Plant
Physiological Ecology

Plant Physiological Ecology

Field methods and instrumentation

Edited by

R.W. PEARCY, J.R. EHLERINGER,
H.A. MOONEY and P.W. RUNDEL

London New York
CHAPMAN AND HALL

First published in 1989 by
Chapman and Hall Ltd
11 New Fetter Lane, London EC4P 4EE
Published in the USA by
Chapman and Hall
29 West 35th Street, New York NY 10001

Softcover reprint of the hardcover 1st edition 1989

Typeset in 10/12 pt Palatino by
Photoprint, Torquay

ISBN-13: 978-94-010-7496-4 e-ISBN-13: 978-94-009-2221-1
DOI: 10.1007/978-94-009-2221-1

British Library Cataloguing in Publication Data

Plant physiological ecology : field methods and instrumentation.
1. Plants. Physiology. Ecological aspects
I. Pearcy, R.W. (Robert W.), 1941–
581.1

Library of Congress Cataloging-in-Publication Data

Plant physiological ecology.

Includes index.
1. Plant physiological ecology – Field work.
2. Plant physiological ecology – Instruments.
I. Pearcy, R. W. (Robert W.), 1941–
QK905.P57 1989 581.5'222 88–30018

Contents

Contributors

William W. Adams III	Department of Environmental, Population and Organismic Biology, University of Colorado, Boulder, CO, USA
J. Timothy Ball	Biological Sciences Center, Desert Research Institute, University of Nevada, Reno, NV, USA
Joseph A. Berry	Department of Plant Biology, Carnegie Institution of Washington, Stanford, CA, USA
Dan Binkley	Department of Forest and Wood Science, Colorado State University, Fort Collins, CO, USA
Arnold J. Bloom	Department of Vegetable Crops, University of California, Davis, CA, USA
Martyn M. Caldwell	Department of Range Science and the Ecology Center, Utah State University, Logan, UT, USA
Gaylon S. Campbell	Department of Agronomy and Soils, Washington State University, Pullman, Washington, USA
F. Stuart Chapin III	Institute of Arctic Biology, University of Alaska, Fairbanks, Alaska, USA
Nona R. Chiariello	Department of Biological Sciences, Stanford University, Stanford, CA, USA

James R. Ehleringer	Department of Biology, University of Utah, Salt Lake City, UT, USA
Christopher B. Field	Department of Plant Biology, Carnegie Institution of Washington, Stanford, CA, USA
John Grace	Department of Forestry and Natural Resources, University of Edinburgh, Edinburgh, Scotland
Carol S. Greitner	Department of General Science, Oregon State University, Corvallis, OR, USA
Wesley M. Jarrell	Department of Environmental Sciences and Engineering, Oregon Graduate Center, Beverton, OR, USA
Roger T. Koide	Department of Biology, Pennsylvania State University, University Park, PA, USA
Harold A. Mooney	Department of Biological Sciences, Stanford University, Stanford, CA, USA
Suzanne R. Morse	Department of Botany, University of California, Berkeley, CA, USA
John M. Norman	Department of Soils Science, University of Wisconsin, Madison, WI, USA
C. Barry Osmond	Department of Botany, Duke University, Durham, NC, USA
Robert W. Pearcy	Department of Botany, University of California, Davis, CA, USA
Robert H. Robichaux	Department of Ecology and Evolutionary Biology, University of Arizona, Tuczon, AZ, USA
Philip W. Rundel	Laboratory of Biomedical and Environmental Sciences and Department of Biology, University of California, Los Angeles, CA, USA

E.-Detlef Schulze	Lehrstuhl für Pflanzenökologie, Universität Bayreuth, D-8700 Bayreuth, Federal Republic of Germany
Celia M. Smith	Department of Botany, Smithsonian Institution of Washington, Washington DC, USA
Stanley D. Smith	Department of Biological Sciences, University of Nevada, Las Vegas, NV, USA
Keith Van Cleve	Institute of Arctic Biology, University of Alaska, Fairbanks, Alaska, USA
Ross A. Virginia	Biology Department and Systems Ecology Research Group, San Diego State University, San Diego, CA, USA
Peter Vitousek	Department of Biological Sciences, Stanford University, Stanford, CA, USA
Kimberlyn Williams	Department of Biological Sciences, Stanford University, Stanford, CA, USA
William E. Winner	Department of General Science, Oregon State University, Corvallis, OR, USA
Reiner Zimmermann	Lehrstuhl für Pflanzenökologie, Universität Bayreuth, D-8700 Bayreuth, Federal Republic of Germany

Introduction

Physiological plant ecology is primarily concerned with the function and performance of plants in their environment. Within this broad focus, attempts are made on one hand to understand the underlying physiological, biochemical and molecular attributes of plants with respect to performance under the constraints imposed by the environment. On the other hand physiological ecology is also concerned with a more synthetic view which attempts to understand the distribution and success of plants measured in terms of the factors that promote long-term survival and reproduction in the environment. These concerns are not mutually exclusive but rather represent a continuum of research approaches. Osmond *et al.* (1980) have elegantly pointed this out in a space–time scale showing that the concerns of physiological ecology range from biochemical and organelle-scale events with time constants of a second or minutes to succession and evolutionary-scale events involving communities and ecosystems and thousands, if not millions, of years. The focus of physiological ecology is typically at the single leaf or root system level extending up to the whole plant. The time scale is on the order of minutes to a year. The activities of individual physiological ecologists extend in one direction or the other, but few if any are directly concerned with the whole space–time scale. In their work, however, they must be cognizant both of the underlying mechanisms as well as the consequences to ecological and evolutionary processes.

IMPORTANCE OF FIELD METHODS TO PHYSIOLOGICAL ECOLOGY

Field research in physiological ecology is heavily dependent on appropriate methods and instrumentation. The writings of early plant geographers and ecologists, such as Schimper (1898), Warming (1896) and Clements (1907), were rich with hypotheses and speculations concerning the role of environment, physiology and morphology in the distributions of plants. Indeed as early as the 1880s attempts were made to measure physiological difference (transpiration) between populations at different elevations (Bonnier, 1890). The theory for water vapor and heat transfer from leaves, including the concept of stomatal resistance, was published in 1900 by Brown and Escombe and porometers capable of providing at least a relative measure of stomatal aperture were first used shortly thereafter (Darwin and Pertz, 1911). The Carnegie Institution of Washington's Desert Research Laboratory in Tucson from 1905 to 1927 was the first effort by plant physiologists and ecologists to conduct team research on the water relations of desert plants. Measurements by Stocker in the North African deserts and Indonesia (Stocker, 1928, 1935) and by Lundegardh (1922) in forest understories were pioneering attempts to understand the environmental controls on photosynthesis in the field.

While these early physiological ecologists were keen observers and often posed hypotheses still relevant today they were strongly limited by the methods and technologies available to them. Their measurements provided only rough approximations of the actual plant responses. The available laboratory equipment was either unsuited or much more difficult to operate under field than laboratory conditions. Laboratory physiologists distrusted the results and ecologists were largely not persuaded of its relevance. Consequently, it was not until the 1950s and 1960s that physiological ecology began its current resurgence. While the reasons for this are complicated, the development and application of more sophisticated instruments such as the infrared gas analyzer played a major role. In addition, the development of micrometeorology led to new methods of characterizing the plant environments. However, most of the instruments such as the infrared gas analyzer and recorders were primarily designed for industrial process control applications and required considerable ingenuity on the part of the physiological ecologist to adapt them for use in the field. There were no instruments manufactured specifically for ecophysiological applications. Physiological ecologists had to be intimately involved with developing instrumentation systems, and be familiar with techniques from many fields, including meteorology, physiology and soil science.

The last 15 years have brought about remarkable changes in the instrumentation available for physiological ecology. This has been in part driven by the electronics revolution that has allowed the development of highly sophisticated 'smart' instruments with microprocessors that control their performance and low-power circuits that allow operation on batteries in the field. In addition, the notebook and analog recorder have been largely replaced by data acquisition systems and solid-state memory, making at least the chore of collecting the numbers less demanding. However, at least as important is the increased demand for ecophysiological and environmental measurements. This has given rise to the development of commercially available instruments specifically designed for ecophysiological measurements and environmental monitoring in the field. For many research problems there is no longer a need for the physiological ecologists to design and assemble the instrumentation systems since commercially available systems function so well. Physiological ecologists have and will continue to play an important role in this development, and the strong cooperation between researchers and instrument companies that has developed in the past few years will lead to further progress.

The increased availability of instruments, however, is a two-edged sword. Physiological ecologists who assembled their gas exchange system were familiar with its limitations. Those who purchase instrumentation systems may not be. It is the responsibility of the investigator to become familiar with the system, its components and their limitations.

METHODOLOGICAL BOOKS FOR PHYSIOLOGICAL ECOLOGY

Since 1960 many important books covering instrumentation of use in physiological ecology have appeared. Environmental measurements were covered by Slatyer and McIlroy (1961), Tanner (1963), Platt and Griffiths (1964) and Fritschen and Gay (1979). Tanner's volume was available only in mimeographed form but was widely utilized because of the practical information it contained on constructing instruments, including such details as the properties of solders and paints. Similarly, Slatyer and McIlroy had only a limited printing but also had enormous impact because of the practical information it provided.

The first treatise covering methods of plant ecophysiology was *Methodology of Plant Ecophysiology. Proceedings of the Montpellier Symposium* edited by F.E. Eckhardt (1965). This volume contained probably the first detailed descriptions of the application of infrared gas analyzers in the field for measuring photosynthesis. *Plant Photosynthetic Production, Manual of Methods* by Sestak, Catsky and Jarvis followed in 1971 and served nearly as a bible for ecophysiologists measuring gas exchange of plants. Although it is clearly dated, the book still contains much valuable information. More recently, several books covering methods have appeared. *Instrumentation for Environmental Physiology* edited by Marshall and Woodward (1985) had the specific objective of covering developments in instrumentation since 1971. *Techniques in Bioproductivity and Photosynthesis* edited by Coombs, Hall, Long and Scurlock (1985) covers a wide array of techniques ranging from growth analysis to enzymology.

SCOPE OF THIS BOOK

This book is intended as a guide to the methodology appropriate for field measurements in physiological ecology, including both measurements of the environment and of the physiological and morphological responses of the plants. In some cases, for example stable isotope measurements (Chapter 13), the measurement is done in the laboratory but on samples collected under field conditions and with the primary objective of understanding the ecological behavior of plants in the field. Other chapters, especially those on soils (Chapter 5) and growth and allocation (Chapter 15), similarly cover analytical techniques on field samples. There has been an attempt to avoid an exhaustive review of all techniques but rather to concentrate on those that have proved most useful to the authors. These techniques provide the basis of evaluating the acquisition of resources (e.g. carbon, light, water, nutrients) and the use of these resources for plant growth. Methods for estimating some of the costs of allocation of resources to particular functions are also covered (Chapter 15). In addition, the methodological framework for extending from organ level (leaf, root) measurements to whole plants is provided in chapters on root systems (Chapter 16) and canopy structure (Chapter 14) and to ecosystem level processes (Chapter 13).

There are many techniques not included in this book that might rightly be an important component of a research program in physiological ecology. The focus is on performance of plants in the field and therefore instrumentation systems and facilities such as growth chambers and phytotrons are not covered. We have also not covered the various biochemically based techniques such as analysis of leaf anatomy, carboxylation enzyme activities, chlorophyll content, electron-transport activities, hormones, or for that matter, root enzyme activities involved in such processes as flooding tolerance. These techniques can clearly be applied to leaf samples collected in the field but are more commonly conducted on greenhouse or growth chamber plants. Methods for biochemical analysis of the photosynthetic apparatus are given in Hipkins and Baker (1986) and Coombs *et al.* (1985) and there are many sources for general information such as the series *Methods in Enzymology*. Problems of experimental design, statistics, etc. are also not covered for reasons of space.

We have also not covered the analytical framework necessary for much research in physiological ecology such as simulation models or cost–benefit analyses. Jones (1983) discusses the important role of models and their interrelationship with experiments in physiological ecology and Loomis *et al.* (1979) provides a discussion of modeling approaches. Biochemically based photosynthetic models such as the one of Farquhar and his colleagues (Farquhar *et al.*, 1980; Farquhar and von Caemmerer, 1982; Farquhar and Sharkey, 1982) have indeed played an important role in linking biochemical events to an understanding of whole-leaf gas exchange. More empirical models have similarly been useful in understanding, for example, limitations on annual carbon gain in specific habitats (e.g. Schulze *et al.*, 1976). Cost–benefit analyses also provide an important analytical framework for understanding plant function and adaptation. The reader is referred to Bloom *et al.* (1985), Givnish (1986) and Gutschick (1987) for thorough discussions of this approach.

REFERENCES

Bloom, A.J., Chapin, F.S. and Mooney, H.A. (1985) Resource limitation in plants – an economic analogy. *Ann. Rev. Ecol. System.*, **16**, 363–92.

Bonnier, G. (1890) Influence des hautes sur les fonctions des vegetaux. *Compt. Rend.*, **111**, 377–80.

Brown, H. and Escombe, F. (1900) Static diffusion of gases and liquids in relation the assimilation of carbon and translocation in plants. *Philos. Trans. R. Soc. Ser. B*, **193**, 223–91.

Clements, F.E. (1907) *Plant Physiology and Ecology*, Holt, New York.

Coombs, J., Hall, D.O., Long, S.P. and Scurlock, J.M.O. (eds) (1985) *Techniques in Bioproductivity and Photosynthesis*, 2nd edn, Pergamon, Oxford.

Darwin, F. and Pertz, D.F.M. (1911) On a new method of estimating the aperture of stomata. *Proc. R. Soc., Ser. B*, **84**, 136–54.

Eckhardt, F.E. (ed.) (1965) *Methodology of Plant Ecophysiology. Proceedings of the Montpellier Symposium*, UNESCO, Paris.

Farquhar, G.D. and von Caemmerer, S. (1982) Modelling of photosynthetic response to environmental conditions. In *Encyclopedia of Plant Physiology*, new series, vol. 12b,

Physiological Plant Ecology (eds O.L. Lange, P.S. Nobel, C.B. Osmond and H. Ziegler), Springer-Verlag, Heidelberg, Berlin and New York, pp. 549–88.

Farquhar, G.D., von Caemmerer, S. and Berry, J.A. (1980) A biochemical model for photosynthetic CO_2 assimilation in leaves of C_3 species. *Planta*, **119**, 78–90.

Farquhar, G.D. and Sharkey, T.D. (1982) Stomatal conductance and photosynthesis. *Annu. Rev. Plant Physiol.*, **33**, 317–45.

Fritschen, L.J. and Gay, L.W. (1979) *Environmental Instrumentation*, Springer-Verlag, Heidelberg.

Givnish, T.J. (ed.) (1986) *On the Economy of Plant Form and Function*, Cambridge University Press, Cambridge.

Gutschick, V.P. (1987) *A Functional Biology of Crop Plants*, Croom Helm, London.

Hipkins, M.F. and Baker, N.R. (1986) *Photosynthesis Energy Transduction a Practical Approach*, IRL Press, Oxford.

Jones, H.G. (1983) *Plants and Microclimate*, Cambridge University Press, Cambridge.

Loomis, R.S., Rabbinge, R. and Ng, E. (1979) Explanatory models in crop physiology. *Annu. Rev. Plant. Physiol.*, **30**, 339–67.

Lundegardh, H. (1922) Zur physiologie und okologie der kohnlensaurassimilation. *Biol. Zbl.*, **42**, 337–58.

Marshall, B. and Woodward, F.I. (eds) (1985) *Instrumentation for Environmental Physiology*, Cambridge University Press, Cambridge.

Osmond, C.B., Björkman, O. and Anderson, D.J. (1980) *Physiological Processes in Plant Ecology: Towards a Synthesis with Atriplex*, Springer-Verlag, Berlin.

Platt, R.B. and Griffiths, J.F. (1964) *Environmental Measurement and Interpretation*, Reinhold, New York.

Schimper, A.F.W. (1898) *Pflanzen-Geographie auf physiologischer Grundlage*, Fischer, Jena.

Schulze, E.D., Lange, O.L., Evenari, M., Kappen, L. and Buschbom, U. (1976) An empirical model of net photosynthesis for the desert plant *Hammada scoparia* (Pomel) Iljin. I. Description and test of the model. *Oecologia*, **22**, 355–72.

Sestak, Z., Catsky, J. and Jarvis, P.G. (eds) (1971) *Plant Photosynthetic Production, Manual of Methods*, W. Junk, The Hague.

Slatyer, R.O. and McIlroy, I.C. (1961) *Practical Microclimatology*, C.S.I.R.O. Plant Industry Division, Canberra.

Stocker, O. (1928) Der wasserhaushalt ägyptischer wüsten-und salzpflanzen von standpunkt einer experimentellen un vergleichenden pflangeographie aus. *Bot. Abh*, **13**, 1–200.

Stocker, O. (1935) Assimilation und atmung westjavanischer tropenbäume. *Planta*, **24**, 402–45.

Tanner, C.B. (1963) *Basic Instrumentation and Measurements for Plant Environment and Micrometeorology*, University of Wisconsin Soils Bulletin 6.

Warming, E. (1896) *Lehrbuch der ökologischen Pflanzengeographie. Eine Einführung in die Kenntnis der Pflanzenvereine*, Borntraeger, Berlin.

1

Principles of instrumentation for physiological ecology

Arnold J. Bloom

1.1 INTRODUCTION

Physiological ecology, as it has adopted more sophisticated techniques, has become increasingly reliant upon instruments to monitor and control physical or chemical parameters. Instrumentation systems, in which concurrent measurements from a range of instruments are needed to derive a secondary quantity of interest, have also proliferated. For example, several gas-exchange systems are now commercially available in which readings from an infrared CO_2 analyzer, flow meters, humidity sensors, a gas-mixing system and temperature sensors are required for the estimation of photosynthetic rates, stomatal conductances and internal CO_2 concentrations. As a consequence, a certain degree of 'equipment-savvy' has become necessary, not only to operate and maintain the tools of the trade, but to select the most appropriate one for a particular task. A thorough treatment of instrumentation such as is available in Horowitz and Hill (1980) is beyond the scope of this chapter; I shall instead touch lightly on the rudiments for the common practitioner.

1.2 MEASUREMENT AND MEASUREMENT ERRORS

Measurement – the quantification of a particular characteristic in standard units – involves a comparison between the characteristic and a reference standard to determine whether they are equal (null or zero detection) or the extent to which they differ (difference or gain detection). Instruments vary greatly in their relative dependence on null and difference detection. For example, a double-pan balance depends predominantly upon null detection and little upon difference detection (deviation of the pointer from the zero mark). By contrast, a modern voltmeter with automatic zero compensation depends predominantly upon difference detection because its input terminals are shorted several times a second (i.e. calibrated to a zero reference standard) and, at other times, the voltage difference between this zero and the input is quantified.

1.2.1 Specifications

Measurement errors for an instrument – including those in the reference standards, null detection and difference detection – are specified in terms of accuracy, precision, noise, stability, sensitivity, resolution and time response.

(a) Accuracy

Accuracy is the amount that a measured quantity differs from its true value. It is often expressed as a percentage of reading (% reading) or full-scale range (% FS). The distinction between these two is important because measurements are often made at values considerably less than the full scale of an instrument. Consequently, the percentage uncertainty in a measurement may be considerably greater than the specified percentage full-scale accuracy.

(b) Precision

Precision is the repeatability of a measurement, that is, the degree to which successive measurements of a constant input will produce the same output. If there is a systematic error in a measurement, such as a calibration error, then accuracy and precision will not be the same: measurements can be quite precise and at the same time not very accurate.

(c) Noise

Noise is any change in output that is unrelated to changes in the measured parameter. It is specified as rms (root mean square) noise, the square root of the sum of the squared deviations from the mean,

$$\text{rms} = \{\sum_{i}[(x_i - \bar{x})^2]\}^{1/2}$$

or as peak-to-peak (pk-to-pk) noise, the range of deviations from the mean. Peak-to-peak noise is roughly six to seven times rms noise. Because noise becomes troublesome only when it is a significant fraction of the transducer signal, a more relevant specification may be the signal-to-noise ratio (SNR) usually expressed logarithmically in decibels (dB):

$$\text{SNR} = 20 \log_{10}[V_{\text{rms}}(\text{signal})/V_{\text{rms}}(\text{noise})]$$

(d) Stability

This refers to changes or drift in the null or difference detection that occur over time or with temperature. It is specified in absolute terms (e.g. mV day^{-1}, mV $°\text{C}^{-1}$) or on a relative basis (e.g. % day^{-1}, % $°\text{C}^{-1}$). If stability of an instrument is low, then it will require more frequent calibration. Difference detection in many electronic instruments is more stable than null detection.

(e) Sensitivity

This is the smallest change in input that will give a discernible change in output.

(f) Resolution

Resolution is the smallest scale division or the last digit that is readable. Clearly an instrument can have no greater accuracy than its sensitivity or resolution, but high sensitivity or resolution do not necessitate high accuracy.

(g) Time response

This is the speed with which an instrument responds to a change in input. It is usually quantified in terms of *settling time,* the time elapsed from the application of a step-change in input to the time when the output remains within a specified error band around its final value. With exponential transients, one defines the *time constant* of an instrument as the time required for the output to reach 67% of its final value. An instrument must have a time response several times faster than the rate at which the phenomenon of interest changes to follow adequately such changes (cf. Chapter 2).

1.2.2 Error analysis

Errors are generally propagated in an additive fashion. Consequently, the total error for an instrument system equals the sum of the independent errors for the components. For example, a switch, amplifier and filter that generate a noise of 1, 6 and 0.5 μV rms, respectively, will in conjunction produce a noise of 7.5 μV rms. Often the errors for one component will dominate (e.g. the amplifier noise); efforts to reduce system errors can focus on that component.

Reducing measurement errors may not be the only consideration in selection of an instrument: a more accurate instrument may require more frequent calibration, cost more, be less portable, or be more difficult to operate than a less accurate counterpart. One should take into account the accuracy required for the particular measurement relative to other sources of variation in the parameters.

1.3 INSTRUMENT ORGANIZATION

An instrument typically consists of four major sections: sensor–transducer, signal conditioner, output and power.

The *sensor* couples the instrument to the environmental phenomenon of interest. The *transducer* converts the signal from the sensor into a usable form such as a voltage, current or mechanical movement that varies in some predictable fashion as the phenomenon changes. Often, but not always, the sensor and transducer are one and the same. For example, a thermocouple junction is thermally coupled to its environment and transduces the temperature into a voltage. However, for a glass thermometer, the sensor is the bulb whereas the mercury is the transducer which changes volume as a function of temperature.

The *signal conditioner* modifies the signal from the sensor–transducer so that it is easily recordable. In electronic instruments this may require (1) signal-to-voltage conversion, (2) multiplexing, (3) amplification, (4) filtering of specific frequencies, (5) transformation and (6) data conversion. For a glass thermometer, the capillary serves as the signal processor because it converts a volume change to a length change.

The *output stage* makes the signal observable through a readout device and/or operative through a process controller. A readout device (e.g. a meter, gauge, or pen on chart paper) displays the processed signal in a quantitative form. The readout for a glass thermometer is the top of the mercury column and the engraved scale. If an instrument serves as a process controller, a readout device may or may not be present. Instead the instrument turns on or off devices such as heaters and valves to bring an environmental factor closer to a desired value.

Finally, the *power section* – be it thermal energy, a wound spring or a battery – provides energy at the proper level to the other sections.

The following text details constraints encountered in performing the function of each section and the procedures to insure compatibility among sections.

1.3.1 Sensors and transducers

Physiological ecologists have at their disposal many different kinds of sensors and electronic transducers. As documented throughout this book, devices for a given physical phenomenon vary greatly in type and magnitude of output signal, precision and accuracy, response time, useful range, power requirements and cost. In the ideal, a transducer would generate a voltage signal that changes linearly with the physical phenomenon of interest; this signal would be 1 V direct current (DC), full-scale, low impedance [i.e. low resistance to alternating current (AC)] and floating (i.e. not grounded). The greater the deviation from this ideal, the more stringent the requirements are for subsequent signal conditioning.

1.3.2 Signal conditioning

(a) Signal-to-voltage conversion

The signal conditioning section usually first converts the transducer signal into a voltage. Current signals are produced by devices such as silicon light sensors, solid-state temperature sensors and polarographic O_2 sensors. Conversion of this current into a voltage usually involves measuring the voltage drop across the shunt resistor ($V = IR_{shunt}$). A small resistor may yield too small a voltage drop; a large one may interfere with the device's generation of current. Alternatively, operational amplifiers in an appropriate configuration (i.e. an inverting amplifier) can convert current to voltage, yielding a high voltage output with a low apparent shunt resistance.

Resistive signals, such as those generated by thermistors, platinum resistance thermometers or strain-gauges, may be converted to a voltage using either a bridge circuit or a constant current source. A Wheatstone bridge (Fig. 1.1(a)) consists of a DC voltage source and three standard resistors. This relatively simple circuit provides a large, but nonlinear, voltage change per resistance change. Circuits that use constant current sources are more difficult to configure, but yield a linear voltage change per resistance change ($\Delta V = IR$). With both these approaches, power dissipation through the sensor should be limited to reduce errors that result from self-heating.

Capacitive signals such as those generated by solid-state humidity sensors or the 'Luft' detectors of infrared gas analyzers are converted to voltages using impedance bridges (Fig. 1.1(b)) or timing circuits. Again, the advantages of bridge circuits are their simplicity and sensitivity; their disadvantage is their nonlinearity. Timing circuits are based on the relationship that, given a constant current, the period required for charging a capacitor to a particular voltage is directly proportional to its capacitance.

Inductive sensors such as linear variable displacement transformers (LVDTs) and some pressure transducers are monitored with a Maxwell bridge (Fig. 1.1(c)). This arrangement is more convenient than strictly using an inductive bridge because standard capacitors are less expensive than standard inductors.

(b) Multiplexing

A multiplexer is a switching system that directs signals from just one device at a time to subsequent stages of signal processing. Multiplexers are used in data acquisition

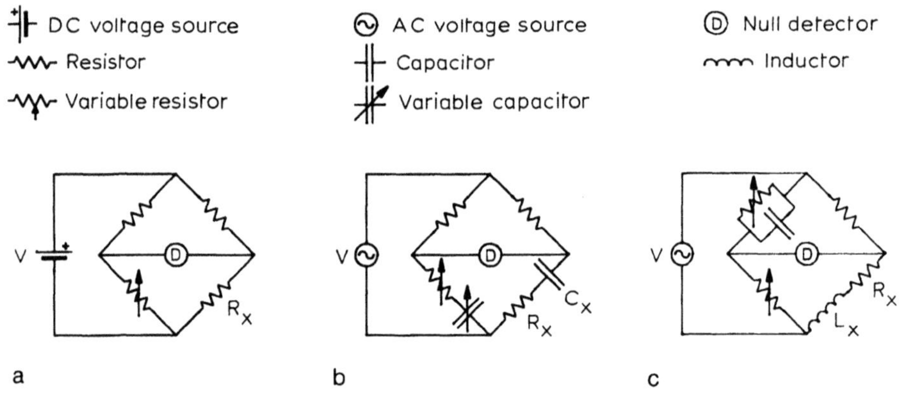

Fig. 1.1 The electronic schematic is shown for a Wheatstone bridge (a), an impedance bridge (b) and a Maxwell bridge (c). The symbols are defined at the top of the figure. R_x, C_x and L_x represent an unknown resistor, capacitor and inductor, respectively. The null detector is a galvanometer or DC voltmeter for the Wheatstone bridge and a set of headphones or AC voltmeter for the impedance and Maxwell bridges. The variable resistors and capacitors are adjusted until no signal is detected, i.e. the bridge is balanced. From the values of the variable resistors and capacitors, the values of the unknown components can be calculated.

systems and data loggers that have one set of signal processing equipment to service many difference sensors and transducers. Traditionally, high-quality multiplexers have employed reed-relays but solid-state switches have improved to the point that they are faster, more reliable, cheaper, consume less power and suitable for low-level (i.e. nanovolt or picoamp) signals.

(c) Amplification

An amplifier is a device that produces an output that is a given multiple of the input signal with minimum disturbance of the input signal. In an electronic amplifier, the voltage applied to the input results in a proportionally larger voltage or current at the output; the proportionality factor is called the gain of the amplifier. For adequate performance the sensor–transducer and the amplifier must be carefully matched. Fortunately, the trend is to encapsulate the sensor–transducer and amplifier together; for example, many solid-state pressure devices have the silicon membrane sensor, piezoelectric transducer, voltage source, bridge resistors, temperature compensation circuitry and amplifiers within a single package. In those cases requiring self-sufficiency, attention should be paid to the impedance, bias current, bandwidth, grounding and input configuration of the devices.

Impedance, bandwidth and bias current. Resistance is the proportionality factor between voltage and current as defined in Ohm's law, $V = IR$. Impedance, the resistance to alternating current (AC), may differ significantly from the resistance to direct current (DC) because current flow through components with capacitance or inductance is frequency dependent. The impedance of the sensor–transducer ($R_{s\text{-}t}$) and that of the amplifier (R_{amp}) act in series to form a voltage divider:

$$V_{amp} = V_{s\text{-}t}R_{amp}/(R_{s\text{-}t} + R_{amp})$$

Only if the impedance of the amplifier is much higher than that of the sensor–transducer ($R_{amp} \geqslant R_{s\text{-}t}$), will the voltage sensed by the amplifier, V_{amp}, accurately reflect the output from the sensor–transducer, $V_{s\text{-}t}$.

Any impedance, including that of the sensor–transducer, generates voltage noise (Johnson's noise) because of the random thermal movement of electrons within it. This noise over a particular frequency range or *bandwidth* equals:

$$V_{rms} = 7.4 \times 10^{-12}[TR_{s\text{-}t}(f_H - f_L)]^{1/2}$$

where V_{rms} = rms noise (V), T = absolute temperature (K), $R_{s\text{-}t}$ = impedance (Ω), f_H = upper frequency limit (Hz) and f_L = lower frequency limit (Hz).

For example, the pk-to-pk noise at 25°C from a 100 Ω thermocouple (about 3 cm of 0.001 inch chromel–constantan wire) over a 20 Hz bandwidth will be about 35 nV, making this thermocouple only marginally useful for thermocouple psychrometry. To achieve an acceptable noise level with a high-impedance sensor–transducer, one must often limit the bandwidth to which the amplifier is sensitive.

All amplifiers require some current (*bias current*) from the sensor–transducer to saturate their input circuits. This bias current I_b, is temperature dependent and contains electronic noise. Interaction between sensor–transducer impedance and input bias current generates an input offset voltage, $V_{off} = I_b R_{s\text{-}t}$, that is perceived as part of the input signal. Consequently, a low-bias current amplifier ($I_b < 10^{-11}$ A) should be used with high-impedance sensor–transducers to reduce this effect.

Grounding and input configuration. Ground is a reference point that is arbitrarily set to zero volts. Theoretically, a voltage ground should be an infinite conduction plane with a constant potential over its entire surface. The surface of the planet Earth approximates such a condition, hence, the term 'ground'. To make contact with the planet, the ground prong in an electrical outlet (wall ground) is connected to a large conductor such as a cold water pipe in moist soil.

Sensor–transducers fall into six classes of signal sources depending on their relationship with wall ground (single-ended/grounded, single-ended/floating, single-ended/driven-off-ground, differential/grounded, differential/floating and differential/driven-off-ground). Single-ended means that only one of the two terminals connected to the source has an active voltage potential; differential means that both terminals are active, that is, each terminal has a voltage potential with respect to a third terminal. Similarly, amplifiers have input configurations which may be classified according to their connection to wall ground (single-ended/grounded, differential/grounded, single-ended/floating and differential/floating). For details on these classes see Nalle (1969) or Aronson (1976). Each class of signal source must be matched with the appropriate class of amplifier to avoid false readings and to reduce sensitivity to external electrical noise. Nonetheless, a differential/floating amplifier will usually handle satisfactorily all kinds of signal sources.

A differential/floating amplifier is typically realized using an instrumentation amplifier. Ideally this type of amplifier responds only to the voltage difference between the two input terminals and not to the common-mode voltage, the voltage common to both terminals. This is analogous to a double-pan balance that shows no deflection when the two weights are equal, no matter how heavy they are. In practice, owing to slightly different gains in the plus and minus inputs, or variations in offset voltage as a function of common-mode levels, common-mode input voltages are not completely eliminated at the output. The common-mode error voltage is the magnitude of this output divided by the gain of the amplifier. The common-mode rejection ratio (CMRR) equals the common-mode input voltage divided by the resulting common-mode error voltage. Common-mode rejection is often expressed logarithmically:

$$\text{CMR (in dB)} = 20 \log_{10} (\text{CMRR})$$

A differential amplifier should have a CMR of 100 dB or greater. In a differential/floating amplifier, the instrumentation amplifier uses a power supply electrically isolated from the wall ground. Usually the ground of this supply is driven at the common-mode voltage. This approach also serves to increase the CMR of the amplifier.

Amplifiers contain input connections, cabling and enclosures, active (i.e. requiring power) devices such as transistors and operational amplifiers, and passive elements such as resistors and capacitors. Each of these parts can be a source of noise and drift.

Input connections. Input connections include plugs and jacks, switches and solder joints. These and, in fact, all junctions between conductors act as thermocouples; thus, they generate small voltages that vary with the temperature difference between themselves and subsequent junctions. For example, lead–tin solder to copper junctions generate about 3 μV °C^{-1} with respect to a junction in an ice bath, copper oxide to copper about 1000 μV °C^{-1} and silicon to copper about 420 μV °C^{-1}, whereas copper to copper, gold to copper, or low-thermal (cadmium–tin) solder to copper generate less than 0.1 μV °C^{-1}. To reduce thermal offsets, one can either use materials such as cadmium–tin solder or minimize the temperature difference between junctions by keeping all input connections in close proximity to one another and thermally insulated.

Cabling and enclosures: grounding. Cabling and enclosures usually become a source of noise as a result of grounding problems. Most sensitive instruments do not make extensive use of the wall ground because its potential fluctuates with the demands of other equipment such as the refrigerators and compressors connected to it. Consequently, instruments usually contain several different internal grounds, that is, several internal voltage reference points. These include one or more analog signal grounds, a digital ground and a ground for the external case which, for safety reasons, is usually connected to wall ground. The different grounds should be either electrically isolated from one another or connected at a single point. Such an arrangement avoids ground loops. A ground loop occurs when a conductor forms a complete circle and thus acts as an antenna which may pick up substantial radio frequency–electromagnetic interference (RFI–EMI). Several symptoms indicate that an instrument suffers from RFI–EMI: (1) a shift in the output signal when either the case is touched, a person walks in front of it or the input and output cables are moved; (2) an increase in output noise during normal work hours or when heavy electrical equipment is being used.

Cables and enclosures can be designed to reduce RFI–EMI noise. A cable consisting of a shielded twisted pair of wires is adequate for most purposes. High-impedance sensors such as pH electrodes require special low-noise coaxial cable which, when bent, produces relatively little piezoelectric current (i.e. the charge induced in crystals under pressure). To minimize RFI–EMI pickup, the internal enclosure for the amplifier and cable shields should be connected at the transducer to the zero-signal-reference-potential (Morrison, 1967). This zero-signal-reference-potential may or may not be identical to the analog ground. For example, the signal from a Wheatstone bridge will have a zero-signal-reference-potential generally higher than the analog ground.

Active devices: op amps. An operational amplifier (op amp) is a direct-coupled high-gain differential voltage amplifier. Today, op amps are fabricated on a single chip of silicon and usually come in an 8-pin package less than 1 cm^2. Op amps vary substantially in cost and quality. They may cost as little as $0.50 or as much as $100. They may generate voltage noise from 0.2 to 20 μV pk-to-pk and current noise from 0.003 to 70 pA pk-to-pk. They also produce voltage and current offsets which depend on temperature (0.1 to 50 $\mu V\ °C^{-1}$; 1.5 to 600 pA $°C^{-1}$) and time (some as low as 0.1 $\mu V\ month^{-1}$). Op amps that use chopper-stabilization and auto-zero compensation trade reduced voltage offsets for increased voltage noise and slower response times.

Passive elements. A resistor, as discussed above, by its very nature generates a voltage noise proportional to the square root of its value. For this reason, high-valued resistors should be avoided when making low-level measurements. In addition, resistors have values that are temperature dependent: carbon or glass resistors change as much as 0.1% $°C^{-1}$ whereas metal film or wire-wound resistors change less than 0.01% $°C^{-1}$. For example, if a silicon light sensor used a carbon resistor as a shunt, its signal output could vary 1% with a 10°C change even under constant light levels.

Capacitors are made from many different types of materials which vary greatly in their sensitivity to temperature and frequency. Critical timing circuits should use capacitors made from mica, NPO ceramic or plastic films such as polycarbonate and polystyrene because they have values that are relatively temperature stable. In higher values ($C > 10$ μF), tantalum capacitors are preferable to aluminium because they suffer less leakage of charge.

(d) Signal filtering and transformation

After the signal from the transducer is converted to a voltage and amplified, it often requires filtering and transformation.

Filtering. Filtering can reduce the amount of background noise in the signal from a sensor–transducer if the periodicity of this signal is distinct from that of the noise. Filters are categorized as low-pass, high-pass, band-pass or notch by how they treat different frequencies. A low-pass filter allows the low-frequency components of a signal to pass through to the output and removes the high-frequency components. Most of the phenomena of interest to physiological ecologists change relatively slowly from an electronic perspective and, thus, low-pass filtering is often employed. A high-pass filter passes high frequencies and removes low frequencies. A band-pass filter passes frequencies only within a narrow frequency range (i.e. band) and removes frequencies lower or higher than this range; techniques such as synchronous modulation–demodulation and lock-in amplification function as highly specific band-pass filters. Lastly, a notch filter removes frequencies only within a narrow band and passes all lower or higher frequencies; for example, a 60 Hz notch filter can be employed to remove power-line interference. Signal averaging serves as a filter in that it diminishes aperiodic or random signals (i.e. white noise) from complex periodic signals by the square root of the number of times the signal is averaged. The effectiveness of any filtering procedure can be evaluated by the signal-to-noise ratio (SNR). Various signal encoding–decoding schemes such as Dolby or synchronous modulation–demodulation can improve the SNR by 20 dB (10-fold) through precisely defining the periodicity of the signal.

Transformation. Transformation, in which the output voltage is set to a mathematical function of the input voltages, can linearize the signal from a nonlinear transducer or combine signals from several transducers. For example, the signal from a thermistor whose resistance changes exponentially with temperature can be transformed so that the instrument reads directly in degrees. The signals from a relative humidity sensor and a platinum resistance thermometer can be combined to indicate vapor pressure.

Analog versus digital signal processing. Analog circuitry deals with continuous signals – signals that can take a wide range of values (e.g. the sweep minute hand on an analog watch can indicate by its position any value between the tick marks). Analog circuitry is less complex and faster for simple processing but becomes unwieldy and inaccurate for sophisticated tasks. Thus the trend in modern instrumentation is to implement filtering and transformation with digital, as opposed to analog, circuitry.

Digital circuitry deals with signals which can only assume discrete values (e.g. the minute digit on a digital watch can indicate only a specific value; 0,1,2,3, . . . 9). Each discrete value which may be either high (1) or low (0) is called a bit. Several parallel digital signals, each defining a bit, can be considered a binary number. In the arcane humour of computer science, 8 parallel digital circuits, or 8 bits, is called a byte; a byte can specify any number between 00000000 and 11111111 in binary or between 0 and $2^8 = 256$ in base 10 numbers. Note that only whole numbers can be represented; a number such as 160.8 cannot be specified. A byte can therefore provide a resolution of 1 in 256 or 0.39% of full scale. While this seems impressive, it is important to remember that most measurements may be at values considerably less than full scale; thus the resulting accuracy is much less. In many scientific applications, 12 bits giving 1 part in 4096 or even 16 bits giving 1 part in 65 536 are required to provide the necessary resolution.

Digital circuitry demands a certain baseline level of complexity that is predominantly associated with the conversion between analog and digital forms and interfacing to the microprocessor and its supporting hardware. Digital circuitry, however, seldom requires additional components to perform sophisticated tasks with high accuracy or to shift from one task to another. For example, an analog logarithmic amplifier whose output voltage is the log of the input voltage contains numerous temperature-sensitive components.

Performing the same transformation with digital circuitry usually entails only programming the microprocessor to perform a logarithm calculation. Once the conversion to a digital signal is made additional transformations will not generally reduce the accuracy. For analog signals, each additional step will introduce new errors. As the cost of digital circuitry continues to plummet, 'smart' instruments, those using digital signal processing, will become the rule.

(e) Data conversion

Digital-to-analog conversion (D/A conversion) and analog-to-digital conversion (A/D conversion) serve as the bridges between the continuous and discrete worlds. D/A converters consist of a voltage reference, a resistor network and solid-state electronic switches. The voltage reference produces a stable voltage which is divided by the resistor network to produce lower values. The high bits of the digital signal turn on solid-state switches that determine which resistors in the network are used and, thus, the extent to which the reference voltage is divided. For example, when all the bits are high, all the switches will be on and the output voltage will equal the reference voltage; when all the bits are low, all the switches will be off and the output voltage will be zero.

For most applications in physiological ecology, analog signals need to be resolved into 12–16 bit digital signals. The two most common techniques for this high-resolution A/D conversion are integrating and successive approximation.

In a dual-slope integrating A/D converter, the unknown incoming voltage charges a capacitor for a period equal to 2^{12} clock pulses for a 12-bit converter, 2^{16} pulses for a 16-bit. A digital counter then times how long in clock pulses it takes for a reference voltage to discharge this same capacitor. If the times for charging and discharging are the same, reflecting that the incoming voltage equals the reference voltage, the counter will have registered 2^{12} or 2^{16} pulses and, thus, all bits will read high. Conversely, if the capacitor discharges immediately, then the incoming voltage is zero and all the bits on the counter will be low. Intermediate discharge times produce intermediate bit patterns. As the name implies, the input signal is integrated through the charging of the capacitor; consequently, this type of converter is relatively insensitive to high-frequency noise.

A successive approximation A/D converter uses a D/A converter and a voltage comparator to conduct a systematic search for the bit pattern which produces a voltage from the D/A converter equal to the incoming voltage. The search proceeds in the following manner: the highest bit is set high and, if the resulting voltage from the D/A converter is greater than the incoming voltage, the bit is set low. This process is then repeated with the second highest bit and so forth. Thus the approach is somewhat analogous to finding the correct weight with a double-pan balance. When the best match is achieved, the bit pattern of D/A converter will be the binary representation of the input voltage as a fraction of the reference voltage. Successive approximation A/D converters are faster but more expensive than integrating converters. They also require additional circuitry because it is essential that the input voltage remain constant during the course of the successive comparisons. This is accomplished via a 'sample and hold' circuit which samples an input voltage and then outputs a constant equivalent voltage until the measurement is complete.

While it is worthwhile for physiological ecologists to understand the basic principles of data conversion, they need not be too concerned with the details. For the brave, converters are available at low cost with all the necessary circuitry so that they can be readily used. Otherwise, they need only be viewed as a basic building block in an instrument. Digital panel meters are essentially A/D converters with some signal-conditioning circuitry, a display driver and a display. These are widely available from electronics suppliers and quite simple to use.

1.3.3 Output

After a signal passes through all the preceding sections of an instrument, it is ready to be revealed to the outside world. For instruments where the signals are read infrequently and can be copied into a notebook, analog (pointer and scale) or digital (numeric display) meters may be satisfactory. Analog displays rapidly communicate trends in the sensor signal. Digital displays provide high accuracy and resolution. For these reasons some multimeters provide both types of displays. Copying down numbers from a display proves inadequate when frequent measurements need to be made, when several different measurements need to be made concurrently or when measurements need to be made with the instrument left unattended. In such situations, analog or digital data recorders play an important role. These are discussed thoroughly in Chapter 2.

(a) Process control

A process controller uses the signal from the sensor–transducer to modify the environmental parameter being monitored. For example, a temperature controller for a water bath may use a signal from a thermistor to control a water heater. The control algorithm is designed to reduce the error signal, the difference between the signal and a desired set point. The four most common approaches are on–off, proportional (P), proportional and integrative (PI) and proportional, integrative and derivative (PID).

An on–off controller like a room thermostat turns on a relay (i.e. an electronic switch) or a valve (i.e. mechanical switch) when the error signal is greater or less than zero.

A P controller achieves a finer degree of control by setting its output proportional to the error signal. This type of controller either proportionally changes the amplitude of the output or maintains a constant amplitude but proportionally varies the time that the output has this amplitude (e.g. a proportional temperature controller either varies the power through a heater or varies the time that the heater is at full power). A simple proportional controller stabilizes at a point different from the desired set point because its output is zero when the error signal is zero. For example, a proportional temperature controller turns the heater off when the error is zero and, therefore, the system stabilizes at a temperature lower than the set point; how much lower is a complex function of the thermal exchange among various parts of the system.

A PI controller has its output proportional to the integral of the error signal over time, thus eliminating the offset from the set point. This integral term, however, produces delays in the controller response time which may reduce its stability.

A PID controller usually gives the best control. It has a derivative term which responds to the rate at which the error term is changing, thus minimizing overshoot. Unfortunately, a PID controller must be carefully tuned to the particular situation in order to avoid instabilities.

Either analog or digital circuitry can be used to effect these controllers. Again, digital circuitry has the advantage that tuning the integral and derivative terms involves only reprogramming. Although process controllers seldom free investigators entirely from the art of tweaking knobs, they do permit greater focus on the biology.

1.3.4 Power

Power to drive field equipment comes from portable generators or batteries. Generators in the USA normally provide 120 V_{rms} AC. Batteries are nominally rated at 1.5, 3, 6, 9 or 12 V DC. Mercury, lead–acid, nickel–cadmium and lithium batteries maintain an output close to their rated voltage throughout their useful working life; the voltages of carbon–zinc and alkaline batteries decline by about one-third of their rated voltage during their useful working life. Nickel–cadmium (Ni–Cd) batteries are particularly sensitive to being overdischarged because the stronger cells that are part of a battery pack will literally destroy the weaker individuals. Temperature has a significant effect on the capacity of batteries; carbon–zinc, alkaline, mercury and Ni–Cd batteries are practically worthless at low temperatures whereas lead–acid and lithium perform acceptably.

A power supply in an instrument converts the output voltages from a generator or battery power source to that needed by the instrument; it may also regulate the supply voltages to compensate for varying source voltage (e.g. spikes from the generator or dying batteries) or varying demands of the instrument. Analog circuitry often requires power at ± 5 to ± 15 V DC whereas digital circuitry usually requires power at +5 V DC. Both types of circuitry are sensitive to fluctuations in power-supply voltage. A measure of this sensitivity for analog circuitry is the power-supply rejection ratio, PSRR, the apparent change in input offset voltage per change in supply voltage.

Regulated power supplies which convert 120 V AC to lower DC voltages fall into two classes, linear or switching. Linear-regulated supplies dissipate voltages in excess of the desired output voltage as heat; they are simpler, less expensive and more precise, but heavier and less efficient. Switching supplies turn on and off the source current as needed to maintain a constant output voltage; they are more efficient and compact, but less precise, more expensive and noisier both electrically and audibly.

DC-to-DC converters transform one DC voltage, such as that from a battery, to one or several other DC voltages. They first must convert the source DC voltage into an AC voltage, then transform this AC voltage into one or several AC voltages, and lastly convert these AC voltages into the appropriate DC voltages. These converters are usually only 70% efficient and produce significant amounts of RFI–EMI noise, but eliminate the need for multiple batteries.

1.4 INSTRUMENT INITIATION

The above foray into the organization of electronic instruments fails to convey that a well-designed piece of equipment is greater than the sum of its parts. For example, packaging of the various sections with an eye to ruggedness and convenience is a highly important characteristic, one which cannot be adequately expressed in a specification sheet; documentation that is intelligible, correct and complete is the exception rather than the rule. Such factors become evident if an instrument is properly initiated.

1.4.1 Initial inspection

Several steps are proper for the inauguration of a new piece of equipment. First read the documentation – seldom are these black boxes self-explanatory. Then gently shake the instrument: in all seriousness, shaking the instrument will serve to warn of loose connections, marginal components or inadequate packaging, faults which may not be obvious on a laboratory bench, but will surely express themselves sooner or later in the great outdoors. Knowing how much force to use in this and in other situations is a matter of experience and good judgment.

Next disconnect the power cord and take the top off the instrument; no reputable company would void a warranty for such an innocent act unless the case is somehow sealed and the documentation contains explicit warnings. Conducting this procedure at this early stage confers familiarity with its intricacies (e.g. the tools which are necessary) under circumstances definitely more serene than the next time this act will have to be performed. Looking inside the instrument alerts one to shoddy workmanship, the position and virgin appearance of critical components and potential trouble spots.

Now test the batteries. A proper test requires that the batteries be under load, that is, conducted when substantial current is being drawn from the batteries: even nearly dead batteries will read close to their proper voltage when not loaded. The easiest procedure to follow is to connect the batteries to the instrument, turn on the instrument for a few minutes, and then read the voltage across the batteries. If possible disconnect one of the battery terminals and insert an ammeter in series to measure the amount of current required by the instrument during its different activities. If the observed current draw is inconsistent with the specifications in the instruction manual, suspect trouble.

1.4.2 Test driving the instrument

First, feed the instrument a null signal: an empty cuvette for a gas-exchange system; a dark environment for a light meter; a desiccant chamber for a relative humidity indicator; an ice-bath for a thermometer; shorted inputs for a voltmeter. With this constant input, determine how much the output from the device varies as the environment around the device changes. Many environmental factors may have major effects. Temperature, of course, influences all components; less obvious factors may include:

(a) Light

Light will produce unpredictable shifts in the output of Ag/AgCl references found in pH and O_2 electrodes and of most silicon solid-state components if not properly encapsulated. Several years ago a digital meter was found to have about a 20% shift in the zero reading when exposed to sunlight because the A/D converter was not properly encapsulated. While this might not be a problem in a dimly lit laboratory, it was disastrous in a field situation.

(b) Humidity

Humidity changes the resistance and capacitance of most materials. High-frequency and low-current circuits are especially sensitive to these changes. In addition, the contacts on high-current switches or relays may experience arcing under high humidity. Bags of silica gel placed inside the instrument case will often alleviate these problems.

(c) Vibration

'Luft' detectors in non-dispersive infrared gas analyzers and LVDTs are vibration sensitive. Also many materials will generate small currents when mechanically stressed as a result of the piezoelectric effect.

(d) EMI–RFI

As discussed above, proper shielding, grounding and filtering can reduce the influence of EMI–RFI.

Then expose your instrument to the range of signals it is likely to experience *in situ*. Again, determine how the conditions the instrument is likely to encounter will influence the output from these signals. Learn to calibrate your instrument with the most accurate devices at your disposal under conditions as close as possible to those to which it will be exposed *in situ*. Do not rely solely on factory calibrations. If you have access to other instruments of the same ilk, compare their readings under the same conditions. Even the same models from a given manufacturer may vary greatly – manufacturer specifications are sometimes based on wishful thinking and 'quality control' is sometimes left to the end user.

1.5 POSTSCRIPT

Although field instrumentation has become

more sophisticated, maintenance has in some ways become simpler. Reliable solid-state devices have supplanted troublesome mechanical parts such as relays, standard voltage battery cells, vacuum tubes and galvanometric panel meters. The modular design of modern instruments permits rapid replacement of integrated circuits or printed circuit boards. 'Smart' instruments, those with built-in microprocessors, often have self-diagnostic capabilities. Unfortunately, the tools and supplies required for maintenance have become increasingly sophisticated: bailing wire is not as useful as it once was.

Perhaps a more serious problem is the distance that modern instrumentation imposes between the biological phenomena and the recorded data. With a glass thermometer, an investigator has an immediate sense of whether the readings are realistic; with a thermocouple attached to a data logger, the validity of the readings is less intuitive. The natural tendency is to treat equipment as a black box, yet to extract a biological signal from noise introduced during measurement – to distinguish between fact and artifact – requires a detailed working knowledge of both the organism and the instrument. Only through this knowledge can investigators exorcise the demons from the black box on which their science depends.

ACKNOWLEDGEMENTS

I thank Robert W. Pearcy and Richard M. Caldwell for their extensive comments on the manuscript.

REFERENCES

Aronson, M.H. (1976) *Low Level Measurements. Part 5. The Input Circuit*, Measurement & Data Corporation Home Study Course, 60/42, C1–C16, Pittsburgh, PA.

Horowitz, P. and Hill, W. (1980) *The Art of Electronics*, Cambridge University Press, Cambridge.

Morrison, R. (1967) *Grounding and Shielding Techniques in Instrumentation*, John Wiley & Sons, New York.

Nalle, D. (1969) *Signal Conditioning*. Brush Instrument Division of Gould, Applications Booklet, Vol. 101, pp. 1–15, Cleveland, Ohio.

2
Field data acquisition

Robert W. Pearcy

2.1 INTRODUCTION

Measurements of environmental or plant parameters must be recorded in some fashion in order to obtain the appropriate data to meet a research objective. For instruments where the signals are read infrequently and can be copied into a notebook, meters of either the analog (pointer and scale) or digital (numeric display) type may be satisfactory. Often, however, values must be recorded over long periods of time, from many sensors at the same time or at a high frequency. Some type of a recording system is essential under any of these circumstances. Thus the requirements for recording systems for different applications in physiological ecology are quite diverse. In this chapter, the various types of recorders, data loggers and data acquisition systems of use in physiological ecology will be discussed along with a consideration of some of the problems involved with spatial and temporal sampling.

The revolution in electronics that has occurred over the last decade or two has had an enormous impact on our ability to record and analyze data. The development of the microprocessor has in particular led to rapid advances in this area. Only 10 years ago, low-cost portable data-acquisition systems suitable for field use were not commercially available. Now, highly sophisticated systems capable of accurate measurements with a variety of transducers are in wide use. These systems have allowed research that was previously impossible, too expensive or too difficult to attempt. It is likely that there will be continued development of even more sophisticated systems, although perhaps not at the pace of the past decade. Nevertheless, because these systems will continue to evolve in terms of capabilities it is necessary that the discussion here focuses on general principles and characteristics, rather than on the specific instruments.

2.2 ANALOG RECORDERS

2.2.1 Mechanical recorders

Mechanical recorders utilizing a spring-driven clock motor and a mechanical arm to move a recording pen have long been available. For example thermographs utilize a bimetalic strip that bends with temperature and hence moves a pivot arm in order to record temperature. Hydrographs utilize the dimensional changes of hairs in response to relative humidity to drive a mechanical pen system. Mechanical recorders are generally of low accuracy and at best provide only a very gross indication of the environment. Another problem is that the sensors are large and the housing even larger so that they are of little use in micrometeorological studies. They must be mounted within instrument shelters in order to avoid radiation errors. Because of these problems they are of little use in ecophysiological studies.

2.2.2 Electronic analog recorders

Electronic chart recorders have motor-driven pens and charts and typically accept an electrical signal (usually volts) from a transducer. Recorders of the *x*/*y* type have separate inputs for the *x* and *y* axis whereas strip-chart recorders rely on a chart drive that transports the paper at a given speed and a pen drive that moves a pen a distance that is proportional to the signal from the transducer in a direction that is perpendicular to the paper movement. Several different mechanisms are in use for positioning the pen, most of which have little relevance to physiological ecology. The most common and useful type is the potentiometric recorder in which the input voltage is compared to a reference voltage derived from a potentiometer or slidewire that is connected to the pen movement mechanism (Fig. 2.1). When there is a difference between the reference and input voltages, the resulting error voltage is amplified and used to drive a servomotor which in turn moves the pen and the potentiometer until it is balanced and no error signal is present. The error signal can be either positive or negative and will drive the servomotor in one direction or the other until a balance is achieved. Typically amplifier and voltage divider circuits on the input allow selection of an appropriate input voltage range to match the transducer output.

Typically recorders have one or two pens, each with an independent input servomechanism, but some have up to six pens for simultaneously recording different inputs. In addition, multipoint models are available that can sequentially sample signals from up to 32 channels and print symbols of different colours or different numbers to correspond to each sensor input. In this case, all of the inputs must be in the same range (e.g. all thermocouples).

Many problems can be encountered when

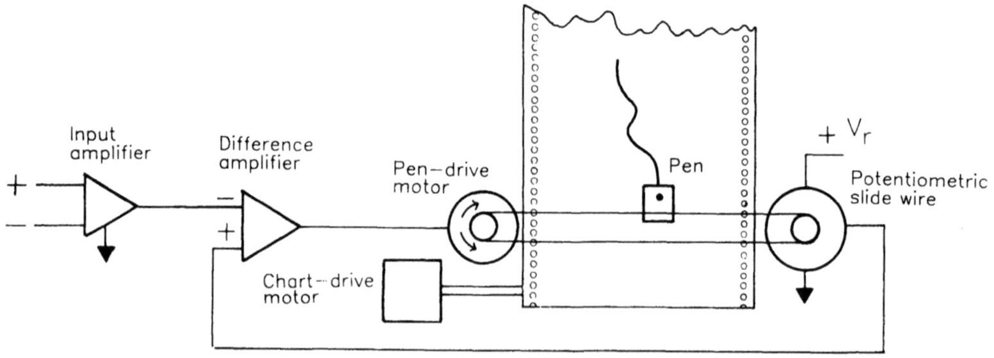

Fig. 2.1 Functional diagram of a strip-chart recorder. The pen is driven by a belt and pulleys on the pen-drive motor and the potentiometric slidewire.

strip-chart recorders are used in the field. Battery powered units are available but chart drives require considerable power and battery life can be short. Ink pens have the propensity to clog, leak or dry out at inopportune times. The mechanical components of the pen and chart drives can be adversely affected by low temperatures in particular. The pen-positioning mechanism can be quite precise but unfortunately the paper is not dimensionally stable and expands and shrinks with changes in humidity and temperature. In practical terms, this limits the accuracy to about 1% of full scale. Finally, retrieving the data from the strip chart further reduces the accuracy and can easily become an impossible burden. In general, strip-chart recorders are most useful where a trend in some variable needs to be observed and decisions are made based on this trend. For example, they are particularly useful in gas exchange systems to indicate when steady conditions (especially CO_2 and water vapor concentrations) have been achieved. For most other applications digital recorders, which are discussed below, offer a better solution.

2.3 DIGITAL RECORDERS

Data loggers are essentially digital recorders that rely on conversion of the analog signal from transducers into a digital signal via an analog to digital (A/D) converter. The digital signal can then be further processed via a microprocessor and stored in a memory, printed, or recorded on magnetic media such as a floppy disk or a magnetic recording tape. Data loggers are capable of operating in a stand-alone mode and those designed for field use may be left to record for periods of weeks to months provided that they are adequately protected from the elements or vandals. By contrast, data-acquisition systems typically require that they be directly interfaced to a computer and thus are for the most part not as suited for long-term recording. However, data-acquisition systems may offer

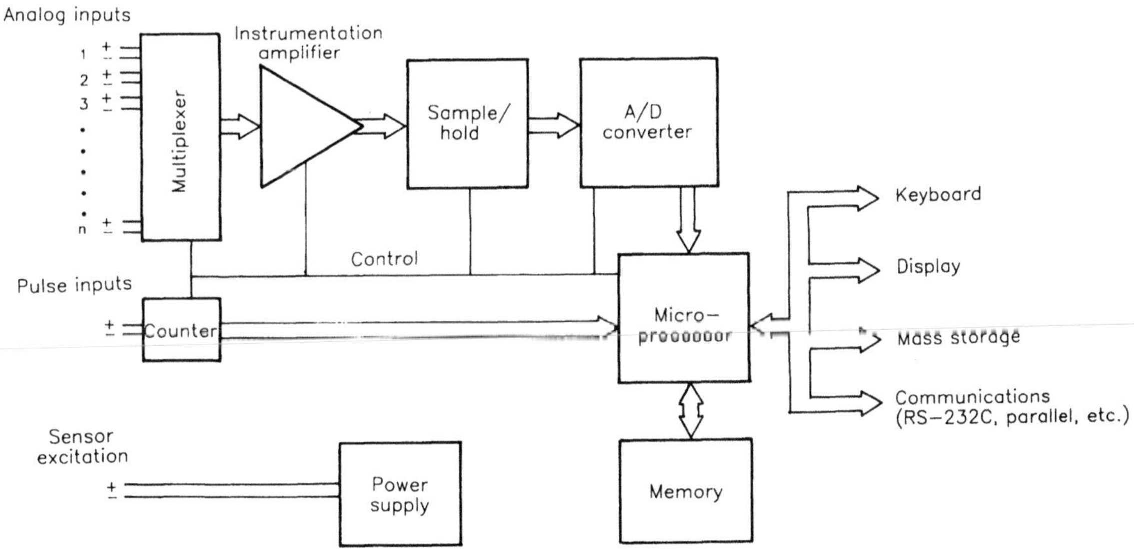

Fig. 2.2 Block diagram of a typical data logger showing the major components. For simplicity, other support components that control the flow of information to and from the microprocessor are not shown.

greater speed and flexibility in manipulation of the data than data loggers.

In addition to the A/D converter and multiplexer, other circuitry is required to make a fully functional data logger (Fig. 2.2). The input signal must be multiplexed to the A/D converter successively. The multiplexer is effectively a switching system that can be set to select just one channel at a time. Small high-quality relays with low thermal offsets have been typically used for the most accurate measurements. However, solid-state multiplexers containing transistor switches have been improved to the extent that they are as good or better than mechanical relays. A sample and hold circuit is usually also required since the A/D converter requires a steady input in order to make a correct conversion. An amplifier may also be required to boost the input signal up to an amplitude where the A/D converter can perform at its best. Programmable gain amplifiers provide the ability to switch ranges so that both high and low voltages can be measured with a high resolution. There may also be other specialized inputs, such as counters, which can handle transducers with either frequency or voltage pulse outputs, such as recording rain gauges or anemometers. Voltage outputs, either AC or DC, may be provided for sensor excitation. The microprocessor provides the control of the multiplexer and A/D converter, and converts and directs the binary output to storage memory, to an external storage medium, or to a computer. Output to the storage medium such as a tape is controlled by an input/output (I/O) device. In many data loggers, the microprocessor along with instructions stored in read-only memory allows for user control of the data logging through commands typically entered via a keyboard.

2.3.1 Communication with computers

Because data loggers are capable of accumulating large amounts of data, some means of transferring it to a computer is needed. When the information is stored on tape for later transfer to the computer, specialized tape readers are required that can translate the magnetic signals on the tape back into electronic digital signals and then transfer them. In some data loggers, solid-state memory modules are used in lieu of tapes. These also require specialized reader interfaces. Memory modules are more expensive than the standard audio cassette tapes used by many systems, but are also less prone to breakdown and use less power. Storage on audio cassettes or solid-state memory are good solutions when the data logger must operate unattended.

There are many applications, however, where it is desirable for the data logger to be directly interfaced with the computer so that further processing can be effected as soon as the measurements are taken. The major problem here is the control of the communication with the computer. Data-acquisition systems that plug directly into the data bus of a computer are widely available for personal computers. However, these are hardly portable systems. Portable personal computers of the type that would be useful in the field generally lack the capacity for addition of an internal data-acquisition card although for some, bus extenders allowing the external addition of a card are available. Alternatively, communication between an external data-acquisition system can be achieved by a communication interface port. Serial (RS-232C) interface ports are commonly found on portable computers and data-acquisition systems with serial interfaces are available. The serial interface sends information one bit at a time and was originally designed to handle communications with printers and terminals. It is relatively simple, requiring as few as three wires. Handshaking, which is control of the flow of information over the interface, can be achieved either via software in the computer and data logger, or by additional wires that provide signals to control the flow of information. While RS-232C interfaces are relatively simple getting them to work can be trying because the standard has been interpreted by manufacturers of computers and data-acquisition systems so loosely. For example, several different kinds of connectors have been used by different manufacturers. Even for the 'standard' connector, the pin assignments may vary from manufacturer to manufacturer. Getting the proper connection generally requires a careful look at the manual for the data logger and computer to find whatever information may or may not be available, plus lots of trial and error. Recently, 'smart cables' containing circuitry that can analyze the signals from the computer and the peripheral device and make the proper connections have become available. While they cost three to four times as much as a standard cable, they can eliminate much frustration. In addition to the lack of a true standard, the major disadvantages of serial interfaces are that they are relatively slow (not a serious problem in most work), require a relatively large amount of power, and allow only one instrument to be connected to the port at a time.

Many of the problems inherent in the serial interface when it is used for data acquisition are overcome in the IEEE-488 interface. This is a parallel interface where 8 bits are sent at once over eight separate wires. Control of the communications is achieved via other lines. Up to 15 instruments can be connected at one time to an IEEE-488 port. Communication with any specific instrument is done by sending it a specific command to either 'listen' or 'talk'. While this interface is much faster than an RS-232C interface, it is quite complicated with a heavy cable and is typically available only as an add-in option on personal computers and in some technical computers. Thus it is not suitable for field applications.

A third interface protocol, called HPIL for Hewlett-Packard Interface Loop, is available that is designed specifically for portable equipment. This interface was designed by the Hewlett-Packard corporation to combine many of the features of the IEEE-488 interface into a simple-to-use low-power interface. It is a serial interface that requires only two wires. Up to 30 peripheral devices, including data-acquisition systems, printers etc., can be 'daisy chained' (connected in a loop arrangement) together. Communication is controlled with listen and talk commands. While it is not as fast as an IEEE-488 interface, depending on the particular task, it can be faster than most serial interfaces. The system is easy to use and offers much more flexibility than RS-232C interfaces and is in many respects nearly ideal for use with portable systems. The major disadvantage is that it is now available only on a limited number of computers and peripherals manufactured by Hewlett-Packard.

2.3.2 Characteristics of field data loggers and data-acquisition systems

There are now a number of commercially available portable data loggers as well as data-acquisition systems for portable computers that are suitable for field use. Table 2.1 lists some manufacturers and currently available models and some of the major characteristics of these models. These systems share many common elements but also differ in important aspects that influence their capabilities for any particular use. Moreover, new models are continually being introduced so it is wise to consult with manufacturers before making decisions about suitability for a particular measurement problem. Rather than describe individual systems in detail, the general characteristics of importance in ecophysiological measurements will be discussed.

(a) Measurement ranges

Many transducers of use in ecophysiology, including light sensors, radiometers and thermocouples, produce low-level (millivolt) voltage signals. Consequently systems for use with these sensors should have an input range capable of measuring signals in the 1–10 mV range with a resolution of at least 5

Table 2.1 Characteristics of data loggers and data-acquisition systems designed for use under field conditions

	Number of channels[a]			Voltage inputs range		Resolution		Pulse counters	Digital outputs	Digital to analog output	Maximum memory	Mass storage	Interface
	Volt (D/S[b])	Current	Resistance	Min.(mV)	Max.(V)	(Bits)	(μV)	(number 8/16 bit)	(number)	switched/ continuous			
Data loggers													
Campbell Scientific (Logan, UT, USA)													
CR 10	6/12[c]	–	12 AC/DC	±2.5	±5	14	0.3	2/1	8	–	64 K	Analog cassette	RS232[j]
CR 21X	8/16[d]	–	16 AC/DC	±5	±5	14	0.3	4/1	6	4/2	64 K	Analog cassette	RS232[j]
CR 7X	14/28	–	28 AC/DC	±1.5	±5	16	0.05	4/0	8	8/2	64 K	Analog cassette	RS232
Omnidata International (Logan, UT, USA)													
Polycorder 516C	5/10[e]	–	10 DC	±50	±5	14	3	–	5	–	30 K	–	RS232
Easylogger	6/12	–	6 AC/DC	±10	±5	12	2.5	0/2	4	–	32 K	128K EProm	RS232
LI-COR Instruments (Lincoln, NE, USA)													
1000	0/8	8	8 AC/DC	±20	±20	14	1.2	2/0	–	–	32 K	–	RS232
Skye Instruments (Isle of Skye, Scotland, UK)													
SBL-600	8/0	–	–	±4	±4	12	1	–	–	–	64 K	–	RS232 Parallel

Grant Instruments (Cambridge, UK)													
Squirrel meter/ logger	4	4	–	±20	±20	12	10	2/0	2	–	64 K	–	RS232 Parallel
Data-acquisition systems													
Hewlett-Packard (Palo Alto, CA, USA)													
Model 3421A	30/56	30	30 DC	±300	±300	16	1	30[g]	30[h]	–	100[i]	–	HPIL HPIB
Remote Measurement Systems (Seattle, WA, USA)													
Model ADC-1	16	–	–	±400[k]	–	12	2	4	4	0/1	–	–	RS232
Acrosystems (Beverley, MA, USA)	16	–	–	±100	±10	16	1.7	–	32	0/1	256K	–	RS232

[a] In most cases, individual channels are user configured for either volt, current or resistance measurements. The numbers given are the maximum possible if all appropriate channels are configured for that measurement.
[b] Differential/single-ended channels.
[c] Up to six 32-channel multiplexers can be added for a total of 192 channels.
[d] The numbers given are for individual cards for analog input, pulse counting and excitation. Up to seven total cards can be added.
[e] A 16-channel multiplexer can be added externally for a total of 26 channels.
[f] Two additional inputs are especially configured for humidity sensors.
[g] Each channel can be configured as a frequency counter or pulse totalizer, only one channel is active at a time.
[h] Each input channel can be configured to close a relay, also a 16-bit digital I/O card can be installed in place of 10-channel input card.
[i] Number of readings.
[j] Requires adapter, no handshaking.
[k] Instrumentation amplifier option can give ±8 mV full scale.

μV and preferably 1 μV. Most data-acquisition boards for personal computers lack this capability. While the input signals can be amplified with instrumentation amplifiers to give more sensitivity and resolution, this often introduces noise unless they are very carefully designed. Systems designed for low-level measurements will typically have an instrumentation amplifier to boost the signal up to a level compatible with the A/D converter. It is much better to pay more for a well-designed system capable of low-level measurements than to attempt to add amplifiers afterwards. Systems should also be capable of measuring high-level (1–5 V) signals since the output of pressure transducers, infrared gas analyzers, flow meters etc. are typically in this range. Differential measurements are preferable over single-ended measurements, especially where electrical isolation of the sensors is not possible (see Chapter 1).

Many systems have the capability of measuring other signals, such as currents, resistances, conductivities, frequencies and pulse outputs, which are often useful for specific types of sensors. In addition, many can provide AC or DC voltage excitations for sensors such as thermistors, soil moisture blocks or conductivity bridges. In each case the primary measurement made is a voltage measurement, with the circuitry within the system prior to the A/D converter doing the conversion to volts. This is convenient since it avoids the problem of the user having to add specialized circuitry to the front end of the system to accommodate a particular sensor. Some systems have built-in thermocouple reference junctions.

(b) Number of channels

The number of different sensors required in any set of measurements differs greatly and will therefore influence the choice of data logger. Smaller systems typically have six to eight input channels whereas the larger systems can be expanded to several hundred channels. Some small data loggers can be expanded to a total of nearly 200 channels by addition of multiplexers.

(c) Data storage

Field data loggers typically have internal memories capable of storing at least a limited number of readings. This may range from as little as 4K of memory allowing storage of a few hundred data values to 64K giving storage of as many as 20 000 data values. Mass storage is usually done with tape recorders or solid-state memory. Transfer of the data from mass storage media to a computer usually requires a specialized reader device. Data-acquisition systems rely on the memory and mass storage system of the host computer.

(d) Programming

Nearly all data loggers have a small keyboard and a display allowing entry of a program or series of commands to control its operation. Some can be programmed via a portable computer attached to an interface port or even programmed remotely via a modem link. Data-acquisition systems rely mostly on the host computer and thus have great flexibility when the system is viewed as a whole.

The programmability of data loggers is, not surprisingly, more limited and less flexible than a computer data-acquisition system but still can be extensive. Because memory must be conserved, the language is usually a 'low-level' one where commands may be cryptic or simply identified by numbers as compared to a more easily understood 'high-level' language such as BASIC or Pascal which are widely used on microcomputers. However, in most cases programming a data logger consists of calling and in effect linking together a series of 'programs' and entering the appropriate parameters as per the instructions to control the operation of the data logger. With good instructions and a little practice, this is not a difficult task.

The programs can be as simple as conversions from voltages to more useful units, and programs for finding means, totals, maximums or minimums over a specified time interval. However, some data loggers may have more specialized programs allowing data reduction and more extensive computational capabilities. For example, the Campbell Scientific data loggers have programs for, among other things, generation of histograms, polynomial curve fitting, conversion of thermistor outputs to temperature and water vapor pressure from temperatures. Programs such as these can reduce the storage capacity needed in the field and later processing requirements. However, it should be remembered that saving more of the data in the field coupled with later processing may be a more flexible approach that can yield more information. The programs within the data logger are necessarily more restrictive than those that can be written for later data reduction on a computer.

(e) Sampling frequency

The sampling frequency is usually chosen by the operator. Some systems allow a choice of several different sampling frequencies ranging from once per min or per 10 min to once per 24 h. Many allow sampling at rates of once per second or even faster, especially for short intervals. It is important to recognize that the realized sampling rate depends on the time required for a single measurement, the number of channels scanned, the processing done with the raw data and the rate at which it is transferred to the storage device. Sometimes the maximum sampling rates given in the specifications do not include times for transfer or the time required for processing or control. The realized rate may be substantially slower.

(f) Environmental conditions

Data loggers for field use need to be able to operate over a wide range of temperatures and humidities and must be carefully engineered to avoid errors due to environmental effects. Polar and high alpine environments impose special problems because of the snow and ice and low temperatures (Walton, 1982). On the other hand, high humidities coupled with warm temperatures typical of tropical forests create another set of problems. Most systems and transducers are capable of operating up to the maximum air temperatures likely to be encountered in terrestrial environments. However, systems should be kept shaded and insulated from changes in air temperature to prevent overheating or thermal gradients that may cause errors. In humid environments, dessicant packages containing silica gel can be placed in the instrument or into an enclosure. Most systems will require some sort of enclosure for operation outdoors. These can have gaskets providing a seal to prevent air and water exchange along with sealed connectors for the wire leads from the instruments. When the enclosure is not hermetically sealed, thermal expansion and contraction of air may result in a significant 'pumping' of moist air into it, leading to a relatively rapid exhaustion of the dessicant.

When a completely sealed enclosure is impractical, an alternative approach is to use a system that allows for thermal expansion. One such approach that has proven to be practical is to attach a tube of dessicant to the enclosure and then connect this to a balloon or a bicycle tyre on the other end to provide a volume for expansion. As the air expands and contracts it passes through the dessicant keeping the enclosure dry. It is still necessary to seal the enclosure as much as possible but not to make it hermetic.

Lightning poses a special hazard in many environments. There is little assurance of survival of a data logger from a direct hit but good earth grounding of the enclosure may help. In the case of a more likely nearby hit, induced voltage surges in the lead wires from sensors can be clamped with an inexpensive spark gap.

Finally, a very real hazard to data loggers is

vandalism. Never underestimate the frequency of passers by even in remote locations.

2.4 INTEGRATORS

Sometimes it is desirable to obtain integrated totals rather than a series of measurements over time as is usually provided by recorders. For example, crop productivity is correlated with the total solar radiation input while growth rates are frequently correlated with degree days.

Integrators are used to provide a measure of the total or integrated signal from a transducer over a given time period. They are useful where hourly, daily or weekly totals are desired and can be used to calculate mean values using the time interval. Various types of integrators have been used, including those based on chemical or electrochemical reactions, physical diffusion and analog or digital electronic circuits.

The most useful types of general purpose integrators are based on digital counting techniques. Generally a voltage-to-frequency converter is used to give pulse frequency that is proportional to the input voltage. A counter circuit then totals the counts and stores it in memory or displays it. Saffell *et al.* (1979) and Woodward and Yaqub (1979) have designed simple integrators useful for light or temperature measurements. A low-cost integrator utilizing a calculator for the display is available from Delta-T instruments. In addition, integrators utilizing solid-state memory modules are available from Omnidata International (Logan, UT). The latter company also manufactures specialized integrators for degree day determinations.

An alternative approach suitable for some types of measurements are electrochemical integrators which utilize some form of electrolysis or electroplating to integrate the output of a sensor. Coulometers or E-cells, which consist of a small silver case (approx. 20 mm by 5 mm) enclosing a silver ion electrolyte and a gold electrode may be especially useful as integrators for light sensors (Newman, 1985). Current from the photocell plates silver out on the electrode. The device is read by passing a constant current through in the opposite direction to deplate the silver. Once all the silver is deplated the resistance rises substantially signalling the end point. The time required is proportional to the integrated total. A circuit diagram for an E-cell reader that interfaces to a microcomputer is given by Newman (1985). If a self-generating sensor like a photocell is used, then no batteries are required. The E cells are inexpensive, allowing many to be used at remote locations.

2.5 SAMPLING CONSIDERATIONS

2.5.1 Temporal sampling

One of the most frequently encountered problems in field measurements is how frequently should measurements be taken and over what spatial scale in order to represent the parameter in question adequately. This of course depends on the variable being measured, the sensor response and nature of the question. Thus few simple rules can be presented. In some cases it is a matter of aesthetics. Where data are being obtained to provide a representation of the environment of a plant, such as might be presented in a figure showing the diurnal temperature or light variation, then sampling must be adequate to show the variation. Measurements every 1 or 2 h may not be adequate whereas if the trends are fairly smooth then sampling every 0.25 to 0.5 h may suffice. When the parameter is changing rapidly such as light within a canopy, then a visually accurate record will require faster sampling. A too frequent sampling schedule simply adds to the burden of handling the data.

When the data are to be used for more than representing general characteristics, problems can arise when a varying signal is sampled at too slow a rate. Shannon's sampling theorem (Shannon, 1948) requires that the frequency of sampling of a signal must be at least twice the highest frequency cycle present in the signal in order to reconstruct it accurately from the data. This sampling frequency is called the Nyquist frequency. If a signal is sampled at a frequency below the Nyquist frequency then aliasing, which is the addition of lower frequency signal components to the original signal, will occur. A familiar example is the appearance of a slow rotation of the stagecoach wheels in western movies because the framing speed of the movie camera undersamples the spoke rotation frequency. Another useful example is that of leaf temperature variations. The thermal time constant of a leaf is on the order of 10 s and the resulting highest frequency variations will have a period of about 3.2 s (Woodward and Sheehy, 1983). Consequently, in order to represent these highest frequency variations faithfully, measurements would have to be made at 1.6 s intervals.

An important consideration in the sampling frequency is the time constant of the sensor. When a sensor is exposed to a new environment some finite amount of time is required before it reaches equilibrium with the environment. For a step change in the environment, the output of a sensor such as a thermocouple or thermistor will first increase rapidly and then asymptotically approach the new equilibrium. In other words the rate of change of the sensor is proportional to the difference between the sensor value and the true value of the environment, or

$$dE/dt = 1/\tau(E_s - E_a)$$

where τ is the time constant, E_s is the sensor value and E_a is the equilibrium value. It can be shown mathematically (see Fritschen and Gay, 1979) that τ is the time required for 63.2% of the total adjustment to be completed.

If the time constant of a sensor is long, the sensor will average the high frequency variations allowing for a lower sampling frequency. Thus sensors with long time constants can be used to advantage to reduce the required sampling frequency provided that they are appropriate for the measurements in other respects. Byrne (1970) and Fuchs (1971) discuss procedures for determining suitable sampling frequencies for micrometeorological measurements based on sensor time constants. Filtering circuits (see Chapter 1) can be applied that increase the time constant of a sensor (Tanner, 1963; Byrne, 1970).

However, a stringent application of Shannon's theorem is only necessary when a precise reconstruction of the signal is required, a rare case in ecophysiology. When a time average is desired then it is only necessary to sample adequately so that the standard error of the mean is within acceptable limits. Sampling need be only relatively infrequent for variables such as temperature for which the higher frequency variations are typically small in magnitude. However, for light within a canopy, which can vary over two orders of magnitude with a frequency of 0.1–100 Hz (Norman and Tanner, 1969), an averaging of hundreds or thousands of measurements may be required. The capacity of some systems to sample frequently but then average the readings for a longer period is important for these kinds of problems.

While means are often useful for characterizing an environment in general terms it should be remembered that they may not be particularly relevant for most biological processes. The inherent nonlinearity of processes such as the light or temperature responses of photosynthesis for example make it difficult to utilize average values in any predictive sense. The same problem also arises with integrated totals. Consequently, other approaches to characterizing the environment may be more useful. One general approach that deserves consideration is to determine the amount of time that an environmental parameter falls within a specified range or within a series of specified ranges, which then allows the data to be presented as a histogram. If the response of the process of interest is known (i.e. the light dependence of photosynthesis), then the ranges of histogram 'bins' can be set to correspond to the regions of importance in this response. This approach has been used to characterize understory environments with respect to photosynthetically available light (Pearcy, 1983; Chazdon *et al.*, 1988). It is also amenable for investigation of the interaction between two variables since the time when both are within certain respective ranges can be determined. Obviously, this results in more data than simple averages, but it still allows a large number of readings to be compressed into a manageable (and presentable) form.

2.5.2 Spatial sampling

The environment within plant communities is also spatially heterogeneous which can present a formidable sampling problem. For example it has been calculated that 50–400 sensors may be required in a forest understory in order to obtain a satisfactory mean solar radiation value (Reifsnyder *et al.*, 1971). For micrometeorological studies where spatial averages are often required, consideration has been given to protocols for adequate sampling. Sensors for air temperature or humidity in fact do spatially average in the horizontal axes, and less so in the vertical axis because of air movement past the sensor. The extent of spatial averaging is dependent on eddy size and wind speed and the sensor time constant. The latter two are in part dependent on position within or above the canopy. Spatial averaging can be increased by moving the sensor, either on a track or a swinging boom (Tanner, 1960, 1963). Tubular or line sensors for solar and photosynthetically active radiation are available that have an output proportional to the average value over their length (1 m in some instances). These line sensors are useful for obtaining averages when the spatial scale of variation is on the order of 1 m or less, such as in row crops, but may be less applicable elsewhere. Woodward and Sheehy (1983) discuss the sensor sizes and distances between sensors that are required for estimating the mean values to within a given error. Ehleringer (Chapter 7) discusses an approach to estimating the number of leaf temperature measurements required for determination of leaf temperatures based on leaf energy balance.

REFERENCES

Byrne, M. (1970) Data logging and scanning rate considerations in micrometeorological experiments. *Agric. Meteorol.*, **9**, 285–6.

Chazdon, R.L., Williams, K. and Field. C. (1988) Interactions between crown structure and light environments in five rainforest *Piper* species. *Amer. J. Bot.*, **75**, 1459–71.

Fritschen, L.J. and Gay, L.W. (1979) *Environmental Instrumentation*, Springer-Verlag, New York.

Fuchs, M. (1971) Data logging and scanning rate determinations in micrometeorological experiments – a discussion. *Agric. Meteorol.*, **7**, 415–18.

Newman, S.M. (1985) Low-cost sensor integrators for measuring the transmission of complex canopies to photosynthetically active radiation. *Agric. For. Meteorol.*, **35**, 243–54.

Norman, J.M. and Tanner, C.B. (1969) Transient light measurements in plant canopies. *Agron. J.*, **61**, 847–9.

Pearcy, R.W. (1983) The light environment and growth of C_3 and C_4 tree species in the understory of a Hawaiian forest. *Oecologia*, **58**, 19–25.

Reifsnyder, W.E., Furnival, G.M. and Horowitz, J.L. (1971) Spatial and temporal distribution of solar radiation beneath forest canopies. *Agric. Meteorol.*, **9**, 12–37.

Saffell, R.A., Campbell, G.S. and Campbell, R.S. (1979) An improved micropower counting integrator. *Agric. Meteorol.*, **20**, 393–6.

Shannon, C.E. (1948) A mathematical theory of communication. *Bell Syst. Tech. J.*, **27**, 379–423, 623–56.

Tanner, C.B. (1960) Energy balance approach to evapotranspiration from crops. *Soil Sci. Soc. Amer. Proc.*, **24**, 1–9.

Tanner, C.B. (1963) Basic instrumentation and measurements for plant environment and micrometeorology. *Department of Soils Bull.* 6, University of Wisconsin, Madison, WI.

Walton, D.W.H. (1982) Instruments for measuring biological microclimates for terrestrial habitats in polar and high alpine regions: a review. *Arc. Alp. Res.*, **14**, 275–86.

Woodward, F.I. and Sheehy, J.E. (1983) *Principles and Measurements in Environmental Biology*, Butterworths, London.

Woodward, F.I. and Yaqub, M. (1979) Integrator and sensors for measuring photosynthetically active radiation and temperature in the field. *J. Appl. Ecol.*, **16**, 545–52.

3

Water in the environment

Philip W. Rundel and Wesley M. Jarrell

In his introduction to a chapter on saturated flow of water through soils, Daniel Hillel (1980a) quoted Galileo Galilei as having written 'I can foretell the way of celestial bodies, but can say nothing about the movement of a small drop of water.' Certainly we can be more confident in our abilities to predict water flow and transformation than Galileo could, but there are still difficulties due to variability and non-ideality of systems in predicting the behavior of water in any system, especially ones as complex as soil and atmosphere.

In this chapter we discuss three aspects of environmental moisture measurement. The first part discusses soil moisture, the second atmospheric humidity and the third environmental moisture fluxes.

3.1 SOIL MOISTURE

In this section we discuss three aspects of soil moisture – the state of water in soils, the relationship between soil water content and soil water potential and methods of soil moisture measurement. A more complete review of these subjects can be found in Hillel (1980a,b) and Klute (1986b).

3.1.1 State of water in soils

The state of water in soils can be described in two ways: in terms of the quantity present and in terms of the energy status of the water. The quantity present is expressed either in gravimetric (mass) terms or volumetric terms. The *gravimetric water content* is the mass of water in a unit mass of dry soil, e.g. kg of water/kg of dry soil. Typically, the wet weight of the soil sample is determined, the sample is dried at 100–110°C to constant weight and reweighed (Gardner, 1986).

The *volumetric water content* is expressed in terms of the volume of water per volume of soil (e.g. litres of water/litres of soil). Volumetric water content can be calculated from gravimetric water content by multiplying the gravimetric by the soil bulk density (kg of dry soil/litre of soil). The bulk density value used should be appropriate for the soil under field conditions, rather than that determined after disturbance.

The energy status of water in soil can be

expressed as follows (Hillel, 1980a; Hanks and Ashcroft, 1980):

$$\Psi_{total} = \Psi_{matric} + \Psi_{solute} + \Psi_{grav.} \quad [MPa] \quad (3.1)$$

where Ψ_{total} = total soil water potential (MPa), Ψ_{matric} = soil matric potential (MPa), Ψ_{solute} = soil solute potential (MPa) and $\Psi_{grav.}$ = pressure potential or gravimetric water potential (MPa). Units of soil water potential are sometimes confusing, since pressure head (in length units), energy/unit volume, energy/unit mass and pressure terms have all been used. This discussion will utilize pressure units, in terms of the pascal (newtons m^{-2}) and megapascal (10^6 pascal, MPa). For a discussion of the interconversion between these units, see Hillel (1980a) or Nobel (1983).

The energy of soil water is attenuated by the hydrophilic surfaces of soil particles. As a result of the attraction of water to these surfaces, the energy of the water is decreased. Water forms films around the particles and fills pores. This fraction of the soil water energy has been termed the 'capillary' suction or matric water potential.

The value of the matric term can be calculated from the capillary rise equation:

$$\Psi_{matric} = -\rho_w \boldsymbol{g} h = \frac{-2\ \gamma \cos\alpha}{r} \quad [MPa] \quad (3.2)$$

where ρ_w = density of water (kg l^{-1}), h = height of rise above a free water (e.g. groundwater) surface (m), $\boldsymbol{g}$ = acceleration due to gravity (m s^{-2}), γ = surface tension (newtons m^{-1}), α = wetting angle (degrees) and r = capillary radius (m) (Hanks and Ashcroft, 1980). The Ψ_{matric} is assigned a negative value because it is expressed as rise above a free water surface.

The pressure potential arises in saturated soil due to the pressure of water above a given point. It can be calculated from the equation

$$\Psi_{grav.} = \rho_w \boldsymbol{g} h \quad [MPa] \quad (3.3)$$

where ρ_w, $\boldsymbol{g}$ and h are as defined in Equation 3.2 above.

The presence of solutes in the soil water further decreases its energy potential. The solute or osmotic potential of soil water is less than or equal to zero, and is directly related to the total solute concentration in the water. van't Hoff's rule allows a first approximation of this term:

$$\Psi_{solute} = c\boldsymbol{R}T \quad [MPa] \quad (3.4)$$

where Ψ_{solute} is the solute potential, $\boldsymbol{R}$ is the universal gas constant (8.3143 J K^{-1} mol^{-1}), T is the absolute temperature, and c is the osmolality of the solution. At low concentrations where the activity coefficient is near 1, c is approximately equal to the total molar concentration of osmotically active species in the water. For example, 1 millimolal sucrose would contribute 1 milliosmolal to the term, while 1 millimolal of a dissociating electrolyte such as NaCl would contribute 2 milliosmolals to the total osmolality of the solution.

The matric and osmotic forces both lower the vapor pressure of the soil water, and thus lower the relative humidity of air in equilibrium with the water. The lower the water potential, the lower the relative humidity of the air in contact with it. This relationship is exploited in the use of psychrometers to estimate the sum of matric and osmotic potentials in the soil water.

3.1.2 Relationship between soil water content and soil water potential

For a given soil, there is expected to be a unique relationship between the soil water content and the soil water potential. This relationship is known as the soil water characteristic curve, soil water retention curve or soil water release curve (Hanks and Ashcroft, 1980; Klute, 1986b). The curve, derived by determining the energy status of water in the soil at several water contents, may vary considerably with changes in soil texture (Fig. 3.1).

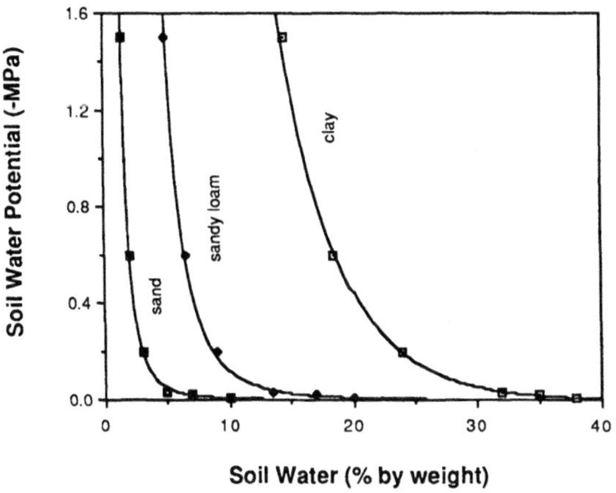

Fig. 3.1 Typical relationships between soil water content and soil water potential in sand, sandy loam and clay soils.

Two general approaches are used to measure the relationship between soil water content and soil water potential. Either a given water content is first established, and the water potential then determined, or conditions are imposed on a soil sample to bring it to a given water potential, and the water content of the sample determined after equilibrium has been reached.

In the latter case, vacuum and pressure plate apparatuses have been used extensively down to −1.5 MPa matric water potentials (Klute, 1986b). These are best applied where the soil solution is dilute and therefore the contribution of solutes to total water potential is minimal. In typical applications, moist soil samples are placed on a ceramic plate (down to −0.08 MPa) or a membrane (to −1.5 MPa), and a fixed suction or pressure is applied to a given potential until no more water is forced out of the sample. In practical terms, a vacuum can be applied to the ceramic plates down to potentials of approximately −0.8 MPa. Below this potential, the soil samples must be housed in a pressure chamber to which constant known air pressure can be applied; water is then forced out of the soil sample and through the ceramic plate or membrane until no more water is lost. At this point, it is assumed that the water potential of the remaining soil water is exactly equal to the negative of the pressure applied. This technique is typically used down to water potentials of −1.5 MPa.

To determine lower water potentials, a dry soil sample may be placed in a dessicator and brought to constant weight in the presence of a solution of known osmotic potential. Water leaves the solution and condenses in the soil or vice versa until equilibrium conditions are reached; the soil water content is determined at that time.

More recently, psychrometric systems have been used to determine the total soil water potential in samples at different soil water contents. Essentially the relative humidity of air in equilibrium with the moist soil sample is determined, and expressed in terms of the corresponding water potential. If the soil is low in salts, only the matric potential component is represented; otherwise, the sum of matric and osmotic potentials results. Because relative humidities near 100% may be difficult to measure accurately, the psychrometric technique is best applied to systems where the soil water potential is less than −0.2 MPa.

3.1.3 Measurements of soil moisture

Six techniques for measurement of soil moisture are described in this section. Three of these [gravimetric analysis, neutron probe measurements and time domain reflectometry (TDR)] measure soil moisture content. Two others, tensiometers and resistance blocks, measure soil matric potential as an analog of soil water potential. The final method, using soil psychrometers, measures soil water potential.

(a) Gravimetric water content

Perhaps the simplest measurement of soil water is that of the *gravimetric soil water content*, often represented by theta (θ). This value is determined by drying the collected soil sample at 100–110°C to constant weight.

Although very simple, it is a destructive sampling procedure and says nothing about the availability of the water to the plant. Availability is thought to be determined primarily by energy status of the water in the soil.

(b) Neutron probe

Fast neutrons slow down when they strike a body of similar mass, such as a hydrogen nucleus. Energy from the neutron is transmitted to the proton, and the rebounding neutron is much slower. This principle has been adapted to estimate the density of hydrogen nuclei in soil. In most soils the bulk of the hydrogen is associated with soil water, although in organic soils or densely rooted soils, this may not be the case. Fast neutrons are emitted from a source lowered into a tube installed in the soil; slow neutrons bouncing back to a detector below the source are counted. The number of slow neutrons detected is proportional to the number of collisions between neutron and hydrogen nuclei, which in turn reflects the soil water content.

Because of its convenience in making repeated measurements at depth over long periods of time, the neutron probe technique has been used in a wide variety of systems, especially agricultural (e.g. Fadl, 1978; Turk and Hall, 1980; Greacen, 1981; Devitt *et al.*, 1983; Hall and Jones, 1983; Hanna *et al.*, 1983; Hodnett, 1986), but natural ecosystems such as forest and grassland as well (e.g. Hanna and Siam, 1980; Kachanoski and DeJong, 1982; Dodd *et al.*, 1984). This technique is best suited for deep finer-textured soils where holes and tubes are easily installed. Sandy soils often present difficulties in installing appropriate tubing because of caving in from the sides. Rocky or stony soils present extreme problems for installation, and often require power-driven drilling rigs.

Depending upon soil properties, soil water content and type of casing used, the neutrons from the probe sample a sphere of soil with a radius of 10–30 cm. This radius makes it difficult to accurately measure soil moisture content in the surface, with erroneously low counts being obtained (Hanna and Siam, 1980). To overcome this problem, special calibration curves can be developed for readings near the surface. Surface samples may also be monitored differently, such as through gravimetric sampling (Hanna and Siam, 1980). Also, fiberglass or aluminum shields have been set in place at the soil surface and filled with surface soil, to keep neutrons from escaping from the soil.

Measurement holes are typically cased with seamless steel, aluminum or polyvinylchloride (PVC) tubing. Diameters used have primarily been 4 cm (1.625 inch), but more recently 5 cm (2 inch) probes have become popular because of easy availability and lower cost for these materials.

In all cases, care should be taken to prevent prolonged human exposure to fast neutrons since they represent a significant health hazard. Operators of a neutron probe must use the instrument properly and wear a proper radiation badge at all times to measure exposure to test neutrons. One should follow the manufacturers' specifications in having the machine tested regularly by a licensed radiation technician.

Calibration. Each individual probe should be calibrated for each general soil type for which it is used (Balbalola, 1978; Hall and Jones, 1983; Bruce and Luxmoore, 1986). Particle size distribution, structure and even mineralogical composition (Balbalola, 1978) can affect the relationship between count rate and soil water content. Gravel layers and stony soils require special attention (Balbalola, 1978; Luxmoore *et al.*, 1981). Different types of casing material can also affect calibration. For these reasons, factory calibrations should not be relied on. Special procedures are also required for clay soils to allow for cracking clay soils (Hodgson and Chan, 1987).

Probes may be calibrated both directly in the field and in the laboratory. Large containers (e.g. 200 litre drums) can be filled with field soil, a tube set in place and the material brought to a known water content and/or potential (Balbalola, 1978). Neutron probe readings are taken, water content is increased and equilibrium again established, then another reading is taken. In this manner, a relationship can be established between the soil water content and the counting rate (Vauclin *et al.*, 1984).

Alternatively, a rapid and generally effective field calibration can be undertaken (e.g. Parkes and Siam, 1979; Field *et al.*, 1984). Soil samples from the field are collected in increments corresponding to those to be sampled with the probe, e.g. 25 cm. Water content and texture of each sample is determined. Since slopes of calibration curves are generally similar even for different textural classes, this provides a means for calibration of the samples.

In soils that have layers of distinct textures, the effective sphere of sampling may encompass two or more of these layers (van Bavel *et al.*, 1954). In this case, accurate calibration is nearly impossible. Other methods may be used to measure soil water status, such as coring and direct determination of gravimetric water content.

Bouma *et al.* (1982) stated that *in situ* measurement of pressure heads, with neutron probe readings of soil water content, gave more realistic curves than those from standard laboratory desorption techniques. Their results clearly show the dangers inherent in transferring results obtained with small disturbed soil samples to field soils with their spatial heterogeneity in pore size distribution.

Data analysis. Typically, ten standard counts are taken with the probe within the shield. This is an internal calibration of the electronics of the instrument. Count rates should vary by less than 5% for the last five determinations. This standard count can then be used to calculated the exact standard curve for that sequence of samples.

The field count results and the calibration curve(s) can be entered into a computer as collected, or subsequent to collection. A microcomputer system (IBM-PC based) has been described which calculates moisture contents for incremental depths, total profile or root zones (Huisman, 1985). Changes in soil water content, seasonal amounts and soil water distributions could be calculated within the same program.

Field studies. Neutron probe access tubes generally are installed to depths over 1 m (Table 3.1), readings taken at regular depth increments and at regular time intervals. Depths may be adjusted to take specific soil horizons into account and time intervals adjusted to account for irrigation or rainfall.

An excellent example of the value of neutron probe studies in assessing field problems of plant water relations can be seen in studies by Dodd *et al.* (1984) of monthly patterns of soil moisture dynamics in *Banksia* woodlands near Perth, Western Australia. The variation in rooting architectures of the dominant woody species allows for equal soil moisture use through the deep profile (Fig. 3.2).

In studies with Carolina forest soils, Hewlett *et al.* (1964) found that instrument and timing variability in their neutron probe data was small compared with spatial variability in soil moisture values. From 30 samples collected over several hectares they found a calibrated standard error of 2.2 mm H_2O for a 75 cm depth. This was the same order of magnitude as the daily evapotranspiration, thus eliminating any value of this technique for evaluating daily cycles of evapotranspiration flux.

The neutron probe has been used to follow the course of water infiltration during irrigation of native desert shrubs (Jarrell and Virginia, unpublished). Soil water profiles, prior to irrigation, are measured, then water content is monitored during irrigation until sufficient water has entered the profile to bring the surface 2 m of soil to approximately

Table 3.1 Comparative aspects of common techniques of measuring soil moisture potential and soil moisture content

Technique	*Range*	*Advantages*	*Disadvantages*
Tensiometer	−0.00 to −0.08 MPa	1. Direct readout of soil water matric potential 2. Inexpensive 3. Can be automated for continuous readout 4. Relatively reliable	1. Requires SMCC* to relate to soil water content 2. Samples small portion of soil, near cup 3. May lose function if drier than −0.20 MPa 4. Measures only matrix potential, not solute 5. Exposed gauge sensitive to disturbance
Psychrometer	−0.20 to −10.00 MPa	1. Measures total water potential 2. Can be automated for continuous monitoring 3. Useful over wide range of potentials, especially in very dry soil	1. Samples very small fraction of total soil volume 2. Relatively sensitive to temperature gradients 3. Requires SMCC to convert value to soil water content 4. May be inactivated by salinity 5. Units fairly expensive
Resistance blocks	−0.10 to −7.0 MPa	1. Calibrated to soil water potential 2. Reliable 3. Inexpensive, readily fabricated 4. Can be automated for monitoring	1. Requires SMCC to relate to soil water content 2. Must be calibrated individually 3. Samples small volume of soil 4. Does not function properly in saline systems
Neutron probe	0.02 to 0.50 kg water kg^{-1} soil	1. Gives direct reading of soil water content 2. Samples large volume of the soil 3. One unit can be used in many locations 4. Relatively precise 5. Small portion above soil surface	1. Individual calibration for each soil type for accuracy 2. Difficult to use in automatic monitoring 3. Use near soil surface requires special technique 4. Instrument initially expensive 5. Calibration with SMCC questionable in layered soils
Soil xylem water potential, pre-dawn	−0.02 to −5.00 MPa	1. Direct relationship to plant root zone 2. Integrated value for root zone 3. Rapid straightforward measurement 4. Can indicate deep sources of water where digging is difficult or impossible	1. Integrates in unknown fashion the rooted soil 2. Cannot be readily related to soil water content 3. May require many plants

Technique	*Range*	*Advantages*	*Disadvantages*
Gravimetric water content	Full range of water contents	1. Low equipment requirements 2. Straightforward interpretation	1. Destructive sampling 2. Collection and transport logistics 3. Sensitive to small-scale spatial variability
Time domain reflectrometry	0.065 to 0.50 kg water kg^{-1} soil	1. Measures water content 2. Samples large soil volume, decreases interference due to heterogeneity 3. May decompose signal to estimate soil water salinity as well as water content 4. Relatively stable over time	1. Requires high initial investment 2. Insertion of probes may be difficult 3. May sample excessively large soil volume

* Soil moisture characteristic curve.

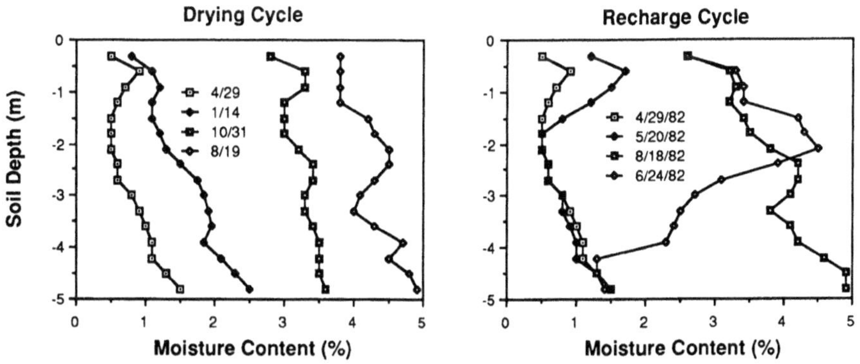

Fig. 3.2 Seasonal changes in soil moisture content with depth over drying and recharge cycles in a *Banksia* woodland community near Perth, Western Australia. Data were collected using a soil neutron probe by Dodd *et al.* (1984).

field capacity (defined as soil water content after 2 days of free drainage). The microemitters for each plant were then closed off separately, depending upon the status of the individual soil.

(c) Tensiometers

Tensiometers are widely used for measuring soil moisture availability where Ψ_{total} is high (Richards, 1965). These devices are simple and inexpensive, making them particularly useful in agricultural studies when many may be required. In operation, tensiometers measure soil suction or matric potential developed as water is pulled out of a ceramic cup into soil by matric forces. The ceramic cup is sealed to a water-filled tube. The vacuum developed is measured using a regular diaphragm-based manometer or mercury manometer. In either case, the vacuum obtained is assumed to be equal to the matric water potential in the soil.

The tensiometer functions best in the range

of 0 to −0.08 MPa (Nnyamah *et al.*, 1978; Devitt *et al.*, 1983). Because air begins to enter the pores of the ceramic cup in soils drier than −0.07 to −0.08 MPa, the device is not useful in even moderately dry soils. Once air enters the tube, values are no longer accurate. Tensiometers are completely insensitive to soil solution osmotic potential, and thus will not provide accurate measurements of soil water potential in soils with significant salinity.

Tensiometers, usually consisting of a simple tube with a ceramic cup glued to one end and a vacuum gauge on the other, have been used for many years to estimate the soil matric potential or suction (Gardner, 1965; Richards, 1965). They are frequently used in combination with the neutron probe, resistance blocks or psychrometers to cover the full range of soil water contents. Much recent research has gone into improving methods for continuous monitoring and data-logging output from tensiometers (Blackwell and Elsworth, 1980; Marthaler *et al.*, 1983). The latter authors described a very inexpensive system that can supply digital signals for a data-logging system.

Tensiometers have been used to estimate water balance (Devitt *et al.*, 1983), follow capillary rise above a water table (McIntyre, 1982) and characterize unsaturated soil hydraulic conductivity (Ward *et al.*, 1983). Estimates based on tensiometer readings were as accurate as pan evaporation measurements in predicting water loss from a weighing lysimeter (Devitt *et al.*, 1983). The significance of tensiometer cup size is well demonstrated by the results of Bouma *et al.* (1982) in describing water movement in soil with vertical worm channels. Large (8 cm long, 2 cm diameter) cups tended to intercept large saturated pores more often during the infiltration phase, falsely suggesting that the soil was at saturation. Large pores were less likely to be encountered with small (0.5 cm length and width) cups, and variability in their readings clearly reflected a more heterogeneous soil moisture regime.

Faiz (1983) has described an approach he used which he termed a 'biological tensiometer.' Water potential 'equilibrium' was induced by covering the plant with a polyethylene bag for 8 h on average, then using a pressure chamber to measure plant water potential. The method was only used with container-grown plants, but the plant potential measurements agreed well with soil water potential measurements taken with soil psychrometers. No comparison was made between pre-dawn water potential and the values derived using this method. Since most plants close stomata in the dark, they are generally assumed to have reached equilibrium with some root-integrated soil water potential some time before dawn. Thus, the measurements should agree well with each other (see Slavik, 1974). Parkes and Waters (1980) report an uncertainty (95% confidence limit) of ±2 cm water for tensions of 25–200 cm water.

(d) Resistance blocks

Resistance blocks provide a second important device for measuring the matric potential of soils as an analog of total soil water potential. Their operation is based on the fact that the electrical conductivity of many materials varies as a function of water content. Most commonly, blocks of solid gypsum containing two embedded electrodes are used. The conductivity through the block increases as the quantity of soil water absorbed by the block increases, while solution salt concentration (saturated gypsum) remains constant.

In operation, measurements are made by connecting an ohmmeter to the electrodes of the resistance block. The resistance is proportional to the quantity of water in the block, which in turn depends upon the extent to which the block competes with the soil constituents for water. Each block must be calibrated to develop a curve relating soil water potential to resistance measure. Preferably, the calibration should be conducted in the same soil in which the field measurements are to be taken. A useful technique is to calibrate blocks in soil on a pressure plate apparatus. In this way, resistance, water content and soil water potential can be determined simultaneously on each sample.

Blocks have also been manufactured of fiberglass and other materials, but since the electrical conductance of soil solution is a function of its composition, they do not work well in saline soils or where saline irrigation water is applied (Colman and Hendrix, 1950). As with gypsum blocks, these must be calibrated to develop relationships between conductivity and soil matric potential.

Resistance blocks work best in soils drier than −0.05 MPa, making them complementary in range of operation to soil tensiometers. They are typically accurate to soil matric potentials as low as −2.0 to −3.0 MPa. Because response time of resistance blocks is slow, they are not useful for following rapid wetting events. Significant hysteresis effects may also be found between wetting and drying calibrations. For these reasons, resistance blocks are most useful for following slow drying cycles. Since gypsum blocks can serve as a local source of calcium and sulfate, they should not be used where these elements affect plant growth.

Resistance blocks have been used routinely in a variety of systems frequently to schedule irrigations. Most systems have historically required an AC Wheatstone bridge circuit to measure the impedance between the imbedded electrodes. New circuitry has been described to make resistance blocks more useful at water potentials as high as −0.03 MPa (Goltz *et al.*, 1981). Other developments include a portable recorder for continuous monitoring (Schlub and Maine, 1979) in the range of plant-available water (−0.03 to −1.50 MPa). After one season, calibration curves were unchanged between −0.1 and −1.50 MPa; at higher potentials, less than a 0.01 MPa shift was noted in the calibration curves. The shift was confirmed through comparison with tensiometer results.

(e) Soil psychrometers

Total soil water potential (i.e. matric plus osmotic potential) can be determined using soil psychrometers. These are typically thermocouple psychrometers encased in a ceramic bulb. The bulb allows equilibration between soil solution and the atmosphere inside the bulb. At equilibrium, the relative humidity inside the bulb is equal to that of the soil atmosphere. Since both matric and solute forces contribute to the equilibrium relative humidity, the device measures total soil water potential (omitting any gravitational or overburden pressures).

The thermocouple psychrometer is effective in moderately dry to dry soils. In wet soils the humidity is too close to saturation for very precise measurements to be made; usually soils above −0.2 MPa cannot be effectively monitored. In fact, over the range of water availability to plants, the soil relative humidity only varies between 0.99 and 1.00 (Rawlins and Campbell, 1986). Where soils are saline, salt may deposit in the ceramic cup as the soil progresses through wetting and drying cycles, effectively inactivating the device. Finally, because relative humidity is affected by temperature, steep temperature gradients in the soil or inaccurate temperature measurements may make results meaningless. This can be a significant problem in surface soil where diurnal heat fluxes occur. In such situations, psychrometer measurements are best made at the two times of the day when thermal gradients are near zero at a particular depth.

Construction and theory of operation. The construction of a typical soil psychrometer is shown in Fig. 3.3. Most commonly, very fine chromel and constantan wires (25 μm diameter) are used to form the thermocouple junction used in the measurement of equilibrium relative humidity, with points of contact at gold pins forming the reference junction. The fine wire gauge is used so that cooling of the thermocouple junction will not significantly change the temperature of the

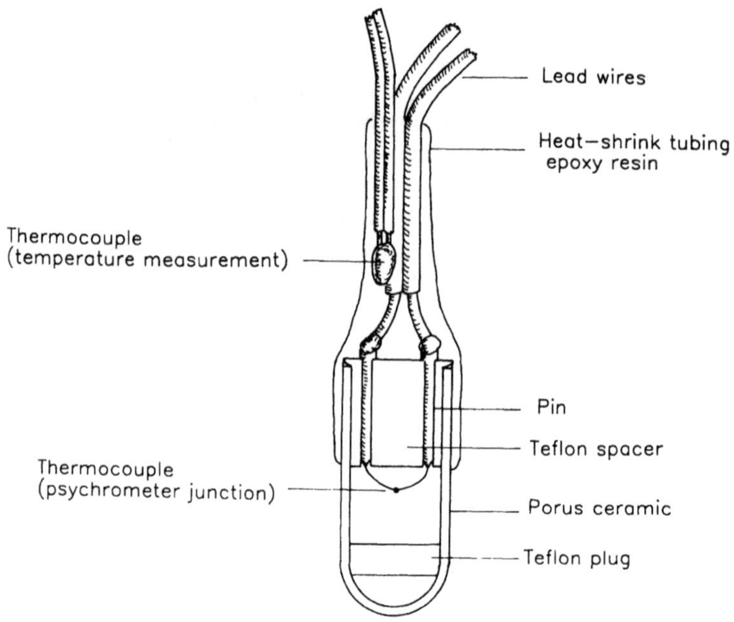

Fig. 3.3 Structure of a typical soil psychrometer. See text for discussion.

larger reference junction. This junction is protected by a porous ceramic or stainless steel screen shield. A second thermocouple formed of larger gauge copper and constantan wire is also included to measure the temperature of the sensor.

There are two modes of operation for soil thermocouple psychrometers. In the *psychrometric mode* the sensing thermocouple is allowed to come into temperature equilibrium with the air space around it, which has been allowed to reach temperature and vapor equilibration with the soil sample enclosed. A small current is then briefly passed through the thermocouple to cause cooling of the junction below the dew point temperature of equilibrated air using the Peltier effect. This cooling will cause water to condense out from the air on to the junction. As the Peltier current is discontinued, the declining microvolt output of the thermocouple is monitored as the junction temperature returns to ambient. The microvolt output drops rapidly in a few seconds before reaching a sloping plateau (point c in Fig. 3.4) when the wet bulb depression temperature is reached. The level of output at this point can be calibrated, with an approximate relationship of 45 $\mu V\ kPa^{-1}$ at 25°C. Readings can be taken until point d is reached (Fig. 3.4) when the condensed water on the junction is depleted.

The *dew-point mode* is somewhat more complex than the psychrometric mode, but offers some advantages in convenience of data readout. In this method of measurement, first described by Neumann and Thurtell (1972) and later refined by Campbell *et al.* (1973), the thermocouple is thermally isolated so that latent heat from the water is the only mechanism for transferring heat to or from the junction. Thus, the wet junction temperature converges to a stable dew-point temperature. The output of the thermocouple quantifies the dew-point temperature and, thus, the relative humidity within the sample

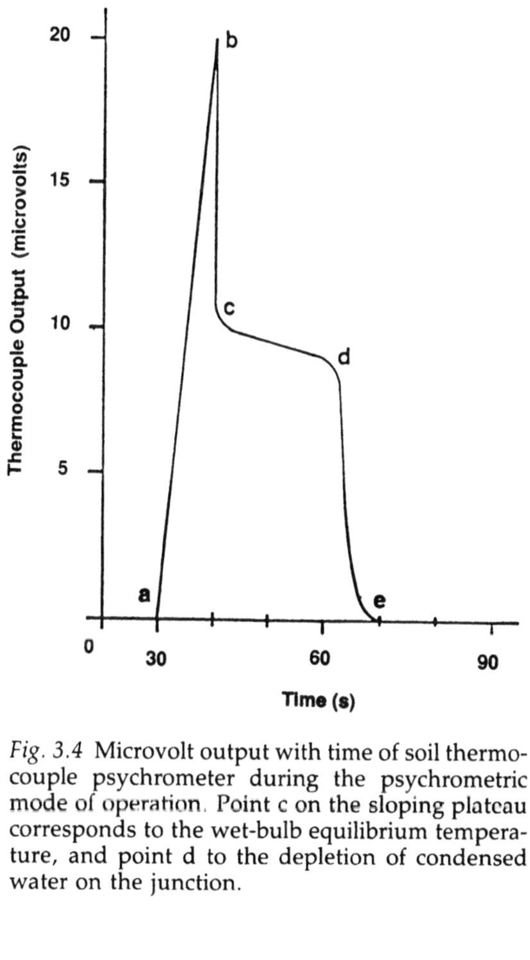

Fig. 3.4 Microvolt output with time of soil thermocouple psychrometer during the psychrometric mode of operation. Point c on the sloping plateau corresponds to the wet-bulb equilibrium temperature, and point d to the depletion of condensed water on the junction.

chamber. This output is proportional to water potential with a relationship of approximately 75 μV kPa^{-1}. Since the thermocouple output in the dew-point mode of reading is constant, the time of reading is not critical as it is in the psychrometric mode (Fig. 3.4).

Installation and calibration. Soil psychrometers should generally be installed so that the axis of the sensor is parallel to the surface of the soil. This orientation minimizes thermal gradients across the sensor. Because of problems of contamination and failure, extra psychrometers should be installed to maximize chances of obtaining continuous data records. Having extra psychrometers in place will also allow occasional removal of sensors to test for changes in calibration.

The calibration of soil psychrometers, as with leaf psychrometers, is generally performed with known solutions of NaCl or KCl (see Appendix Table A14). Each individual psychrometer should be calibrated. Since soil psychrometers include no large heat sink, it is impossible to calibrate these accurately unless the temperature of the sensors is equilibrated under isothermal conditions. If insulated containers are used, at least 50 cm of lead wire should be included within the container to prevent heat fluxes along this wire (Briscoe, 1986). Water baths are a common means of stabilizing psychrometer temperatures, but care must be taken to minimize thermal gradients in such baths. Turning off the heat source of the water bath for several minutes prior to calibration measurement is a useful technique for briefly reducing or eliminating such thermal gradients. Wescor, a major producer of psychrometric instrumentation, suggests that the calibration of soil psychrometers will most nearly approximate field conditions if the sensors are immersed directly in the calibrated solution (Briscoe, 1986). The pore size of the ceramic or stainless-steel shield surrounding the thermocouple is sufficiently small to prevent liquid water from entering the cavity.

Because accurate measurement of soil water potential with thermocouple psychrometers requires a uniform rate of water evaporation from the thermocouple junction and a uniform concentration of water vapor throughout the equilibration cavity, contamination presents a significant problem. The presence of salts on the junction changes the rate of evaporation and thus invalidates the calibration. Ceramic shields over the thermocouple have a pore diameter of 3 μm, while stainless-steel shields have a pore diameter of 20–30 μm. While soil particles will not pass through these shields, dissolved salts may pass through and precipitate on the inner wall of the shield or the junction. In addition to calibration problems, precipitated salts and high humidities often lead to corrosion problems which can affect the precision of water potential measurements. It is not unusual for corrosion to lead to destruction of the fine thermocouple wires. For these reasons a single calibration of soil psychrometers may not be sufficient to insure accurate readings over time.

Field studies. There have been many field studies covering a wide range of ecosystems in which soil psychrometers have been successfully used to monitor soil water potential. For example, Fonteyn *et al.* (1987) and Schlesinger *et al.* (1987) have very carefully documented the value of such psychrometers in field studies of soil moisture potential in a desert grassland.

Roundy (1984) used *in situ* electrical conductivity sensors and Peltier psychrometry to estimate the total, solute and, by difference, the matric potentials of saline soils. Electrical conductivity of the soil solution had to be determined and a calibration established between the electrical conductivity and the solute potential. Field salt concentrations were calculated on the basis of soil water content and saturation extract electrical conductivity. With the availability of vapor pressure osmometers based on psychrometry (e.g. Wescor, Logan, Utah), it is simple to determine the solute potential of any liquid sample directly. Also, laboratory psychrometer chambers now available can be used to measure the total water potential of any sample which will fit into the sample wells, and work well on a routine basis (Decagon Devices, SC-10A).

Nnyamah *et al.* (1978) have used psychrometers with tensiometers to follow soil potential changes under a Douglas fir forest, to estimate contact resistance between roots and soil. Root xylem water potentials were determined with hygrometers.

(f) Time-domain reflectrometry (TDR).
Measurement of bulk soil water content is most desirable for water-balance studies in the field. The larger the volume of soil sampled with a given technique, the smaller the effect of small-scale heterogeneity on measured values. For these reasons, a new technique called time-domain reflectrometry (TDR) shows promise for measurements of soil water content (Dasberg and Nadler, 1987; Topp, 1987; Dalton, 1987).

The relative dielectric constant of soil materials varies directly with soil water content (Topp *et al.*, 1980, 1982; Dasberg and Dalton, 1985; Topp and Davis, 1985). The dielectric constant is proportional to the length of time required for a voltage pulse to pass through a known length of soil (Topp *et al.*, 1980). The soil dielectric constant has been measured with coaxial cables oriented vertically (Topp *et al.*, 1980), or brass or stainless-steel rods oriented vertically or horizontally (Dasberg and Dalton, 1985; Topp and Davis, 1985). Topp *et al.* (1980) developed an empirical relationship between the dielectric constant and the soil water content.

Horizontal lines have been found to have lower standard errors than the vertical lines (Topp and Davis, 1985). Both 3 mm and 6 mm diameter rods appear adequate, with the advantage of not requiring any pilot hole preparation compared with 12.7 mm diameter rods. Rods 0.125–1 m or more in length have been used (Topp and Davis, 1985). Distances between rods are on the order of 0.05–0.5 m (Dasberg and Dalton, 1985; Topp and Davis, 1985). The TDR trace may be obtained with a cable tester (Topp and Davis, 1985) or a combination of oscilloscope, pulse generator and sampling head (Dasberg and Dalton, 1985).

Soil texture, density, temperature and salt content have relatively little effect on the measurement. However, sufficient consistent variation in signal has been found that soil water electrical conductivity can also be estimated using this technique (Dalton *et al.*, 1984; Dasberg and Dalton, 1985). Again, this provides an in-field estimate of the soil salinity averaged over a substantial soil volume, thus making it less sensitive to local variations. Presumably this could correlate better with plant response, since the value should relate more directly to conditions to which the plant is exposed in the field.

The major advantages of this technique include (1) the soil volume tested is relatively large; (2) opportunity for frequent measurements, throughout the growing season; (3) possibly general relationships that hold across several soils, with relatively little calibration required; (4) ability to estimate soil salinity as well as water content. Disadvantages include (1) lack of good database for understanding how general the above relationships are; (2) no standardized equipment; (3) some inconsistent results near the surface; (4) difficulties in inserting the horizontal probes. With further development, this technique may fulfill its promise for larger-scale accurate water measurement, especially in plant communities where plant spacing is small relative to the probe lengths, such as grasslands or crop lands.

3.1.4 Solute potential

High salt concentrations in the soil solution decrease plant growth (Greenway and Munns, 1980). Although specific ion toxicities may occur, decreased water availability is also a factor. Methods have been developed for measuring extractable salts (Page *et al.*, 1982) as well as determining soil salt content *in situ* (Rhoades and Oster, 1986).

Solute potential is commonly determined by direct extraction of soil solution, often after the water content of the sample has been raised to saturation. Soil solution can be extracted from saturated soil using a simple vacuum apparatus. The solute potential can be determined directly using a vapor pressure osmometer or a thermocouple psychrometer system. More frequently, the specific conductivity of this saturation extract is determined using a conductivity cell. Units of conductivity are decisiemens m^{-1}, dS m^{-1} (historically mmhos cm^{-1}). Conductivity is generally proportional to the solute water potential, although the precise relationship depends upon the specific salt ions present.

Techniques for assessing soil salinity *in situ* include porous matrix salinity sensors and bulk electrical conductivity sensors (Rhoades and Oster, 1986). Porous matrix sensors (Richards, 1966) are units with soil-sensor contact areas on the order of 1 cm^2. Salt in the soil solution enters the pores of the matrix and electrical conductivity is determined. The device therefore provides a point measurement of soil salinity. Bulk soil electrical conductivity sensors, on the other hand, include time-domain reflectrometry and the four-electrode technique (Rhoades and van Schilfgaarde, 1976; Rhoades and Corwin, 1980). These techniques typically measure soil salinity characteristics over an area of meters-wide dimension. Depending on spacing and current inputs, different depths of soil may be sampled (Rhoades and Oster, 1986). Smaller four-probe sensors have also been tested (Rhoades, 1979).

3.2 ATMOSPHERIC MOISTURE

3.2.1 Expressions to describe water vapor in the atmosphere

Water vapor in air, like other gases, exerts a pressure termed *vapor pressure*. If a sample of air is sealed over a surface of pure water under conditions of constant temperature, an equilibrium is eventually reached at which the rates of water molecules entering and leaving the liquid are the same. The vapor pressure of water molecules at this stage is termed the *saturation vapor pressure*. This equilibrium value is temperature dependent, as shown in Fig. 3.5 and in Appendix Table A7.

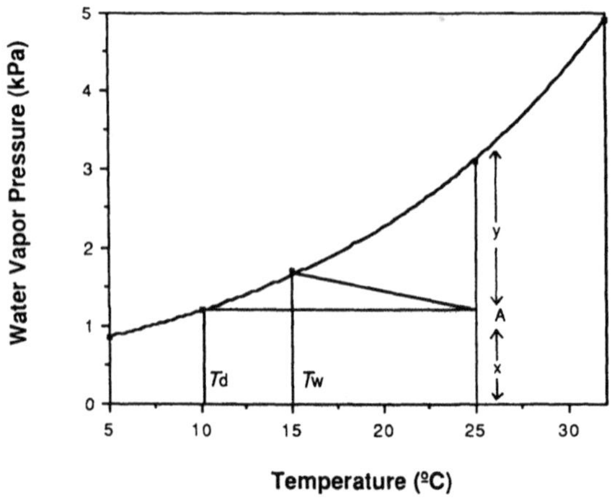

Fig.3.5 Relationship of water vapor pressure to temperature, illustrating the nature of various measures of humidity. At the point indicated by A, the vapor pressure (e) = x, the vapor pressure deficit (Δe) = y, the relative humidity = $x/(x+y)$, and the dew-point temperature (T_d) = the temperature where saturation vapor pressure equals x. Also shown is T_w, the wet-bulb temperature. Adapted from Jones (1983).

Saturation vapor pressures over ice are somewhat lower than those over water at the same temperature. Although the difference is relatively small, accurate measurements of vapor pressures at temperatures at or below freezing must consider the existence of these two possible states for water. Moist air at a given temperature cannot be in equilibrium with liquid water and ice at the same time.

In analyses of atmospheric humidity, water vapor is considered as a partial pressure compared to the total pressure of the air–vapor mixture. Following Dalton's law, the total pressure of a mixture of ideal gases or vapors is the sum of the pressures which would exist if the water pressure of each individual component were considered individually. Provided condensation is not taking place, water vapor behaves very nearly as an ideal gas.

$$PV = n\boldsymbol{R}T \qquad (3.5)$$

where P is the partial pressure of water vapor (Pa), V is the volume of the system, n is the number of moles of water present, $\boldsymbol{R}$ is the universal gas constant (8.3143 J K^{-1} mol^{-1}) and T is the temperature (K).

Equation 3.5 can be modified to describe the density (ρ) of water vapor or other gas, that is the mass per unit volume, by expressing n as the ratio of the mass of gas (m) to the molecular weight of the gas (M). Thus by making this substitution and dividing both sides by V, the ideal gas law becomes:

$$P = \rho \frac{\boldsymbol{R}}{M} T \quad \text{[Pa]} \qquad (3.6)$$

or for density:

$$\rho = \frac{P}{\boldsymbol{R}T} M \quad [\text{g m}^{-3}] \qquad (3.7)$$

This density of water vapor per unit volume of atmosphere is termed the *absolute humidity*. Measurements of absolute humidity are useful in calculations of diffusion (see Penman, 1948).

When the ideal gas laws are applied specifically to water vapor, the vapor pressure of water is termed e, the molecular weight equals 18 and the density is expressed in units of g m^{-3}. Therefore the vapor density can be calculated from the vapor pressure according to:

$$\rho = \frac{2.17}{T} e \quad [\text{g m}^{-3}] \qquad (3.8)$$

Another expression of water vapor mass in the atmosphere is *specific humidity* (q) which is the ratio of density of water vapor to the density of moist air. This can be derived from Equation 3.7 (see Gates, 1980) to give:

$$q = 0.622 \frac{e}{P - 0.379e} \quad [\mathrm{g\ kg^{-1}}] \quad (3.9)$$

where 0.622 is the ratio of the molecular weights of water (M_w) to dry air (M_a), and 0.379 equals $(M_a - M_w)/M_a$, specific humidity is commonly expressed in units of g kg^{-1}. For cold and dry air masses, typical values of q are 1–3. For warm humid air masses they may range from 12 to 25 (Gates, 1980). Approximate conversions between absolute and relative humidity can be easily made since below 3000 m elevations at normal temperature ranges, a cubic meter of air weighs about 1000 g (Riehl, 1965).

Some literature, particularly in meteorology, also refers to a related measure, the *mixing ratio* (Penman, 1948). This value is the ratio of water vapor mass to the density of dry air. Since the vapor pressure of water vapor (e) is generally about two orders of magnitude lower than the atmospheric pressure (p), mixing ratio and specific humidity are numerically almost equal.

The ratio of the actual water vapor pressure present in the atmosphere (e_a) and the water vapor pressure in a saturated atmosphere (e_s) at the same temperature is termed the *relative humidity* (h):

$$h = \frac{e_a}{e_s} \times 100 \quad [\%] \quad (3.10)$$

Relative humidity can also be expressed as the ratio of the absolute or specific humidities of an ambient level of water vapor and saturated air at the same temperature and pressure. It is important to note the obvious importance of temperature to relate relative humidities to other quantitative measures of atmospheric water vapor. As can be seen in Appendix Table A8, there can be a great range of absolute humidities, and thus water vapor pressures at the same relative humidity if temperature varies.

The difference between the ambient water vapor pressure in the atmosphere (e_a) and the saturated water vapor pressure (e_s) is termed the *vapor pressure deficit* or *saturation deficit*. This value has an important influence as a driving gradient for calculations of evaporation.

3.2.2 Dew-point and wet-bulb temperatures

If a sample of air is cooled without any net loss or gain of water vapor, it will eventually reach a temperature at which the air becomes saturated with water vapor and the relative humidity reaches 1.0. Any cooling below this point will cause condensation or dew formation. This level, termed the *dew-point temperature,* is of great value in measurements of atmospheric humidity since it can be produced by forced cooling of ambient air. The relationship between air vapor pressure, vapor pressure deficit, dew point and relative humidity are shown in Fig. 3.5. In studies below 0°C, this point where saturation is reached, is termed the *frost-point temperature*.

Another very useful term in humidity studies is the *wet-bulb temperature*. This is the temperature that a moist surface reaches when it is cooled by evaporation into the atmosphere. If the atmosphere is unsaturated with water vapor, then the amount of cooling of the surface which may occur is a function of the atmospheric vapor pressure, temperature and the rate of air movement over the evaporating surface. This wet-bulb temperature is an easily measured value which can be determined experimentally by blowing air rapidly across a thermometer bulb which is continuously wet by a wick. Used in conjunction with a paired dry-bulb thermometer, the atmospheric vapor pressure can be determined by the relationship:

$$e_a = e_s\ \gamma\ (T_a - T_w) \quad (3.11)$$

where γ is the psychrometric constant of 0.66 kPa K^{-1} for a well-ventilated surface at 100 kPa and 20°C, and $(T_a - T_w)$ is the temperature depression between the dry-bulb (T_a) and wet-bulb (T_w) thermometers. The psychrometric constant varies from 0.649

at 0°C to 0.678 at 45°C (see Appendix Table A6). The theoretical derivation of this constant is discussed by Monteith (1973).

3.2.3 Measurement of atmospheric humidity

(a) Mechanical hygrometers

One of the most common devices in use today for the measurement of atmospheric humidity is the mechanical hygrometer which measures changes in the length of hair with humidity (Davey, 1965). This type of device has been used for centuries, and has the obvious advantages of low cost and simplicity. Relative changes in length of human hair are somewhat logarithmically related to relative humidity, subject to a significant hystereses (about 3% pH) and temperature response. Response time is about 5 min. There are a number of manufacturers of commercial hygrometers or hydrothermographs which use the calibrated change in hair length to move a mechanical pen on a motor-driven drum to record diurnal and weekly changes in relative humidity.

Despite their widespread use, mechanical hygrometers are neither accurate nor reproducible. Furthermore, if the hairs are exposed to relative humidities below 15% or above 90%, the calibration may be permanently shifted (Day, 1985). Aging or other forms of hair degradation may also alter the calibration. Because of these problems, mechanical hygrometers are only suitable for approximate measurements where low cost is the primary concern. Since hydrothermographs are often much in demand as small rodent condominiums, it is advisable to screen off access to the exposed hair elements.

(b) Wet- and dry-bulb psychrometer

Taking advantage of the simple principles of wet-bulb temperature depression described above, wet- and dry-bulb psychrometers are the most widely used device for measuring humidity today. Even when more complex instrumentation is used to measure humidity, wet- and dry-bulb psychrometers are often used as a standard for calibration.

In operation, two identical thermometers are used in this technique. The dry-bulb thermometer is exposed to the atmosphere in a normal manner, while the wet-bulb thermometer has a close fitting cloth sheath over its bulb. This sheath is kept wet by a wick connecting it to a reservoir of distilled water. Under nonsaturated atmospheric conditions, a well-ventilated wet bulb will cool by latent heat of evaporation as water evaporates into the atmosphere. When an equilibrium point for cooling is reached (usually after several minutes), this wet-bulb temperature and the adjacent dry-bulb temperature can be utilized to calculate vapor pressure from Equation 3.11.

For convenience, psychrometric tables are widely available for relative humidities and vapor pressure deficits at measured values of wet- and dry-bulb temperature (Smithsonian Institution, 1968, and Appendix Table A8). Nomograms are also available to determine these values (see Slavik, 1974), and sliderule calculators for relative humidity are commonly included with commercial wet- and dry-bulb psychrometers.

There are a variety of means used to ventilate wet- and dry-bulb psychrometers. The simplest and least expensive form is called a sling psychrometer which provides a swivel handle or rope to manually whirl the thermometers. In an Assman-type psychrometer, the air is moved across the thermometer bulbs by forced ventilation, commonly at a rate of about 4 m s^{-1}. These types of psychrometers can be used in recording meteorological stations if battery power is available to turn on the airflow system before each measurement.

While thermometers are the common temperature-sensing element used in these psychrometers, any type of accurate sensing device can be used. Thermocouple, thermistor and resistance thermometers can all be made to work well.

The accuracy of relative humidity measurements made with a wet- and dry-bulb psychrometer is a function of the accuracies of both the wet- and dry-bulb temperatures recorded and the algorithm used to solve the equation. When properly designed and operated, a wet- and dry-bulb psychrometer is capable of providing reasonably accurate data. Accuracy is highest at high relative humidities, and then drops off as wet-bulb depression increases.

As a general rule, wet- and dry-bulb psychrometers should not be used for measuring humidity or vapor pressure in small enclosed spaces (White and Ross, 1983). In such circumstances there is a risk that the added water vapor evaporated from the wet bulb may significantly impact these values.

(c) Resistance hygrometers

A variety of instruments have been designed which fall into the general category of resistance hygrometers. Carbon resistance elements, lithium chloride or other saturated salts, are commonly used in sensors whose electrical resistance changes with humidity. In a lithium chloride sensor, a fabric bobbin is covered with a winding coated with a dilute solution of lithium chloride. As an alternating current passes through the winding and heats it, water is evaporated from this coating. The rate of evaporation is a function of the vapor pressure of the surrounding air. As the bobbin dries out, the resistance of the salt solution increases, resulting in less current flow and a cooling. Alternative heating and cooling of the bobbin leads to an equilibrium temperature and moisture content which can be calibrated to the ambient water vapor pressure, and through this, the dew-point temperature.

Between dew-point temperatures of −10 to 38°C, lithium chloride sensors provide accuracies of only ±1°C (White and Ross, 1983). Another major disadvantage of this type of sensor is the slow response time, commonly on the order of 5–10 min to reach equilibrium, and associated hysteresis effects. An additional limitation is the lower limit of measurement imposed by the chemical behavior of lithium chloride which has a saturation vapor pressure of about 11% relative humidity (White and Ross, 1983).

Instrument designs using lithium chloride sensors for monitoring air humidity have been published by many authors (e.g. Dunmore, 1938; Steiger, 1951; Brastad and Borchardt, 1953; Lieneweg, 1955; Tanner and Suomi, 1956). Descriptions of other types of resistance hygrometers have also been published, although these have not been widely utilized. Wexler (1957) described a sulfonated polystyrene hygrometer and other types of electric hygrometers with rapid response times. Jones and Wexler (1960) have designed a barium fluoride film electrode on a glass substrate with sensitivity from 20 to 100% relative humidity and a response time of seconds.

(d) Dew-point hygrometers

Dew-point hygrometers, which can operate over the whole range of environmental humidity measurements, provide one of the most accurate methods for measuring relative humidity. In this instrument an air stream is passed through a chamber containing a metallic mirror whose temperature can be controlled very accurately by cooling with a small Peltier block, or alternatively heated by a resistance wire. The mirror is cooled, in operation, until light reflected off of the mirror's surface changes in response to dew formation. A photodetector senses this point of light scattering and controls the mirror temperature through a feedback system so that the dew-point temperature is just maintained. The temperature of the mirror, measured accurately with a thermistor or platinum resistance thermometer, is equal to the dew-point temperature of the surrounding air.

Because the dew-point temperature is

measured directly, and the corresponding relative humidity or vapor pressure found from tables, this type of instrument is very accurate and stable. Dew-point mirrors can resolve ±0.1°C with an accuracy of ±0.2°C. The principal source of error in this measurement is the contamination of the mirror surface with soluble salts which would shift the point of condensation to slightly higher levels. Response time is moderate, but sufficient to provide accurate measurements of relative humidity in flowing gas volumes.

Dew-point hygrometers are largely laboratory instruments and require line power, but systems suitable for field use in gas exchange systems have now been developed. These are discussed in more detail in Chapter 8. The principal negative feature of these instruments is their relatively high cost.

(e) Solid-state humidity sensors

In the last fifteen years, there has been the development of solid-state capacitance sensors for atmospheric humidity. The broad sensitivity, accuracy and rapid response time of these sensors has led to their widespread use in environmental physiology and micrometeorological studies. The most common of these solid-state devices is the Vaisala dielectric polymer sensor which was developed in 1973 (see also Chapter 8). This small sensor consists of a polymer layer coated on to a metalized glass plate, with a second gold electrode vacuum-evaporated on to the polymer surface (Day, 1985). This upper electrode is so thin that water vapor diffuses through it rapidly. In operation, the electrical capacitance of the dielectric polymer changes virtually linearly over a relative humidity range of 0–75%, with an absolute accuracy of ±2% RH. Over this range the response time to humidity change is very rapid, although it slows somewhat at low temperatures. Above 75% relative humidity, measurements suffer from problems of slower response, increasing nonlinearity and long-term drift (Field and Mooney, 1984; Day, 1985).

As advances continue to be made in dielectric polymer sensors, improved solid-state humidity sensors will be produced. While other designs are now available (Day, 1985), there is a tendency for a tradeoff between broader range of response on one side and lower absolute accuracy and slower response on the other hand.

(f) Spectroscopic hygrometers

While spectroscopic methods for measuring atmospheric humidity are widely used in laboratory-based instrumentation, these are only now beginning to be adopted for field use. Infrared gas analyzers have been used for many years for this purpose (Slavik, 1974) and offer the strong advantages of very rapid response in a flowing air mass and extreme sensitivity at low relative humidities. These devices are discussed in more detail in Chapter 8.

Remote sensing of atmospheric moisture can be accomplished spectroscopically using ultraviolet spectroscopy. These instruments commonly distinguish the absorption of water vapor from O_2 by illuminating with 121.6 nm radiation and measuring the resultant fluorescence at 310 nm from activated water molecules (Kley and Stone, 1978; Stone, 1980). These approaches may have increasing interest to physiological ecologists in the future.

The disadvantage of spectroscopic hygrometers is their high cost and need for direct calibration. Accuracy is highly dependent on good calibration.

3.2.3 Choice of an appropriate humidity sensor

As described above, there are a great variety of commercial sensors that are available for measurements of atmospheric humidity. In addition to these common sensors, there are many other types of humidity sensors that have not been widely used for studies of environmental physiology. These include electrolytic hygrometers using phosphorus pentoxide films, piezoelectric sorption hygrometers, cellulose crystallite sensors, refraction hygrometers, thermal conductivity gas analysis and pneumatic bridge hygrometers. These devices are discussed in more detail by Wexler (1965), Slavik (1974) and White and Ross (1983).

In choosing a sensor, there are a variety of factors to consider: price, accuracy, response time, humidity range, temperature range and susceptibility to contamination. For flexibility and high accuracy, the dew-point hygrometer is generally the instrument of choice, although the cost of these is high. The small size, moderate cost and simple electronics of dielectric capacitance sensors such as the Vaisala chip make these very suitable for use in portable battery-operated humidity sensors for micrometeorological stations and physiological instruments for measurements of photosynthesis and transpiration.

Wet- and dry-bulb psychrometers are still widely used to make spot measurements of mesoscale relative humidity. Their low cost and simplicity of operation makes them very attractive. Care must be taken with these, however, to insure that adequate ventilation of the wet bulb is made, as described above. Mechanical hair hygrometers are also widely used because of their convenience in making continuous records of humidity. These have serious problems in accuracy, however, and are not an adequate choice for physiological relationships to microclimatic parameters.

Regardless of the instrument used, the need for regular and careful calibration is of critical importance. The use of instruments in a field environment where dust or gaseous pollutants may be present increases the importance of frequent calibration. A laboratory dew-point hygrometer can serve as a local source of routine calibration. Many dew-point hygrometers are sold with a National Bureau of Standards (NBS) or equivalent calibration certificate.

3.3 MOISTURE FLUX

While hydrologic approaches to measurements of water flux between and within ecosystem components are beyond the scope of the chapter, it is important to recognize that there are increasing numbers of cooperative studies underway between ecosystem modelers and physiological plant ecologists. These interactions are directed toward the goal of better understanding the nature of water fluxes through the soil–plant–atmosphere continuum.

3.3.1 Soil water flux

Water moves through soils in response to gradients in soil water potential (Klute, 1986a). The rate of movement is proportional to the intensity of the gradient in potential in the soil, expressed as $d\Psi/dx$, where Ψ = soil water potential and x = distance. The proportionality factor between flux rate and potential gradient is termed the hydraulic conductivity, K. The value of K depends upon soil characteristics, environment and the water content of the soil itself. A typical relationship between hydraulic conductivity and soil water potential in a loam soil is shown in Fig. 3.6.

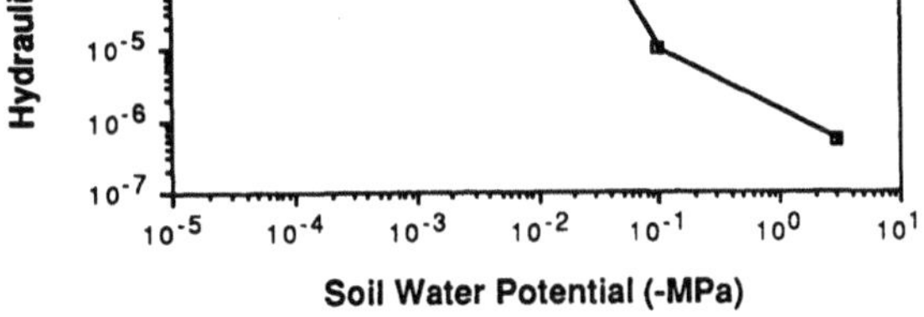

Fig. 3.6 The relationship between hydraulic conductivity and soil water potential in a loam soil (redrawn from Slatyer, 1967).

Saturated hydraulic conductivity, because it is expressed at a single water content, is directly related only to soil properties, particularly total porosity and pore size distribution. In simplest terms this may be expressed as:

$$F = K(\mathrm{d}\Psi/\mathrm{d}x) \qquad [\mathrm{cm}^3\ \mathrm{s}^{-1}] \qquad (3.12)$$

where F = flux. Under conditions of unsaturated flow, K changes as a function of water content, making the analysis much more difficult.

Various laboratory and field methods have been employed to estimate hydraulic conductivities (Fluhler *et al.*, 1976; Parkes and Waters, 1980; Daniel, 1982; Ward *et al.*, 1983). Because conductivity is so sensitive to the three-dimensional geometry of soil solids and pores, measurements based on disturbed samples in the laboratory have proven relatively useless in predicting field-scale processes. In-field estimates have used tensiometers (Fluhler *et al.*, 1976; Ward *et al.*, 1983), neutron probe (Parkes and Waters, 1980) and thermocouple psychrometers (Daniel, 1982).

(a) Infiltration

Water movement into soils is termed infiltration, as opposed to water movement through soils which is termed permeability.

Soil water infiltration in initially dry soil is frequently found to be a two-step process (Hillel, 1980a). In the first stage, gravitational and matric forces combine to produce rapid initial inflow. As soil water content increases, inflow rates slow to a nearly constant rate, as the matric potential gradient reaches very low values (Bertrand, 1965).

While laboratory measurements of infiltration can be made using intact or reconstituted soil cores, it is very difficult to relate results of such analyses to field soil conditions. As a result, simple infiltrometers have been widely used in field studies.

Single- and double-ring infiltrometers are frequently used in a variety of ecological field studies. The single ring is simply a circular metal band perhaps 1.5 m in diameter. Water is rapidly introduced into the ring, a constant head established and the rate of water entry into soil measured. A 'falling head' infiltrometer can be operated simply by measuring the depth of water over time after initial filling, without maintaining a constant depth.

The double-ring system is similar to the single ring, except two concentric rings are placed on the soil, and water dynamics in the inner ring only are monitored. The outer ring is filled with water and offers a barrier against lateral movement of water out of the inner ring.

Ring geometry is important in the design of infiltrometers. Ahuja (1982) found no lateral flow effects with an inner ring of 30 cm and an outer ring diameter of 120 cm. Outer rings with smaller diameters generally affected infiltration rates.

(b) Evaporation

Evaporation from soil occurs in three stages, the first, a constant rate nearly equal to that from a free surface of water, the second a decreasing rate limited by diffusion from within the soil and the third, a slow rate stage (van Bavel and Hillel, 1976; Hillel, 1980b). The rate of water loss during the first stage is limited primarily by the boundary layer thickness of air above the soil, and the atmospheric relative humidity. At some point, during surface soil drying, the rate of loss becomes progressively more limited by the rate of liquid water flux from deeper within the soil. Finally, loss is controlled by rates of water movement to the surface in the vapor phase.

Rose (1966) modeled evaporation in terms of a simple relationship between time and water loss:

$$W_E = st^{1/2} + bt \qquad (3.13)$$

where W_E is evaporation, s is sorptivity (Philip, 1957), t is time and b is a constant <0. Hillel (1975) also developed models to describe evaporation under bare soil.

Measuring or predicting evaporation under

vegetated soil has proven to be very complex; most reports lump evaporation and transpiration together as evapotranspiration, as discussed later in this chapter. When leaf-area indices are relatively high (over 4), soil evaporation from well-watered sites may be only 5% or less of total evapotranspiration. When leaf-area indices are less than 2 in well-watered soil, soil evaporation may comprise as much as 50% of total evapotranspiration. As soil surfaces dry, soil evaporation is a negligible component of this total.

Seymour and Hsiao (1984) describe a thermocouple psychrometer apparatus capable of measuring humidity right at the soil surface. The psychrometer is used to monitor air equilibrated with the soil surface, in a chamber placed briefly on the soil. Cumulative vapor pressure differences between surface humidity and humidity at 1.5 m were highly correlated with water loss from microlysimeters.

(c) Internal soil transport

If the water potential gradient and soil hydraulic conductivities are known, rates of water transport through the soil can be predicted (Parkes and Waters, 1980). Large differences have been found between measured and estimated (laboratory-based) unsaturated hydraulic conductivities, suggesting that only a small portion of the soil pore space contributes to flow through the whole profile.

Field *et al.* (1984) found difficulty in obtaining 'representative' undisturbed core samples for laboratory evaluation of hydraulic properties. As a result, laboratory values did not agree with measurements made *in situ* using the tensiometer and neutron probe. Inter-aggregate porosity and root channels were both said to contribute to nonuniform drainage. A soil-water flux sensor has recently been described (Kawanishi, 1983). Using a heat-pulse source (10 cm metal tube wound with nichrome wire) and a differential temperature sensor arrangement (a differential thermocouple with one couple on each side of the source, about 2 cm from it), soil-water flow rates between 0.01 mm h^{-1} and several mm h^{-1} could be measured. The sensor must be calibrated under controlled flux conditions in the laboratory. Rainfall infiltration and capillary rise above a water table were both measured with the device, which gave satisfactory results.

3.3.2 Atmospheric moisture flux

(a) Precipitation

Precipitation measurements are commonly made with a standard US Weather Bureau rain gauge. This device is 20 cm (8 inches) in diameter and 51 cm (20 inches) tall, with a funnel built into the upper part which transfers precipitation into an inner cylinder with 10% of the cross-sectional area of the outside opening. For weather stations, a Belfort-type recording rain gauge is commonly used in addition to measure the time and rate of rainfall.

Although standard rain gauges are desirable for permanent weather stations, it is possible to obtain satisfactory records using small and cheaper collecting devices. Size of the opening does not apparently affect measurements significantly so long as wind speeds are moderate. Various forms of baffles are often used around rain gauges in windy microsites. On steep slopes, there is evidence that the most accurate measurements are obtained with tilted rain gauges with their rims parallel to the ground surface (Hayes and Kittredge, 1949).

With increasing interest in atmospheric transport of pollutants in recent years, there has been a resurgence of interest in the chemistry of precipitation. Many meteorological stations are now equipped with wetfall collectors and bucket collectors which remain closed under dry conditions and open automatically with rainfall. The sampling efficiencies of these collectors compare fairly well with standard precipitation collectors

(Bogen *et al.*,1980; Schroder *et al.*, 1985). Open bulk-precipitation collectors of other designs (e.g. funnel and bottles, plastic buckets with polyethylene bag inserts) may vary slightly from standard collectors in catchment efficiency and significantly in chemical characteristics of precipitation if dryfall deposition is significant (Galloway and Likens, 1976; Chan *et al.*, 1984).

Studies of pollution ecology have also led to new interest in chemical analyses of throughfall precipitation and stemflow. These measurements of canopy effects on precipitation are beyond the scope of this chapter. Parker (1983) reviews the literature on throughfall and stemflow in forest nutrient cycles.

(b) Cloudwater collectors

Recent interests in the environmental quality of atmospheric moisture has led to the development of cloudwater collectors by the Institute of Ecosystem Studies of the New York Botanical Garden. Since windspeeds during fog events are often too low for sample collection with a passive collector, the cloudwater collectors utilize an active system with a battery-powered fan to draw atmospheric water droplets through a vertical opening and across a cartridge of Teflon strands where they condense and drip into a collection bottle.

(c) Evaporation

Standard measurements of evaporation for Weather Bureau purposes are generally made with an evaporation pan. A standard pan in the United States is circular, 25.4 cm (10 inches) deep by 120.6 cm (47.5 inches) inside diameter, and is placed on a platform at ground level. Pan evaporation data are commonly used as a general climatic index incorporating temperature, humidity, wind and irradiance. The coefficient of conversion of evaporation from such a pan to the free water surface of a shallow lake is approximately 0.7 (Haddock, 1981).

Recording evaporation pans are commercially available which provide a mechanical chart drive to record weight change due to evaporation from a 250 cm^2 surface. A Piche evaporimeter is a device providing another measure of evaporation. It consists of a cylindrical tube of glass filled with water, with one end closed and the other end covered with a disk of porous paper held in place by a brass spring clip. When held vertically, the amount of water evaporated from the paper surface of 11 cm^2 can be read on a scale on the tube.

(d) Evapotranspiration

There is a vast literature describing a variety of methods to estimate evapotranspiration from moist or vegetated surfaces using standard meteorological data for the calculations. The models used in these calculations generally use one of three approaches in utilizing meteorological data. These are water vapor flux measurements, energy balance approaches and empirically derived relationships between evaporation and one or more standard meteorological factors (Federer, 1970).

One of the most simple approaches is the Thornthwaite method (1948) for calculating both evapotranspiration and potential evapotranspiration. The simplicity of this approach is that it relies on an empirical relationship between potential evapotranspiration and mean air temperature, and thus can be utilized for any site at which daily high and low temperatures are recorded. Although the theoretical basis for this empirical fit is somewhat lacking, the Thornthwaite method does surprisingly well in predicting evapotranspiration. It may also miss badly on occasions (Palmer and Havens, 1958).

The basic Thornthwaite formula for computing monthly potential evapotranspiration (PE) is:

$$\mathrm{PE} = 1.6\,\frac{(10T)^{a}}{I} \tag{3.14}$$

Where T is the mean monthly temperature

(°C) and I is a heat index, constant for any site, made up of monthly index values of i which are a function of the long-term mean temperatures. The final exponent a, is an empirical derived value which is a function of I. Graphical solutions to this formula make the calculations very simple (Palmer and Havens, 1958). Evapotranspiration is estimated by combining daily PE and percent soil moisture depletion on a simple nomogram.

A more elegant approach to evapotranspiration estimation comes from the Penman–Monteith equation using a energy-balance approach. This method was first developed by Penman (1948) to describe evaporation from a free water surface, and later modified to apply to leaf surfaces (Penman, 1953). It was refined by Monteith (1965) to apply to whole plant canopies. In operation, this equation combines estimates of the vertical flux of water vapor derived by energy-balance and aerodynamic approaches to develop a relationship which describes water vapor flux (E) in terms of atmospheric meteorological measurements.

$$E = \frac{[s(R_n - G) + \rho_a c_p g_h \Delta e]}{\lambda\ [s + \gamma(g_h/g_w)]} \qquad (3.15)$$

which utilizes values for the slope of vapor pressure to temperature relationship from soil to air (s), the available energy of net radiation (R_n) and soil heat flux (G), the density of air (ρ_a), the specific heat of air (c_p), total thermal conductance (g_h), vapor pressure deficit (Δe), latent heat of vaporization (λ), the psychrometric constant (γ) and total pathway conductance (g_w). For a plant leaf, g_w is the sum of stomatal and boundary layer conductance, but for a whole canopy the sum conductances comprising g_w are more complex. In operation, this equation can be solved with data inputs for net radiation balance, ambient temperature, vapor pressure deficit, the conductance at the soil or canopy surface and the leaf or canopy conductance. This latter factor provides a biological component to the calculations.

A variety of simplifications have been applied to the Penman–Monteith equation to provide greater ease of calculation. These simplifications, discussed by Thom and Oliver (1977), Thompson (1982) and Jones (1983), can provide acceptable results. A more complete discussion of this and other modeling approaches to measurements evapotranspiration are beyond the scope of this chapter.

(e) Interception

Evaporation from plant canopies includes not only losses derived from transpiration, but from interception processes following wet canopy conditions. This may be a very significant flux, especially in humid areas where interception losses may exceed transpirational losses (Calder, 1976). A considerable degree of technical interest has been given to the interception process, particularly for forest canopies (Roberts, 1983). Work on canopy interception by Corsican pine (Rutter *et al.*, 1971) led to a detailed development of a predictive model of the rate and magnitude of inputs, stores, interchanges and losses from a forest canopy during and following a rainstorm. This model demonstrated that the forest structure had a relatively minor role in influencing interception loss in comparison to variables of the duration and frequency of the rainstorm (Rutter and Morton, 1977). This finding has led to simplified rainfall interception models (see Gash, 1979) in which annual interception loss is predicted by multiplying annual precipitation by a locally determined coefficient. This interception fraction is reasonably constant for forests at a given annual rainfall, but decreases with increasing annual total precipitation (Roberts, 1983).

(f) Dewfall

The condensation of water from the atmosphere termed dewfall is governed from the same physical principles as evaporation (Monteith, 1957). For this reason, Equation 3.15 is applicable, with dew forming whenever E is negative. In calculating dewfall, however, there is no physiological resistance to water transfer and thus $g_h = g_w$. Therefore:

$$E = \frac{[s(R_n - G) + \rho_a c_p g_h \Delta e]}{\lambda\,(s+\gamma)} \qquad (3.16)$$

This equation predicts that dewfall will occur when $-s(R_n - G)$ is greater than $\rho_a c_p \Delta e$ (Monteith, 1963). Assuming that the net radiation balance $(R_n - G)$ falls to about -100 W m^{-2} on a clear night, Equation 3.16 predicts a maximum rate of dewfall of approximately 0.1 mm h^{-1} for saturated air (Jones, 1983). Typical rates of dewfall that have been measured are commonly only 0.1–0.2 mm night^{-1} (Tuller and Chilton, 1973), with maximum rates up to about 0.4 mm night^{-1} (Monteith, 1963; Burrage, 1972). These field values thus are well below theoretical limits. The causes of this are undoubtedly many, including nonsaturated air masses which require cooling air masses to the dew-point temperature and the sensitivity of dewfall to wind speed.

Because dewfall is almost always an order of magnitude or more less than evapotranspiration, dew is not generally an important factor in community water balance. There are exceptions, however, such as coastal fog deserts in Baja California, northern Chile and Namibia. The ecological significance of atmospheric dewfall and fog moisture are reviewed in detail by Rundel (1982). Mooney *et al.* (1980) describe use of dew moisture by *Nolana mollis* in the Atacama Desert of northern Chile. This situation is remarkable in that hygroscopic salts on the leaf surfaces condense water out of the air masses with relative humidities no higher than 82%.

There have been a variety of approaches used to measure dewfall. These range from weighing lysimeters (Jennings and Monteith, 1954; Collins, 1961) to artificial surfaces positioned on a recording balance. Commercial dew sensors of this latter type are manufactured by Adolf Thies and Company in West Germany and Belfort in the United States. Data collected by such balance devices are only ecologically realistic if the radiative and aerodynamic properties of the collector resemble those of real leaves. Qualitative records of dew presence (i.e. duration) can be measured by electrical resistance between two electrodes attached to the surface of a leaf.

REFERENCES

Ahuja, L.R. (1982) Determining unsaturated hydraulic conductivity of soil during drainage under a ring infiltrometer. *J. Hydrol.*, **58**, 167–73.

Balbalola, O. (1978) Field calibration and use of the neutron moisture meter on some Nigerian soils. *Soil Sci.*, **126**, 118–24.

Bertrand, A.E. (1965) Rate of water intake in the field. In *Methods of Soil Analysis. Part 1 Physical and Mineralogical Properties, Including Statistics of Measurement and Sampling* (ed. C.A. Black), American Society of Agronomy, Madison, pp. 197–209.

Blackwell, P.S. and Elsworth, M.J. (1980) A system for automatically measuring and recording soil water potential and rainfall. *Agric. Water Manag.*, **3**, 135–41.

Bogen, D.C., Nagourney, S.J. and Torquato, C. (1980) A field evaluation of the HASL wet–dry deposition collector. *Water Air Soil Pollut.*, **13**, 453–8.

Bouma, J., Belmans C.F.M. and Dekker, L.W., (1982) Water infiltration and redistribution in a silt loam subsoil with vertical worm channels. *Soil Sci. Soc. Am. J.*, **46**, 917–21.

Brastad, W.A. and Borchardt, L.F. (1953) Electric hygrometer of small dimensions. *Rev. Sci. Instrum.*, **24**, 1143–4.

Briscoe, R. (1986) Thermocouple psychrometers for water potential measurements. In *Advanced Agricultural Instrumentation Design and Use* (ed. W.G. Gensler), Martins Nighoff, Dordrecht, pp. 193–209.

Brown, R.W. and van Haveren, B.P. (1972) *Psychrometry in Water Relations Research. Proceedings of the Symposium on Thermocouple Psychrometers*, Utah Agriculture Experimental Station, Logan, 342 pp.

Bruce, R.R. and Luxmoore, R.J. (1986) Water retention: field methods. In *Methods of Soil Analysis Part 1 Physical and Mineralogical Methods*, 2nd edn, Agronomy, Number 9 (Part 1), (ed. A. Klute), American Society of Agronomy, Madison, pp. 635–62.

Burrage, S.W. (1972) Dew on wheat. *Agric. Meteorol.*, **10**, 3–12.
Calder, I.R. (1976) The measurement of water losses from a forested area using a 'natural' lysimeter. *J. Hydrol.*, **30**, 311–25.
Calder, I.R. (1978) Transpiration observations from a spruce fir forest and comparisons with predictions from an evaporation model. *J. Hydrol.*, **38**, 33–47.
Campbell, E.C., Campbell, G.S. and Barlow, W.K. (1973) A dew-point hyprometer for water potential measurement. *Agric. Meteor.*, **12**, 113–21.
Chan, W.H., Lusig, M.A., Stevens, R.D.S. and Vet, R.J. (1984) A precipitation sampler intercomparison. *Water Air Soil Pollut.*, **23**, 1–13.
Collins, B.C. (1961) A standing dew meter. *Meteorol. Mag.*, **90**, 114–17.
Colman, E.A. and Hendrix, T.M. (1950) The fiberglass electrical soil-moisture instrument. *Soil Sci.*, **77**, 425–38.
Dalton, F.N. (1987) Measurement of soil water content and electrical conductivity using time domain reflectrometry. In *Proceedings of the International Conference on Measurement of Soil and Plant Water Status*, vol. 1, Utah State University, pp. 95–8.
Dalton, F.N., Herkelrath, W.N., Rawlins, D.S. and Rhoades, J.D. (1984) Time domain reflectrometry: simultaneous assessment of the soil water content and electrical conductivity with a single probe. *Science*, **224**, 989–90.
Daniel, D.E. (1982) Measurement of hydraulic conductivity of unsaturated soils with thermocouple psychrometers. *Soil Sci. Soc. Am. J.*, **46**, 1125–9.
Dasberg, S. and Dalton, F.N. (1985) Time domain reflectrometry field measurements of soil water content and electrical conductivity. *Soil Sci. Soc. Am. J.*, **49**, 293–7.
Dasberg, S. and Nadler, A. (1987) Field sampling of soil water content and bulk electrical conductivity with time domain reflectrometry. In *Proceedings of the International Conference on Measurement of Soil and Plant Water Status*, vol. 1, Utah State University, pp. 99–101.
Davey, F.K. (1965) Hair humidity elements. In *Humidity and Moisture* (ed. A. Wesler), Reinhold, New York, Vol. 1, pp. 571–3.
Day, W. (1985) Water vapor measurement and control. In *Instrumentation for Environmental Physiology* (eds B. Marshall and F.I. Woodward), Cambridge Press, Cambridge, pp. 59–78.
Devitt, D., Jury, W.A., Sternberg, P. and Stolzy, L.H. (1983) Comparison of methods used to estimate evapotranspiration for leaching control. *Irrig. Sci.*, **4**, 59–69.
Dodd, J., Heddle, E.M., Pate, J.S. and Dixon, K.W. (1984) Rooting patterns and their functional significance. In *Kwongan: Plant Life of the Sandplain* (eds J.S. Pate and J.S. Beard), University of Western Australia Press, Nedlands, pp. 146–77.
Dunmore, J.M. (1938) An electric hygrometer and its application to radiometeorography. *Bull. Am. Meteorol. Soc.*, **19**, 225–43.
Fadl, O.A.A. (1978) Evapotranspiration measured by a neutron probe on Sudan Gezira vertisols. *Exp. Agric.*, **14**, 341–7.
Faiz, S.M.A. (1983) Use of a pressure in the determination of soil water potential. *Plant and Soil* **73**, 257–64.
Federer, C.A. (1970) *Measuring Forest Evapotranspiration – Theory and Problems*, USDA Forest Service Research Paper NE-165, 25 pp.
Field, C. and Mooney, H.A. (1984) Instrumentation for photosynthetic research. In *Physiological Ecology of Plants of the Wet Tropics* (eds E. Medina, H.A. Mooney and C. Vasquez-Yanez), Junk, The Hague.
Field, J.A., Parker, J.C. and Powell, N.L. (1984) Comparison of field- and laboratory-measured and predicted hydraulic properties of a soil with macropores. *Soil Sci.*, **138**, 385–96.
Fluhler, H., Ardakani, M.S. and Stolzy, S.H. (1976) Error propagation in determining hydraulic conductivities from successive water content and pressure head profiles. *Soil Sci. Soc. Am. J.*, **40**, 830–6.
Fonteyn, P.J., Schlesinger, W.H. and Marion, G.M. (1987) Accuracy of soil thermocouple hygrometer measurements in desert ecosystems. *Ecology*, **68**, 1121–4.
Galloway, J.W. and Likens, G.E. (1976) Calibration of collection procedures for determination of precipitation chemistry. *Water Air Soil Pollut.*, **6**, 241–58.
Gardner, W.H. (1965) Water content. In *Methods of Soil Analysis. Part 1 Physical and Mineralogical Properties, Including Statistics of Measurement and Sampling* (ed. C.A. Black), American Society of Agronomy, Madison, pp. 82–127.
Gardner, W.H. (1986) Water content. In *Methods of Soil Analysis Part 1 Physical and Mineralogical Methods*, 2nd ed., Agronomy Number 9 (Part 1) (ed. A. Klute), American Society of Agronomy, Madison, pp. 493–544.
Gash, J.H.C. (1979) An analytical model of rainfall interception by forests. *Q. J. R. Meteorol. Soc.*, **105**, 43–55.
Gates, D.M. (1980) *Biophysical Ecology*, Springer Verlag, New York. 611 pp.
Goltz, S.M., Benoit, G. and Schimmelpfennig, H.

(1981) New circuitry for measuring soil water matric potential with moisture blocks. *Agric. Meteorol.*, **24**, 25–82.

Greacer, E.L. (1981) *Soil Water Assessment by the Neutron Method*, CSIRO, East Melbourne, 140 pp.

Greenway, H. and Munns, D.R. (1980) Mechanism of salt tolerance in non-halophytes. *Annu. Rev. Plant Physiol.*, **31**, 149–90.

Haddock, D. (1981) Pan evaporation: guide to crop water use and reservoir evaporation. *Weekly Weather and Crop Bulletin*, US Department of Commerce, US Department of Agriculture, Vol. 68, No.23.

Hall, D.G.M. and Jones R.J.A., (1983) Soil moisture changes under grassland as measured by neutron probe in midland England. *J. Agric. Sci.*, **101**, 481–4.

Hanks, R.J. and Ashcroft, G.L. (1980) *Applied Soil Physics*, Springer-Verlag, Berlin, 159 pp.

Hanna, L.Y., Harlan, P.W. and Lewis, D.T. (1983) Effect of landscape position and aspect on soil water recharge. *Agron. J.*, **75**, 57–60.

Hanna, G.W. and Siam, N. (1980) The estimation of moisture content in the top 10 cm of soil using a neutron probe. *J. Agric. Sci.*, **94**, 251–4.

Hayes, G.L. and Kittredge, T. (1949) Comparative rain measurements and raingauge performances on a steep slope adjacent to a pine stand. *Trans. Am. Geophys. Union*, **30**, 295–301.

Hewlett, J.D., Douglas, J.E. and Clutter, J.L. (1964) Instrument and soil moisture variance using the neutron scattering method. *Soil Sci.*, **97**, 19–24.

Hillel, D. (1975) Evaporation from bare soil under steady and diurnally fluctuating evaporativity. *Soil Sci.*, **120**, 230–7.

Hillel, D. (1980a) *Fundamentals of Soil Physics*, Academic Press, New York, 413 pp.

Hillel, D. (1980b) *Applications of Soil Physics*, Academic Press, New York, 385 pp.

Hodgson, A.S. and Chan, K.Y. (1987) Field calibration of a neutron moisture meter in a cracking grey clay. *Irrig. Sci.*, **8**, 233–44.

Hodnett, M.G. (1986) The neutron probe for soil moisture measurement. In *Advanced Agricultural Instrumentation, Design and Use* (ed. W.G. Gensler), Martins Nighoff, Dordrecht, pp. 148–92.

Hulsman, R.B. (1985) The neutron probe and the microcomputer. *Soil Sci.*, **140**, 153–7.

Jennings, G. and Monteith, J.L. (1954) A sensitive recording dew-balance. *Q. J. R. Meteorol. Soc.*, **80**, 222–6.

Jones, F.E. and Wexler, A. (1960) A barium chloride film hygrometer element. *J. Geophys. Res.*, **65**, 2087–95.

Jones, H.G. (1983) *Plants and Microclimate*, Cambridge University Press, Cambridge.

Kachanoski, R.G. and DeJong, E. (1982) Comparison of the soil water cycle in clear-cut and forested sites. *J. Environ. Qual.*, **11**, 545–9.

Kawanishi, H. (1983) A soil-water flux sensor and its use for field studies of transport processes in surface soil. *J. Hydrol.*, **60**, 357–65.

Kley, D. and Stone, E.J. (1978) Measurement of water vapor in the stratosphere by photodissociation with Ly (1216°A) light. *Rev. Sci. Instrum.*, **49**, 691–7.

Klute, A. (ed.) (1986a) *Methods of Soil Analysis. Part 1 Physical and Mineralogical Methods*, 2nd edn, Agronomy Number 9 (Part 1), American Society of Agronomy, Madison, 1188 pp.

Klute, A. (1986b) Water retention: laboratory methods. In *Methods of Soil Analysis. Part 1 Physical and Mineralogical Methods*, 2nd edn, Agronomy, Number 9 (Part 1), (ed. A. Klute), American Society of Agronomy, Madison, pp. 635–62.

Lieneweg, F. (1955) Absolut und relative Feuchtebestimmungen mit dem Lithiumchlorid-Feuchtemesser. *Siemens-Zeitschr.*, **29**, 212–18.

Luxmoore, R.T., Grizzard, T. and Patterson, M.R. (1981) Hydraulic properties of Fullerton cherty silt loam. *Soil Sci. Soc. Am. J.*, **45**, 692–8.

McIntyre, D.S. (1982) Capillary rise from saline groundwater in clay soil cores. *Austr. J. Soil Res.*, **20**, 305–13.

Marthaler, H.P., Vogelsanger, W., Richard, F. and Wierenga, P.J. (1983) A pressure transducer for field tensiometers. *Soil Sci. Soc. Am. J.*, **47**, 624–7.

Monteith, J.L. (1957) Dew. *Q. J. R. Meteorol. Soc.*, **83**, 322–41.

Monteith, J.L. (1963) Dew: facts and fallacies. In *The Water Relations of Plants* (eds A.J. Rutter and F.H. Whitehead), Blackwells, Oxford, pp. 37–56.

Monteith, J.L. (1965) Evaporation and environment. *Symp. Soc. Exp. Biol.*, **19**, 205–34.

Monteith, J.L. (1973) *Principles of Environmental Physics*, Arnold, London.

Mooney, H.A., Gulmon, S.L. Ehleringer, J. and Rundel, P.W. (1980) Atmospheric water uptake by an Atacama Desert shrub. *Science*, **209**, 693–4.

Neumann, H.H. and Thurtell, G.W. (1972) A Peltier-cooled thermocouple dewpoint hygrometer of *in situ* measurement of water potential. In *Psychrometry in Water Relations Research* (eds P.W. Brown and B.P. van Haveren), Utah

Agricultural Experiment Station and Utah State University, Logan, pp. 103–12.

Nnyamah, J.U., Black, T.A. and Tan, C.S. (1978) Resistance to water uptake in a Douglas fir forest. *Soil Sci.*, **126**, 63–76.

Nobel, P.S. (1983) *Biophysical Plant Physiology and Ecology*, W.H. Freeman, San Francisco, 603 pp.

Page, A.L., Miller, R.H. and Keeney, D.R. (eds) (1982) *Methods of Soil Analysis*, 2nd edn, Agronomy, Number 9 (Part 2), American Society of Agronomy, Madison, 1159 pp.

Palmer, W.C. and Havens, A.V. (1958) A graphical technique for determining evapotranspiration by the Thornthwaite method. *Monthly Weather Rev.*, **86**, 123–8.

Parker, G.G. (1983) Throughfell and stemflow chemistry in the forest nutrient cycle. *Adv. Ecol. Res.*, **13**, 57–133.

Parkes, M.E. and Siam, N. (1979) Errors associated with measurement of soil moisture change by neutron probe. *J. Agric. Engineer. Res.*, **24**, 87–93.

Parkes, M.E. and Waters, P.A. (1980) Comparison of measured and estimated hydraulic conductivity. *Water Resour. Res.*, **16**, 749–54.

Penman, H.L. (1948) Natural evaporation from open water, bare soil and grass. *Proc. R. Soc. London, Ser. A*, **193**, 120–45.

Penman, H.L. (1953) The physical basis of irrigation control. *Rep. 13th Int. Horticult. Congr.*, **2**, 913–14.

Philip, J.R. (1957) Evaporation, moisture and heat fields in the soil. *J. Meteorol.*, **14**, 354–66.

Rawlins, S.L. and Campbell, G.S. (1986) Water potential: thermocouple psychrometry. In *Methods of Soil Analysis. Part 1 Physical and Mineralogical Methods*, 2nd edn, Agronomy Number 9 (Part 1), (ed. A. Klute), American Society of Agronomy, Madison, pp. 597–618.

Rhoades, J.D. (1979) Inexpensive four-electrode probe for monitoring soil salinity. *Soil Sci. Soc. Am. J.*, **43**, 817–18.

Rhoades, J.D. and Corwin, D.L. (1980) Determining soil electrical conductivity–depth relations using an inductive electromagnetic soil conductivity meter. *Soil Sci. Soc. Am. J.*, **45**, 255–60.

Rhoades, J.D. and Oster, J.D. (1986) Solute content. In *Methods of Soil Analysis. Part 1 Physical and Mineralogical Methods*, 2nd edn, Agronomy Number 9 (Part 1), (ed. A. Klute), American Society of Agronomy, Madison, pp. 985–1006.

Rhoades, J.D. and van Schilfgaarde, J. (1976) An electrical conductivity probe for determining soil salinity. *Soil Sci. Soc. Am. J.*, **40**, 647–51.

Richards, L.A. (1965) Physical condition of water in soil. In *Methods of Soil Analysis. Part 1 Physical and Mineralogical Properties, Including Statistics of Measurement and Sampling* (ed. C.A. Black), American Society of Agronomy, Madison, pp. 128–52.

Richards, L.A. (1966) A soil salinity sensor of improved design. *Soil Sci. Soc. Am. Proc.*, **30**, 333–37.

Richards, S.J. (1965) Soil suction with tensiometers. In *Methods of Soil Analysis. Part 1 Physical and Mineralogical Properties, Including Statistics of Measurement and Sampling* (ed. C.A. Black), American Society of Agronomy, Madison, pp. 153–63.

Riehl, H. (1965) *Introduction to the Atmosphere*, McGraw-Hill, New York, 365 pp.

Roberts, J. (1983) Forest transpiration: a conservative hydrological process? *J. Hydrol.*, **66**, 133–41.

Rose, C.W. (1966) *Agricultural Physics*, Pergamon, Oxford.

Roundy, B.A. (1984) Estimation of water potential components of saline soils of Great Basin rangelands. *Soil Sci. Soc. Am. J.*, **48**, 645–50.

Rundel, P.W. (1982) Water uptake by other organs. In *Encyclopedia of Plant Physiology. Physiological Ecology (Water Relations and Photosynthetic Productivity*, Vol. 12B) (eds O.L. Lange, P. Nobel, B. Osmund and H. Ziegler), Springer-Verlag, Berlin, pp. 111–34.

Rutter, A.J., Kershaw, K.A., Robins, P.C. and Morton, A.J. (1971) A predictive model of rainfall interception in forests. I. Derivation of the model from observations in a plantation of Corsican pine. *Agric. Meteorol.*, **9**, 367–84.

Rutter, A.J. and Morton, A.J. (1977) A predictive model of rainfall interception in forests. III. Sensitivity of the model to stand parameters and meterological variables. *J. Appl. Ecol.*, **144**, 567–88.

Schlesinger, W.H., Fonteyn, P.J. and Marion, G.M. (1987) Soil moisture content and plant transpiration in the Chihuahuan Desert of New Mexico. *J. Arid Environ.*, **12**, 119–26.

Schlub, R.L. and Maine, J.W. (1979) Portable recorder for the continuous monitoring of soil moisture resistance blocks. *J. Agric. Engineer. Res.*, **24**, 319–23.

Schroder, J.L., Linthurst, R.A., Ellson, J.E. and Vozzo, F. (1985) Comparison of daily and weekly precipitation sampling efficiencies using automatic collectors. *Water Air Soil Pollut.*, **24**, 177–89.

Seymour, V.A. and Hsiao, T.C. (1984) A soil surface psychrometer for measuring humidity and studying evaporation. *Agric. For. Meteorol.*, **32**, 61–70.

Slatyer, R.O. (1967) *Plant–Water Relationships.* Academic Press, London.
Slavik, B. (1974) *Methods of Studying Plant Water Relations,* Springer-Verlag, Berlin.
Smithsonian Institution (1968) *Smithsonian Meteorological Tables,* 6th edn, Smithsonian Institution Press, Washington, D.C.
Steiger, W.R., (1951) A new development in the measurement of high relative humidities. *Science,* **114**, 152–3.
Stone, R.J. (1980) Ultraviolet fluorescence water vapour instrument for aircraft. *Rev. Sci. Instrum.,* **51**, 677–8.
Tanner, C.B. and Suomi, V.E. (1956) Lithium chloride dew cell properties and use for dewpoint and vapor-pressure gradient measurements. *Trans. Am. Geophys. Union,* **37**, 413–20.
Thom, A.S. and Oliver, H.R. (1977) On Penman's equation for estimating regional evaporation. *Q. J. R. Meteorol. Soc.,* **103**, 345–58.
Thompson, N. (1982) A comparison of formulae for the calculation of water loss from vegetated surfaces. *Agric. Meteorol.,* **26**, 265–72.
Thornthwaite, C.W. (1948) An approach toward a rational classification of climate. *Geogr. Rev.,* **38**, 55–94.
Topp, G.C. (1987) The application of time-domain reflectrometry (TDR) to soil water content measurement. In *Proceedings of the International Conference on Measurement of Soil and Plant Water Status,* vol. 1, Utah State University, pp. 85–93.
Topp, G.C. and Davis, J.L. (1985) Measurement of soil water content using time-domain reflectrometry (TDR): a field evaluation. *Soil Sci. Soc. Am. J.,* **49**, 19–24.
Topp, G.C., Davis, J.L. and Annan, A.P. (1980) Electromagnetic determination of soil water content: measurement in coaxial transmission lines. *Water Resour. Res.,* **16**, 579–82.
Topp, G.C., Davis, J.L. and Annan, A.P. (1982) Electromagnetic determination of soil water content using TDR: I. Application to wetting fronts and steep gradients. *Soil Sci. Soc. Am. J.,* **46**, 672–8.
Tuller, S.E. and Chilton, R. (1973) The role of dew in the seasonal moisture balance of a summer-dry climate. *Agric. Meteorol.,* **11**, 135–42.
Turk, K.J. and Hall, A.H. (1980) Drought adaptations of cowpea. II. Influence of drought on plant water status and relations with seed yield. *Agron. J.,* **72**, 421–7.
van Bavel, C.H.M. and Hillel, D.I. (1976) Calculating potential and actual evaporation from a bare soil surface by simulation of concurrent flow of water and heat. *Agric. Meteor.,* **17**, 453–76.
van Bavel, C.H.M., Hood, E.C. and Underwood, N. (1954) Vertical resolution in the neutron method for measuring soil moisture. *Trans. Am. Geophys. Union,* **35**, 595–600.
Vauclin, M., Haverkamp, R. and Vachaud, G. (1984) Error analysis in estimating soil water content from neutron probe measurements. 2. Spatial standpoint. *Soil Sci.,* **137**, 141–8.
Ward, A., Wells, L.G. and Phillips, R.E. (1983) Characterizing unsaturated hydraulic conductivity of Western Kentucky surface-mine spoils and soils. *Soil Sci. Soc. Am. J.,* **47**, 847–54.
Wexler, A. (1957) *Electric Hygrometers,* US Bureau of Standards, Circular 586.
Wexler, A. (1957) *Humidity and Moisture,* Vols. 1–3, Reinhold, New York.
White, G.M. and Ross, I.J. (1983) Humidity. In *Instrumentation and Measurement for Environmental Sciences* (ed. B.W. Mitchell), American Society of Agricultural Engineering, Special Publ. 13–82, pp. 8–01–8–12.

4

Measurement of wind speed near vegetation

John Grace

4.1 INTRODUCTION

Wind affects plants and animals in numerous ways (Grace, 1977, 1981) and consequently ecologists require to measure air movement for a variety of reasons, ranging from the field survey of shelter to the ventilation of leaves in assimilation chambers. The range of sensors available is considerable, and the selection made must be appropriate to the task in hand.

It should be clear that wind speed is a vector quantity with attributes of direction and magnitude. In some studies, it may be necessary to resolve the vector quantity on *x*, *y* and *z* Cartesian coordinates, resulting in *u*, *v* and *w* components (Fig. 4.1). Usually this is not the case, and more often interest centers on measuring only the horizonal component *u* or any other quantity which can be related empirically to heat or mass transfer through the boundary layer of an organism. However,

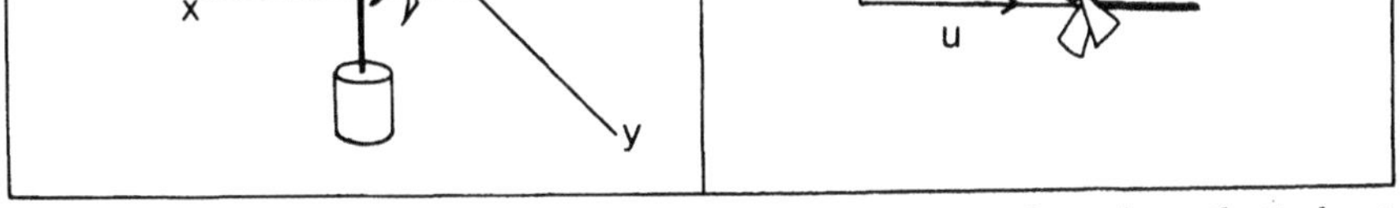

Fig. 4.1 Resolution of the wind vector into *u*, *v* and *w* components by a hypothetical set of sensors pointing in the *x*, *y* and *z* directions (into the wind, across and horizontal to the wind, and across and vertical to the wind). (b) shows the meaning of the term 'cosine response'. The anemometer is said to have a perfect cosine response if it responds to $U \cos \theta$ where θ is the angle between the wind and the propeller axis.

it must always be realized that anemometers differ greatly in their directional responses and so different types of anemometer cannot be expected to register the same wind speed when exposed alongside each other in a turbulent flow.

Anemometers differ from each other in several important respects. All mechanical anemometers have a starting *threshold* wind speed, below which the frictional forces in their bearings prevent working. Not all anemometers display *linearity* of response with respect to wind speed. This is inherently true of anemometers which rely on the principle of heat transfer. Nonlinear responses are especially troublesome when integration is required, and so such instruments are often equipped with a linearizer circuit.

The speed of response is another important attribute. The *time constant,* normally defined as the time taken for a sensor to reach 63% of its complete response, is not very useful when applied to anemometers, as the time of response is itself dependent on wind speed. In anemometry it is usual to quote the analogous *distance constant*: the length of air that must pass the sensor to elicit 63% of its complete response (Mazzarella, 1972).

In addition, the *size* of the instrument, its *endurance* and *cost* are all important factors in affecting choice.

Most of what follows is concerned with various *sensors* for the measurement of wind speed. It should however be realized that an operational *system* will usually require not only sensors but the equipment necessary to convert the signal into a form which can be displayed or stored for subsequent analysis.

4.2 FLOW IN WIND TUNNELS, GROWTH CABINETS AND DUCTS

The standard instrument used in wind tunnels for calibration purposes is the *pitot static tube* (Ower and Pankhurst, 1966; Bryer and Pankhurst, 1971). When constructed according to the standard pattern, it does not itself require calibration and has no moving parts to deteriorate. It essentially measures the force per area which is generated when moving air is halted by a nozzle pointing into the wind. The modern instrument consists of a concentric arrangement of two tubes connected to opposing arms of a manometer (Fig. 4.2). The pressure recorded on the manometer, P, is related to the wind speed according to the equation

$$P = 0.5\ \rho u^2 + p$$

where u is the wind speed, ρ is the density of the air and p is the prevailing *static pressure* which would be sensed by a measuring device at rest relative to the fluid. The manometer effectively subtracts p from P and so reads a pressure that can be equated to $0.5\rho u^2$. The device does not require especially careful alignment into the wind: 12° errors can be tolerated.

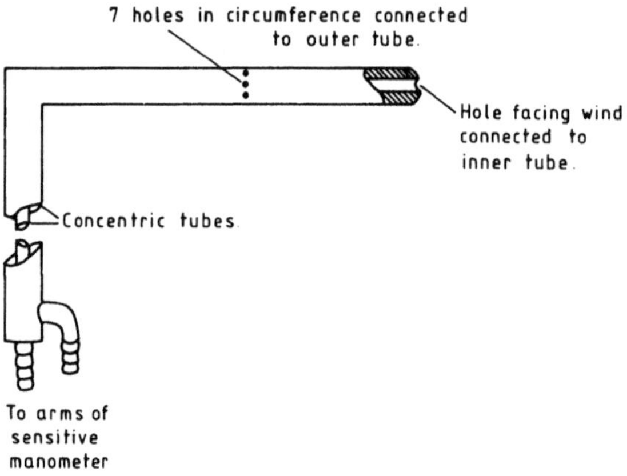

Fig. 4.2 The pitot static tube.

The dynamic pressure developed at the wind speeds that interest plant scientists (0–10 m s^{-1}) is small, and highly sensitive manometers or electronic pressure transducers must be used to measure wind speeds below 5 m s^{-1}. Below 1 m s^{-1} the pressure response is in any case influenced by poorly understood viscous effects (Bryer and Pankhurst, 1971). Even though corrections are available, it is probably wise to regard 1 m s^{-1} as the lowest usable limit of the pitot tube.

For very accurate calibration work it must be realized that air density is a function of temperature, pressure and water content, so these parameters must be recorded.

Pitot tubes should not be used in turbulent flow to determine mean flow, as the pressure is not a linear function of wind speed. To overcome this problem, some electrical pressure sensors incorporate a 'square root extractor' so that the output is proportional to the square root of pressure. However, the turbulent fluctuations recorded at the pressure sensor depend partly on the volume of air within the tubing connections, and so small-scale turbulence will not be recorded.

The principle of the pitot-static tube has been employed in field sensors. An interesting example is the three-dimensional system of tubes described by Thurtell *et al.* (1970), suitable for measuring u, v and w components of wind over vegetated surfaces.

Another instrument which is preferred in cases where flow is turbulent and where the mean flow rate of air is required is the vane anemometer. It is essentially a miniature windmill, made of flat vanes with some means of measuring rotations (Ower and Pankhurst, 1966). It responds in a practically linear manner to wind speed and in one

Fig. 4.3 Vane anemometer suitable for use as a hand-held general-purpose instrument for the laboratory, and as a three-dimensional sensor in the field (Leda, Lowne Instruments Ltd., London).

commercial version known as the Leda anemometer, the starting threshold is about 0.11 m s^{-1}. The cosine response is quite good, and the instrument is fairly small (Fig. 4.3). Sensitive vane anemometers are particularly useful in ascertaining flow rates in growth rooms, through ventilator fans in glasshouses and in general-purpose installations like fume cupboards and sterile-transfer units.

4.3 WEATHER STATIONS AND FIELD SURVEY

The classical instrument is the *cup-counter anemometer*. It consists of a set of cups, usually three, connected by arms to a vertical spindle. Its main advantage is that the rotation rate is a linear function of wind speed over a wide range, so the mean value can be found by simply summing the number of rotations. The sensor is omnidirectional in the horizontal plane, but the directional response is complex and variable in the vertical plane (Wyngaard, 1981). Large versions used in permanent weather stations are robust but have a fairly high starting threshold (up to 0.4 m s^{-1}), and a long distance constant (5 m). Small lightweight versions are available for micrometeorology, with typical distance constants about 1 m (Fig. 4.4).

The original design of cup anemometer (Patterson, 1926) has been modified to minimize the inertial tendency of the cups to 'run-on' and thus overestimate a fluctuating wind (Sheppard, 1940; Scrase and Sheppard, 1944; Deacon, 1951; Rider, 1960; Fritschen, 1967).

Historically, several other anemometers which also depend on drag forces between

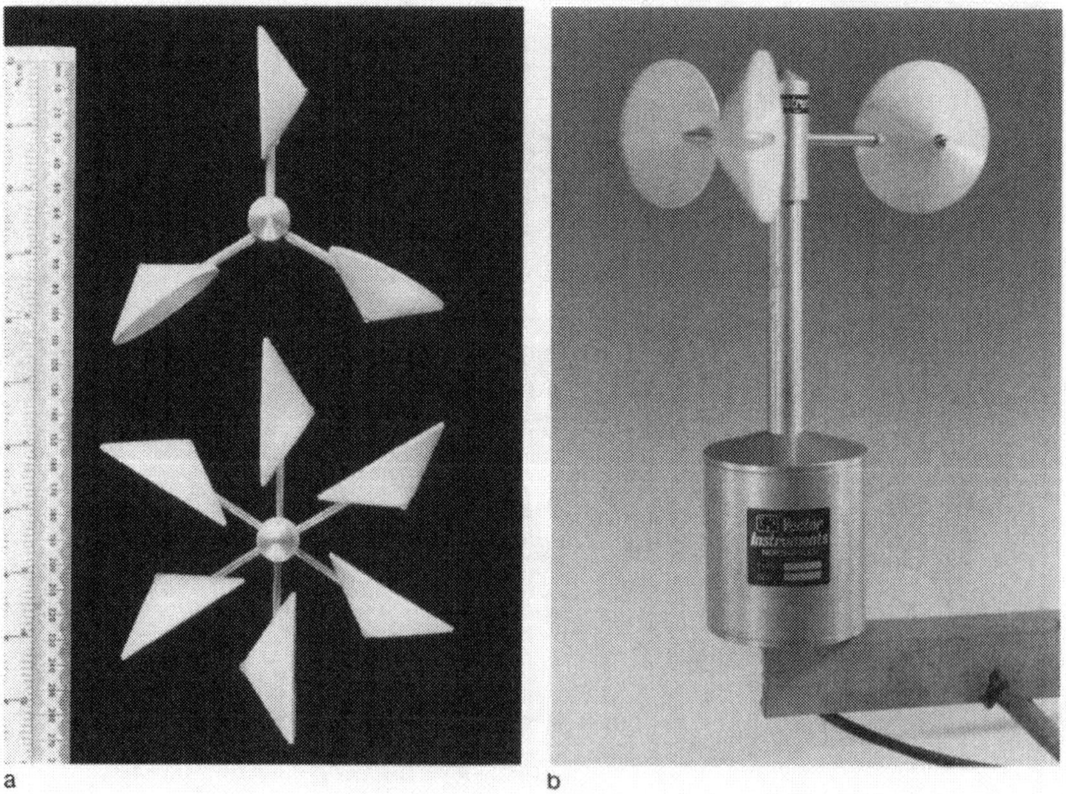

a b

Fig. 4.4 Lightweight cup anemometer set, showing three-cup and six-cup rotors (Vector Instruments, Rhyl, Wales).

the air and a body have been used. The simplest is the swinging plate (Jensen, 1954). The modern equivalents employ strain gauges or displacement transducers. An interesting modern example of one of these is the *thrust anemometer* described by Smith (1980), consisting of a sphere on a vertical stalk. The stalk pivots on a fulcrum half-way along its length. Wind forces acting on the sphere deflect the rod and the deflection is sensed by displacement transducers.

The *wind vane* is an essential feature of all weather stations, either simply to indicate wind direction or to orientate anemometers into the wind. In some studies, it is important to characterize the dynamic response of these units to a fluctuating wind direction. This is discussed at length by Fritschen and Gay (1979) and Wyngaard (1981).

Complete stations incorporating cup anemometer, wind vane as well as other environmental sensors and data logger are commercially available. Many workers prefer to assemble their own system. For a recent example, the reader is referred to Duncan (1985).

For general survey work, there is a need for semiquantitative methods that have the advantage of being very cheap and therefore capable of being used at many stations spread over an area of landscape. *Flags* have been used by foresters to survey the degree of exposure (Lines and Howell, 1963). The rate of loss of area from standard flags can be used to estimate the windiness of a site over periods of weeks. The main limitation of this technique is that the relationship between wind speed and loss of flag is markedly non-linear when the flags are wet (Rutter, 1965). Naphthalene evaporation is another possibility, though the measurement of wind speed is inevitably confounded by variations in temperature (Bernier, 1988).

4.4 WIND PROFILES ABOVE VEGETATION

The instruments for micrometeorological profiles must be very small and sensitive to much lower wind speeds than those used for general survey. The most demanding application is the evaluation of momentum transfer and the roughness parameter of the vegetation z_0. Here, calibration must be very carefully done as small errors (<1%) in the calibration of individual sensors can cause surprisingly large errors in z_0 and other derived quantities (Tanner, 1963).

The usual instrument in such studies is a miniaturized version of the *cup anemometer* with an electrical output (e.g. Jones, 1965; Bradley, 1969). In this form the starting threshold is as low as 0.1 m s^{-1} and the distance constant is about 1 m. Some versions have six cups made of expanded polystyrene, and are extremely delicate (Fig. 4.4). Because of the necessity to work with very exact data, many authors have discussed the errors caused by the supporting masts and the imperfect cosine response in the vertical plane (Moses and Daubek, 1961; MacCready, 1966; Gill *et al.*, 1967; Bernstein, 1967; Izumi and Barad, 1970). However, it is probably true that the absolute accuracy of cup anemometers in this application is limited by the dynamic response of the instrument, especially its tendency to 'run on' as the wind speed is reduced. Such errors probably increase with height above the vegetation, and so the observed wind speed may be seriously distorted. Limited observational data and theoretical calculations suggest that the error in the mean wind speed caused by overspeeding could sometimes be as much as 10% (Kaganov and Yaglom, 1976; Wyngaard, 1981).

4.4.1 Turbulent flows near plants

Close to plant surfaces the air movement is often less than 1 m s^{-1} and the wind direction is to some degree chaotic. Such flows are of great practical interest, yet very difficult to quantify. Mechanical anemometers are too large (though vane anemometers can be as small as 15 mm in diameter), and their directional characteristics are unsuitable. Interest in such flows occurs because the movement of air within a centimeter or so of the plant parts determines to a large extent the local aerodynamic resistance for transfers of heat and mass. Mechanical anemometers, even if they could be made small enough, would underestimate such flows because they do not respond to flows in all directions and might turn first one direction and then another, underestimating flow.

In studies of heat or mass transfer it is intuitively appropriate to use an anemometer which itself depends on heat or mass transfer. The instrument first developed according to this principle was the hot-wire anemometer. The characteristics of heat loss from a hot wire are well known (Compte-Bellot, 1976; Freymuth, 1978), and are fully exploited in this device to produce an extremely sensitive instrument.

In this sensor, current is passed through a very fine wire to raise the temperature considerably above the ambient. When air flows across this wire, the latter is cooled, and the consequent change in electrical resistance can be detected as a change in the voltage across the wire. A calibration curve can be obtained relating voltage output to wind speed. The sensor is incorporated into a Wheatstone bridge to permit small changes in resistance to be measured. In this form, the wire runs at a temperature which varies with the wind speed.

The useful range of the sensor is increased (and its response time made even shorter) when employed in the constant temperature mode. Here, the change in resistance unbalances the bridge and the error voltage activates a fast servo amplifier. This applies a voltage to restore the balance, thus keeping the wire temperature constant. The applied voltage is also the output signal.

Hot wire anemometers are inherently nonlinear, but for quantitative work the raw signal is passed through a linearizing circuit, designed to produce a voltage that is directly proportional to velocity over a wide range (e.g. $0.1-5$ m s^{-1}). Below about 0.1 m s^{-1} the output from the sensor remains nonlinear because at such low speeds natural convection takes over from forced convection as the main mode of heat transfer.

A typical hot wire probe is 5–25 μm in diameter and 1–3 mm long, mounted between prongs that have fine tapers to minimize conduction of heat away from the wire (Fig. 4.5). Fairly good probes can be made in the laboratory using Wollaston wire which is fine platinum coated with silver (Perry, 1982). Hot-wire anemometers have sometimes been used in the field but only for short periods as they are fragile and their calibration changes when they are dirty. They are best used in a clean indoor environment (Grace and Wilson, 1976; Raupach *et al.*, 1980).

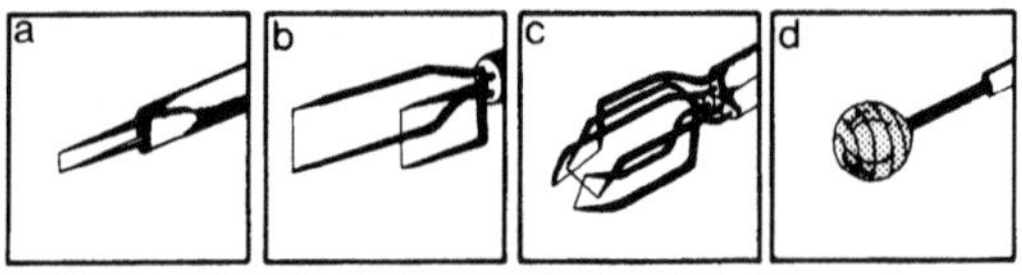

Fig. 4.5 Configurations of hot-wire anemometer probes (Disa Instruments, Bristol). a, 5 μm platinum-plated tungsten wire; b, two-wire probe, where second wire provides temperature compensation; c, triple-wire sensors for directional resolution; d, omnidirectional film-type probe.

Much more robust probes than these are now available, from companies marketing hot-wire anemometers, in which the probe has a film configuration. Film probes are made commercially by depositing a thin layer of metal on a quartz substrate and covering it with a protective ceramic layer. Robustness is thus achieved with some loss in speed of response. Many possibilities are opened up by this technique of depositing metal in a thin layer. The substrate may be spherical to achieve a nearly omnidirectional response – a useful characteristic for ecologists and physiologists interested in 'general windiness and ventilation'. There is also a film probe made with sticky tape which can be glued to leaves or other surfaces (Grace, 1978). There are exciting possibilities of miniaturization: at least one company is developing a flow sensor with associated signal processing available as a single microchip.

Hot wires and hot films have an especially fast response time (to about 0.001 s) and are therefore widely used to detect and measure turbulence. There is very extensive literature on the fluid dynamics of turbulent flows (Goldstein, 1957). For an introduction to turbulence in relation to vegetation the reader is referred to Bradshaw (1971), Raupach and Thom (1981) and Finnigan (1985).

There are several other types of sensor which depend on heat transfer, but which are larger and respond more slowly than the hot wire. They may be made using a thermocouple wound round a heater (Fritschen and Shaw, 1961) or with a heated thermistor bead. Sometimes corrections for ambient temperature may be required, especially in the simplest 'home-made' circuits (Bergen, 1971; Unwin, 1980).

4.4.2 Turbulence in the atmosphere

The hot-wire anemometer senses the arrival of eddies which may be only a few millimeters in diameter, but which may nevertheless be important in fundamental studies of heat transfer from biological surfaces at the microscale. Above the canopies of plants, most of the movement of gases and water vapor occur in eddies which are one or more orders of magnitude larger than this. The extreme sensitivity of the hot-wire anemometer may therefore be practically redundant. Moreover, the hot wire is inherently deficient in one respect: it cannot distinguish the direction of any component of flow. For these reasons, lightweight propellers have been used instead.

The propeller anemometer described by Gill (1975) has a four-blade helicoid propeller of expanded polystyrene. The original model was mounted on a lightweight bi-vane which pointed the propeller into the wind, aligning it to both azimuth and elevation angle. The calibration coefficient is fixed by the pitch of the propeller and on the sensitivity of the tachometer-generator. Gill (1975) describes the development of this single-propeller model to a three-propeller version capable of measuring u, v and w components of the wind.

The Gill anemometer, because of its low inertia, has been frequently used for eddy correlation (Dyer *et al.*, 1967; Hicks, 1972; Thompson, 1979; Milne, 1979). It does not respond fully to frequency fluctuations greater than about 1 Hz and thus in this particular application it underestimates the heat and mass transfer (McBean, 1972; Hicks, 1972; Garratt, 1975). Its distance constant at the most favorable angles is 1 m but much greater at oblique angles, and the calibration becomes nonlinear below 1 m s^{-1}. Corrections for poor cosine response have been described (see Horst, 1973; Pond *et al.*, 1979).

The Leda vane anemometer (Grace, 1986) is an eight-bladed version of the flat-vane anemometer, and may be marginally superior to the Gill anemometer in terms of its linearity, cosine response and distance constant (Figs 4.3 and 4.6). In its original form it generates a train of pulses which may be integrated or converted to a voltage using a frequency-to-voltage converter. The signal does not convey the direction of the flow past the vanes (unlike in the Gill anemometer) but recently a modified version has been developed which does do this.

All mechanical anemometers have deficiencies such as starting threshold, time lag and imperfect directional responses. Even when these deficiencies are minimized by careful design, information regarding small-scale turbulence is lost (Garratt, 1975). These problems are overcome by the use of sonic anemometers. Though conceptually simple, sonic instruments are considerably more expensive than mechanical anemometers be-

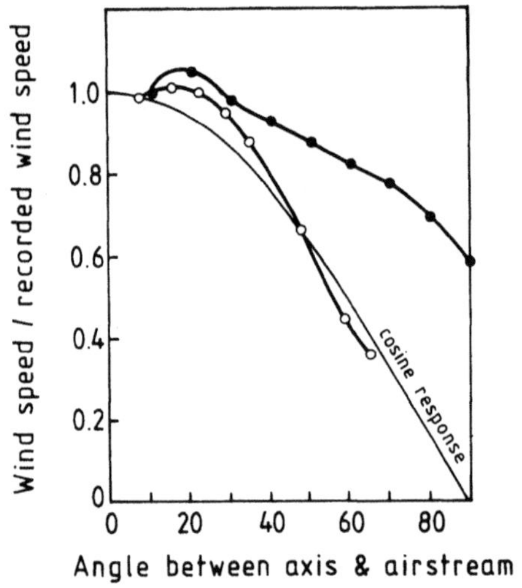

Fig. 4.6 Directional response of Gill propeller anemometer (●) and Leda vane anemometer (○). The line denotes the ideal cosine response. Leda data from unpublished data of M. Holland.

cause of the need for rapidly responding data recorders and signal-processing electronics which can be considerable.

The basic principle is the measurement of the flight time of a pulse of ultrasound between two points (Fig. 4.7). The flight time depends on the speed of sound in still air plus the speed at which the air is moving. In one mode of operation, ultrasound pulses are sent in both directions at a constant frequency, and the difference in flight time is measured from the phase shift between pulses arriving at each end of the flight path. Practical instruments are described by Mitsuta (1966), Campbell and Unsworth (1979), Shuttleworth *et al.* (1982) and Coppin and Taylor (1983), and the theory is presented by Kaimal (1979) and Wyngaard (1981).

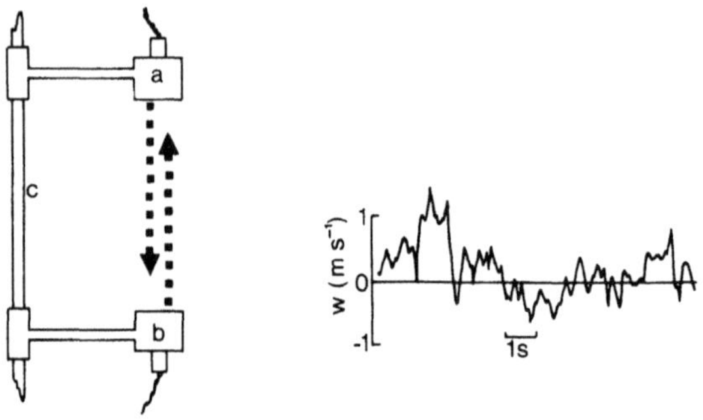

Fig. 4.7 Sonic anemometer (Campbell and Unsworth, 1979). Ultrasonic transducers (a and b), mounted on mast c are used as both transmitters and receivers. The resulting record of the *w* component of wind is shown on the right.

The sonic anemometer is commercially available and rapidly replacing the propeller anemometer in measurements of fluxes over and within vegetation canopies, including forests, by eddy correlation. The characteristics of the instrument are, by now, well documented, though there is still some uncertainty about the extent to which the supporting arms of the sensors may affect the turbulence, especially when the length of the flight path is reduced in an attempt to sense even smaller-scale turbulence.

The ultimate environmental sensor is one which does not affect the environment which it is sensing. The laser-Doppler anemometer achieves this as the laser unit can be remote from the place where the flow is being sensed. Flow is detected from the Doppler shift in the frequency of radiation scattered by small particles which happen to occur naturally in the air. As the optics are focused on a tiny spot, the spatial resolution is very small indeed. The equipment is very expensive, and so far has been used only as a laboratory instrument (Durst and Zare, 1974; Abbiss *et al.*, 1974; Durst *et al.*, 1975). It is especially useful in the measurement of wind speed and turbulence within a millimeter or so of surfaces, where hot-wire anemometers do not perform satisfactorily because of radiative coupling between the surface and the probe.

4.5 BOUNDARY LAYER RESISTANCE (r_a)

It is often required to measure the boundary

layer resistance of leaves in assimilation chambers, growth cabinets or elsewhere. The most common method has been to make a paper model of the leaf, wet it with water, then measure the evaporation rate by weighing (Jarvis, 1971). Calculation of the resistance from the evaporation rate requires a knowledge of the water vapor pressure at the evaporating surface. This value is very sensitive to the considerable variations in temperature which may develop over the paper as a result of differential evaporative cooling, and which are usually unknown. In one such experiment, it was estimated that the errors in the resistance could be as much as 45% (Grace and Wilson, 1976). A much better technique involves measurement of the transfer of heat from a metal model to the air (Grace *et al.*, 1980; Jones, 1983). The metal model should be brightly polished to minimize radiative heat transfer. The rate of heat transfer is conveniently determined by warming the model to a few degrees above ambient and then letting it cool. The resistance may be determined by analysis of the cooling curve described in detail elsewhere (Grace *et al.*, 1980; Jones, 1983). The technique requires only a set of differential thermocouples and a flat-bed recorder. Another method, with the advantage that it does not require a model to be made, is to derive r_a from the energy balance of the leaf (Parkinson, 1985). In assimilation chambers this is convenient as the necessary environmental parameters are usually being measured in any case, and do not fluctuate very much. In the field, the technique would be more awkward to apply.

There is scope for the use of diffusing species other than heat and water vapor for the measurement of aerodynamic resistance or for mapping the local variation in this value over the surface. Any material which diffuses through the boundary layer will ö: organic solvents (Thom, 1975), a radio isotope of lead (Chamberlain, 1974) and naphthalene (Neal, 1975) have all been used with success.

The boundary layer resistance of leaves in assimilation chambers and porometer chambers should always be determined for the leaf or shoot concerned. It is not meaningful to state that the air flow in the chamber 'was 4 m s^{-1}' or 'was turbulent'. Preferably the fan speed should be varied in a preliminary experiment, to ensure that a minimal r_a has been reached. The aim of this procedure is to enable stomatal resistance to be measured in conditions of low r_a so that the CO_2 and water vapor concentrations at the leaf surface are not very different from those in the bulk of air inside the chamber.

The dependence of boundary layer resistance on leaf size, wind speed and on the leaf-to-air temperature difference can be estimated using relationships gleaned from engineering textbooks (Fig. 4.8). When applying these formulae it must be remembered that heat transfer occurs from both sides of the leaf, and that experimental determinations on leaves usually yield rather lower resistances than those calculated (Grace and Dixon, 1985).

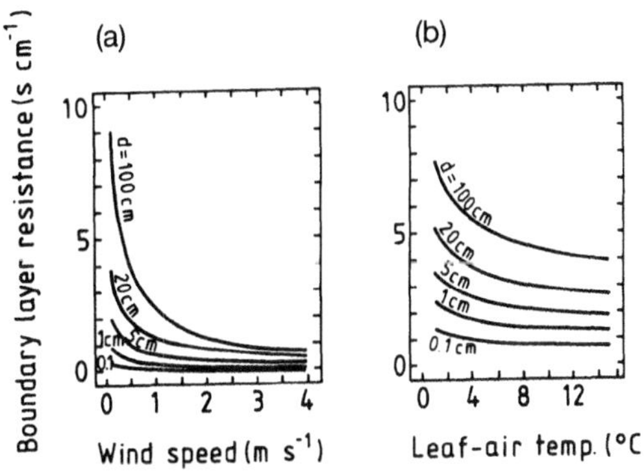

Fig. 4.8 Boundary layer resistances for water vapor flux from leaves with dimensions $d = 0.1$ to 100 cm (from Grace, 1983). In (a) the resistance for a forced convection is plotted as a function of wind speed. (b) The resistance for a free convection regime as a function of the difference in temperature between leaf and air. In both cases the resistance is for *one* side of the leaf only.

4.6 CALIBRATION

Rigorous calibration requires a high-grade wind tunnel, providing laminar flow in the range 0.5 – 20 m s^{-1} and equipped with a pitot static tube as the standard against which other sensors can be calibrated (see Pankhurst and Holder, 1952; Pope and Harper, 1966; Vogel, 1969; Grace, 1977). Moreover, the density of air depends on temperature, pressure and water content, so these should be recorded before and after a calibration run. If it is intended to calibrate a set of anemometers for use in measurements of micrometeorological profiles, the set should be calibrated before and after use. Over the course of a field season it is wise to keep a record of the location in the profile of each anemometer, reassorting the set each day, so that a faulty or miscalibrated instrument will be readily apparent when the data are analyzed. Also, as an additional precaution, the sensors can occasionally be mounted in a horizontal line in the field to compare outputs and hence to identify 'rogues'.

For small sensors, especially hot wires, miniature wind tunnels are available commercially, or can be made using wide glass tubing connected to a falling head of water which is allowed to draw air through the tube at a known rate. Some workers prefer to calibrate hot wires dynamically by mounting them on a suspended sledge which is undergoing simple harmonic motion (Perry, 1982).

4.7 AERODYNAMIC INFLUENCE BY MASTS

The supporting structures to which anemometers are attached inevitably disturb the flow (Moses and Daubek, 1961; Gill *et al.*, 1967; Cormack and Horn, 1968; Wucknitz, 1977). Dyer (1981) presents visualization tests which give some general idea of the extent of aerodynamic influence. Masts should be rigid, and the booms pointed into the flow.

Large anemometers should not be mounted too close to each other. It is a simple matter to test the magnitude of any aerodynamic interference between adjacent anemometers if a large wind tunnel is available.

4.8 VISUALIZATION

Air flow can be traced and photographed using smoke-generating equipment which is commercially available. Smoke often reveals features of the flow which might otherwise go unnoticed (Oliver, 1973; Merzhirch, 1974; Arkin and Perrier, 1974; Bergen, 1975; Wilson and Crowther, 1985). For photography, very careful illumination is required, and some workers suggest the use of lightweight particles (Vogel and Feder, 1966) or small bubbles filled with helium (Moen, 1974). However, even the smallest bubbles do not trace the small-scale turbulence.

In relation to individual leaves, the classical technique of Schlieren photography is useful in detecting the development of a thermal boundary layer, especially in natural convection (Barnes and Bellinger, 1945). Naphthalene sublimation can be used to map local variation in air flow over a leaf surface (Grace, 1983).

4.9 PRESSURE MEASUREMENTS

Atmospheric pressure varies considerably from day to day, as well as with altitude above sea level (Table 4.1). At most places on the earth the atmospheric pressure varies ±5% from week to week according to the meteorological situation. The basic properties of air such as density and the diffusivity of constituent gases vary appreciably with temperature and pressure. Such variations may be important enough to require correction to standard conditions, and especially where comparisons of gas exchange are being made at sites of different altitude. The effect of pressure on gas volume may be calculated

Table 4.1 Effect of altitude on atmospheric pressure, based upon the American Standard Atmosphere (see Weast, 1985)

Altitude (m)	0	500	1000	1500	2000	3000	4000
Pressure (kPa)	101	95	90	85	80	70	62

using the ideal gas equation, but the effects of small changes in pressure on diffusivity and on the psychrometric constant are less obvious (Table 4.2). Calibration of anemometers should be accompanied by measurements of pressure, temperature and humidity, as the force of the moving air on the sensor clearly depends on air density.

There are other applications where it is necessary to measure differential pressure. One of these has already been referred to in this chapter: the pitot static tube used in wind tunnels for calibration purposes must be connected to a differential pressure transducer. For the low wind speeds of interest to plant scientists (1 m s^{-1}), the differential pressure is very small, and so very sensitive sensors are called for. Similar requirements occur in infrared gas analysis, where, as in many gas flow systems, the pressure in the reference cell is different from that in the analysis cell.

Although the SI unit of pressure is the Pascal (1 N m^{-2}), meterologists still use bars and millibars (1 bar = 10^5 Pa), and instruments are still calibrated in these units. Moreover, electrical manometers designed for use in industry are sometimes calibrated in inches or millimeters of water or mercury.

For occasional measurement of atmospheric pressure in the laboratory, a *mercury barometer* may be permanently fixed to the wall in a place which is unlikely to experience strong temperature fluctuations or draughts. The use of these instruments in their various patterns (Fortin, Patterson or Kew) is described in meteorological handbooks (Meteorological Office, 1956). As an alternative, the *aneroid barometer* should be considered. The reading on this barometer depends on the expansion and contraction of an aneroid cell – a sealed elastic bellows-like structure. These changes in dimension are linked mechanically to a needle on a dial or can be sensed electrically thus lending to automatic recording. The

Table 4.2 Effect of pressure and temperature on air density (kg m^{-3}), D/D_o and the psychrometric constant for aspirated psychrometers, γ (kPa °C^{-1}). D_o is the diffusion coefficient of any gas in air at standard conditions of 0 °C and 100 kPa and D is its value at the stated temperature and pressure. The psychrometric constant is defined as $\gamma = C_pP/\lambda\epsilon$ where C_p is the specific heat of air at constant pressure, P is the atmospheric pressure, λ is the latent heat of vaporization of water and ϵ is the ratio of the molecular weight of water vapor to that of air (= 0.622). Calculated from formulae in Kay and Labey (1973) and Weast (1985)

Pressure (kPa)	*Atmospheric density at 50% relative humidity (kg m^{-3})*			*D/D_o*			*γ (kPa °C^{-1})*		
	10°C	*20°C*	*30°C*	*10°C*	*20°C*	*30°C*	*10°C*	*20°C*	*30°C*
80	0.98	0.94	0.91	1.33	1.41	1.50	0.052	0.053	0.053
90	1.10	1.06	1.02	1.18	1.26	1.33	0.059	0.060	0.060
95	1.17	1.12	1.08	1.12	1.19	1.26	0.062	0.063	0.063
100	1.23	1.18	1.14	1.06	1.13	1.20	0.066	0.066	0.067
105	1.29	1.24	1.20	1.01	1.08	1.14	0.069	0.069	0.070

aneroid barometer is considerably more portable than the mercury barometer and can be transported to field stations or used outside, but is somewhat less accurate. Absolute accuracy claimed for mercury barometers is about 0.2 mbar (0.02%), whilst tests on the standard Meteorological Office MkII aneroid suggest a figure of around 0.5 mbar. This difference is unimportant for most biological investigations. Both types of barometer are virtually maintenance free. In the *aneroid barograph* the change in linear dimensions of the cell is coupled by levers to a pen which writes on a moving chart to record variations in pressure.

As mentioned above, it is important to have in the laboratory a manometer for measuring differential pressure, especially if the pitot static tube is to be used as a calibration standard for a wind tunnel or anemometers. In this case it is necessary to measure differential pressures in the range 0.5–100 Pa. Direct-reading manometers suitable for this incorporate a U-tube filled with a colored low-density liquid.

The U-tube is held at an angle only slightly displaced from the horizontal so that a small displacement in height of the fluid results in a large lateral displacement on the meniscus. The meniscus is read using a microscope, and

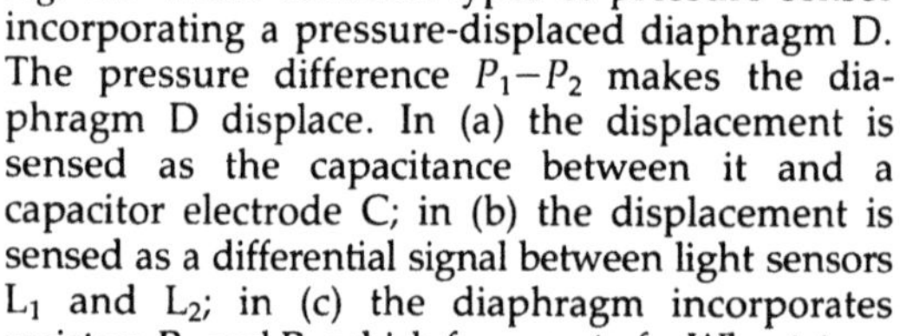

Fig. 4.9 These common types of pressure sensor incorporating a pressure-displaced diaphragm D. The pressure difference $P_1 - P_2$ makes the diaphragm D displace. In (a) the displacement is sensed as the capacitance between it and a capacitor electrode C; in (b) the displacement is sensed as a differential signal between light sensors L_1 and L_2; in (c) the diaphragm incorporates resistors R_1 and R_2 which form part of a Wheatstone bridge circuit set up to generate a voltage which can be proportional to the displacement.

the most sensitive instruments incorporate the null principle whereby the user adjusts a calibrated screw which adjusts the angle of the U-tube to achieve a null point.

Direct-reading manometers, both absolute and differential, are now largely replaced by electrical transducers. These can be small, sensitive, very portable and cheap. Most of these work on the principle that a pressure difference across an elastic diaphragm causes a displacement which can be measured and calibrated in terms of pressure (Fig. 4.9).

In the most expensive cells, the displacement is sensed as a change in capacitance between the diaphragm and a capacitor electrode plate (Fig. 4.9a). Such cells have very good long-term stability and can be used as transfer standards against which other devices can be calibrated. The most sensitive, but not the most accurate, type incorporates an optical system to detect the displacement of the diaphragm (Fig. 4.9b). This high sensitivity is useful in physiological work and when measuring differential pressures developed at low wind speed by a pitot static tube. A third method is to sense the distortion of the diaphragm using strain gauges or displacement transducers. This is the method that lends itself to miniaturization: the diaphragm can be reduced to a silicon chip with piezo resistors to detect the elastic distortion. The overall unit can thus be small enough to be included in a gas line or in an assimilation chamber. In other respects, the performance of this type may be not quite as good as the larger types (Table 4.3).

Most electrical transducers are available in forms suitable for differential or absolute pressure measurements. In the latter, the volume on one side of the diaphragm is closed and filled with gas at a suitable pressure.

4.10 SOME APPLICATIONS

Wind speed is often measured routinely or

Table 4.3 Typical specifications of pressure sensors, based on data from suppliers: Chell Instruments, Norfolk NR28 9JH; Mercury Electronics, Glasgow; Farnell, Leeds LS12 2TU; Radiospares, Corby, NN17 9RS

Type	*Variable capacitance cell with signal conditioner*	*Diaphragm electromanometer*	*Silicon chip with integral sensing diaphragm*
Size of sensor	< 100 mm	100 mm	5 – 30 mm
Price range	£500 – £1000	£500 – £1000	£30 – £300
Linearity	Good	2%	1%
Ranges	Differential from 100 Pa FSD† Absolute from 100 Pa FSD to 10^6 Pa	Differential down to to 10 Pa FSD	Many available
Temperature error 0–50°C in			
(a) Sensitivity	*0.02% reading °C^{-1}	Unknown	0.02% reading °C^{-1}
(b) Zero shift	*0.005% FSD† °C^{-1}	Unknown	0.04% FSD† °C^{-1}
Response time	< 16 ms	< 80 ms	Fast
Absolute accuracy	0.15% of reading, stable over 3 years	2%	Depends on calibration Stability over 3 years 1.5%
Recommended for	General purpose laboratory use, gas control systems	Very small differential pressures, as in physiological experiments, pitot static tubes	Portable equipment including portable weather stations

* Available from one company in transfer standard series with temperature stability ten times greater than the standard range quoted.
† Full scale deflection.

incidentally as part of the site description in an ecological study. There are, however, many other instances where the measurement is central to the purpose of the investigation. Beginners may find it useful to peruse such case studies to set wind measurement in the context of specific biological problems. Some examples are mentioned in the following paragraphs.

The agronomic importance of windbreaks and shelterbelts has lead to much descriptive work on the extent of reduction of the wind speed in the lee of barriers. This general field is the subject of a recent book (Brandle *et al.*, 1988). Natural shelter afforded by topography and the roughness of the vegetation itself may be very significant in understanding behavior patterns of animals (Grace and Easterbee, 1979).

There have been many attempts to relate heat and mass transfer to wind speed, sometimes in the field but more often in a wind tunnel (Landsberg and Thom, 1971; Pearman *et al.*, 1972; Proctor, 1981; Grace *et al.*, 1980). At very low wind speeds there are formidable problems, as flows are difficult to measure and there is little guidance in the literature regarding the treatment of natural convection (Grace and Dixon, 1985). Measurements of surface temperature of plants in extreme environments such as deserts and mountains have demonstrated the importance of convective processes for survival and growth (Nobel, 1989; Wilson *et al.*, 1987).

Wind also causes direct mechanical damage to leaves, especially important in horticulture but significant also in forestry (Wilson, 1980). The large-scale destruction of plantations during storms appears to depend on a resonance between gustiness and the natural period of sway, but data on this point are rather scarce (Holbo *et al.*, 1980; Mayer, 1987).

The boundary layer over leaves is frequently turbulent (Grace and Wilson, 1976), and the air immediately above and within the canopy space of crops displays complex patterns of flow (Cionco, 1972; Hicks *et al.*, 1975; Hutchison and Hicks, 1985).

The behavior of airborne materials in the form of mists and small particles (aerosols) near vegetation is influenced by the wind regime. Studies in wind tunnels suggest they may behave in the same way as gases when the particles are very small (Chamberlain and Little, 1981). Dispersal of spores and other propagules from plant surfaces normally requires a threshold wind speed (Grace and Collins, 1976), but in the field wind-induced vibration may have a more important effect on dispersal than wind acting directly.

ACKNOWLEDGEMENTS

Thanks are due to my colleague John Moncrieff for his comments on an earlier draft, and Michael Holland for permission to use the result of the directional test of the Leda anemometer.

REFERENCES

Abbiss, J.B., Chubb, T.W. and Pike, E.R. (1974) Laser Doppler anemometry. *Opt. Laser Technol.*, **1974**, 249–61.

Arkin, G.F. and Perrier, E.R. (1974) Vorticular air flow within an open row crop canopy. *Agric. Meteorol.*, **13**, 359–74.

Barnes, N.F. and Bellinger, S.L. (1945) Schlieren and shadowgraph equipment for air flow analysis. *J. Opt. Soc. Am.*, **35**, 497–509.

Bergen, J.D. (1971) An inexpensive heated thermistor anemometer. *Agric. Meteorol.*, **8**, 395–405.

Bergen, J.D. (1975) Air movement in a forest clearing as indicated by smoke drift. *Agric. Meteorol.*, **15**, 165–79.

Bernier, P.Y. (1988) Low-cost wind speed measurements using naphthalene evaporation. *Journal of Atmospheric and Ocean Technology*, **5**, 662–5.

Bernstein, A.B. (1967) A note on the use of cup anemometers in wind profile experiments. *J. Appl. Meteorol.*, **6**, 280–6.

Bradley, E.F. (1969) A small, sensitive anemometer system for agricultural meteorology. *Agric. Meteorol.*, **6**, 185–93.

Bradshaw, P. (1971) *An Introduction to Turbulence and its Measurement*, Pergamon, Oxford.

Brandle, J., Hintz, J. and Sturroch, H. (eds) (1988) *Windbreaks*, Elsevier, Amsterdam.

Bryer, D.W. and Pankhurst, R.C. (1971) *Pressure-Probe Methods for Determining Wind Speed and Flow Direction*, National Physical Laboratory, Her Majesty's Stationery Office, London.

Campbell, G.S. and Unsworth, M.H. (1979) An inexpensive sonic anemometer for eddy correlation. *J. Appl. Meteorol.*, **18**, 1027–77.

Chamberlain, A.C. (1974) Mass transfer to bean leaves. *Bound. Layer Meteorol.*, **6**, 477–86.

Chamberlain, A.C. and Little, P. (1981) Transport and capture of particles by vegetation. In *Plants and their Atmospheric Environment* (eds J. Grace, E.D. Ford and P.G. Jarvis), Blackwell Scientific Publications, Oxford, pp. 147–73.

Cionco, R.M. (1972) Intensity of turbulence within canopies with simple and complex roughness elements. *Bound. Layer Meteorol.*, **2**, 453–65.

Compte-Bellot, G. (1976) Hot-wire anemometry. *Annu. Rev. Fluid Mech.*, **8**, 209–31.

Coppin, P.A. and Taylor, K.J. (1983) A three component sonic anemometer/thermometer system for general micrometeorological research. *Bound. Layer Meteorol.*, **27**, 27–42.

Cormack, J.E. and Horn, J.D. (1968) Tower shadow effect. *J. Geophys. Res.*, **73**, 1869–76.

Deacon, E.L. (1951) The over-estimation error of cup anemometers in fluctuating winds. *J. Sci. Instrum.*, **28**, 231–4.

Duncan, C.N. (1985) Metdads: a microcomputer-controlled weather display station. *Weather*, **40**, 68–76.

Durst, F. and Zare, M. (1974) *Bibliography of Laser-Doppler-Anemometry Literature*, Unpublished manuscript.

Durst, F., Zare, M. and Wigley, G. (1975) Laser-Doppler anemometry and its application to flow investigations related to the environment. *Bound. Layer Meteorol.*, **8**, 281–322.

Dyer, A.J. (1981) Flow distortion by supporting structures. *Bound. Layer Meteorol.*, **20**, 243–51.

Dyer, A.J., Hicks, B.B. and King, K.K. (1967) The

Fluxatron – a revised approach to the measurement of eddy fluxes in the lower atmosphere. *J. Appl. Meteorol.*, **6**, 408–13.

Finnigan, J.J. (1985) Turbulent transport in flexible plant canopies. In *The Forest–Atmosphere Interaction* (eds B.A. Hutchinson and B.B. Hicks), Reidel, Dordrecht, pp. 443–80.

Freymuth, P. (1978) A bibliogaphy of thermal anemometry. *TSI Q.*, **4**, 3–26.

Fritschen, L.T. (1967) A sensitive cup-type anemometer. *J. Appl. Meteorol.*, **6**, 695–8.

Fritschen, L.J. and Gay, L.W. (1979) *Environmental Instrumentation*, Springer-Verlag, New York, Heidelberg and Berlin.

Fritschen, L.T. and Shaw, R.H. (1961) A thermocouple-type anemometer and its use. *Bull. Am. Meteorol. Soc.*, **42**, 42–6.

Garratt, J.R. (1975) Limitations of the eddy-correlation techniques for the determination of turbulent fluxes near the surface. *Bound. Layer Meteorol.*, **8**, 255–9.

Gill, G.C. (1975) Development and use of the Gill UVW Anemometer. *Bound. Layer Meteorol.*, **8**, 475–95.

Gill, G.C., Olsson, L.E., Sela, J.I. and Suda, M. (1967) Accuracy of wind measurements on towers or stacks. *Bull. Am. Meteorol. Soc.*, **48**, 665–74.

Goldstein, S. (1957) *Modern Developments in Fluid Dynamics*, Clarendon, Oxford.

Grace, J. (1977) *Plant Response to Wind*, Academic Press, London.

Grace, J. (1978) The turbulent boundary layer over a flapping *Populus* leaf. *Plant, Cell Environ.*, **1**, 35–8.

Grace, J. (1981) Some effects of wind on plants. In *Plants and their Atmospheric Environment* (eds J. Grace, E.D. Ford and P.J. Jarvis), Blackwell Scientific Publication, Oxford, pp. 32–56.

Grace, J. (1983) *Plant–Atmosphere Relationships*, Chapman and Hall, London.

Grace, J. (1986) The measurement of wind speed and turbulence. In *Instrumentation in Environmental Physiology* (eds B. Marshall and F.I. Woodward), Cambridge University Press, Cambridge, pp. 101–21.

Grace, J. and Collins, M.A. (1976) Spore liberation from leaves by wind. In *The Microbiology of Leaf Surfaces* (eds C.H. Dickenson and T.F. Preece), Academic Press, London and New York, pp. 185–98.

Grace, J. and Dixon, M. (1985) Convective heat transfer from leaves. In *Effects of Shelter on the Physiology of Plants and Animals* (ed. J. Grace), Swets & Zeitlinger, Lisse, pp. 1–16.

Grace, J. and Easterbee, N. (1979) The natural shelter for red deer (*Cervus elaphus*) in a Scottish glen. *J. Appl. Ecol.*, **16**, 37–48.

Grace, J., Faschun, F.E. and Dixon, M. (1980) Boundary layer conductance of the leaves of some tropical timber tress. *Plant, Cell Environ.*, **3**, 443–50.

Grace, J. and Wilson, J. (1976) The boundary layer over a *Populus* leaf. *J. Exp. Bot.*, **27**, 231–41.

Hicks, B.B. (1972) Propeller anemometers as sensors of atmospheric turbulence. *Bound. Layer Meteorol.*, **3**, 214–28.

Hicks, B.B., Hyson, P. and Moore, C.J. (1975) A study of eddy fluxes over a forest. *J. Appl. Meteorol.*, **14**, 58–66.

Holbo, H.R., Corbett, T.C. and Horton, P.J. (1980) Aeromechanical behaviour of selected Douglas-fir. *Agric. Meteorol.*, **32**, 81–91.

Hutchison, B.A. and Hicks, B.B. (eds) (1985) *The Forest–Atmosphere Interaction*, Reidel, Dordrecht.

Horst, T.W. (1973) Corrections for response errors in a three component propeller anemometer. *J. Appl. Meteorol.*, **12**, 716–25.

Izumi, Y. and Barad, M.L. (1970) Wind speeds as measured by cup and sonic anemometers and influences by tower structure. *J. Appl. Meteorol.*, **9**, 851–6.

Jarvis, P.G. (1971) The estimation of resistances to carbon dioxide transfer. In *Plant Photosynthetic Production, Manual of Methods* (eds Z. Sestak, J. Catsky and P.G. Jarvis), Junk, The Hague, pp. 556–631.

Jensen, J.M. (1954) *Shelter Effect*, Danish Technical Press, Copenhagen.

Jones, J.I.P. (1965) A portable sensitive anemometer with proportional d.c. output and a matching velocity-component resolver. *J. Sci. Instrum.*, **42**, 414–17.

Jones, H.G. (1983) *Plants and Microclimate*, Cambridge University Press, Cambridge.

Kaganov, E. and Yaglom, A.M. (1976) Errors in wind-speed measurements by rotation anemometers. *Bound. Layer Meteorol.*, **10**, 15–34.

Kaimal, J.C. (1979) Sonic anemometer measurements of atmospheric turbulence. *Proceedings of Dynamic Flow Conference 1978*, Skovlunde, Denmark.

Kay, G.W.C. and Labey, T.H. (1973) *Tables of Physical and Chemical Constants and some Mathematical Constants*, 14th edn, Longman, London.

Landsberg, J.J. and Thom, A.S. (1971) Aerodynamic properties of a plant of complex structure. *Q. J. R. Meteorol. Soc.*, **97**, 565–70.

Lines, R. and Howell, R.S. (1963) The use of flags to estimate relative exposure of trial plantations. *For. Comm., For. Rec.*, **51**, 1–31.

McBean, G.A. (1972) Instrument requirements for

eddy correlation measurements. *J. Appl. Meteorol.*, **11**, 1078–84.
MacCready, P.B. (1966) Mean wind speed measurements in turbulence. *J. Appl. Meteorol.*, **5**, 219–25.
Mayer, H. (1987) Wind-induced tree sways. *Trees*, **1**, 195–206.
Mazzarella, D.A. (1972) An inventory of specifications for wind measuring instruments. *Bull. Am. Meteorol. Soc.*, **53**, 860–71.
Merzhirch, W. (1974) *Flow Visualization*, Academic Press, New York.
Meteorological Office (1956) *Handbook of Meteorological Instruments*, H.M.S.O., London.
Milne, R. (1979) Water loss and canopy resistance of a young Sitka spruce plantation. *Bound. Layer Meteorol.*, **16**, 67–81.
Mitsuta, Y. (1966) Sonic anemometer–thermometer for general use. *J. Meteorol. Soc. Jpn*, **44**, 12–24.
Moen, A.N. (1974) Turbulence and the visualisation of wind flow. *Ecology*, **55**, 1420–4.
Moses, H. and Daubek, H.G. (1961) Errors in wind measurements associated with tower-mounted anemometers. *Bull. Am. Meteorol. Soc.*, **42**, 190–4.
Neal, S.B.H.C. (1975) The development of the thin-film naphthalene mass-transfer analogue technique for the direct measurement of heat-transfer coefficients. *Int. J. Heat Mass Transfer*, **18**, 559–67.
Nobel, P.S. (1989) Principles underlying the prediction and analysis of temperature and its influences on plants, with particular reference to hot climates. In *Plants and Temperature* (eds S. Long and F.I. Woodward), Cambridge University Press, Cambridge.
Oliver, H.R. (1973) Smoke trails in a pine forest. *Weather*, August 1973, 345–7.
Ower, E. and Pankhurst, R.C. (1966) *The Measurement of Air Flow*, 4th edn, Pergamon, Oxford.
Pankhurst, R.C. and Holder, D.W. (1952) *Wind Tunnel Technique*, Pitman, London.
Parkinson, K.J. (1985) A simple method for determining the boundary layer resistance in leaf cuvettes. *Plant, Cell Environ.*, **8**, 223–6.
Patterson, J. (1926) The cup anemometer. *Trans. R. Soc. Can.*, **20**, 1–56.
Pearman, G.I., Weaver, W.L. and Tanner, C.B. (1972) Boundary layer heat transfer coefficients under field conditions. *Agric. Meteorol.*, **10**, 83–92.
Perry, A.E. (1982) *Hot-wire Anemometry*, Clarendon Press, Oxford.
Pond, S., Large, W.G., Miyake, M. and Burling, R.W. (1979) A Gill twin propeller-vane anemometer for flux measurements during moderate and strong winds. *Bound. Layer Meteorol.*, **16**, 351–64.
Pope, A. and Harper, J.J. (1966) *Low-speed Wind Tunnel Testing*, Wiley, New York.
Proctor, M.C.F. (1981) Diffusion resistances in Bryophytes. In *Plants and their Atmospheric Environment* (eds J. Grace, E.D. Ford and P.G. Jarvis), Blackwell Scientific Publications, Oxford, pp. 219–29.
Raupach, M.R. and Thom, A.S. (1981) Turbulence in and above plant canopies. *Annu. Rev. Fluid Mech.*, **13**, 97–129.
Raupach, M.R., Thom, A.S. and Edward, I. (1980) A wind tunnel study of turbulent flows close to regularly arranged rough surfaces. *Bound. Layer Meteorol.*, **18**, 373–97.
Rider, N.E. (1960) On the performance of sensitive cup anemometers. *Meteorol. Mag., London*, **89**, 209–15.
Rutter, N. (1965) Tattering of flags under controlled conditions. *Nature, Lond.*, **205**, 168–9.
Scrase, F.J. and Sheppard, P.A. (1944) The errors of cup anemometers in fluctuating winds. *J. Sci. Instrum.*, **21**, 160–8.
Sheppard, P.A. (1940) An improved design of cup anemometer. *J. Sci. Instrum.*, **17**, 218.
Shuttleworth, W.J., McNeil, D.D. and Moore, C.J. (1982) A switched continuous-wave sonic anemometer for measuring surface heat fluxes. *Bound. Layer Meteorol.*, **23**, 425–48.
Smith, S.D. (1980) Evaluation of the Mark 8 Thrust Anemometer-thermometer for measurement of boundary-layer turbulence. *Bound. Layer Meteorol.*, **19**, 273–92.
Tanner, C.B. (1963) Basic instructions and measurements for plant environment and micrometeorology. *Soils Bull.*, No. 6, College of Agriculture, University of Wisconsin.
Thom, A.S. (1975) Momentum, mass and heat exchange in plant communities. In *Vegetation and the Atmosphere* (Vol. 1 *Principles*) (ed. J.L. Monteith), Academic Press, London and New York, pp. 57–109.
Thompson, N. (1979) Turbulence measurements above a pine forest. *Bound. Layer Meteorol.*, **16**, 293–310.
Thurtell, G.W., Tanner, C.B. and Wesely, M.L. (1970) Three dimensional pressure-sphere anemometer system. *J. Appl. Meteorol.*, **9**, 379–85.
Unwin, D.M. (1980) *Microclimate Measurement for Ecologists*, Academic Press, London and New York.
Vogel, S. (1969) Low-speed wind tunnels for biological investigations. In *Experiments in*

Physiology and Biochemistry (ed. G.A. Kerkut), Academic Press, London, Vol. 2, pp. 295–325.

Vogel, S. and Feder, N. (1966) Visualisation of low-speed flow using suspended plastic particles. *Nature, Lond.* **209**, 186–7.

Weast, R.C. (1985) *CRC Handbook of Chemistry and Physics*, CRC Press, Florida.

Wilson, C.E. and Crowther, J.M. (1985) Flow visualisation and the study of shelter effects for vegetation at the microscale. In *Effects of Shelter on the Physiology of Plants and Animals* (ed. J. Grace), Swets & Zeitlinger, Lisse, pp. 17–36.

Wilson, C., Grace, J., Allen, S. and Slack, F. (1987) Temperature and stature: a study of temperatures in montane vegetation. *Funct. Ecol.*, **1**, 405–14.

Wilson, J. (1980) Macroscopic features of wind damage of leaves of *Acer pseudoplatanus* L. and its relationship with season, leaf age and windspeed. *Ann. Bot.*, **46**, 303–11.

Wucknitz, J. (1977) Disturbance of wind profile measurements by a slim mast. *Bound. Layer Meteorol.*, **11**, 155–69.

Wyngaard, J.C. (1981) Cup, propeller, vane and sonic anemometers in turbulence research. *Annu. Rev. Fluid Mech.*, **13**, 399–423.

5
Soil nutrient availability

Dan Binkley and Peter Vitousek

5.1 INTRODUCTION

Many methods have been developed for assessing the availability of soil nutrients, but for a variety of reasons none are universally applicable. In this chapter, we discuss the conceptual basis for measuring nutrient availability and describe the strengths and limitations of some of the methods for assessing nonagricultural soils. We also discuss methods for characterizing soil acidity, salinity and redox potential because they often control nutrient cycling and availability.

The concept of nutrient availability can be viewed from two distinct points of view. From a soils perspective, the flux of nutrients from unavailable pools into pools accessible by plants represents a *nutrient supply rate*. Ecosystems differ widely in nutrient supply rates because of differences in rates of decomposition, mineral weathering and other processes. Alternatively, the productivity of individual plants or whole ecosystems can be affected by *nutrient limitation*, defined as the extent to which productivity is reduced by an inadequate rate of nutrient supply (Chapin *et al.*, 1986). In agricultural situations, nutrient supply rate and nutrient limitation are closely linked, although some species have a greater flexibility in growth rates and respond better to increasing nutrient supply (cf. Bray, 1961). In natural ecosystems, however, differences in species composition, plant life stage, and moisture supply can uncouple low nutrient supply rates from nutrient limitation.

This distinction is illustrated in Fig. 5.1. At low nutrient supply rates, Species T_3 would dominate the site. If fertilization increased nutrient supply rate, productivity would at first increase only to level 3′ (in the absence of species replacement). Fertilization of a site dominated by species T_2 or T_1 could yield a greater increase in productivity than fertilization of T_3, despite the greater initial nutrient supply. In the longer term, T_2 or T_1 could replace T_3 on a fertilized site, but on shorter time scales nutrient limitation would not be closely correlated with nutrient supply (Fig. 5.1).

We focus here on methods for estimating rates of nutrient supply, which we will use interchangeably with nutrient availability. Nutrient supply rate is based on soil characteristics, and provides information that can be compared across a range of sites (or treatments) that vary in species composition,

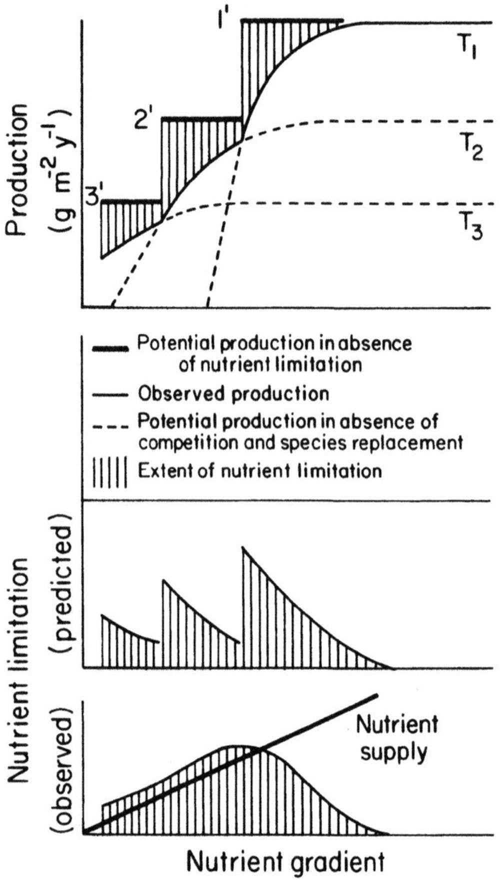

Fig. 5.1 Nutrient limitation of plant growth depends upon the supply rate of nutrients and the ability of a species to increase growth in response to increases in the supply rate (see text) (after Chapin *et al.*, 1986).

plant age or soil fertility. Without a great deal of background information, nutrient limitation can only be determined by fertilization.

5.2 DIFFICULTIES IN MEASURING NUTRIENT AVAILABILITY

No soil assay determines nutrient availability precisely, as even the best methods cannot account for all of the factors that influence fluxes of nutrients in natural ecosystems. Soil nutrients are found in an array of pools that vary in availability to plants, and plants affect nutrient availability in a number of ways. Nutrient absorption creates concentration gradients that favor diffusion of more ions towards roots. Plants vary the amount of soil they exploit by changing the production of fine roots, mycorrhizal fungi, and altering the chemistry of the rhizosphere. On a longer time scale, plants also affect nutrient availability by producing organic litter of varying chemical and physical properties (see Pastor and Post, 1986).

The choice among soil assay methods involves weighing the strengths and limitations of each method in relation to research objectives. In many cases, the extent of nutrient limitation is of primary interest, and fertilization experiments or foliar analysis (see Chapter 10) are most useful. In other cases, an index of differences in soil fertility among sites is desired, and simple laboratory measurements serve as well as more complex designs.

5.3 NITROGEN AVAILABILITY

Nitrogen is added to ecosystems primarily as ammonium, nitrate and organic nitrogen dissolved in precipitation, by gas and aerosol inputs to the forest canopy, and as ammonia fixed biologically by procaryotes. Annual inputs in precipitation usually range from 1 to 5 kg ha^{-1} in unpolluted regions, up to 10 to 40 kg ha^{-1} in areas with nitric acid rain. Ecosystems dominated by symbiotic N-fixing plants may have input rates of 100 kg ha^{-1} or more annually through this pathway. The total N content of most ecosystems falls within the range of 1500–15 000 kg ha^{-1} (see Post *et al.*, 1986). Over 98% of this nitrogen is bound in organic detritus and soil humus, and becomes available for plant uptake through microbial mineralization (decomposition). Some organic N pools are fairly labile and decompose readily, while others are very recalcitrant due to complex chemical structures and physical occlusion in soil microaggregates (Sollins *et al.*, 1984). The annual uptake of N in most nonagricultural ecosystems falls within the range of 20–200 kg ha^{-1} annually.

Approximately 2–5% of soil organic nitrogen is released to inorganic forms annually as a byproduct of microbial oxidation of organic matter. A large fraction of this released nitrogen (50–99%) is taken up by microbes and is called 'immobilization' (Rosswall, 1976, 1982). The total amount of N liberated from organic matter is 'gross mineralization'; the quantity remaining after subtraction of microbial immobilization is 'net mineralization'. Although microbes are generally effective competitors for inorganic N (especially ammonium), mycorrhizae can increase the proportion obtained by plants (see Barber, 1984; Tinker, 1984).

The rate of mineralization is affected by substrate chemistry (N content and types of organic compounds), biology (comminuters, microbes, grazers on microbes) and environmental factors (temperature, moisture and aeration). For a review see Binkley and Hart (1989). Mineralization is generally high for substrates with high N contents and under environmental conditions favorable for microbial activity. Microbial immobilization tends to be high (lowering net mineralization) when N concentrations in organic matter are low, because a relative abundance of carbon compounds provides energy for using N to synthesize new microbial biomass. Fluctuations in environmental conditions promote mineralization by altering microbial populations, which makes the N contained in microbial biomass one of the more labile of soil N pools.

Ammonia (NH_3) is the first product of microbial mineralization (ammonification). In all but the most alkaline soils, ammonia rapidly acquires one H^+ to form ammonium (NH_4^+). Ammonium may be taken up and utilized by plants and microbes, retained in the soil, leached from the soil or oxidized by autotrophic bacteria. The last process, nitrification, releases 2 H^+ and forms nitrate (NO_3^-). Ammonium is retained in the soil on cation-exchange sites which are present in most soils, while the anion-exchange capacity of temperate-zone soils is usually negligible. This difference in ion-exchange properties results in nitrate mobility (diffusivity) in most soils exceeding that of ammonium by 10 to 100-fold (Barber, 1974). Therefore, when the concentrations of the two ions are similar, the supply of nitrate to plant roots should be much greater than the supply of ammonium. Leaching losses of nitrate can be large, but losses of ammonium are negligible except in very sandy soils with little cation exchange capacity. Nitrate may also be lost through dissimilatory reduction as N_2O or N_2 (denitrification).

5.3.1 Methods for assessing N availability

(a) Total N content

The most common method used for measuring the total N content of soils was developed by J. Kjeldahl in 1883. In this procedure, soil samples are boiled (digested) in sulfuric acid (along with potassium sulfate to raise the boiling point, and selenium or mercury to act as catalysts) until all organic matter has been oxidized and all organic N is converted to free ammonium. Most nitrate present in the digest is not recovered as ammonium unless a catalyst (DeVarda's alloy, a mixture of copper, aluminum and zinc) is included to reduce it to ammonium. Nitrate generally accounts for much less than 1% of the total N content. Classically, the ammonium was distilled by adding a strong base to convert ammonium to gaseous ammonia, which was then trapped in an acidic solution. Ammonia captured could be determined by titrating the acid back to the original pH. Most laboratories now use semiautomated systems for determining ammonium content in the digest by colorimetric reactions. Variations on the Kjeldahl method are described in detail in Hesse (1971), Allen *et al.* (1974), and Bremner and Mulvaney (1982). Total N can be a good index for ranking sites according to N availability across large gradients, but is often a poor index over smaller gradients and where site disturbance is important.

(b) Inorganic N

Although ammonium and nitrate represent the available inorganic forms of N in soils, the sizes of these pools are generally small relative to annual nutrient supply rates and hence turnover rapidly. A small pool of nitrate may indicate either a low nitrification rate, a high rate of nitrate uptake by plants, or rapid denitrification. In general, high levels of ammonium and nitrate (>25–50 mg N kg^{-1} soil) probably indicate high N availability, but lower levels may not indicate infertile soils.

Ammonium and nitrate are usually determined by extraction with strong salt solutions (usually 5–20 g of soil in 50–200 ml of 2 M KCl), followed by filtration or centrifugation. Ammonium and nitrate in the filtrate are usually determined colorimetrically, although selective ion electrodes are also used (see Hesse, 1971; Keeney and Nelson, 1982; Smith, 1983). Some soil ammonium cannot be extracted by salt solutions because it is fixed in the interlattice of certain clays. The dynamics and turnover of this ammonium are uncertain (Mengel and Scherer, 1981; Keeney and Nelson, 1982).

(c) Mineralizable N

Methods for assessing mineralizable N provide an index of N availability by determining the release of inorganic N from organic matter over a given time period. Many techniques are used; their purposes range from providing a rapid comparable index of the N-supplying power of soil, through determining the effects of field temperature and moisture on the process, to estimating the capacity for N release in the soil. The proportion of the total N in soils that is released by the different methods varies widely (Table 5.1).

Table 5.1 Proportion of total N recovered by several N-availability methods (n = 9 soils from a Douglas-fir/Sitka alder stand; Binkley, unpublished)

Method	*Mean (mg kg^{-1})*	*Proportion of total N recovered*
Total nitrogen	2400	1
Inorganic N	24	0.01
Anaerobic incubation	83	0.04
Aerobic incubation	74	0.03
Boiling water total N	32	0.01
Autoclaved total N	120	0.05
Bioassay seedling uptake	48	0.02

Anaerobic incubations. A simple method for estimating N availability involves incubating water-saturated soil for 7 or 14 days at 35°C or 40°C (called 'anaerobic' incubations, see Keeney, 1982). The soils are extracted with a strong salt solution (usually 2 or 4 M KCl) after incubation. Only ammonium need be determined, as any nitrate initially present is lost through denitrification, and nitrate production does not occur under anaerobic conditions. Although the incubation conditions do not mimic normal environments, the values often correlate well with plant uptake in agricultural fields (Keeney and Bremner, 1966), and with foliar nutrient levels and fertilization response in some forests (Shumway and Atkinson, 1978; Powers, 1980).

Aerobic incubations. Aerobic incubations are more time consuming, but may also be more realistic. The simplest approach to aerobic incubation involves 10–30-day incubations of 5–20 g samples at 20°C or 25°C with moisture held constant near field capacity (moist but still aerobic). Both ammonium and nitrate are measured in the salt solution extracts (100 ml of 2 M KCl), and both net mineralization (ammonium+nitrate after incubation minus ammonium+nitrate before incubation) and net nitrification (nitrate after incubation minus nitrate before incubation) can be calculated

(Robertson and Vitousek, 1981). Aerobic incubations in low-fertility soils often yield negative rates due to microbial immobilization of N during the course of incubation.

Nitrogen mineralization potential. The most intensive laboratory approach involves determining the time course of N mineralization by sequential extractions over several months (Stanford and Smith, 1972; Keeney, 1982). This procedure allows the calculation of potentially mineralizable nitrogen (N_0) in a soil. Samples (5 g) are leached weekly with 50 ml of very dilute (0.02 – 0.10 M $CaCl_2$) salt solution, sometimes amended with an N-free nutrient solution. Leaching removes the products of mineralization (ammonium and nitrate), and the leachate is measured for inorganic N content. The pattern of nitrogen mineralization over time often resembles first-order kinetics, although different processes are involved because of the complicated interactions among various N pools and microbes. With appropriate curve fitting, the size of a potentially mineralizable pool can be calculated.

Once N_0 is determined, it can be used to calculate N mineralization under varying field conditions. First-order kinetics (rate proportional to the amount of mineralizable N remaining) commonly are assumed, and measured temperature and soil moisture levels are used together with information on the influence of temperature and moisture on mineralization. For example, Burger and Pritchett (1984) measured N_0 for soils from plots in loblolly pine (*Pinus taeda* L.) stands that were unharvested, harvested, or harvested and plowed. The mineralization potential decreased with increasing level of disturbance: control 5.5 mg kg^{-1} 21 day^{-1}, harvested 4.5, and harvested plus plowed 3.3. However, both temperature and moisture increased with increasing disturbance, so the net effect was higher calculated rates in disturbed plots: control 2.2 mg kg^{-1} 21 day^{-1}, harvested 2.3, harvested and plowed 3.3.

In field incubations. Some of the effects of varying environments can be included by incubating soil cores on-site in plastic bags (Eno, 1960) or other containers. Polyethylene bags allow gas transfer, but are impermeable to water and prevent any leaching of nutrients. Field temperature regimes are reflected in the bags, but moisture content is constant throughout the incubation. The presence of freshly dead fine roots may alter the balance between immobilization and net mineralization, but sieving to remove roots also disrupts soil structure and may alter mineralization values. Generally, enough cores are collected so that initial ammonium and nitrate concentrations can be determined on a subset, with the remainder incubated intact for about a month. Samples are extracted with KCl and analyzed as described in Section 5.3.1.b.

Seasonal and treatment-related differences in mineralization can be determined with sequential incubations (Ellenberg, 1977; Nadelhoffer *et al.*, 1983; Pastor *et al.*, 1984). For example, Vitousek and Matson (1985) examined the effects of clearcutting followed by various site-preparation treatments on nitrogen mineralization from 0–15 cm depth soils under both laboratory and field conditions. The reference (uncut control) stand mineralized about 24 kg N ha^{-1} in the growing season of 1983, versus an estimate of 56 kg N ha^{-1} under laboratory conditions. The site with harvesting plus minimal site preparation (crushing residual vegetation with a roller behind a tractor) mineralized about 69 kg N ha^{-1} in field incubations and 100 kg N ha^{-1} in the laboratory. The most intense site preparation (piling of slash into windrows, disking the soil and applying herbicide to minimize nonpine vegetation) yielded 87 kg N ha^{-1} in the field, and 123 kg N ha^{-1} in the laboratory.

A recent variation of the in-field incubation approach uses open cores of soil in tubes with a layer of ion-exchange resin in a mesh bag at the bottom of the tube. Water can enter from the top and flow through the tube, and dissolved nutrients are retained on the resin (DiStefano and Gholz, 1986). After incubation, both the soil in the core and the resin are extracted and analyzed for ammonium and nitrate. This method is an improvement, but it still kills roots and prevents water uptake by roots (allowing the water content of the cores to remain somewhat higher than in the surrounding soil). Sasser and Binkley (1988) used this method to determine changes in mineralization across a mortality wave in a Fraser fir [*Abies fraseri* (Pursh) Poir.] ecosystem. During 2.5 summer months, mineralization in the disturbed zone was about 65 kg N ha^{-1}, while the young regeneration and sapling zones averaged 30 kg N ha^{-1}. Mineralization in the mature stand (70 kg N ha^{-1}) was similar to the disturbed zone.

(d) Chemical extraction of labile N

Another approach to estimating N availability involves empirical indexes of the pool of N that is most susceptible to mineralization. Mild extraction by chemicals such as sodium hydroxide or simply boiling in water liberates only the N contained in fairly simple compounds. These methods provide reasonable indexes of N availability in a variety of situations (see Keeney, 1982). The boiling water method is especially simple and has proved as useful as most other methods: a 5 g sample is boiled in water (or 2 M KCl) for 10–60 min, followed by determination of ammonium (or sometimes organic N) in the filtered extract (Keeney and Bremner, 1966; Hart and Binkley, 1985; Gianello and Bremner, 1986).

(e) Chloroform fumigation/incubation

The chloroform fumigation/incubation (CFI) technique is becoming widely used as a method for estimating microbial biomass and its nitrogen and phosphorus content (Jenkinson and Powlson, 1976). Microbial biomass is the major pool of labile (readily mineralizable) N in many soils, so the CFI method may also be used as an index of N availability. Soil samples are fumigated with chloroform in a vacuum dessicator, then repeatedly flushed with air and evacuated, and incubated for 10 days. Samples are then extracted with 2 M KCl and analyzed for ammonium and nitrate. This chloroform-labile fraction can be used alone as an index of available N, or converted to microbial-biomass nitrogen if the carbon released by fumigation is known (Voroney and Paul, 1984).

(f) Ion-exchange resin bags

A simple approach to estimating nutrient availability in the field uses ion-exchange resin beads in nylon mesh (Binkley and Matson, 1983). This approach has many advantages, but careful quality assurance is needed to develop confidence in resin bag values obtained under specific field conditions. Resin bags are prepared by adding a known volume of resins to fine-mesh nylon bags. The quantity of resin per bag needs to have sufficient exchange capacity to retain all ions reaching the bag (i.e. the resin must not become saturated with ions from the soil). Several kinds of resins have been used, and no standard guidelines are available. The resins can be obtained with various ions saturating the exchange sites (such as OH^-, Cl^-, H^+, Na^+). Anion and cation resins can be mixed or kept in separate bags. Mixed resin bags are simpler, but may present complications as described below.

As a general guideline, we use mixed resins saturated with OH^- and H^+, and add about 14–28 ml of resin with a measuring spoon to each stocking. A knot is tied in the stocking, and the excess cut off. The exchange sites are then saturated with Na^+ and Cl^- by putting each bag in 100 ml of 1 M NaCl for 1 h; this minimizes problems later in chemical analysis. The bags are buried in the soil and ions are removed from water flowing through the bag (Binkley, 1984). After a given period, the bags are collected and the resins extracted (either in or out of the bag) with 100 ml of 2 M KCl. This extraction recovers about 0.80–0.90%

of the ammonium and nitrate on the resins. Some problems have arisen with the pH of the extract (especially if H^+ and OH^- loaded resins are used), and other interferences; we have found preloading with NaCl coupled with dilution of extracts (1:30 or more with distilled water) removes any interference.

Resin-bag estimates of nutrient availability may be difficult to interpret in terms of processes, because ion capture by the bags is affected by mineralization rate, uptake rate and ion mobilities in the soil (Binkley, 1984). However, resin bags provide a simple way to develop information on relative in-field patterns of nutrient availability that appears to relate well to other measures. For example, Binkley *et al.* (1986) compared resin-bag estimates in 17 forests in Wisconsin with values from buried-bag incubations in earlier years. Nitrate accumulated on the resin correlated highly with net nitrate production in buried bags (r = 0.87–0.92), as did resin ammonium+nitrate with the net production of ammonium + nitrate (r = 0.73–0.82).

(g) ^{15}N to trace N flux

Nitrogen has two stable (nonradioactive) isotopes: ^{14}N comprises about 99.6338% of atmospheric N_2, and ^{15}N accounts for the remaining 0.3662%. Nitrogen dynamics can be traced through soils and ecosystems by adding various compounds enriched or depleted in ^{15}N.

The use of N tracers in research on soil N dynamics often involves the addition of small amounts of very highly enriched ^{15}N materials. The recovery of the labeled N in various pools can then be used to calculate transfer rates among pools. For example, the net mineralization of N from organic pools is the balance between microbial release of ammonium and microbial immobilization. If small amounts of highly enriched $^{15}NH_4^+$ are added to a soil prior to incubation, the proportion remaining as ammonium and nitrate can be followed as a measure of microbial immobilization. The use of ^{15}N can also be valuable in the field; it is the only method that allows direct determination of the proportion of fertilizer N actually taken up by plants. Methods of analyses for N isotopes are covered in Chapter 13; recent references include Vose (1980) and Hauck (1982).

5.3.2 Limitations of assessing N availability

No one method provides a precise estimate of N availability. As mentioned earlier, laboratory indexes cannot account for variations in field conditions among sites. In-field incubations are also subject to unavoidable artifacts, such as constant moisture contents and exclusion of plant uptake. Experiments with ^{15}N may be most powerful, but are more expensive and still limited in resolution by differential mixing of label with soil N pools and other possible artifacts. All methods should be considered as indexes of availability, rather than as measures of actual supply rates.

5.4 PHOSPHORUS AVAILABILITY

Phosphorus inputs from the atmosphere are usually small, commonly ranging from 0.1 to 0.5 kg ha^{-1} annually. Inputs from weathering vary greatly among ecosystems, but are often of the same order of magnitude. Annual plant uptake is commonly 2–15 kg ha^{-1} annually. Phosphorus is always present as the phosphate anion; it does not undergo oxidation or reduction as does nitrogen. Between pH 4 and 6, most phosphate in the soil solution occurs as $H_2PO_4^-$. The concentration of phosphate in soil solutions is usually very low (often less than 5 μeq l^{-1}).

The availability of P to plants is tightly constrained by inorganic chemical reactions. Phosphate forms barely soluble precipitates in the presence of aluminum, iron or calcium, and solubility depends strongly on pH. A shift of one pH unit may change P solubility by 10-fold (Fig. 5.2). Many soils (especially Spodosols, Ultisols and Oxisols) have high contents of iron and aluminum oxides that adsorb phosphate tightly, keeping the concentration in soil solution very low (Uehara and Gillman, 1981). This explains in part why many tropical forests appear limited more by phosphorus than nitrogen (Vitousek and Sanford, 1986).

The amount of organic phosphate exceeds inorganic phosphate (excluding P contained in rocks) in most soils, and turnover of organic P pools provides a large portion of the P taken up by plants. Whereas ammonium is released as a byproduct of microbial oxidation of carbon compounds, the release of phosphate is often the direct result of enzymes produced by microbes and higher plants (McGill and Cole, 1981). The importance of these phosphatase enzymes is currently debated. For example, Emsley (1984) claimed that in the absence of phosphatases, the half-life of organic P molecules in soils would be on the order of centuries. However, Barber (1984) cautions that very little evidence supports the idea that P supply rate increases with increasing enzyme activity.

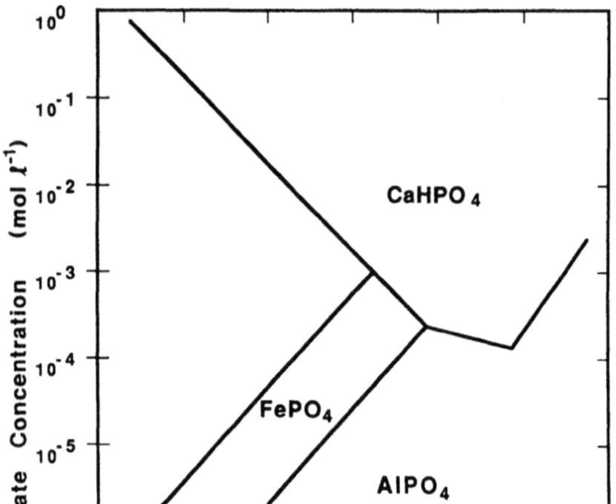

Fig. 5.2 Phosphorus availability depends in part on the solubility of P salts with calcium, iron and aluminum. At low pH, aluminum salts are the least soluble and regulate P concentrations in soil solution (after Lindsay and Vlek, 1977).

5.4.1 Extraction indexes of P availability

The total P content of soils can be determined following carbonate fusion or digestion with strong perchloric or hydrofluoric acid, but total P content is not well related to P availability. Various chemical extractants have been developed to measure the quantity of potentially available P; each method has proven useful in some areas, but none works well in all soil types. Most methods have been developed and tested for agricultural use, but some have been applied to forest and grassland soils.

(a) Acid ammonium fluoride extraction

With this extractant, the acid dissolves phosphate associated primarily with calcium, and the fluoride replaces some of the P adsorbed by iron and aluminum oxides (Olsen and Sommers, 1982). The method appears suited for many types of acid soils, but provides low estimates of P availability on some alkaline soils, and in some cases high estimates where the pool of P precipitated with calcium carbonate is large (W. Schlesinger, personal communication). In this procedure, 1 g of soil is shaken with 7 ml of extractant (0.03 M NH_4F and 0.5 M HCl) for 1 min and then filtered. The P concentration of the filtrate may be determined colorimetrically in a spectrophotometer or by automated techniques. For agricultural soils, extractable P levels (by this method) of less than 3 mg of P kg^{-1} of soil indicate low P availability, and values above 20 mg of P kg^{-1} are high. Little calibration work has been done for nonagricultural soils, but soils with less than 3 mg kg^{-1} extractable P may well be P limited, and soils far above this level probably have ample P for plant growth.

(b) Double acid extraction

The combination of hydrochloric and sulfuric acids appears well suited for estimating P availability in old highly weathered soils where P availability is regulated by sorption with iron and aluminum. This method is commonly used in the Southeastern US and in tropical regions. The basic procedure calls for shaking a 5 g sample in 20 ml of extractant (0.05 M HCl and 0.0125 M H_2SO_4) for 5 min, followed by filtration and analysis of the filtrate for P (Olsen and Sommers, 1982). Extractable P levels below 10 mg of P kg^{-1} of soil may indicate a low P supply; many P-deficient forest soils have less than 3 mg of extractable P kg^{-1} of soil.

(c) Sodium bicarbonate extraction

This Olsen method uses alkaline sodium bicarbonate (pH 8.5) to extract labile phosphorus. The solubility of P associated with aluminum and iron tends to increase with increasing pH, so extraction at pH 8.5 tends to recover P precipitated with these metals. The bicarbonate also precipitates calcium, increasing the phosphate concentration of the solution (Olsen and Sommers, 1982). This method has had some success in evaluating P availability for acid soils (e.g. Wells *et al.*, 1973, but see Kadeba and Boyle, 1978), but is generally more useful for evaluating soils with pH near 7 (see Jones, 1974). The procedure involves shaking 5 g of soil in 100 ml of extractant (0.5 M $NaHCO_3$ at pH 8.5) for 30 min, followed by filtration and P determination. Levels below 10 mg of P kg^{-1} of agricultural soil may indicate low P availability, but critical levels for nonagricultural soils have not been established.

(d) Anion-exchange resin extraction

The anion-exchange resin method has proven more useful across a broad range of soil types than any method not employing radioactive ^{32}P (Amer *et al.*, 1955; Kadeba and Boyle, 1978; Smith, 1979; van Raij *et al.*, 1986). The basic method (Sibbeson, 1977, 1978, 1983) involves shaking 5 g of soil in 100 ml of deionized water with 2 g of anion-exchange resin (the amount of resin varies with resin types). The resin may be free in the solution (and later separated by sieving), or encased in a mesh bag. The resin adsorbs P from solution, creating a diffusion gradient from adsorbed pools into the solution. After a set time, the resins are removed from the sample and the adsorbed P is removed by equilibration with a replacing salt solution (such as 1 or 2 M NaCl).

The anion-exchange resin method appears to be a good index of plant-available P, and it correlates well with results from ^{32}P isotope experiments (see below). However, resins provide an index of the pool of readily available P, and the long-term supply of P in nonagricultural soils may derive from less-readily available pools. A unique long-term experiment by Barber (1979) illustrates the importance of the less labile pool in agricultural soils. Plots were fertilized annually with 22 or 44 kg of P ha^{-1} and cropped for 17–21 years, and then cropped without fertilization for 8 more years. The P removed in crops was measured for each of the 8 years, as was the resin-extractable P. Expressing both P measures on a kg ha^{-1} basis, the resin-extractable pool of the less-fertilized plot showed a decline of about 1 kg of P ha^{-1} for every 20 kg ha^{-1} removed by crops (Fig. 5.3). In the heavily fertilized plots, a reduction of 1 kg ha^{-1} in resin-extractable P occurred for each 6 kg of P ha^{-1} removed in crops. Apparently a large pool of nonlabile P gradually resupplied the pool of labile P that was extractable by the anion-exchange resins.

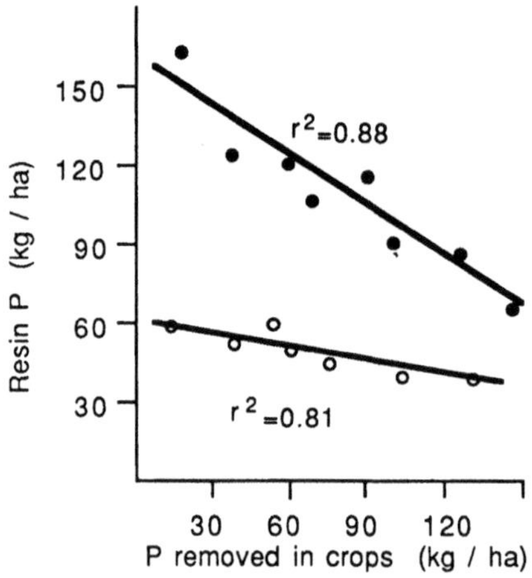

Fig. 5.3 Resin-extractable P correlated well with plant-available P over 8 years, but the different slopes between two soil types indicates a different rate of resupply from non-extractable pools (after Barber, 1984).

5.4.2 Phosphate sorption isotherms

When phosphate is added to a soil sample, a small amount will remain in soil solution, but most is absorbed by microbes, adsorbed by iron and aluminum compounds, or precipitated as barely soluble salts. Soils that strongly 'fix' P (through adsorption or other processes) generally have lower P availability than soils that can maintain higher levels of P in solution. This feature allows P availability to be assessed through development of phosphate sorption isotherms. The term 'isotherm' is used because the concentrations of ions vary but the total energy of the system remains constant.

These isotherms can be derived by adding 25 ml of solution of varying P concentration (often ranging from 10 to 300 mg of P l^{-1}) to a series of suspensions containing 2 g of soil (Sommers *et al.*, 1977). Samples are sometimes sterilized before the experiment to prevent microbial immobilization of P. With increasing P concentration, the soil's capacity for further adsorption decreases and a larger portion remains in solution. A soil that maintains a greater portion of added P in solution should have higher P availability than a soil that adsorbs most of the added P.

5.4.3 Isotope equilibration

The most accurate approach to estimating P availability involves adding small quantities of P highly labeled with ^{32}P (a strong beta emitter) to a soil suspension and allowing a set time for equilibration between P in solution and P adsorbed (or precipitated) to the soil. The isotopically exchangeable (or labile) pool of non-radioactive ^{31}P is calculated:

$$\text{Labile } ^{31}\text{P} = \frac{(^{31}\text{P} + ^{32}\text{P})}{F_{\text{sol}}}$$

where F_{sol} is the fraction of the ^{32}P that remains in solution after the equilibration period. For example, if the solution contained 10 μg of P (both isotopes), and 10% of the radioactive P remained in solution, then the labile pool would equal 100 μg per weight of soil in the suspension. Similar approaches can be used to examine P availability patterns under field conditions, improving upon the information obtained with the anion-exchange resin methods.

A variety of methods has been used to calculate plant-available P from the results of ^{32}P equilibration experiments. The most widely used approach calculates an '*A*' value that represents the quantity of P available in a field soil:

$$A = \frac{B\,(1-y)}{y}$$

where *B* is the amount of labeled fertilizer added and *y* is the proportion of P in plants obtained from the fertilizer. Other calculations, such as '*L*' and '*E*' values, evaluate the pool of P that directly interchanges with added ^{32}P rather than with a 'plant available' pool (see Vose, 1980).

^{32}P is not the only isotope useful in soil–plant studies; the less-energetic ^{33}P has been widely used to measure P depletion zones around roots (see Nye and Tinker, 1977). Caldwell *et al.* (1985) used double-labeling with ^{32}P and ^{33}P to evaluate the below-ground competitive ability of two species of *Agropyron* with a shrub (see Chapter 10).

5.5 SULFUR AVAILABILITY

The biogeochemical cycle of sulfur shares features of the cycles of both N and P. The S cycle involves important oxidation and reduction transformations (as does the N cycle), and also has a substantial pool that remains in the oxidized (sulfate) form (similar to P). Inputs of S from the atmosphere in the form of sulfate (SO_4^{2-}) range from 1 to 10 kg of S ha^{-1} annually in unpolluted regions to over 50 kg ha^{-1} in regions with very acidic precipitation. Sulfur is also released into soils through the weathering of various inorganic minerals, but estimates of weathering rates are rare (Mitchell *et al.*, 1986). Reduced forms of S are found in unweathered minerals (such as pyrites), in anaerobic soils and in organic compounds. Oxidized forms dominate in the atmosphere, in some sedimentary rocks, in the solution of aerobic soils and on adsorption surfaces in soils.

Sulfur availability does not limit ecosystem production in the vast majority of nonagricultural ecosystems. Although some tree species have low levels of S in foliage on some sites, clear increases in productivity with S fertilization alone are very rare (Turner, 1979; Turner and Lambert, 1986; Blake *et al.*, 1988). Sulfate is the dominant form used by plants, although in some cases direct uptake of SO_2 from the atmosphere appears important (e.g. Hoeft *et al.*, 1973; cited in Tabatabai, 1982; Mengel and Kirkby, 1982).

Available sulfur comes from readily soluble inorganic pools, from pools adsorbed on iron and aluminum oxides, and from the mineralization of organic sulfur. Sulfur may be mineralized as a byproduct of decomposition, or more directly through the action of sulfatases. As with P, a range of methods have been used to assess S availability, and none has been consistently superior in predicting S response to fertilization (see Ensminger and Freney, 1966; Tabatabai, 1982). In agricultural soils, S availability has been assayed by extraction with water, acids, acetate and bicarbonate. Biological methods have used aerobic incubations under controlled conditions, and assays of sulfatase activity. Tracer studies have used ^{35}S (a weak beta emitter). Sorption isotherms have also been developed for a variety of soils from the perspective of sulfate retention and soil acidification by acid deposition. No method can be recommended highly for nonagricultural soils, as little calibration work has been done (but see Blake *et al.*, 1988).

5.6 AVAILABILITY OF ESSENTIAL CATIONS

Input rates of essential cations (Ca, Mg, K and others) in precipitation and mineral weathering are sufficient to support plant productivity in most nonagricultural ecosystems. However, deficiencies of cation nutrients are observed on organic soils, very sandy soils and very old, highly weathered soils. The major pools of cations in soils are: unweathered minerals, cation-exchange sites and organic matter. In alkaline soils, precipitated salts (such as calcium carbonate) can form major pools (Schlesinger, 1985). The total content of minerals can be determined by fusion in furnaces at 600–1000°C, or by treatment with perchloric and hydrofluoric acids (see Allen *et al.*, 1974; Lim and Jackson, 1982). Total mineral content has little relation to the availability of cation nutrients, and is rarely measured by ecologists.

Extractable cations are adsorbed on negatively charged sites derived from charge imbalances in clay minerals and from dissociated organic acids. The potential cation storage of a soil is termed the cation-exchange capacity, and adsorbed ions are called extractable or exchangeable cations. In acid soils, a substantial portion of the exchange sites are occupied by nonnutrient cations such as H^+ and Al^{3+}. The availability of a nutrient cation such as calcium held on exchange sites can be low if cation exchange capacity is small, or if calcium occupies a small portion of the exchange sites (see Reuss and Johnson, 1986). The proportion of nutrient cations on the exchange complex is termed 'percent base saturation' by soil scientists, although these ions are not bases in a chemical sense.

The extractable pool of cations can be a poor indicator of long-term availability, because the pool is at least partially resupplied from atmospheric inputs, mineral weathering, sparingly exchangeable pools and mineralization of organic matter. For example, a Hawaiian soil with about 500 kg ha^{-1} of exchangeable K supplied 15 successive crops of Napier grass (*Pennisetum purpureum* Schum.) with a total of 4000 kg of K ha^{-1}, without a substantial decline in the exchangeable K^+ pool (Fig. 5.4).

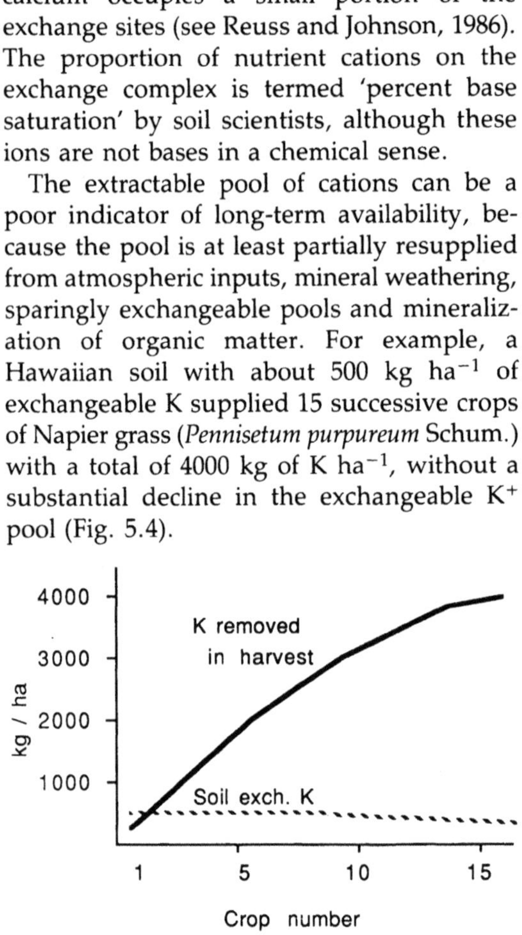

Fig. 5.4 Fifteen harvests of Napier grass removed about eight times the quantity of potassium present on exchange sites, with no significant decrease in the exchangeable pool (after Ayers *et al.*, 1947).

The supply of essential cations limits ecosystem production in some regions, but little work has focused on calibrating analyses of nonagricultural soils with plant responses to fertilization. In one example, Adams (1973) examined indexes of Mg^{2+} availability and the concentration of Mg^{2+} in foliage of radiata pine (*Pinus radiata* D. Don) at eight locations in New Zealand. Three sites were thought to be severely deficient in Mg^{2+} availability, two were moderately deficient, and three were considered nondeficient. Exchangeable Mg^{2+} did not correlate significantly with foliar concentration, and Adams speculated that the exchangeable pool related more to the organic matter content (and cation-exchange capacity) than to the Mg^{2+}-supplying ability of the soil. Extractions with nitric and hydrochloric acids dissolved some of the Mg^{2+} contained in minerals, and correlated fairly well with foliar concentrations (correlation coefficients of 0.7–0.8, $P<0.05$). Surprisingly, the total Mg content (determined by digestion in hydrofluoric acid) gave the best correlation with foliar Mg^{2+} ($r = 0.95$, $P<0.001$).

5.6.1 Extractable cations and cation-exchange capacity

As noted above, the pool of extractable cations is usually determined following extraction with a strong salt solution. Various salts have been used, such as 2 M KCl, NaCl, NH_4NO_3 and ammonium acetate. The choice of extracting salt has little effect on the quantity of ions extracted, because the very high concentration of the added cation forces almost all exchangeable cations into solution. Ammonium nitrate and acetate interfere less with the subsequent measurement of the cations by atomic absorption.

The general sequence for determining extractable cations involves replacing all extractable cations in a 5 or 10 g soil sample by shaking with 25 ml of a strong extractant, such as 1 M NaCl, for 30 min. The sample (plus extractant) is transferred to a Buchner funnel and filtered. The concentrations of cations in the filtered extract are then determined by atomic absorption, flame emission spectrophotometry, or inductively coupled plasm (ICP) emission (see Allen *et al.*, 1974; Page *et al.*, 1982). An estimate of the cation-exchange capacity (CEC) can be obtained by rinsing the soil sample in the Buchner funnel with a series of five additions of 20 ml of alcohol to remove all of the nonadsorbed Na^+. In the next step, 100 ml of another salt solution (typically 1 M KCl) is drawn through the sample to replace all the adsorbed Na^+. The quantity of Na^+ recovered from the second extraction gives the CEC of the sample.

A large portion of the CEC of many soils derives from dissociated organic acids, and the degree of dissociation of these acids is affected by soil pH. This pH-dependent CEC decreases as soil pH drops (as soils become more acidic), and makes the choice of extractant important in determining CEC. Extractants such as NaCl and KCl have only small effects on the pH of the soil suspension, whereas ammonium acetate buffered at pH 7 will change the pH of the soil suspension to 7. Because most agricultural soils fall in the range of pH 5.5 to 7.0, the ammonium acetate method provides a reasonable standard for comparing CEC across soil types. Many grassland soils fall in the same pH range, so again the ammonium acetate method may be suitable. However, most forest soils fall between pH 3.5 and 5.5, and in this region the pH-dependent CEC is large and the difference in CEC measured at pH 7 and at the normal pH of forest soils may differ by more than twofold.

Overestimates of CEC can result from the use of highly concentrated salt solutions, because the measured CEC tends to increase with the square root of the salt concentration (Gillman, 1981; Kalisz, 1986). In arid and semiarid soils, concentrated extractants also may dissolve precipitated salts (see Gupta *et al.*, 1985). Methods using solutions of low ionic strength are discussed by Rhoades (1982a) and Hendershot and Duquette (1986). Alternatively, the effective CEC of a sample can be estimated by summing all exchangeable ions, including aluminum and H^+.

5.6.2 Cation-exchange resins

Cation-exchange resins can also be used to estimate cation availability in the same manner that anion resins have been used for N and P availability (van Raij *et al.*, 1986). For example, Gibson (1986) examined spatial and temporal patterns in cation availability in grasslands that had been grazed or left ungrazed. He found that availability of all cations differed between summer and autumn, and that calcium and potassium were significantly higher in the ungrazed site in both seasons.

5.7 MICRONUTRIENT AVAILABILITY

The availability of micronutrients is plentiful in most nonagricultural ecosystems, but deficiencies have striking effects where they occur. Among the micronutrients, boron probably limits forest production most commonly. The availability of B is often measured simply by boiling a 5 g sample in water and determining the B concentration in the filtered extract by colorimetry (Bingham, 1982). Soils with less than 0.5 mg of B kg^{-1} may be B-limited (Stone, 1968). The availability of some other micronutrients is often assayed with organic chelators. For example, the availability of copper, zinc and iron may be estimated by extracting a 10 g soil sample with 20 ml of chelating extractant [0.005 M diethylenetriaminepentaacidic acid (DTPA), 0.1 M triethanolamine (TEA), and 0.01 M $CaCl_2$], followed by determination of the metals in the extract by atomic absorption spectrophotometry (Baker and Amacher, 1982). The critical levels for these micronutrient assays generally have not been calibrated for nonagricultural soils. See Cox and Kamprath (1972) for a review of methods used in agricultural soils.

5.8 SOIL CLASSIFICATION

A great deal of research has focused on identifying nutrient limitation in commercial forests, and in some regions soil classification provides an index of nutrient limitation. For example, the Cooperative Research in Forest Fertilization program at the University of Florida has developed a classification scheme based on drainage and characteristics of the B horizon (Kushla and Fisher, 1980) that allows prediction of response to fertilization with N or P.

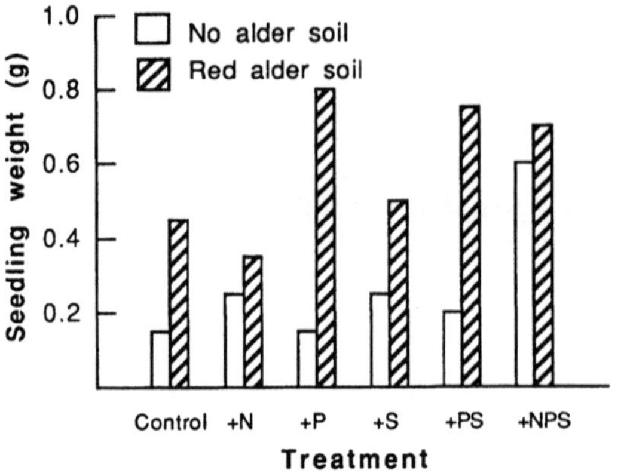

Fig. 5.5 A bioassay with Douglas-fir seedlings indicated that soils from a plot without alder were low in N, P and S, whereas soils from an alder plot appeared limited only by P.

5.9 BIOASSAY OF NUTRIENT AVAILABILITY

Field fertilization trials provide the most powerful information on nutrient limitations (for protocols, see Binkley, 1986), but are expensive and usually require several years to complete. A simpler approach uses bioassays of soil fertility under controlled conditions. The nutrient content of the plant biomass provides an index of a soil's ability to supply nutrients, and growth response to fertilization identifies limiting nutrients and gives some idea of the potential response to fertilization. For example, analysis of Douglas-fir [*Pseudotsuga menziesii* (Mirb.) Franco] foliage from a 23-year-old plantation suggested severe nitrogen limitation (Binkley, 1983). Foliage from Douglas fir mixed with N-fixing red alder appeared to have adequate N, but to be deficient in P and S. A greenhouse bioassay with Douglas-fir seedlings also indicated an N limitation in soils from the no-alder plantation; P and S were also limiting when N was added (Fig. 5.5). The seedlings in soils from the mixed Douglas-fir/alder plantation appeared limited only by P availability.

Physiological characteristics of bioassay-grown plants can also be utilized. Harrison and Helliwell (1979) demonstrated that $^{32}PO_4$-uptake capacity per unit of root was greatest in P-deficient plants and declined linearly with P supply. Although this measurement provides an estimate of nutrient limitation rather than nutrient supply, it can be used in the interpretation of bioassays (see also Chapter 10).

Bioassays are subject to several limitations. Disturbance of the soil prior to the assay (such as sieving and removing roots) may alter nutrient availability. The relative size of bioassay containers and plants is important. Small plants may take a long time to exploit the soil fully, and, where nitrate is present, young plants utilize a much larger fraction of the soil's N than P (Peace and Grubb, 1982). Greenhouse conditions are usually more conducive to nutrient mineralization and plant growth than are field environments, and bioassays cannot address differences in nutrient availability due to differences in field microclimates. Finally differences between the nutrition and physiology of juvenile and adult plants may introduce artifacts.

5.10 SOIL ACIDITY

The availability of most nutrients is affected directly or indirectly by soil acidity. In the short term, the concentration of H^+ in soil solution is probably the most useful index of soil acidity. Because the H^+ concentration commonly varies across sites by orders of magnitudes, a logarithmic scale is used. The pH of a soil solution is the negative of the logarithm (base 10) of the H^+ concentration:

$$pH = -\log_{10} [H^+]$$

Therefore, a pH of 6 indicates a H^+ concentration of 10^{-6} mol l^{-1}, and a pH of 4 means that H^+ is 100 times more concentrated (10^{-4} mol l^{-1}).

The H^+ concentration of the soil solution is buffered by very large pools of H^+ on cation-exchange sites, on undissociated organic acids, and in the waters of hydration on aluminum ions. The H^+ that can be displaced from exchange sites by salt cations is termed exchangeable acidity, and the H^+ that can be dislodged from the soil only by adding a base (such as OH^-) is called titratable acidity. The dynamics of H^+ in ecosystems are reviewed by Binkley and Richter (1986) and Reuss and Johnson (1986).

5.10.1 Measurement of soil pH

A variety of methods are common for the measurement of soil pH. Most involve mixing a soil sample with a solution of water or dilute salt (such as 0.01 M $CaCl_2$) in a ratio of one part soil weight to one part solution. The sample is stirred and allowed to equilibrate for 10 min, and the pH is measured by an electrode inserted into the suspension (McLean, 1982). The choice of soil/solution ratios is somewhat subjective, and soils high in organic matter often need ratios of 1:5 or more to form a suspension.

5.10.2 Lime potential, exchangeable acidity and titratable acidity

In agricultural soils, the exchangeable and titratable pools of H^+ often are assessed by measuring a 'lime potential' which estimates the amount of lime needed to raise soil pH. The Shoemacher, McLean and Pratt (SMP) method is one of the most common (McLean, 1982), and involves equilibrating a 5 g soil sample with 10 ml of a buffer solution ($CaCl_2$, K_2CrO_4 and *p*-nitrophenol buffered at pH 7.5) for 30 min. A very acidic soil with a high lime potential will produce a large drop in the pH of the buffer solution, whereas a soil with a low lime potential will have less effect on the pH of the buffer. The amount of lime required to raise soil pH to a desired level is then read from a table (McLean, 1982).

Lime potential measures are empirical; they provide little insight into the components of soil acidity. The size of the exchangeable H^+ pool itself can be measured by titrating a salt extract (extraction as described in Section 5.6.) with a dilute base to an end point of pH 7 or 8.2 (Thomas, 1982). This pool is especially important in explaining and predicting changes in soil pH (see Reuss and Johnson, 1986).

The classic approach for measuring the titratable pool of H_+ (Thomas, 1982) is similar to lime potential measures: a 10 g sample is equilibrated with 25 ml of a buffer solution ($BaCl_2$ and triethanolamine buffered to pH 8.2), and then rinsed with three washes of 25 ml of 0.25 M $BaCl_2$. The extract/leachate is then titrated with a weak acid to a colorimetric end point. A blank sample of the buffer (plus $BaCl_2$ rinses) is also titrated, and the titratable acidity is the difference between the acid needed to titrate the blank versus the sample extract. An extract of a soil high in titratable acidity requires less acid to titrate than a soil low in titratable acidity.

Much of the acidity neutralized in a titration to pH 7 or 8.2 does not dissociate near the ambient pH of soils, and some measure of buffer capacity near ambient pH may be helpful. Titration curves allow comparisons across a pH range (Mehlich, 1941), but no standard methodology has been developed for preparing titration curves for soils. We currently weigh seven samples of 5 g each into cups, and add 25 ml of 0.5 M KCl to buffer the ionic strength of the suspensions. Then 2 ml of water is added to the control cup, three concentrations of acid to three of the cups, and three concentrations of base to the remaining cups. The range of concentration of the acids and bases needed to span the desired range of pH is determined in pilot measurements for each soil type; 0.001 M – 0.01 M is the common range. The cups equilibrate for 24 h, and then pH is measured and plotted as a function of added base or acid. The buffer capacity across any pH range can be calculated from the curves.

5.11 SOIL SALINITY

Soil salinity affects soil water potentials, pH and microbial activity. High concentrations of salts are expected in arid and semiarid ecosystems, but salinity is also important in some unexpected situations such as primary succession on river floodplains in Alaska (G. Marion and K. Van Cleve, personal communication). The simplest method involves determination of the salts that dissolve in distilled water [other methods are described in Rhoades (1982b) and Hesse (1971)]. An aqueous extract of 200 g of soil is prepared by saturation with distilled water to form a moist paste with no excess water. After 5 h, the sample is filtered and the extract is analyzed for individual cations by atomic absorption or other methods. This method estimates the soluble salts; the actual concentration of ions in soil solutions varies with the water content of the soil. Direct measurements of salinity can be obtained by sampling soil solution with lysimeters, or with potentiometric sensors buried in the soil (Rhoades, 1982b).

5.12 SOIL REDOX POTENTIAL

Soil redox potential relates to the activity of electrons as acidity relates to the activity of H^+. Redox and acidity largely determine the context of geochemical and biological activity in soils. Similar to soil acidity, the redox state of a soil can be described by an intensity factor (the potential, E_h, or the logarithmic form pe) and a buffer capacity (referred to as 'poising'). Oxygen serves as an electron acceptor and dominates the redox potential in well-aerated soils; they have a high E_h (above 400 mV at pH 5) and pe near 15 (at pH 5), and are well-poised to resist changes in potential. Changes in pH and pe are inverse linear functions of one another. Bartlett (1986) provides an excellent synthesis of the redox behavior of soils.

Soil redox potentials are commonly measured potentiometrically with platinum electrodes. Under aerobic conditions, the mV readings are highly variable and not very meaningful. Under reducing conditions, readings are more stable and useful, but many observations are needed to characterize a site (see McIntyre, 1970; Hess, 1971). Standard methods have not been developed for measuring the poise of a soil, but the common approach is to titrate reduced soils with oxidized titrants (such as $Fe(III)Cl_3$) and measure changes in potential (see Hesse, 1971).

5.13 COMMENTS ON SAMPLING

The value of all assay methods is limited by the handling of samples and by the sampling scheme used to characterize the field site. The design of soil sampling requires sound statistical planning based on a few important criteria.

5.13.1 Sample preparation

Many indexes of nutrient availability are affected by the treatment of samples, especially in relation to temperature and moisture content. Uniform guidelines are not appropriate because the best procedures vary with the analytical methods to be used. For example, total nutrient content of soils is usually determined on air-dried or oven-dried samples, and dried samples can be stored in air-tight containers indefinitely. At the other extreme, samples collected for analysis of extractable ammonium and nitrate usually are transported in a cooler and immediately extracted in the laboratory. Quality-assurance procedures should include comparisons of fresh and air-dried values to check for possible problems (see Bartlett and James, 1980; Gilliam and Richter, 1985; Phillips *et al.*, 1986). Results are often expressed on an oven-dried basis after determining moisture content on separate subsamples.

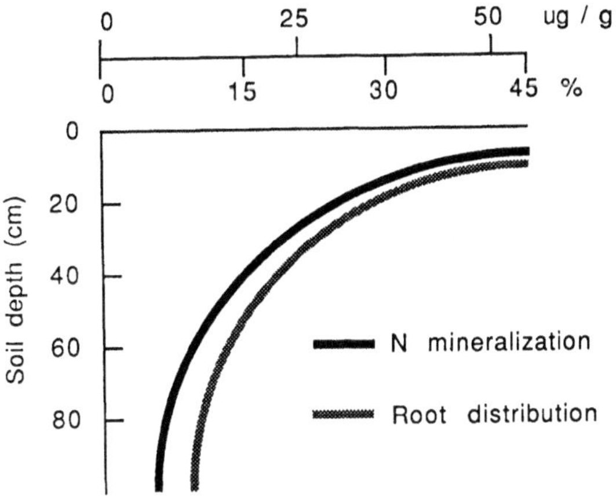

Fig. 5.6 The distribution of fine roots of ponderosa pine parallel the mineralizable N profile (after Powers, 1984).

5.13.2 Sampling depth

To what depth should soils be sampled? The fine roots responsible for absorbing most nutrients are usually concentrated in the upper soil, so most assays focus on nutrient availability in the top 10 or 15 cm. Although this may be the most active zone, deeper horizons may also supply substantial quantities of nutrients. For example, Powers (1984) found that about half of the fine roots and half of the mineralizable N in a mixed-conifer forest were located in the top 20 cm of soil (Fig. 5.6). Lower horizons appeared more important in a slash pine (*Pinus elliottii* Engelm.) ecosystem, where Van Rees and Comerford (1986) found one-third of the fine roots were deeper than 1 m, with taproots extending below 3 m. They applied strontium (a calcium analog) to the forest floor and to the soil at a depth of 1 m. No increase in the concentration of Sr was found in needles from the plot where the forest floor had been labeled, which the authors attributed to competition with the shallow roots of understory plants. Significant uptake of Sr occurred in the plot labeled at 1 m depth.

The upper soil may be the best location for diagnosing current nutrient limitations, but estimates of total nutrient supply and long-term nutrient availability require consideration of the entire rooting zone. Deeper roots could 'mine' soils of phosphorus or cations that weather at depth, or tap nutrients from ground water.

5.13.3 Variability, replication and quantitative spatial analysis

Table 5.2 gives ranges for coefficients of variation (standard deviation divided by mean) commonly found for forest soils. Some measures, such as bulk density (the weight of dry soil per unit volume) of sandy soils, have a low variability and a few samples may provide a precise estimate of the mean. The pool of extractable nitrate is much more variable, and precise estimates may require very large numbers of samples. The number of samples required in an experiment can be estimated based on the expected variability, the levels of confidence desired, and the desired sensitivity of the comparisons among sites (see Armson, 1977; Lloyd and McKee, 1983).

The variability within samples from replicates within one ecosystem can be reduced by compositing many samples before laboratory analysis. Soil sampling is usually less expensive (in labor and money) than laboratory work, and compositing reduces within-site variability and allows greater precision in the estimates of site means.

Careful attention to replication is needed to insure efficient allocation of effort and to avoid pseudoreplication. For example, if two 1 ha plots dominated by different species were to be tested for differences in nutrient availability, replicate samples within plots would provide an error estimate that would allow testing for differences between plot means. However, lack of replication of the plots prevents any statistical inference on the effect of species on nutrient availability. If four replicate (1 ha) plots for each species were available, what would be the optimal allocation of sampling effort? Multiple samples within each plot would be superior to taking just one sample per plot only to the degree that multiple samples gave a better estimate of the true mean of each plot. The number of subsamples provides no degrees of freedom in the test for the effect of species.

Table 5.2 Some common ranges for coefficients of variation (standard deviation divided by the mean) for various soil measures; from a variety of sources

Property	*Common coefficients of variation*
Bulk density	0.15 – 0.30
Forest floor biomass	0.35 – 0.80
Extractable calcium	0.40 – 0.75
Total nitrogen concentration	0.25 – 0.45
Extractable ammonium	0.50 – 0.75
Extractable nitrate	0.80 – 1.30
Mineralizable N	
Anaerobic incubations	0.40 – 0.70
Aerobic incubations	0.60 – 0.90
Boiling water extraction	0.25 – 0.35
Ion-exchange resin bags (greenhouse)	0.15 – 0.20
Ion-exchange resin bags (in field)	0.60 – 0.80
In-field buried-bag incubations	0.50 – 0.90

Therefore, multiple samples within plots should be bulked to provide a single composite sample for each plot, minimizing laboratory effort with no sacrifice of statistical power.

In some situations, the spatial variability of nutrient availability may be of interest. Statistical methods developed for mineral exploration have been used for examining variability in spatial patterns of soil properties (see Riha *et al.*, 1986; Robertson *et al.*, 1988). These techniques are summarized by Vieira *et al.* (1983) and Webster (1977, 1985).

5.14 INDEX UNITS

The results of soil assays can be expressed in a variety of forms. The simplest is the weight of the available or labile pool per g of dry soil (mg g^{-1} for large pools, $\mu g\ g^{-1}$ for small pools). However, roots exploit soil on a volume basis rather than a mass basis, and comparisons among soils of different bulk densities (mass/volume) may be more meaningful if nutrient availability is calculated per volume of soil. The choice of the most appropriate units needs to be made with consideration of the objectives of each study, keeping in mind that these methods are indexes of nutrient availability rather than measures of actual availability.

ACKNOWLEDGEMENTS

This chapter was substantially improved by comments from Bill Schlesinger, Bob Pearcy, Dave Valentine, an anonymous reviewer and the graduate students in Pearcy's 1986 class in ecophysiological methods.

REFERENCES

Adams, J.A. (1973) Critical soil magnesium levels for radiata pine nutrition. *N.Z. J. For. Sci.*, **3**, 390–4.

Allen, S.E., Grimshaw, H.M., Parkinson, J.A. and Quarmby, C. (1974) *Chemical Analysis of Ecological Material*, Wiley, New York.

Amer, F.D., Bouldin, D., Black, C. and Duke, F. (1955) Characterization of soil phosphorus by anion exchange resin adsorption and P-32 equilibration. *Plant and Soil*, **6**, 391–408.

Armson, K.A. (1977) *Forest Soils*, University of Toronto Press, Toronto.

Ayers, A.S., Takahashi, M. and Kanehiro, Y. (1947) Conversion of non-exchangeable potassium to exchangeable forms in a Hawaiian soil. *Soil Sci. Soc. Am. Proc.*, **38**, 985–61.

Baker, D.E. and Amacher, M.C. (1982) Nickel, copper, zinc, and cadmium. In *Methods of Soil Analysis. Part 2 Chemical and Microbiological Properties* (eds A.L. Page, R.H. Miller and D.R. Keeney) American Society of Agronomy, Madison, pp. 323–36.

Barber, S.A. (1974) Influence of the plant root on ion movement in soil. In *The Plant Root and Its Environment* (ed. E.W. Carson), University Press of Virginia, Charlotte, pp. 525–64.

Barber, S.A. (1979) Soil phosphorus after 25 years of cropping with five rates of phosphorus application. *Commun. Soil Sci. Plant Anal.*, **10**, 1459–68.

Barber, S.A. (1984) *Soil Nutrient Bioavailability*, Wiley, New York.

Bartlett, R. (1986) Soil redox behavior. In *Soil Physical Chemistry* (ed. D.L. Sparks), CRC Press, Boca Raton, pp. 179–207.

Bartlett, R. and James, B. (1980) Studying dried, stored soil samples – some pitfalls. *Soil Sci. Soc. Am. J.*, **44**, 721–4.

Bingham, F.T. (1982) Boron. In *Methods of Soil Analysis. Part 2 Chemical and Microbiological Properties* (eds A.L. Page, R.H. Miller and D.R. Keeney), American Society of Agronomy, Madison, pp. 431–48.

Binkley, D. (1983) Interaction of site fertility and red alder on ecosystem production in Douglas-fir plantations. *For. Ecol. Manag.*, **5**, 215–27.

Binkley, D. (1984) Ion exchange resin bags: factors affecting estimates of nitrogen availability. *Soil Sci. Soc. Am. J.*, **48**, 1181–4.

Binkley, D. (1986) *Forest Nutrition Management*, Wiley, New York.

Binkley, D., Aber, J., Pastor, J. and Nadelhoffer, K. (1986) Nitrogen availability in some Wisconsin forests: comparisons of resin bags and on-site incubations. *Biol. Fertil. Soils*, **2**, 77–82.

Binkley, D. and Matson, P. (1983) Ion exchange resin bag method for assessing forest soil N availability. *Soil Sci. Soc. Am. J.*, **47**, 1050–2.

Binkley, D. and Richter, D. (1986) Nutrient cycles and H^+ budgets of forest ecosystems. *Adv. Ecol. Res.*, **16**, 1–51.

Binkley, D. and Hart, S. (1989) The components of nitrogen availability in forest soils. *Adv. Soil Sci.* (in press).

Blake, J., Webster, S. and Gessel, S. (1988) Soil sulfate-sulfur and growth responses of nitrogen-fertilized Douglas-fir to sulfur. *Soil Sci. Soc. Am. J.*, **52**, 1141–7.

Bray, R.H. (1961) You can predict fertilizer needs with soil tests. *Better Crops with Plant Food*, **45**, 18–27.

Bremner, J.M. and Mulvaney, C.S. (1982) Nitrogen – total. In *Methods of Soil Analysis. Part 2 Chemical and Microbiological Properties* (eds A.L. Page, R.H. Miller and D.R. Keeney), American Society of Agronomy, Madison, pp. 595–624.

Burger, J.A. and Pritchett, W.L. (1984) Effects of clearfelling and site preparation on nitrogen mineralization in a Southern pine stand. *Soil Sci. Soc. Am. J.*, **48**, 1432–7.

Caldwell, M.M., Eissenstat, D.M., Richards, J.H. and Allen, M.F. (1985) Competition for phosphorus: differential uptake from dual-isotope-labeled soil interspaces between shrub and grass. *Science*, **229**, 384–6.

Chapin, F.S.H. III, Van Cleve, K. and Vitousek, P. (1986) The nature of nutrient limitation in plant communities. *Am. Nat.*, **127**, 148–58.

Cox, F.R. and Kamprath, E.J. (1972) Micronutrient soil tests. In *Micronutrients in Agriculture* (eds J.J. Mortvedt, P.M. Giordano and W.C. Lindsay), Soil Science Society of America, Madison, pp. 289–318.

DiStefano, J. and Gholz, H.L. (1986) A proposed use of ion exchange resin to measure nitrogen mineralization and nitrification in intact soil cores. *Commun. Soil Sci. Plant Anal.*, **17**, 989–98.

Ellenberg, H. (1977) Stickstoff als Standortsfactor, inbesondere für mitteleuropaische Pflanzengesellschaften. *Oecol. Plant.*, **12**, 1–22.

Emsley, J. (1984) The phosphorus cycle. In *The Natural Environment and the Biogeochemical Cycles*, Vol. 1, Part A (ed. O. Hutzinger), Springer-Verlag, Berlin, pp. 147–62.

Eno, C.F. (1960) Nitrate production in the field by incubating the soil in polyethylene bags. *Soil Sci. Soc. Am. Proc.*, **24**, 277–9.

Ensminger, L.E. and Freney, J.R. (1966) Diagnostic

techniques for determining sulfur deficiencies in crops and soils. *Soil Sci.*, **101**, 283–90.

Gianello, C. and Bremner, J.M. (1986) A simple chemical method of assessing potentially available organic nitrogen in soil. *Commun. Soil Sci. Plant Anal.*, **17**, 195–214.

Gibson, D.J. (1986) Spatial and temporal heterogeneity in soil nutrient supply measured using *in situ* ion-exchange resin bags. *Plant and Soil*, **96**, 445–50.

Gilliam, F.S. and Richter, D.D. (1985) Increases in extractable ions in infertile Aquults caused by sample preparation. *Soil Sci. Soc. Am. J.*, **49**, 1576–8.

Gillman, G.P. (1981) Effects of pH and ionic strength on the cation exchange capacity of soils with variable charge. *Austr. J. Soil Res.*, **19**, 93–6.

Gupta, R.K., Singh, C.P. and Abrol, I.P. (1985) Determining cation exchange capacity and exchangeable sodium in alkali soils. *Soil Sci.*, **139**, 326–31.

Harrison, A.F. and Helliwell, D.R. (1979) A bioassay for comparing phosphorus availability in soils. *J. Appl. Ecol.*, **16**, 497–505.

Hart, S.C. and Binkley, D. (1985) Correlation among indices of forest soil nutrient availability in fertilized and unfertilized loblolly pine plantations. *Plant and Soil*, **50**, 230–3.

Hauck, R.D. (1982) Nitrogen–isotope–ratio analysis. In *Methods of Soil Analysis. Part 2 Chemical and Microbiological Properties* (eds A.L. Page, R.H. Miller and D.R. Keeney), American Society of Agronomy, Madison, pp. 735–80.

Hendershot, W.H. and Duquette, M. (1986) A simple barium chloride method for determining cation exchange capacity and exchangeable cations. *Soil Sci. Soc. Am. J.*, **50**, 605–8.

Hesse, P.R. (1971) *A Textbook of Soil Chemical Analysis*, Chemical Publishing, New York.

Hoeft, R.G., Walsh, L.M. and Keeney, D.R. (1973) Evaluation of various extractants for available soil sulfur. *Soil Sci. Soc. Am. Proc.*, **37**, 401–4.

Jenkinson, D.S. and Powlson, D.S. (1976) The effects of biocidal treatments on metabolism in soil. I. Fumigation. *Soil Biol. Biochem.*, **8**, 167–77.

Jones, M.B. (1974) Fertilization of annual grasslands of California and Oregon. In *Forage Fertilization* (ed. D.A. Mays), American Society of Agronomy, Madison, pp. 255–76.

Kadeba, O. and Boyle, J.R. (1978) Evaluation of phosphorus in forest soils: comparison of phosphorus uptake, extraction method and soil properties. *Plant and Soil*, **49**, 285–97.

Kalisz, P.J. (1986) Effect of ionic strength on the cation exchange capacity of some forèst soils. *Commun. Soil Sci. Plant Anal.*, **17**, 999–1007.

Keeney, D.R. (1982) Nitrogen – availability indices. In *Methods of Soil Analysis. Part 2 Chemical and Microbiological Properties* (eds A.L. Page, R.H. Miller and D.R. Keeney), Agronomy Society of America, Madison, pp. 711–34.

Keeney, D.R. and Bremner, J.M. (1966) A chemical index of soil nitrogen availability. *Nature (London)*, **211**, 892–3.

Keeney, D.R. and Nelson, D.W. (1982) Nitrogen – inorganic forms. In *Methods of Soil Analysis. Part 2 Chemical and Microbiological Properties* (eds A.L. Page, R.H. Miller and D.R. Keeney), Agronomy Society of America, Madison, pp. 643–98.

Kushla, J.D. and Fisher, R.F. (1980) Predicting slash pine response to nitrogen and phosphorus fertilization. *Soil Sci. Soc. Am. J.*, **44**, 1301–6.

Lim, C.H. and Jackson, M.L. (1982) Dissolution for total elemental analysis. In *Methods of Soil Analysis. Part 2 Chemical and Microbiological Properties* (eds A.L. Page, R.H. Miller and D.R. Keeney), American Society of Agronomy, Madison, pp. 1–12.

Lindsay, W.L. and Vlek, P.L.G. (1977) Phosphate minerals. In *Minerals in Soil Environments* (eds J.B. Dixon and S.B. Weed), Soil Science Society of America, Madison, pp. 639–72.

Lloyd, F.T. and McKee, W.H. Jr. (1983) Replications and subsamples needed to show treatment responses on forest soils of the Coastal Plain. *Soil Sci. Soc. Am. J.*, **47**, 587–90.

McGill, W.B. and Cole, C.V. (1981) Comparative aspects of cycling of organic C, N, S and P through soil organic matter. *Geoderma*, **26**, 267–86.

McIntyre, D.S. (1970) The platinum microelectrode method for soil aeration measurement. *Adv. Agron.*, **22**, 235–81.

McLean, E.O. (1982) Soil pH and lime requirement. In *Methods of Soil Analysis. Part 2 Chemical and Microbiological Properties* (eds A.L. Page, R.H. Miller and D.R. Keeney), American Society of Agronomy, Madison, pp. 199–224.

Mehlich, A. (1941) Base unsaturation and pH in relation to soil type. *Soil Sci. Soc. Am. Proc.*, **6**, 150–6.

Mengel, K. and Kirkby, E.A. (1982) *Principles of Plant Nutrition*, International Potash Institute, Berne.

Mengel, K. and Scherer, H.W. (1981) Release of nonexchangeable (fixed) soil ammonium under field conditions during the growing season. *Soil Sci.*, **131**, 226–32.

Mitchell, M.J., David, M.B., Maynard, D.G. and Telang, S.A. (1986) Sulfur constituents in soils and streams of a watershed in the Rocky Mountains of Alberta. *Can. J. For. Res.*, **16**, 315–20.

Nadelhoffer, K.J., Aber, J.D. and Melillo, M.J. (1983) Leaf-litter production and soil organic matter dynamics along a nitrogen-availability gradient in southern Wisconsin. *Can. J. For. Res.*, **13**, 12–21.

Nye, P.H. and Tinker, P.B. (1977) Solute movement in the soil–root system. *Studies in Ecology*, No. 4, University of California Press, Berkeley.

Olsen, S.R. and Sommers, L.W. (1982) Phosphorus. In *Methods of Soil Analysis. Part 2 Chemical and Microbiological Properties* (eds A.L. Page, R.H. Miller and D.R. Keeney), American Society of Agronomy, Madison, pp. 403–48.

Page, A.L., Miller, R.H. and Keeney, D.R. (eds) (1982) *Methods of Soil Analysis. Part 2 Chemical and Microbiological Properties*, American Society of Agronomy, Madison.

Pastor, J., Aber, J.D., McClaugherty, C.A. and Mellilo, J.M. (1984) Aboveground production and N and P cycling along a nitrogen mineralization gradient on Blackhawk Island, Wisconsin. *Ecology*, **65**, 256–68.

Pastor, J. and Post, W.M. (1986) Influence of climate, soil moisture and succession on forest carbon and nitrogen cycles. *Biogeochemistry* **2**, 197–210.

Peace, W.J.H. and Grubb, P. (1982) Interaction of light and mineral nutrient supply in the growth of *Impatiens parviflora*. *New Phytol.*, **90**, 127–50.

Phillips, I.R., Black, A.S. and Cameroon, K.C. (1986) Effects of drying on the ion exchange capacity and cation adsorption properties of some New Zealand soils. *Commun. Soil Sci. Plant Anal.*, **17**, 1243–56.

Post, W.M., Pastor, J., Zinke, P.J. and Stangenberger, A.G. (1986) Global patterns of soil nitrogen storage. *Nature, London* **317**, 613–16.

Powers, R.F. (1980) Mineralizable soil nitrogen as an index of nitrogen availability to forest trees. *Soil Sci. Soc. Am. J.*, **44**, 1314–20.

Powers, R.F. (1984) Sources of variation in mineralizable soil nitrogen. In *Nitrogen Assessment Workshop May 19–20*, 1982, RFNRP Report No. 2, College of Forest Resources, University of Washington, Seattle, pp. 25–31.

Reuss, J.O. and Johnson, D.W. (1986) *Acid Deposition and Acidification of Soils and Waters*, Springer-Verlag, New York.

Rhoades, J.D. (1982a) Cation exchange capacity. In *Methods of Soil Analysis. Part 2 Chemical and Microbiological Properties* (eds A.L. Page, R.H. Miller and D.R. Keeney), American Society of Agronomy, Madison, pp. 149–58.

Rhoades, J.D. (1982b) Soluble salts. In *Methods of Soil Analysis. Part 2 Chemical and Microbiological Properties* (eds A.L. Page, R.H. Miller and D.R. Keeney), American Society of Agronomy, Madison, pp. 167–80.

Riha, S.J., James, B.R., Senesac, G.P. and Pallant, E. (1986) Spatial variability of soil pH and organic matter in forest plantations. *Soil Sci. Soc. Am. J.*, **50**, 1347–52.

Robertson, G.P. and Vitousek, P.M. (1981) Nitrification in primary and secondary succession. *Ecology*, **62**, 376–86.

Robertson, G.P., Houston, M., Evans, F. and Tiedje, J. (1988) Spatial variability in a successional plant community: patterns of nitrogen availability. *Ecology*, **69**, 1517–24.

Rosswall, T. (1976) The internal cycle between vegetation, microorganisms, and soils. In *Nitrogen, Phosphorus, and Sulfur – Global Cycles* (eds B.H. Svennson and R. Soderlund), *Ecol. Bull. NFR*, **22**, 157–67.

Rosswall, T. (1982) Microbiological regulation of the biogeochemical nitrogen cycle. In *Nitrogen Cycling in Ecosystems of Latin America and the Caribbean* (eds G.P. Robertson, R. Herrera and T. Rosswall), *Plant and Soil*, **67**, 15–34.

Sasser, C. and Binkley, D. (1988) Nitrogen dynamics across a mortality wave in balsam fir and Fraser fir ecosystems. *Biogeochemistry* (in press).

Schlesinger, W. (1985) The formation of caliche in soils of the Mojave Desert, California. *Geochim. Cosmochim. Acta*, **49**, 57–66.

Shumway, J. and Atkinson, W.A. (1978) Predicting nitrogen fertilizer response in unthinned stands of Douglas-fir. *Commun. Soil Sci. Plant Anal.*, **9**, 529–39.

Sibbeson, E. (1977) A simple ion-exchange resin procedure for extracting plant-available elements from soil. *Plant and Soil*, **46**, 665–9.

Sibbeson, E. (1978) An investigation of the anion-exchange resin method for soil phosphorus extraction. *Plant and Soil*, **50**, 305–21.

Sibbeson, E. (1983) Phosphate soil tests and their suitability to assess the phosphate status of soils. *J. Food Sci. Agric.*, **34**, 1368–74.

Smith, V.R. (1979) Evaluation of a resin-bag procedure for determining plant-available P in organic, volcanic soils. *Plant and Soil*, **53**, 245–9.

Smith, K.A. (ed.) (1983) *Soil Analysis*, Marcel Dekker, New York.

Sollins, P., Spycher, G. and Glassman, C.A. (1984) Net nitrogen mineralization from light- and heavy-fraction soil organic matter. *Soil Biol. Biochem.*, **16**, 31–7.

Sommers, L.E., Nelson, D.W., Owens, L.B. and Floyd, M. (1977) *Technical Bulletin 99*, Purdue University, West Lafayette, Indiana.

Stanford, G., and Smith, S.J. (1972) Nitrogen mineralization potentials of soils. *Soil Sci. Soc. Am. Proc.*, **36**, 465–72.

Stone, E.L. (1968) Microelement nutrition of forest trees: a review. In *Forest Fertilization: Theory and Practice* (ed. G.W. Bengtson), Tennessee Valley Authority, Muscle Shoals, AL, pp. 132–79.

Tabatabai, M.A. (1982) Sulfur. In *Methods of Soil Analysis. Part 2 Chemical and Microbiological Properties* (eds A.L. Page, R.H. Miller and D.R. Keeney), American Society of Agronomy, Madison, pp. 501–38.

Thomas, G.W. (1982) Exchangeable cations. In *Methods of Soil Analysis. Part 2 Chemical and Microbiological Properties* (eds A.L. Page, R.H. Miller and D.R. Keeney), American Society of Agronomy, Madison, pp. 159–66.

Tinker, P.B. (1984) The role of microorganisms in mediating and facilitating the uptake of plant nutrients from soil. *Plant and Soil*, **76**, 77–91.

Turner, J. (1979) Interactions of sulfur with nitrogen in forest stands. In *Forest Fertilization Conference* (eds S.P. Gessel, R.M. Kenady and W.A. Atkinson), Institute of Forest Resources Contribution No. 40, University of Washington, Seattle, pp. 116–25.

Turner, J. and Lambert, M.J. (1986) Nutrition and nutritional relationships of *Pinus radiata*. *Annu. Rev. Ecol. System.*, **17**, 325–50.

Uehara, G. and Gillman, G. (1981) *The Mineralogy, Chemistry, and Physics of Tropical Soils with Variable Charge Clays*, Westview Press, Boulder.

van Raij, B., Quaggio, J.A. and da Silva, N.M. (1986) Extraction of phosphorus, potassium, calcium, and magnesium from soils by an ion-exchange resin procedure. *Commun. Soil Sci. Plant Anal.*, **17**, 547–66.

Van Rees, K.C.J. and Comerford, N.B. (1986) Vertical root distribution and strontium uptake of a slash pine stand on a Florida Spodosol. *Soil. Sci. Soc. Am. J.*, **50**, 1042–6.

Vieira, S.R., Hatfield, J.L., Nielsen, D.R. and Biggar, J.W. (1983) Geostatistical theory and application to variability of some agronomical properties. *Hilgardia*, **51**, 1–49.

Vitousek, P. and Matson, P. (1985) Disturbance, nitrogen availability, and nitrogen losses in an intensively managed loblolly pine plantation. *Ecology*, **66**, 1360–76.

Vitousek, P. and Sanford, R.L. Jr., (1986) Nutrient cycling in moist tropical forests. *Annu. Rev. Ecol. System.*, **17**, 137–67.

Voroney, R.P. and Paul, E. (1984) Determination of Kc and Kn *in situ* for calibration of the chloroform fumigation-incubation method. *Soil Biol. Biochem.*, **16**, 9–14.

Vose, P.B. (1980) *Introduction to Nuclear Techniques in Agronomy and Plant Biology*, Pergamon, Oxford.

Webster, R. (1977) *Quantitative and Numerical Methods in Soil Classification and Survey*, Clarendon Press, Oxford.

Webster, R. (1985) Quantitative spatial analysis of soil in the field. *Adv. Soil Sci.*, **3**, 1–70.

Wells, C.G., Crutchfield, D.M., Berenyi, N.M. and Davey, C.B. (1973) *Soil and Foliar Guidelines for Phosphorus Fertilization of Loblolly Pine*, USDA Forest Service Research Note SE-110, Asheville, NC.

6

Radiation and light measurements

Robert W. Pearcy

6.1 INTRODUCTION

Measurements of radiation in physiological ecology are of primary importance because of their role in energy balance determinations (see Chapter 7) and in photosynthesis measurements (see Chapter 11). Although less frequently made in physiological ecology, light quality measurements are important with respect to photomorphogenesis, or for understanding the effects of specific wavelengths, such as UV radiation, on physiological processes. Each of these measurements requires a different approach. In energy balance studies we are primarily interested in the energy incident on a leaf or plant, and how much of this energy is absorbed. Photosynthesis, on the other hand, is a photochemical process that is driven by absorption of photons by chlorophyll in the wavelength band from 400 to 700 nm. Here the absorptance of most leaves is quite high (Fig. 6.1). According to Planck's law, the energy content of a 700 nm photon (171 kJ mol^{-1}) is only 57% of that of a 400 nm photon (299 kJ mol^{-1}). However, photosynthetic rate should be essentially independent of whether it was driven by the absorption of a photon of 400 or 700 nm light, despite the large difference in energy content. Any excess energy in the absorbed photon is dissipated as heat or fluorescence. Thus photosynthetic rate should more closely follow the number of photons absorbed rather than the amount of energy absorbed.

Light action in photomorphogenesis is photochemical in nature and thus the same considerations apply. However, there is a need to measure specific wavelengths such as red or far-red light or to determine the spectral characteristics. These needs have quite different requirements for measurement and instrumentation systems.

6.2 DEFINITIONS AND UNITS

The radiant energy incident on a unit surface from all directions is called the *incident radiant flux density* or *irradiance* which has the appropriate SI units of W m^{-2}. If the energy per unit wavelength is specified, it is termed *spectral irradiance* and has the units of W m^{-2} nm^{-1}. *Radiant intensity*, on the other hand, is the flux from a point source through a unit solid angle and has the units of W $steradian^{-1}$.

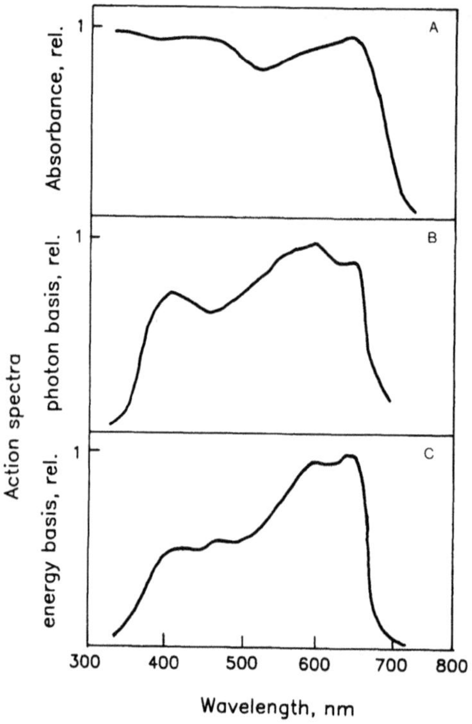

Fig. 6.1 Spectral absorptance of leaves (A) and action spectra for photosynthesis plotted on the basis of CO_2 uptake per absorbed photon (B) or on the basis of CO_2 uptake per unit of incident energy (C). The action spectra are plotted on a relative basis. The action spectra show that rates are more dependent on wavelength when they are expressed on a per unit of energy basis than on a per number of photons basis. The curves are averages from 22 species. Redrawn from McKree (1972b).

Intensity thus refers to a source and is often misused in place of the correct terms irradiance or radiant flux density.

Electromagnetic radiation has both the properties of a wave and of discrete particles. A *photon* is a discrete packet of electromagnetic radiation. A *quantum* is the amount of energy contained in a photon. In photobiology, *quantum* and *photon* are often used interchangeably with reference to the numbers of particles. An *Einstein* is 6.02×10^{23} photons (=1 mol of photons). The use of the unit Einstein has been criticized since in photochemistry it has sometimes been used in reference to the quantity of energy in 1 mol of photons and at other times to 1 mol of photons itself (Incoll *et al.*, 1977). In photobiology, the definition has apparently always been 1 mol of photons. *Photon flux* is the net number of photons per unit of area emitted or absorbed per unit of time and has the units of mol s^{-1}. *Incident photon flux area density* is the number of photons incident per unit of plane surface per unit of time with the units of mol m^{-2} s^{-1}. This is usually just referred to as *photon flux density*. The term *fluence rate*, derived from the latin verb for flow, is sometimes used in place of flux density. In fact, the two are not quite identical since fluence rate refers to the integral of photon flux over all angles about that point whereas flux density refers to a unit surface.

In an attempt to standardize measurement terminology and units, a committee of the Crop Science Society of America proposed the following definitions (Shibles, 1976).

1. *Photosynthetically active radiation* (PAR): radiation in the 400–700 nm waveband;
2. *Photosynthetic photon flux density* (PPFD): incident photon flux density of PAR: the number of photons (400–700 nm) incident per unit time on a unit surface. The SI unit is mol m^{-2} s^{-1}. Usually the units of μmol m^{-2} s^{-1} or μEinsteins m^{-2} s^{-1}, which is not part of the SI system, are used;
3. *Photosynthetic irradiance* (PI): radiant energy flux density of PAR: the radiant energy (400–700 nm) incident per unit of time on a unit surface. Appropriate units are W m^{-2}.

Units and definitions for light measurement are likely to undergo further modification. For example, two papers (Mohr and Schäfer, 1979; Bell and Rose, 1981) provide extensive derivation of the units and definitions but make different recommendations as to terminology. Biggs (1986) on the other hand generally follows the Crop Science Society of America recommendations with some extensions. One difficulty with the very precise definitions is that often several terms with slightly different definitions result. These, in turn, have multiletter abbreviations further adding to the confusion. For example, PFD is a more readable and pronounceable abbreviation than PPFD even if the definition is less precise. By defining the wavelength range once in the methods, the reader or listener can be spared at least part of the agony later on. The terminology used in this chapter is generally consistent with that of Bell and Rose although the more commonly used photon flux density and its abbreviation, PFD, will be used in place of photon flux area density. They also define many other terms that would be rarely applicable in physiological ecology.

6.3 ENERGY VERSUS PHOTONS AS A MEASURE OF PAR

The above considerations have led to the suggestion that sensors for PAR should transduce the number of photons in the 400–700 nm waveband, rather than the amount of energy. McKree (1972a) has suggested that, ideally, the sensor should have a sensitivity similar to that of the photosynthetic process, but this is practically unattainable. Furthermore, the measurement would then include properties of the plant rather than just properties of the active radiation. The actual waveband considered to be photosynthetically active has differed, with Russian workers in particular using 380 and 720 nm as the limits (McKree, 1979). However, efforts to standardize units, measurements and definitions, plus the availability of commercial sensors for the 400–700 nm waveband, dictate that this range be accepted.

McKree (1972a,b, 1973) has exhaustively investigated the consequences of using energy or photon flux units in measurements of photosynthetically active radiation. Based on measured action spectra of photosynthesis for a wide variety of crop species (Fig. 6.1) and the spectral properties of various light sources, he calculated the expected initial slope of the photosynthesis (*P*) versus light (*I*) curves for an average leaf under the different light sources (Table 6.1). An ideal sensor corresponding to the sensitivity of the leaf would give a value of 1 under the different light sources. In other words, the sensor would measure only the radiation that could actually be utilized in photosynthesis. A sensor based on a photon response (Fig. 6.2) gives a variation of 8% between various light sources, while one based on an energy

Table 6.1 Photosynthetic rate per unit of light flux calculated for an 'average plant' for light from different sources and for different units of light flux measurement. Values are normalized to 1.00 for sun and sky light. Adapted from McKree (1972b)

Light source	*PFD* ($\mu E\ m^{-2}\ s^{-1}$)	*PAR* ($W\ m^{-2}$)	*Photometric* (*lux*)
Sun and sky	1.00	1.00	1.00
Blue sky	0.98	0.93	1.14
High-pressure sodium vapor	1.06	1.14	0.80
Metal halide	1.0	1.02	0.80
Mercury vapor	1.01	1.03	0.78
Warm white fluorescent	1.02	1.01	0.65
Cool white fluorescent	0.99	0.97	0.67
Quartz–iodine	1.04	1.15	1.14
Range	8%	22%	49%

response gives 22% variation. Calculations were also made for a photometric sensor having a sensitivity maximum of 550 nm, the same as that of the human eye, since, in the past these have often been used in photosynthesis research. Here, the variation in the sensor reading was 49%. Thus photon flux measurements are not ideal in terms of McKree's definition but are clearly better than those based on energy. Measurements based on photometric sensors and the units of lux or foot-candles are unacceptable today.

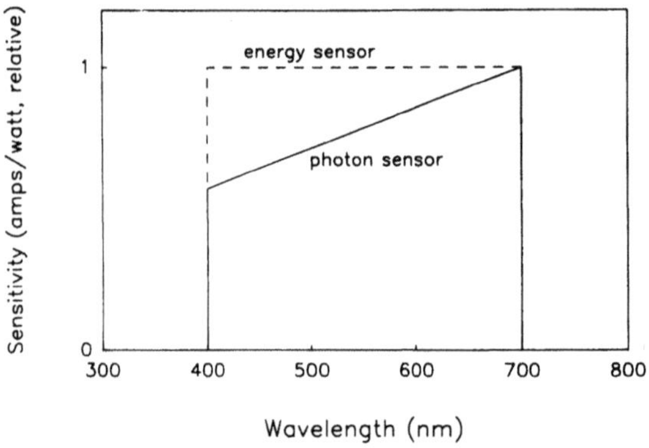

Fig. 6.2 Spectral response curves showing the output of (1) a photon sensor and (2) an energy sensor for equal energy input as a function of wavelength. The greater sensitivity in terms of energy for the photon sensor at the longer wavelengths compensates for the lower energy per photon.

6.4 RADIATION SENSORS: GENERAL CHARACTERISTICS

Two types of sensors, thermoelectric and photoelectric, are commonly used in radiation measurements in physiological ecology. A third kind of sensor, photochemical [a light-sensitive paper (blueprint paper, for example) or a dye solution that changes color in proportion to the exposure to light], has occasionally been used (Friend, 1961). Photochemical sensors, however, should be avoided since they are highly wavelength dependent.

Thermoelectric sensors are commonly used for solar and longwave radiation measurements and are designed to respond equally to incident energy over their wavelength range. Typically, a thermoelectric sensor consists of a temperature transducer such as a thermopile or thermistor attached to a surface that is painted with a highly absorbent black paint. The reference temperature transducer is attached to either the instrument housing, which is shielded from solar radiation or to a white reflective surface in the same radiation field as the black surface. The difference in temperature between the black surface and the reference is then a function of the difference in absorbed radiation. It is important that the other energy exchanges from the two surfaces such as convection and conduction be similar. Typically the surfaces are covered with glass or polyethylene domes to minimize convective losses and keep them similar.

Photoelectric sensors are solid-state detectors that are very useful because of their low cost and spectral response characteristics. The most common types that are sensitive in the visible region are silicon cells (Si), gallium arsenide phosphide (GaAsP), selenium cells (Se) and cadmium sulfide (CdS) cells. Lead sulfide (PbS) and lead selenide (PbSe) cells are most sensitive in the infrared. The most useful are Si and GaAsP which make use of the photovoltaic effect to generate a voltage at a p/n junction when it is exposed to light. Both consist of a bare plate coated with an n-type layer of silicon (or GaAsP) covered by a thin p-type layer. Absorption of a photon of sufficient energy (short enough wavelength) excites an electron, creating a positively charged 'hole'. Electrons in the n layer and holes from the p layer drift towards the p/n layer (barrier layer or junction) where they combine, resulting in a positively charged n layer, a negatively charged p layer, and an electric field across the junction. This voltage potential cannot continue to build up with continued absorption of light, however, since ultimately there is back diffusion. Consequently the voltage output is very nonlinear.

If a short-circuit is created between the electrodes attached to p and n layers, current will flow in proportion to the number of photons absorbed (holes created) in the p layer. This short-circuit current is then a linear function of the absorbed light, often over six to eight orders of magnitude of light. Alternatively, if a voltage is applied across the cell in a reverse bias, then the apparent electrical conductance is a function of the number of photons absorbed. Reversed biased operation decreases the response time but has the disadvantage of causing a significant dark current that must be subtracted from the signal. The temperature coefficient of a photocell is a wavelength dependent and greater in the red and infrared than at shorter wavelengths. In general, the temperature coefficient is quite small (-0.1–$0.2\%\ °C^{-1}$) and for all practical purposes the effect of temperature can be ignored.

Spectral responses of the various types of photoelectric cells differ widely (Fig. 6.3) because different energies are required for excitation of the electrons into the conduction band. For Si photocells the bandgap energy is 1.12 eV (electron volts) whereas for GaAsP cells it is 1.8 eV. Thus the long wavelength cutoffs are 1100 and 700 nm respectively. The peak sensitivity of Si cells is near 900 nm whereas for GaAsP cells it is at 610 nm. At short wavelengths, the photons are absorbed near the surface of the p layer and the electron–hole pairs can recombine before reaching the p/n layer, limiting the response. Alterations in manufacturing can, however, shift the spectral response characteristics to a limited degree. For example, the short-wavelength sensitivity of Si cells can be enhanced (silicon-blue cells) resulting in a peak response at 680 nm.

Because of the inherent nonlinearity of the voltage output from a photocell, measurement of either the current or a conversion of the current to a voltage is required. The simplest system is to convert the current to a voltage with a shunt resistor placed across

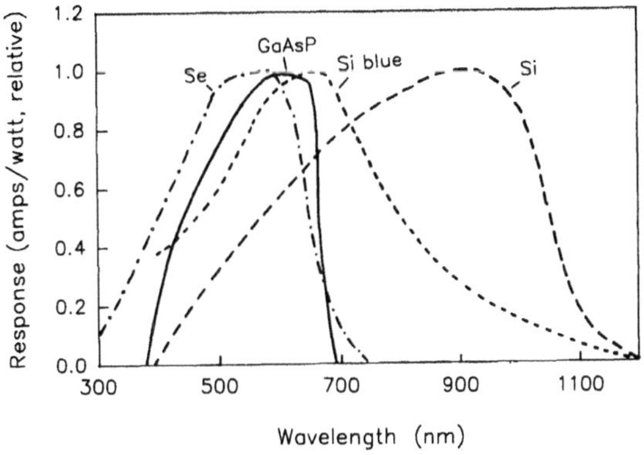

Fig. 6.3 Spectral response curves for several common types of photocells. The curves are adapted from literature from Optical Coatings Laboratories and Hanamatsu.

the output. If the resistance is kept small enough (1–3 ohms) so that the maximum voltage developed is not more than 10 mV, then the response will be nearly linear with flux density up to full sunlight. This is a convenient technique if a number of voltage-producing sensors, such as thermocouples, are to be used in addition to a photocell since a single microvoltmeter or recorder can then be used for all sensors. The disadvantage is that the resistance of the wire leads can have a significant influence, particularly if they are more than 1 or 2 m long. The resistor should be a precision wire-wound or metal film type having a low temperature coefficient.

Alternatively, a current-to-voltage converter constructed from an operational amplifier can be used to condition the signal from a photocell (Fig. 6.4). This has the advantage of giving a higher signal and a lower temperature coefficient for the output than operation with a resistor. The required circuit is very simple but certain precautions must be taken, especially when measurements are to be made in low light. Under low light when the current output of the photocell is low, the input bias current, which in effect is a current that leaks between the inputs of an op-amp, can be significant as compared to the photo-

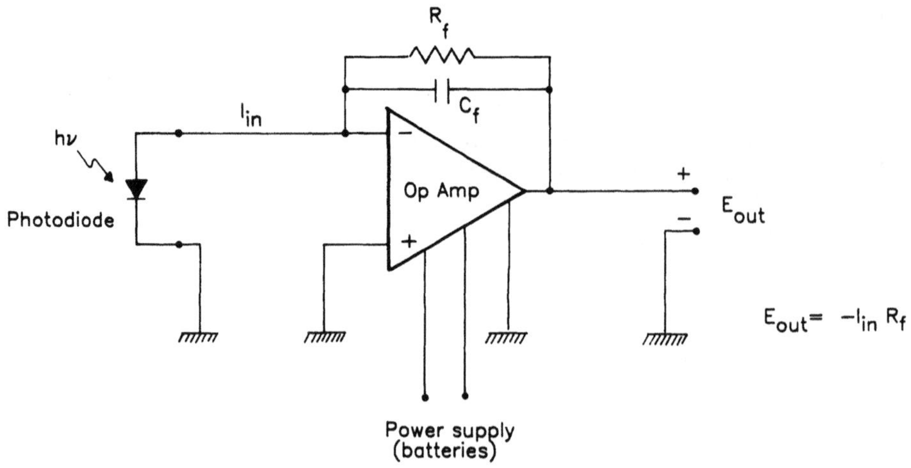

Fig. 6.4 A simplified circuit for a current-to-voltage converter that will function with photocells to give a voltage output. A low bias current operational amplifier should be used in this circuit. The output voltage (E_{out}) as a function of input current (I_{in}) is a function of the feedback resistor (R_f). The capacitor (C_f) acts as a noise filter at the penalty of a slightly reduced time constant. A bipolar (±) power supply is required for most amplifiers. This can be two 9 V batteries connected to give a ± 9 V supply.

cell current output for standard low-cost bipolar op-amps. The photocell and bias currents are additive so that it appears as an offset voltage in the dark, causing the apparent response to be nonlinear at low light. To avoid this problem, op-amps with very low bias currents (field effect transistor input or FET op-amps) should be utilized. Suitable low bias current FET op-amps are available from most semiconductor manufacturers. These manufacturers can often supply data sheets showing examples of their use in circuits for measuring light with photocells.

Unfiltered GaAsP, Si and Se cells are inexpensive ($1–$10), highly useful light sensors for use in physiological ecology, provided that certain precautions are taken. Because the spectral response of GaAsP cells is from 300 to 680 nm they are particularly useful for PAR measurements. However, for a given light input the current output is considerably less for GaAsP than for Si cells. This limitation is offset when they are used with a shunt resistor since they can tolerate a large resistance before deviating from a linear response. The spectral sensitivity of Se cells is better than that of Si cells for PAR measurements but they suffer from fatigue and have poor long-term stability when exposed to high light. The fatigue can be overcome by adding neutral density filters but at the penalty of reduced sensitivity. When used above a canopy, either type of cell can be adequately calibrated to read in PI, PFD or irradiance, since spectral shifts with changes in sky conditions are not large or tend to compensate for one another. Similarly, photocells can be calibrated for use under artificial lights, such as those used with photosynthesis chambers, provided that flux densities are altered with neutral density filters or in other ways that do not shift the lamp spectrum. Reducing the flux density from an incandescent lamp source, for example, by reducing the voltage causes large spectral shifts and a significant error in the measurements. The shiny surface of photocells also causes a poor response at low angles of incidence such as occur at dawn or dusk. In most cases, these errors are usually of little consequence.

Unfiltered photocells cannot be used under canopies without errors (McPherson, 1969). Selenium cells give too high a reading under canopies because of the higher transmission of green than red wavelengths. Silicon cells, on the other hand, are highly sensitive to wavelengths beyond 700 nm where the canopy transmission is especially high. Serious errors in estimates of PFD can result even if a sensor has a relatively low sensitivity beyond 700 nm. The reduced sensitivity in the 650–700 nm band and at wavelengths less than 450 nm causes a spectral error in GaAsP cells. However, GaAsP cells are generally the best choice, as will be discussed below, since the spectral errors are less than those for Si or Se cells and because of the lack of a response to wavelengths greater than 700 nm.

6.4.1 Solar radiation sensors

Solar radiation instruments of most use in physiological ecology are *pyranometers* which measure both direct and diffuse solar radiation and *pyrradiometers* which measure both solar and longwave thermal radiation. In contrast, *pyrheliometers* are narrow-aperture instruments that are designed to measure only the direct-beam component of solar radiation. There are many different kinds of pyranometers, pyrradiometers and pyrheliometers and they are described in detail in Fritschen and Gay (1979). No attempt will be made to review all kinds here and only a brief description of representative types will be given.

The total solar radiation received on a horizontal surface is referred to as the *global solar radiation*. The *reflected solar radiation* is that component that is reflected from the earth's surface. *Total radiation* includes the global radiation and the longwave radiation emitted principally by the atmosphere.

(a) Pyranometers based on thermoelectric sensors
Most pyranometers employ thermoelectric sensors (thermopiles) to detect the difference in temperature of alternate black and white surfaces when exposed to solar radiation. Alternatively the temperature of a black surface is referenced to some other surface or the base of the instrument which is not exposed to solar radiation. Hemispherical quartz or glass shields over the surfaces minimize convection losses, preventing changes in temperature due to wind. They also limit the response to wavelengths shorter than 4000 or 3000 nm, respectively, so that solar but not longwave radiation is sensed. An example of a pyranometer is the Eppley Black and White Pyranometer manufactured by Eppley instruments (Fig. 6.5). A thermopile made up of 15 copper–constantan junctions is mounted on the bottom of concentric black and white wedges which constitute the sensor. The sensor is covered by two glass domes that minimize convective heat transfer and restrict the wavelength response to 300–3000 nm. The output for this type of instrument is in the range of 5–15 $\mu V\ W^{-1}\ m^{-2}$.

(b) Pyranometers based on photocells
Pyranometers based on thermopiles are typically quite expensive and often have relatively small voltage outputs. As an alternative, pyranometers utilizing silicon cells have been developed. While these have spectral responses that are much narrower than solar radiation, the relatively small shifts in the wavelength distribution of solar radiation with changes in sky conditions such as cloudiness still allow satisfactory accuracy for many purposes. Some models utilize bare Si photocells mounted under glass domes to provide protection. However, the cosine response of the bare Si photocell is rather poor. Newer versions have the photocells mounted in small weatherproof plastic or metal housings similar to those of quantum sensors (see Section 6.4.3.a) that give a good cosine response. These sensors are available from LI-COR Instruments (model LI-200SA), Skye Instruments (model SKS 110) and Delta-T Devices (type ES). They are satisfactory for use under open sky conditions but should

Fig. 6.5 An Eppley black and white pyranometer (top) and a net pyrradiometer (bottom).

never be used under canopies or artificial light sources or for measuring reflected radiation.

6.4.2 Net and total radiation measurements

Total radiation includes both the short wavelengths derived from solar radiation as well as long wavelengths emitted by the atmosphere, the earth's surface or surrounding vegetation etc. Pyrradiometers for measurement of total radiation typically employ black thermoelectric sensors with polyethylene shields to minimize convection and to protect the surfaces. A thin layer of polyethylene is only about 5% less transparent to long than to short wavelengths. This difference, as well as the slightly lower absorptance in the infrared of the black paint used on the sensor surface, can be corrected for by painting a small part white so that it reflects shortwave but still absorbs longwave radiation.

Net pyrradiometers, which are more commonly called net radiometers, consist of two flat sensors facing in opposite directions and well insulated from each other which measure the difference between the incoming and outgoing total radiation. The most commonly used types consist of black sensors covered by hemispherical polyethylene domes (Fig. 6.5). The difference in temperature between the two surfaces is then proportional to the difference in absorbed radiation and is typically sensed by either a thermopile or by matched thermistors. Total radiometers differ in that the radiation received by each surface is measured independently. This is a distinct advantage since studies of leaf energy balance require total radiation measurements. Unfortunately, total radiometers are considerably more expensive than net radiometers.

Net radiometers can be converted into total radiometers by covering one side with a metalic or polyethylene shield that is painted black on the inside so that it acts as a good emitter of longwave radiation. The radiation emitted from this shield (R_1) can be calculated from its temperature according to the Stefan–Boltzmann equation:

$$R_1 = e\sigma T^4 \quad (6.1)$$

where σ is the Stefan–Boltzmann constant (5.67×10^{-8} W m^{-2} K^{-4}), e is the emissivity (0.99) and T is the temperature in K. By adding R_1 to the apparent net radiation measured by the sensor, the total radiation received by the other side can be determined. Idso (1971, 1972) discusses modifications to a net radiometer for measurement of hemispherical radiation, including the necessary precautions and the potential sources of error.

6.4.3 Photosynthetically active radiation sensors

Many different attempts to come up with suitable photocell–filter combinations to minimize spectral errors in measuring PFD have been published. Federer and Tanner (1966) and McPherson (1969) tested a wide variety of sensor–filter combinations and analyzed the errors associated with each.

(a) Quantum sensors

Biggs *et al.* (1971) designed a filtered silicon cell light sensor with a spectral response that closely approximates the ideal photon response. A modified version of this sensor is available from LI-COR Instruments under the name Quantum Sensor (model LI-190SA). Other manufacturers produce sensors with similar response characteristics (Skye Instruments model SKP 215; Delta-T Instruments Quantum Sensor type QS). The commercial availability of this rugged convenient sensor at a relatively low cost has made it a standard for measurement of PFD.

The original design consisted of a silicon-blue photocell mounted under a 400–700 nm broad-band inteference filter, a piece of heat-absorbing glass and a Wratten 85A colored gelatin filter. The housing was designed to control the angle of incidence of the light on the interference filter. The opaque white plastic disk on the top acts as a diffuser which in combination with the raised edge of the housing provides a cosine correction that compensates for the increased reflectance of the diffuser at low angles of incidence. The heat-absorbing glass provides further blockage beyond 800 nm where transmission of the interference filter is significant. The Wratten filter corrects the spectral sensitivity so that the sensor response corresponds more closely to a true photo response. The small holes in the housing are designed to restrict the angle of incidence of the light on the interference filter to within 15° of vertical. At greater angles of incidence, the filter cut-off shifts to shorter wavelengths causing an error. Overall, the sensor gives a good photon response over the 400–700 nm range (Fig. 6.6). While the LI-COR Quantum Sensor still has the same principle of operation, the components have improved stability. The sensors are supplied calibrated either in μA per μmol photons m^{-2} s^{-1} for use with a current-to-voltage converter or in mV per μmol photons m^{-2} s^{-1} with a load resistor installed across the lead wires. In the current mode, the quoted temperature coefficient is quite low (0.15% $°C^{-1}$) and the long-term stability is excellent (2% $year^{-1}$). Various other models are now manufactured, including a 1 m long-line sensor designed for spatial averaging under canopies and a spherical sensor designed for underwater use. Several meters, integrators, etc., designed to operate with the Quantum Sensor and calibrated to read out directly in PFD, are also available.

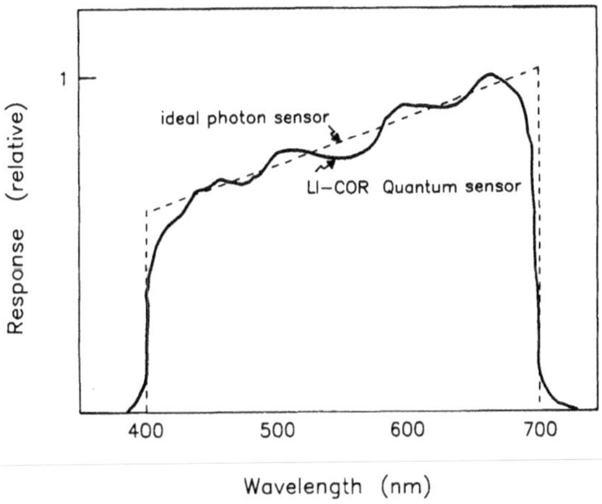

Fig. 6.6 Spectral response curve for a LI-COR Quantum Sensor as compared to the ideal photon response curve. Adapted from the manual for the LI-COR model 190S Quantum Sensor.

(b) GaAsP photocells

Gutschick *et al.* (1985) used an array of small GaAsP photocells mounted directly on to leaves to record histograms of PFD at different levels in a canopy. The photocells were interfaced to a portable computer via amplifiers and analog-to-digital converters. The photocells, originally designed for use in cameras, were sufficiently light (10 mg) that when connected to the leads they presented only a small dynamic load (15–20 mg) and did not disturb leaf position. The original sensor (NEC electronics model PH 201A) is no longer manufactured but other similar models are available from Hamamatsu Photonics or International Rectifier Corp. These photocells cost less than $10.

The low cost of the GaAsP photocells like those used by Gutschick and his colleagues makes them an attractive choice when large numbers of sensors are needed to measure light in a canopy or at many sites in an understory. The spectral error is a potential problem, but the errors may be within a tolerable range. To examine this, the expected outputs of a LI-COR Quantum Sensor and a NEC PH 201A sensor were calculated from the wavelength sensitivities in the manufacturers' specification sheets and the spectral distribution of PFD in a clearing and in the understory in a Costa Rican rain forest (Chazdon and Fetcher, 1984). These calculations showed that if the NEC PH 201A sensor was calibrated against a Quantum Sensor under sunlight in the open then it should give a PFD reading 3.2% higher than the Quantum Sensor for measurements in an understory. Of course there is an additional transfer error plus the spectral error in the Quantum Sensor itself so that the total error may be closer to 10–15%. The cosine response of the NEC PH 201A photocells appears to be quite good up to angles of 70°. Where only a few sensors are needed, GaAsP photocells are not appropriate substitutes for Quantum Sensors. However, where multiple sensors are needed, the cost of Quantum Sensors may be prohibitive. The advantage gained by using multiple GaAsP sensors may outweigh the disadvantages introduced by the spectral errors.

(c) PI sensors

For some purposes, such as the estimation of energy conversion efficiencies, PI may be a more appropriate measure than PFD. Above a canopy, PI can be directly estimated from pyranometer readings, since the 0.4–0.7 waveband radiation is an almost constant fraction of total shortwave radiation (300–3000 nm) under widely varying sky conditions (Anderson, 1971). Measurements by Efimova (1967; cited in Anderson, 1971) throughout the Soviet Union give a relationship of:

$$\mathrm{PI} = 0.43\, S_{\mathrm{dir}} + 0.57\, S_{\mathrm{dif}} \qquad (6.2)$$

where S_{dir} is direct solar radiation and S_{dif} is diffuse radiation. PI can also be measured with a pyranometer with a difference technique by first measuring total solar radiation and then using colored-glass UV and IR filters that block wavelengths shorter than 400 and 700 nm respectively. The difference in the readings with these two filters gives the PI. Filters designed to fit on Eppley pyranometers are available, but these plus the instrument are quite expensive.

Photocells can also be filtered to respond to PI. In general, the filtering is more successful with selenium than with silicon cells because the former are limited in their response to wavelengths less than 730 nm. With silicon cells the residual transmission through filters at wavelengths greater than 700 nm can cause significant errors, particularly when used within canopies. A filtered photocell in a quantum sensor-type housing having a reasonably flat energy response in the 400–700 nm band and apparently little residual response above 700 nm is manufactured by Skye Instruments (model SKE 510).

McKree (1972) has calculated factors for converting between PI and PFD for various light sources based on published lamp, diffuse sky and solar spectra. He, however, remarks that these factors must be used cautiously since lamp spectra can differ from those used in the calculation. Furthermore, relative errors due to the nonideal photon or irradiance response of sensors add to the uncertainties associated with the use of these factors.

6.5 DETERMINATION OF THE DIFFUSE AND DIRECT COMPONENTS OF RADIATION

For many studies it is necessary to separate the diffuse from the direct solar-beam radiation. Diffuse radiation can be measured by shading the sensor either with a disk or with a shadow band that blocks the direct solar radiation from striking the sensor. Direct solar radiation can then be determined by subtracting the diffuse radiation from the total. When only an occasional measurement is needed the sensor can be shaded with a disk on a stick. Sufficient time must be allowed before recording the readings so that a steady signal is achieved. This can be 1 min or more for pyranometers with thermopiles. If one sensor is available for measurement of total solar radiation and another for diffuse radiation, then a motor-driven system that moves the disk so that it continuously obscures the sun can be used. However, shadow bands are more convenient and do not require power. The shadow band consists of a semicircular metal band mounted over the sensor so that it obscures the solar track. The mounting of the band must accommodate the complex geometry of the earth's movement. For example the sun appears to rise and set from the north of east and west, respectively, during the summer in the northern hemisphere. The points of sunrise and sunset shift to the south in the winter. Shadow bands that can be gradually shifted to accommodate these seasonal changes can be easily constructed. Instructions for construction of a shadow band are given by Horowitz (1969).

6.6 CALIBRATION OF RADIATION SENSORS

The calibration of radiation sensors is a most difficult proposition. Pyranometers are usually calibrated by comparison with a standard pyranometer whose calibration can be traced to a standard pyrheliometer (Fritschen and Gay, 1979). For PI sensors, a calibration could be achieved by comparison against PI values measured with a pyranometer by the difference method. Calibration of PFD sensors is more difficult because of the need to know spectral irradiances over the 400–700 nm range. The most practical approach is to use a standard lamp (usually a tungsten–halogen lamp) of known spectral irradiance distribution operated on a highly regulated and calibrated constant current source. The spectral irradiance from standard lamps has an uncertainty of ±3% in most cases. Irradiance from this lamp can then be determined with a thermopile detector, with PI determined by the difference method. Then, knowing the spectral irradiance (E_λ) from the lamp at the standard distance as supplied by the calibration curve, the photon flux density can be calculated from:

$$\mathrm{PFD} = \frac{\mathrm{Na}}{hc} \int_{400}^{700} E_\lambda \lambda \, d\lambda \tag{6.3}$$

where h is Planck's constant, Na is Avogadros number (6.02×10^{23}) and c is the velocity of light. If the sensor has a nonideal photon response, then the calibration will strictly hold only under this light source, and there will be a relative error depending on the spectral distributions of light from the source. Spectral responses of sensors can be checked by comparison with thermopile detectors under monochromatic light from interference filters or monochrometers.

The above procedure is obviously too difficult, complicated and perhaps unnecessary for most ecophysiological studies; it is intended only to illustrate the complexities of light calibrations. A more practical approach is to compare the measurement to that made with a standard sensor. Thus, a Quantum Sensor, for example, which is preferably stored in the laboratory and used only for calibrations serves as the standard for PFD measurements. Calibration of a photocell within a leaf chamber used for photosynthesis measurements can then be achieved by placing the Quantum Sensor in the position of the leaf. This calibration procedure contains the absolute error of the Quantum Sensor, the relative error due to a nonideal source and a transfer error related to the uncertainties in the readout instruments. Biggs (1986) comprehensively discusses the errors involved in calibration and use of radiation sensors.

6.7 SAMPLING CONSIDERATIONS

The general considerations regarding sampling that are discussed in Chapter 1 hold for radiation measurements. In understories or under partly cloudy conditions frequent sampling may be required. When PAR measurements are being made in conjunction with photosynthesis measurements the slow response of the gas-exchange equipment relative to the changes in light and the response of the light sensor presents a special problem. Under these circumstances, it is appropriate to integrate the PAR signal so that the time constants are similar. This procedure was followed by Björkman and Ludlow (1972), and allowed a much closer comparison of photosynthetic responses of a rain forest understory plant with the variable light regime. If the primary interest is response to sunflecks, then the problem is clearly not how to sample PAR fast enough, but how to decrease the time constant of the photosynthesis apparatus and how to handle the increasingly rapid data sampling and amount of data. Weekly or monthly integrated totals may be most appropriate when the parameter of interest is plant growth or phenology.

Above a canopy, one sensor will usually suffice to characterize PAR over a relatively broad area of similar exposure. Within a canopy, however, sampling with a single sensor is inadequate because of the large spatial variations (Anderson, 1966, 1971) and in some cases 50–400 sensors may be required (Anderson, 1966; Reifsnyder *et al.*, 1971). Line sensors, such as the 1 m long-line Quantum Sensor, can be used under crop canopies to provide a spatial average as long as proper attention is paid to orientation. The cosine and azimuth responses of line sensors are not as good as the typical small circular sensors. Under taller canopies, such as in shrublands or forests, the spatial sampling of line sensors may not be large enough. Single small sensors mounted on tracks that allow movement back and forth under the canopy can be used to provide a spatial average. With the proper data-acquisition system, this procedure could be adapted to provide both a spatial average and spatial and time distributions of flux densities. Spatial averages are useful for characterizing in a general way light penetration into a canopy, but they may not be that closely related to photosynthetic response. Photosynthesis does not respond linearly to PAR and therefore the average PAR could not be expected to predict CO_2 uptake closely. This problem has been considered in some detail relative to canopy transmission models by Monsi *et al.* (1973). Sunflecks on the forest floor of relatively short duration are of great importance to the carbon balance of understory plants (Pearcy, 1988) and averages could clearly obscure this relationship (Gross, 1982). Moreover, photosynthetic utilization of sunflecks depends not only on the PFD but also on their frequency and duration (Chazdon and Pearcy, 1986). Both spatial and time distributions of PFD are needed to characterize the light environment of an understory or in a canopy adequately where sunflecks are important. An effective way to analyze these distributions is with frequency histograms of PFD, duration or time between sunflecks. Data loggers such as the Campbell Scientific CR21x can be programmed to reduce the large numbers of observations into histograms so that large data-storage capacities are not needed. This approach was used by Pearcy (1983) to reduce 2.7×10^6 light measurements from 18 homemade PFD sensors based on the design of Biggs *et al.* (1971) to obtain an understanding of the contribution of sunflecks to the understory light regimes in a Hawaiian forest. Records from portable stripchart recorders were also used to obtain further information on data on frequency, PFD and duration of sunflecks, but this is clearly much more laborious and limited to at most a few sensors.

6.8 PHOTOGRAPHIC ESTIMATIONS OF LIGHT CLIMATE

Superwide-angle lenses attached to a camera can give a 180° hemispherical 'fisheye' picture of a forest or shrub canopy if the picture is taken upwards from within the canopy. Photographs of this type can be used to estimate indirectly the radiation received at the point where the photograph was taken (Evans and Coombe, 1959; Anderson, 1964a, 1970, 1971). Furthermore, provided that there are no significant seasonal changes in the canopy, a single photograph can be used to estimate the potential radiation received at different times during the year. The technique appears particularly useful for providing relative comparisons between sites.

Superwide-angle 'fisheye' lenses are available for a variety of cameras. The most convenient types are 35 mm cameras fitted with 7.5 mm lenses which give circular fisheye images covering a 180° hemispherical field of view in the center of the film. Lenses of this type are available for several of the common 35 mm single-lens reflex cameras, but they are quite expensive. Conversion lenses designed to fit in front of a normal 50 mm lens to give a nearly 180° hemispherical image are much less expensive, but have slightly poorer resolution (Evans *et al.*, 1975; Anderson, 1971). For most purposes, however, the resolution is adequate. These lenses can be used on larger-format cameras and the larger image may more than compensate for the lower lens resolution. It is important, in terms of the analysis, that the lens gives an equiangular rather than a rectilinear projection, as do most camera lenses. Wide-angle lenses that are not 'fisheye' are of the latter type. An equiangular projection, or Hill projection named after the inventor of the first equiangular lens (Hill, 1924), has a directly proportional relationship between angular altitude and radial distance, whereas rectilinear lenses do not. However, most fisheye lenses do have some angular distortion in the outer one-third of the image that requires correction. The procedures for determining the optical center of the lens and the angular corrections are given by Evans *et al.* (1975).

To take a hemispherical canopy photograph, the camera, attached to a tripod, is oriented with the film plane horizontal and the top of the camera pointing either north or south. Alignment of the camera should be done carefully with a level and a compass, remembering to correct magnetic north to true north. A sight attached to the camera can aid considerably in alignment. In addition, a marker that is visible in the field of view of the photograph is also helpful in correct alignment of the photographs during the analysis. This can be done by attaching a horizontal platform under the camera with white vertical markers oriented north–south and extending 10° above the optical center of the lens. For best resolution, the markers should be at least 0.5 m from the camera. In densely shaded understories a small light on the marker such as a light-emitting diode or illumination with a small flashlight is helpful.

The film used in hemispherical photography can be either black or white or color transparency with the former having higher resolution but requiring more effort to develop suitable photographs for analysis. Relatively fast black and white films are the most suitable, since they allow for faster shutter speeds which are required if there is any wind. The film is developed in a high-speed developer that gives a 2- to 3-fold increase in film speed. Contrast between the sky and canopy can be increased by using a red filter on overcast days or a blue filter on clear days. Overcast days are best because of the relatively uniform sky illumination, but early mornings or evenings on clear days are also suitable, since the sun will not interfere with the photograph. The exposure that gives good contrast between the sky and the canopy, yet does not lose detail, must be arrived at by trial and error. Generally, two to three photographs at different shutter speeds will give at least one suitable image. About 5 min per site is required so many sites can be sampled in a single day.

The analysis of the photograph can be done manually with overlays showing either solar tracks (Fig. 6.7(a)) or a 'spider web' grid (Fig. 6.7(b)). For manual analysis, these can be drawn on to clear acetate film or, if the image is to be projected as with a color slide, on to a large sheet of drafting film which can then be mounted on a clear plastic sheet and used as a rear projection screen. Solar tracks for a specific latitude and date can be drawn on polar coordinate paper using relationships giving the azimuth and elevation angle of the sun,

$$\sin\alpha = \sin\psi \sin\delta + \cos\psi \cos\delta \cos\eta \quad (6.4)$$

and

$$\sin\beta = -\cos\delta\,(\sin\eta/\cos\alpha) \quad (6.5)$$

where α is the solar altitude or the angular elevation above the horizon, β is the solar azimuth, δ is the solar declination at the respective date, ψ is the latitude, and η is the hour angle. The hour angle is equal to $\pi/180\ (H-12)\ 15$, where H is the mean solar time in hours. The geometrical relationships between the solar track on the hemisphere and on the Hill projection are shown in Fig. 6.8. All angles are expressed in radians. A detailed discussion of the computation is given in Evans and Coombe (1959). Measurement of the lengths of the track transversing holes in the canopy gives an estimate of the relative amount of time the site could receive direct sunlight (direct site factor). The frac-

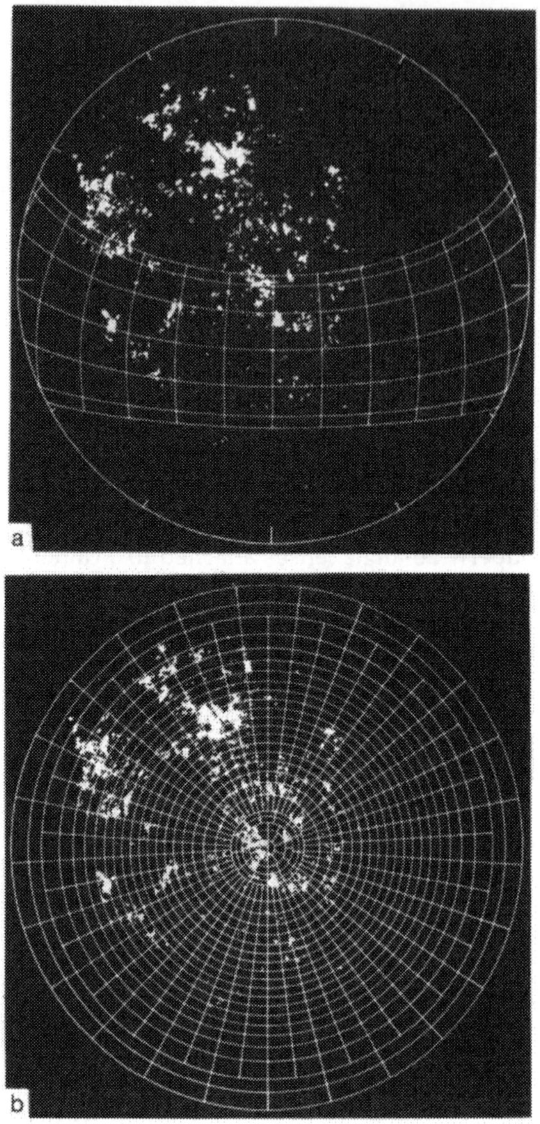

Fig. 6.7 Hemispherical 'fisheye' photograph overlayed with solar tracks for analysis of the direct site factor (a) or a grid (b) for estimation of the diffuse site factor according to the procedures of Anderson (1971). The picture is from a Hawaiian forest site at latitude 21°N.

tional transmission of diffuse light (diffuse site factor) can be estimated from spider-web grids (Fig. 6.7(b)) constructed according to instructions given in Anderson (1964b). Areas within the 1000 segments vary so that each potentially contributes 0.1% of the total sky illuminance from a standard overcast sky. An estimate of the fractional transmittance is obtained scoring each segment as to whether it is 0, 25, 50, 75 or 100% open. Summation of the score yields the diffuse site factor.

The very labor-intensive requirements of manual analysis have led to several attempts to automate it via computerized image techniques. Bonhomme and Chartier (1972) utilized a rotating arm with photocells mounted on it to determine gap frequencies from images projected on to a screen. This method is not readily adaptable to obtaining solar tracks, however. Image analysis techniques can be used to digitize the image and store it as an array of pixel values, representing different grey scales or in some cases colors of individual points. Image analysis systems currently available typically provide resolutions of 256 × 256 (65 536 pixels) or 512 × 512 (262 144 pixels) although some of the newer systems have resolutions of 1024 × 1024 (1.04 × 10^6 pixels). Since the circular fisheye image must fit within the square, the actual number of pixels digitized are about 78% of these values.

Since the location of each pixel on the image is known, it is a relatively simple matter to calculate the pixels to be sampled along a solar track using Equations 6.4 and 6.5. The coordinates of the pixels to be sampled are calculated from the solar azimuth and elevation using the Pythagorean theorem. It is probably best to sample pixels at a given time increment (usually 2–4 min) along the solar track. Once the values of the pixels along a solar track are determined, a cutoff must be established so that a pixel can be assigned as being either 'black' indicating canopy or 'white', indicating a hole. Then the number of open pixels times the time increment will give an estimate of the potential number of minutes of sunflecks. Alternatively, empirical relationships for solar radiation at the earth's surface, such as those given by

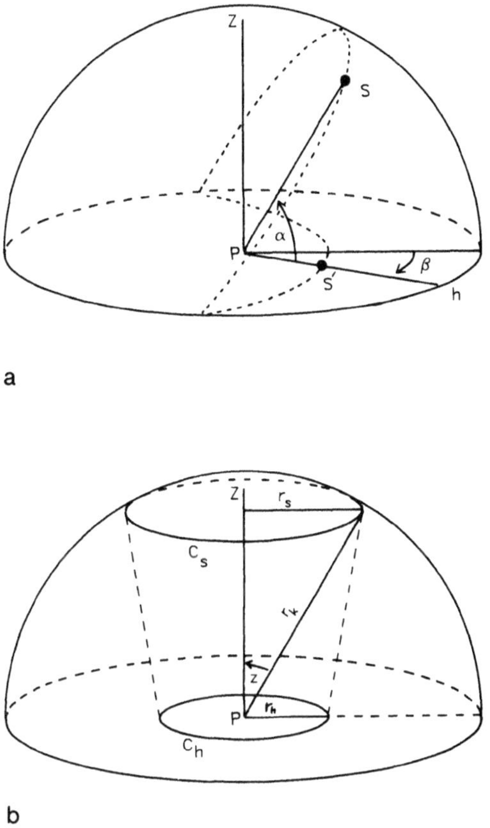

Fig. 6.8 Geometrical relationships between a hemisphere and a Hill projection taken at point P which are used to derive the solar tracks (a) and the weighting factor for the diffuse site factor (b). For (a) the short dashed lines give the solar tracks. The solar images on the hemisphere and Hill projection are given by *S* and *S'*, respectively, and are described by the azimuth (ß) and solar elevation (α). The distance from *S'* to a point on the horizon of the Hill projection (*h*) is proportional to α. The zenith is given by *Z*. For (b), C_s and r_s are the circumference and radius of a slice across the hemisphere taken at a zenith angle z. The corresponding circumference and radius on the Hill projection are C_h and r_h. r_k is the radius of the hemisphere.

Kreith and Kreider (1978) and shown below, can be used with the solar track information to predict the direct radiation penetrating the canopy. Direct solar radiation or PFD at the top of the canopy can be calculated by first estimating the value of the solar radiation at the outside of the atmosphere (S_{out}) from

$$S_{out} = C(1+\{0.34 \cos[360(d/365)(\pi/180)]\}) \quad (6.6)$$

where C is the solar constant for either energy (1353 W m^{-2}) or the estimated value for PFD (2510 μmol of photons m^{-2} s^{-1}), d is the Julian day and $\pi/180$ converts degrees to radians. The direct beam solar radiation at the earth's surface can now be calculated by first determining the airmass the beam traverses from

$$M = [1229 + (614 \sin\alpha)^2]^{0.5} - 614 \sin\alpha \quad (6.7)$$

where M is the airmass and α is the solar elevation from Equation 6.4. This relationship is valid for sea level; at higher elevations it should be corrected by the factor P/P_o where P is the atmospheric pressure and P_o is standard pressure at sea level. Direct radiation on a surface normal to the beam (S_{no}) is then given by

$$S_{no} = S_{out}\, 0.56\, (e^{-0.65M} - e^{-0.095M}) \quad (6.8)$$

Finally, the radiation on a horizontal surface at the top of the canopy (S_{dir}) is given by

$$S_{dir} = S_{no} \sin\alpha \quad (6.9)$$

which can be determined for each open pixel on the solar track. Summation will then give the potential direct radiation received in sunflecks on the forest floor. This empirical approach ignores penumbral and cloudiness effects as well as variations in atmospheric transmission but on clear days can give a reasonable estimate of the actual values (Chazdon and Field, 1987). Cloudiness effects could be included if the fractional cloud cover were known. This value will vary over a day in most locations and on a seasonal basis.

For the diffuse site factor, either the values of all pixels or a sample of the pixels can be determined, and then the pixels that are 'white' can be multiplied by a weighting factor that scales each for its contribution to the total illumination from the sky. The weighting factor is similar in principle to the different sampling areas on the spider-web grid of Anderson, and must take into account the differences in the relative projection of area on the photograph and on a hemisphere (Fig. 6.8). In addition, if only part of the pixels are sampled, a sampling scheme must be established so that they are evenly distributed over the hemisphere. Chazdon and Field (1987) derived a scheme for calculating diffuse radiation by sampling along 90 evenly spaced radii.

If all pixels are sampled then a weighting factor can be calculated by considering the area of a pixel on the Hill projection relative to the area it represents on the hemisphere. Along the radius the angular 'height' of a pixel on the Hill projection is constant but the angular 'width' decreases from the zenith to the horizon. The radius on the Hill projection (r_h) is directly proportional to the angle from vertical (z). If r_k is the radius to the edge of the sphere, which equals r_h at $z = \pi/2$ radians (90°), then

$$r_h = (2\, r_k/\pi)z \quad (6.10)$$

It follows then that the circumference on the Hill projection (C_h) is given by

$$C_h = 2\,(2\, r_k/\pi)z = 4r_k z \quad (6.11)$$

The radius (r_s) of a horizontal slice through the hemisphere at an angular height z is given by

$$r_s = r_k \sin z \quad (6.12)$$

Therefore, the circumference of the slice is given by

$$C_s = 2\pi r_k \sin z \quad (6.13)$$

and the weighting factor for a pixel at angle z [$w(z)$] is given by

$$w(z) = C_s/C_h \quad (6.14)$$

To calculate the diffuse light transmitted through gaps in the canopy, the distribution of radiation over the sky must be considered. On clear days, most diffuse radiation comes from within 10° of the solar disk (Hutchison *et al.*, 1980). However, the calculation of the radiation distribution for clear skies is complex and for most purposes the simpler approximation of either a uniform overcast sky distribution (UOC) or the distribution for a standard overcast sky (SOC; Moon and Spencer, 1942) can be used. First an estimate of the diffuse radiation (D_h) at the top of the canopy is obtained from

$$D_h = 0.5 \text{ sinz } (0.91\ S_{out} - S_{no}) \quad (6.15)$$

Each pixel is assigned a value of one (p=1) if it is open and zero (p=0) if it is closed. Assuming a uniform distribution, the canopy openness (θ) is determined from the sum of the weighting factor for the open pixels divided by the sum of the weighting factor for all pixels, or

$$\theta = \sum_i^m \sum_j^n p_{ij} w_{ij} \Big/ \sum_i^m \sum_j^n w_{ij} \quad (6.16)$$

and diffuse radiation on a horizontal surface at the point of the photograph (D_u) is given by

$$D_u = \left(\sum_i^m \sum_j^n p_{ij} w_{ij} (\text{sinz})_{ij} \Big/ \sum_i^m \sum_j^n w_{ij}\right) D_h \quad (6.17)$$

where the term sinz provides a cosine correction. A further correction can be made for a SOC distribution but for most purposes this is probably unnecessary. Comparison with values calculated for a uniform overcast sky distribution suggest that it yields acceptable results for periods of a day or longer, but a more complex clear sky approximation would be required for predictions at any particular time of the day (Hutchison *et al.*, 1980). However, given the uncertainties introduced by clouds, etc., more complicated procedures seem unwarranted.

Image analysis systems differ widely in their configuration. Systems using video cameras are available for use with personal microcomputers. Chazdon and Field (1987) used a photocell-based digitizer (Thunderscan, Thunderware, Orinda, CA) that fits on to the carriage of an Apple Imagewriter printer, and an Apple Macintosh computer to digitize hemispherical photographs. This system appears to have quite high resolution and is relatively low cost – the Thunderscan attachment costs less than $300. Dedicated image-analysis systems typically use a video camera and are much more expensive, even for relatively small systems designed to be used with personal computers. Some systems provide for color analysis. For analysis of solar tracks, a high resolution is desirable, since at 256 pixels, each one covers approximately 3 min on the track. Moreover, pixels that overlap the edge of the canopy and a gap give an intermediate gray value, requiring assignment to either 'black' or 'white' for analysis. The Thunderscan appears to be capable of a resolution in excess of 1000 × 1000 pixels. The primary disadvantages of the Thunderscan system is that it requires that a print be made and that it is slow. Video-based systems can with a macro or close-up lens directly digitize a negative or color slide. Digitizing and analysis of an image with the Macintosh-Thunderscan system takes approximately 20 min but only about 3 min on a dedicated system running with a minicomputer. Most systems allow manipulation of the image so that, for example, dark clouds can be lightened or reflective leaves darkened. The potential advantages of color in helping to distinguish sky from canopy have not been explored.

6.9 SPECTRAL RADIOMETRY

Measurement of the spectral distribution of light requires a spectroradiometer. These consist of a monochromator to separate and isolate individual wavelengths, a light sensor and system for correcting for the variation in sensitivity of the light sensor and the optical transmittance of the system with wavelength (Fig. 6.9). The monochrometer can consist of a diffraction grating, a prism or a wedge interference filter. Secondary filters are usually required to reduce stray light transmitted at other wavelengths than the one of interest. Alternatively, a series of individual wavelength interference filters mounted on a wheel can be used to provide measurements at a series of discrete wavelengths. In most modern instruments, the detector consists of a photodiode. Correction of the sensitivity at different wavelengths can be done by a cam attached to the wavelength drive that adjusts a potentiometer. Alternatively, microprocessor-based instruments can store corrections for each wavelength in nonvolatile memory. The output is calibrated in units of W m^{-2} s^{-1} nm^{-1} or in μmol m^{-2} s^{-1} nm^{-1} and is typically plotted on an XY recorder as a function of wavelength.

Most commercially manufactured spectroradiometers are primarily designed for laboratory use but a battery-powered version for field use is available from LI-COR (model LI-1800 portable spectroradiometer). Because of their complexity, spectroradiometers, typically cost in the range of $10 000–20 000. A relatively simple spectroradiometer using a wedge interference filter and cam system for correcting the sensitivity to photon flux density has been developed by Halldal (1969) for underwater measurements.

Frequently, measurements are not needed for the whole spectrum but rather for specific wavelengths. This is especially true for studies of photomorphogenesis where measurements are needed of red and far-red wavelengths centering around 660 and 730 nm where the two forms of phytochrome absorb. This is most frequently done with combinations of filters and a photocell or photodiode. Colored glass filters that exhibit a sharp cutoff at specific wavelengths are available from Schott Glass as well as other manufacturers. These filters are highly absorbant at shorter wavelengths but transmit wavelengths longer than the cutoff point. Typically the transition from 1% to 90% transmittance occurs within a 25–50 nm wavelength range. Measurements of relatively narrow spectral bands can be obtained from the difference between readings taken with two filters.

Interference filter–photodetector combinations are another approach to measuring light of specific wavelengths. Interference filters are multilayer thin film devices that transmit a desired wavelength interval while blocking both longer and shorter wavelengths. The interference between the multiple reflec-

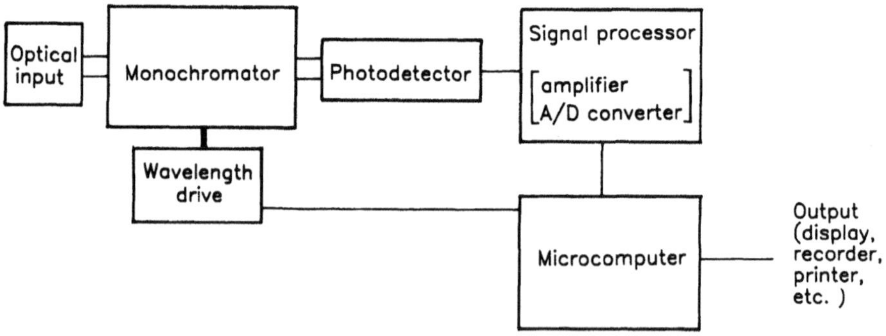

Fig. 6.9 Functional diagram of a spectroradiometer showing the various components. Redrawn from Duffey (1986).

tions within the filter are destructive for most wavelengths but allow transmission of the peak wavelength. By combining layers of different thicknesses, broader or narrower bandpasses can be achieved. The advantage of interference filters is that only a single filter is necessary. They are, however, more expensive than colored glass filters and also more delicate. In particular, moisture can destroy the multilayer construction.

REFERENCES

Anderson, M.C. (1964a) Light relations of terrestrial plant communities and their measurement. *Biol. Rev.*, **39**, 425–86.

Anderson, M.C. (1964b) Studies of the woodland light climate. I. The photographic computation of light conditions. *J. Ecol.*, **52**, 27–41.

Anderson, M.C. (1966) Some problems of simple characterization of the light climate in plant communities. In *Light as an Ecological Factor* (eds R. Bainbridge, G.C. Evans and O. Rackham), Blackwell, Oxford, pp. 77–90.

Anderson, M.C. (1970) Interpreting the fraction of solar radiation available in forest. *Agric. Meteorol.*, **9**, 191–216.

Anderson, M.C. (1971) Radiation and crop structure. In *Plant Photosynthetic Production Manual of Methods* (eds Z. Sestak, J. Catsky and P.G. Jarvis), Junk, The Hague, pp. 412–67.

Bell, C.D. and Rose, D.A. (1981) Light measurement and the terminology of flow. *Plant Cell Environ.*, **4**, 89–96.

Biggs, W.W. (1986) Radiation measurement. In *Advanced Agricultural Instrumentation* (ed. W.G. Gensler), Martinus Nijhoff, Dordrecht, pp. 3–20.

Biggs, W.W., Edison, A.R., Eastin, J.W., Brown, J.W., Maranville, J.W. and Clegg, M.D. (1971) Photosynthesis light sensor and meter. *Ecology*, **52**, 126–31.

Björkman, O. and Ludlow, M.M. (1972) Photosynthetic performance of two rainforest species in their native habitat and analysis of their gas exchange. *Carnegie Inst. Wash. Yearb.*, **71**, 94–102.

Bonhomme, R. and Chartier, P. (1972) The interpretation and automatic measurement of hemispherical photographs to obtain sunlit foliage area and gap frequency. *Isr. J. Agric. Res.*, **22**, 53–61.

Chazdon, R.L. and Fetcher, N. (1984) Photosynthetic light environments on a lowland tropical forest in Costa Rica. *J. Ecol.*, **72**, 553–64.

Chazdon, R.L. and Field, C.B. (1987) Photographic estimation of photosynthetically active radiation: Evaluation of a computerized technique. *Oecologia*, **73**, 525–32.

Chazdon, R.L. and Pearcy, R.W. (1986) Photosynthetic responses to light variation in rainforest species. II Carbon gain and photosynthetic efficiency during lightflecks. *Oecologia*, **69**, 524–31.

Duffey, B.L. (1986) Possible errors involved in the dosimetry of solar UV-B radiation. In *Stratospheric Ozone Reduction, Solar Ultraviolet Radiation and Plant Life* (eds R.C. Worrest and M.M. Caldwell), Springer-Verlag, Berlin, Heidelberg and New York, pp. 75–86.

Evans, G.C. and Coombe, D.E. (1959) Hemispherical and woodland canopy photography and the light climate. *J. Ecol.*, **47**, 103–13.

Evans, G.C., Freeman, P. and Rackham, O. (1975) Developments in hemispherical photography. In *Light as an Ecological Factor II* (eds R. Bainbridge, G.C. Evans and O. Rackham) Blackwell, Oxford, pp. 549–56.

Federer, C.A. and Tanner, C.B. (1966) Sensors for measuring light available for photosynthesis. *Ecology*, **47**, 654–7.

Friend, D.J.C. (1961) A simple method for measuring integrated light values in the field. *Ecology*, **42**, 577–80.

Fritschen, L.J. and Gay L.W. (1979) *Environmental Instrumentation*, Springer-Verlag, Berlin, New York and Heidelberg.

Gross, L. (1982) Photosynthetic dynamics in varying light environments: A model and its application to whole leaf carbon gain. *Ecology*, **63**, 84–93.

Gutschick, V.P., Barron, H.P., Waechter D.A. and Wolf, M.A. (1985) Portable monitor for solar radiation that accumulates irradiance histograms for 32 leaf-mounted sensors. *Agric. For. Meteorol.*, **33**, 281–90.

Halldal, P. (1969) Automatic recording of action spectra of photobiological processes, spectrophotometric analyses, fluorescence measurements and recording of the first derivative of the absorption curve in one simple unit. *Photochemistry and Photobiology*, **10**, 23–34.

Hill, R. (1924) A lens for whole sky photographs. *Q. J. R. Meteorol. Soc.*, **50**, 227–35.

Horowitz, J. (1969) An easily constructed shadowband for separating direct and diffuse solar radiation. *Sol. Energy*, **12**, 543–5.

Hutchison, B.A., Matt, D.R. and McMillen, R.T.

(1980) Effects of sky brightness distribution upon penetration of diffuse radiation through canopy gaps in a deciduous forest. *Agric. For. Meteorol.*, **22**, 137–47.

Idso, S.B. (1971) Transformation of a net radiometer into a hemispherical radiometer. *Agric. Meteorol.*, **9**, 109–21.

Idso, S.B. (1972) Simplifications in the transformation of net radiometers into hemispherical radiometers. *Agric. Meteorol.*, **10**, 473–6.

Incoll, L.D., Long, S.P. and Ashmore, M.R. (1977) SI units in publications in plant science. *Curr. Adv. Plant Sci.*, **9**, 331–43.

Kreith, F. and Kreider, J.F. (1978) *Principles of Solar Engineering*, McGraw-Hill, New York.

McKree, K.J. (1972a) The action spectrum, absorptance and quantum yield of photosynthesis in crop plants. *Agric. Meteorol.*, **9**, 191–216.

McKree, K.J. (1972b) Test of current definitions of photosynthetically active radiation against leaf photosynthesis data. *Agric. Meteorol.*, **10**, 443–53.

McKree, K.J. (1973) A rational approach to light measurements in plant ecology. *Curr. Adv. Plant Sci.*, **3**, 39–43.

McKree, K.J. (1979) Radiation. In *Controlled Environment Guidelines for Plant Research* (eds T.W. Tibbits and T.T. Kozlowski), Academic Press, New York, pp. 11–27.

McPherson, H.G. (1969) Photocell-filter combinations for measuring photosynthetically active radiation. *Agric. Meteorol.*, **6**, 347–56.

Mohr, H. and Schäfer, E. (1979) Uniform terminology for radiation: a critical comment. *Photochem. Photobiol.*, **26**, 1061–2.

Monsi, M., Uchijima, Z. and Oikaga, T. (1973) Structure of foliage canopies and photosynthesis. *Annu. Rev. Ecol. System.*, **301**–27.

Moon, P. and Spencer, D.E. (1942) Illumination from a nonuniform sky. *Illumin. Eng.*, **37**, 707–26.

Pearcy, R.W. (1983) The light environment and growth of C_3 and C_4 tree species in the understory of a Hawaiian forest. *Oecologia*, **58**, 19–25.

Pearcy, R.W. (1988) Photosynthetic utilization of lightflecks by understory plants. *Australian J. Plant Physiol.*, **15**, (in press).

Reifsnyder, W.E., Furnival, and Horowitz, J.L. (1971) Spatial and temporal distribution of solar radiation beneath forest canopies. *Agric. Meteorol.*, **9**, 21–37.

Shibles, R. (1976) Committee report: Terminology pertaining to photosynthesis. *Crop Sci.*, **16**, 437–9.

7
Temperature and energy budgets

James R. Ehleringer

7.1 INTRODUCTION

Temperature is of fundamental importance in affecting rates of metabolic activity in plant tissues. In this chapter, we will focus on methods for temperature measurement under field conditions and on the energy budget equation, which basically describes the influences of abiotic/biotic factors in affecting a deviation in plant tissue temperature from the surrounding ambient air temperature. As such the primary emphases of this chapter will be to describe (1) the principles behind the energy budget approach, (2) the leaf parameters which will influence leaf energy balance and thus need to be measured, (3) how leaf, air and soil temperatures are most commonly measured and (4) the precautions necessary to minimize errors in leaf, air and soil temperature measurements.

Although temperature is the emphasis of this chapter, a number of topics will not be considered because of space limitations. These include factors influencing air/soil temperature profiles (such as Richardson's numbers and damping depths), Bowen ratios and degree-day concepts. These as well as other microclimatological topics are discussed in depth in recent texts by Monteith (1973), Campbell (1977), Jones (1983) and Rosenberg *et al.* (1983).

7.2 ENERGY BUDGET APPROACH

All organisms and objects interact with their physical environment through energy-exchange processes. For leaves these processes include radiation absorption, reradiation, convection and transpiration. Different leaf temperatures arise because of a combination of changes in the air temperature surrounding the leaf and changes in leaf energy balance. An understanding of leaf energy balance or, as it is more often called, the energy budget equation, will allow the quantitative prediction of how leaf temperatures will change in response to abiotic factors such as solar radiation intensity and wind speed or to biotic factors such as leaf angle and stomatal conductance to water vapor.

7.2.1 Energy budget equation

At equilibrium, the rate of energy absorbed by a leaf equals the rate of energy loss. If these energy components are not in balance, then leaf temperature will change (increase or decrease) until an equilibrium situation is achieved. Because of the very high surface to volume ratios of most leaves, energy exchange is rapid and equilibrium in energy exchange is usually approached within a matter of seconds.

$$\text{Absorbed radiation} = \text{reradiation} + \text{convection} + \text{transpiration} \quad (7.1)$$

While this equation describes the major components of energy exchange in leaves, it ignores three other possible forms of energy exchange: energy fixation into stable carbon compounds during photosynthesis, energy production via respiration, and energy transfer via conduction. These three components of the leaf energy budget equation can usually be ignored because they are of low magnitude in comparison with the major forms of leaf energy exchange as shown in Equation 7.1.

Before proceeding to describe the energy budget equation further, it may be useful to describe briefly the energy-exchange processes being considered. Radiation absorption involves the absorption by leaf tissues of both shortwave radiation (solar radiation) and longwave radiation (sky and terrestrial infrared radiation). Reradiation is the process by which longwave radiation is reemitted by leaves. Convection is a process by which heat is exchanged between the leaf surface and the surrounding bulk air masses through physical contact. Finally, transpiration is the process by which water moves by diffusion from the inner leaf surfaces through stomatal pores to the outside air. Although transpiration is a mass movement, energy is also being transferred because of the kinetic energy content of that water leaving and because water movement from the leaf involves a state conversion of water from liquid to a gaseous phase. This state conversion requires a significant amount of energy.

Equation 7.1 can be expanded for a single leaf to

$$aS + \epsilon IR = 2\varepsilon\sigma(T_l + 273.15)^4 + 2h_c(T_l - T_a) + 2L(\rho_l - \rho_a)/(r_s + r_a) \quad (7.2)$$

where a is the leaf absorptance to total solar radiation (400–3000 nm), S is the total amount of direct, diffuse and reflected solar radiation incident on a leaf, ϵ is the emissivity of the leaf to infrared radiation, IR is the sum of infrared radiation from the ground and the sky, σ is the Stefan–Boltzmann constant (5.67×10^{-8} W m^{-2} K^{-4}), T_l and T_a are the leaf and air temperatures, h_c is the convection coefficient, L is the latent heat of vaporization (2441 J g^{-1} at 25°C), ρ_l and ρ_a are the water vapor densities of the inner leaf air spaces and the outside air, and r_s and r_a are the stomatal and boundary layer resistances (inverse of the conductances) to water vapor transfer. Note that the factor two appears in several places in Equation 7.2. This is because energy is being absorbed and lost by both upper and lower leaf surfaces. The leaf temperature can be solved for iteratively using the previous equation or by using a linear solution (Miller, 1972):

$$T_l = T_a + \frac{E_{abs} - \varepsilon\sigma(T_a + 273.15)^4 \times (L)(\rho_{s,ta} - \rho_a)/(r_l + r_a)}{4\varepsilon\sigma(T_a + 273.15)^3 + h_c + (L)(d\rho/dT)/(r_l + r_a)} \quad (7.3)$$

where E_{abs} is one-half the total energy absorbed by the leaf (solar and infrared radiation), $\rho_{s,ta}$ is the saturated water vapor density at air temperature, and $d\rho/dT$ is the slope of the saturation water vapor density versus temperature curve at air temperature.

The linear solution of the energy budget equation is approximate and reliable when leaf and air temperatures do not deviate greatly from each other. However, as leaf temperature deviates from air temperature, the error in this linear solution will increase proportionally. Given the speed of today's computers, a more satisfying solution is to use the Miller linear solution to obtain the approximate leaf temperature and then to iterate incrementally to the final solution (you specify the solution limits).

7.2.2 Leaf coupling factors

The linkages between leaf temperature and the physical environment as shown in the energy budget equation are made by leaf coupling factors. These coupling factors represent either physical properties of the leaf or are parameters based specifically on leaf characteristics. The interrelationships between energy exchange processes, environment, coupling factor and leaf properties are presented in Table 7.1. It is through changes in leaf coupling factors that plants adapt and adjust to different physical environments.

Leaf coupling factors involved with absorbed radiation are the leaf absorptance to total solar radiation and leaf emissivity. Since the molecular composition of leaves is essentially the same, there are only small differences in leaf emittance to longwave radiation (Table 7.2). However, large variations in leaf absorptances do occur and representative values and how they are measured are discussed in Section 7.7. Changes in leaf absorptance need not be the only mechanism for reducing the amount of solar radiation absorbed. An alternative would be to reduce the amount of solar radiation incident on the leaf. This can be accomplished by increasing the leaf angle or by orienting the leaf away from the sun. Methods for measuring leaf orientation and quantifying the total amount of solar radiation incident on the leaf are presented in Sections 7.5 and 7.6.

The stomatal and boundary layer resistances serve to regulate rates of water loss from the leaf and, therefore, provide the coupling for latent energy exchange. Stomatal resistance is a function of the diameter and density of stomatal pores. The boundary layer resistance depends in part on leaf size and shape and arises because of the presence of a thin layer of laminar air flow above the leaf which increases the path length for water vapor diffusion.

The convection coefficient for sensible heat transfer is related to the boundary layer resistance as

$$h_c = V/r_a' \qquad (7.4)$$

Table 7.1 Energy-exchange processes and leaf coupling factors. Based on Collier *et al.* (1973)

Energy-exchange process	*Environmental factors involved*	*Coupling factor*	*Organism properties involved in coupling*	*Organism response*
Convection	Wind speed, air temperature	h_c	Size of leaf, shape of leaf	Temperature
Evaporation (transpiration)	Wind speed, water content of air	r_a	Size of leaf, shape of leaf	Temperature, water loss
		r_s	Stomatal density, stomatal opening, cuticular resistance	
Reradiation	–	ε	Molecular composition	Temperature
Energy absorption	Solar	a	Leaf color, leaf orientation	Temperature, photosynthesis
	Terrestrial radiation	ε	Molecular composition	Temperature

Table 7.2 Emittances to longwave radiation for various materials determined at approximately 25°C. Data are from Birkebak (1966), Holman (1968) and Idso *et al.* (1969)

	Emissivity
Non-biological materials	
Brass (polished)	0.04
Brass (dull)	0.22
Brick	0.93
Copper (oxidized)	0.78
Glass	0.94
Iron (new cast)	0.44
Iron (oxidized surface)	0.80
Silver	0.02
Water	0.95
Soils	
loam	0.97
sand	0.95
Biological materials	
Animals	
Glaucomys volans (flying squirrel)	0.95
Peromyscus sp. (deer mouse)	0.94
Scalopus carolinensis (mole)	0.97
Sylvilagus floridanus (rabbit)	0.97
Leaves	
Gossypium hirsutum (cotton)	0.96
Nicotiana tabacum (tobacco)	0.97
Opuntia basilaris (cactus)	0.98
Phaseolus vulgaris (bean)	0.96
Populus fremontii (poplar)	0.98

where V is the volumetric heat capacity of air (the product of the heat capacity of air and the density of air) and r_a' is the boundary layer resistance for heat transfer (approximately $r_a^{1.5}$). A more detailed description of boundary layer and stomatal resistances and the procedures used to measure them are presented in Chapters 4 and 8 respectively.

7.2.3 Parameters to be measured for the energy budget equation

A number of parameters must be measured as inputs into the energy balance equation in order to calculate leaf temperature and transpiration rate correctly. These parameters and a sequence for their calculations are presented in Table 7.3. The measurements described in Table 7.3 allow the determination of the individual paths of energy absorption and loss by the leaf. However, for simulation or calculation purposes when leaf temperatures are not measured directly, it is more convenient to specify what environmental parameters must be measured in order to calculate the leaf temperature. In short, the environmental parameters are total solar radiation, total infrared radiation, wind speed and atmospheric humidity. The corresponding leaf parameters which must be measured are the total conductance to water vapor, absorptance and width.

7.2.4 Energy budget simulations

Perhaps the greatest utility of the energy budget equation is that it allows simulations

Table 7.3 Components of the energy budget equation and parameters that go into its calculation

Component	*Measurement*	*Instrument required*	*Chapter*	*Comments*
Solar radiation absorbed	Horizontal incident	Thermopile pyranometer	6	Measure on horizontal surface
	Diffuse	Thermopile pyranometer	6	Measure using shadow band
	Direct	Thermopile pyranometer	6	Subtract diffuse from solar total
	Reflected	Thermopile pyranometer	6	Measure by inverting pyranometer

Component	Measurement	Instrument required	Chapter	Comments
	Total incident solar			Multiply direct by cos(*i*) for the leaf angle and azimuth. Then add the diffuse and reflected radiation
	Leaf absorptance	Integrating sphere	7	–
	Total absorbed solar			Multiply the total incident solar times the leaf absorptance
Longwave (infrared) radiation absorbed				
	Sky	Total radiometer or infrared thermometer	6, 7	Subtract solar from the total or calculate from the 'effective' sky temperature using the Stefan–Boltzmann equation
	Soil	As above or thermocouple		Subtract reflected solar from total measured with the total radiometer inverted or calculate from the soil surface temperature
	Total longwave			Add sky and soil together, then multiply by the emittance
Longwave (infrared) radiation loss				
	Leaf	Thermocouple or infrared thermometer	7	Calculate using Stefan–Boltzmann equation. Multiply by two to account for both surfaces. Then multiply by the leaf emittance
Latent heat loss				
	Leaf resistance	Porometer	8	Leaf resistance is inverse of leaf conductance
	Boundary layer to water vapor	Anemometer, ruler	4	Calculate boundary layer knowing leaf width and wind speeds using equations from Chapter 4
	Transpiration		8	Calculate from leaf and boundary layer resistances, and measure leaf to air water vapor gradient
	Latent heat			Multiply transpiration by latent heat of vaporization
Sensible heat loss				
	Leaf–air temperature difference	Thermocouples	7	Mostly easily measured using differential thermocouples
	Boundary layer	Anemometer, ruler	4	Calculate boundary layer knowing leaf width and wind speeds using equations from Chapter 4
	Convection coefficient			Divide volumetric heat capacity of air by boundary layer resistance
	Sensible heat			Multiply convection coefficient by leaf to air temperature difference

or calculations of what leaf temperatures and transpiration rates would be if one of the leaf or environmental parameters were to change. Such simulations are extremely valuable in assessing the sensitivity of leaf temperature and/or transpiration rate to changes in a single variable. The two simulations which follow illustrate the significance of changes in leaf coupling factors on the leaf energy budget and therefore on the temperature and water loss rates of the leaf.

Let us assume a reasonable set of environmental conditions such as might be observed at midday during a mild summer day: an air temperature of 25°C, a soil temperature of 30°C, a relative humidity of 30%, a wind speed of 1 m s^{-1}, and an incident total solar radiation load of 1000 W m^{-2}. Let us also assume reasonable nonstress values for the leaf coupling factors: a leaf width of 2 cm, a leaf conductance to water vapor of 0.2 mol m^{-2} s^{-1} on both upper and lower leaf surfaces.

Let us now ask the first question, how will leaf temperature and transpiration rate change solely as a function of changing leaf absorptance to total solar radiation? From Fig. 7.1, we see that leaf temperature will decrease from 27.6°C to 23.5°C when leaf absorptance decreases from 50% (typical for most green leaves) to 20% (quite reflective leaf surface). Solely as a consequence of this decrease in leaf temperature, the transpiration rate will decrease 28%, from 9.0 to 6.5 mmol m^{-2} s^{-1}.

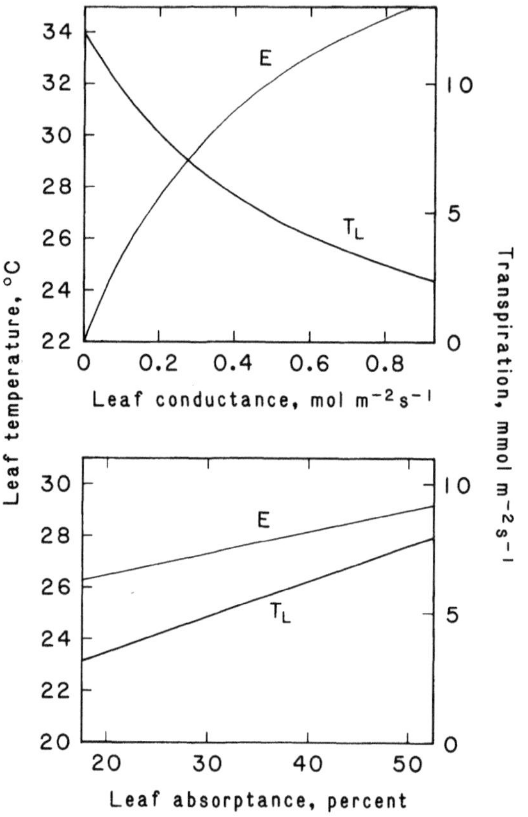

Fig. 7.1 Leaf energy budget calculations of the effects of changes in leaf absorptance (400–3000 nm) and in leaf conductance on leaf temperature (T_c) and transpiration rate (E). For these simulations T_a = 25°C, relative humidity = 30%, total solar radiation = 1000 W m^{-2}, leaf width = 2 cm, and wind speed = 1 m s^{-1}.

For the second leaf energy budget simulation, let us ask how the temperature and transpiration rate of the green leaf will change as the leaf conductance is varied. For simplicity, the leaf conductance of the upper and lower surfaces will be equal, and the results are plotted as a function of total leaf conductance. From Fig. 7.1, we see that conductance changes have a profound effect on both leaf temperature and on transpiration rate. Leaf temperatures increase from 25.0°C to 33.9°C as leaf conductance decreases from 0.8 to 0.0 mol m^{-2} s^{-1}. At the same time, the transpiration rate decreases from 13.7 to 0 mmol m^{-2} s^{-1}, although the decrease is not linear with conductance because of the compounding effect of changes in leaf temperature with leaf conductance.

In both of these simulations it is clear that changes in the leaf coupling factors, even under moderate environmental conditions, will play a significant role in affecting both leaf temperature and transpiration rate. Thus, it should be evident that in approaching leaf temperature studies and in understanding the relationships between leaf and air temperatures, it is imperative to have a complete consideration of both the abiotic environment as well as the leaf coupling factors. Additionally, since air temperatures vary with height (see Section 7.3), the air and leaf temperatures should be measured at the same position within the microclimatic profile.

7.3 VARIATIONS IN AIR AND LEAF TEMPERATURES WITH HEIGHT

7.3.1 Horizontal variation and sample size

At any time there may be considerable spatial variation in air temperatures on a horizontal as well as vertical scale. Horizontal spatial variation in air temperatures results from nonuniformity in the surrounding landscape. For instance, a mixture of shrubs, trees and open areas often results in small differences in air temperature of 1–4°C. These differences arise because of differential surface heating (depending on surface spectral and water loss characteristics), and thus result in differences in the rate of sensible heat transfer to upper air layers. On a larger scale, substantial variations in air temperature may arise if two adjacent slopes have different exposures (angles and azimuths). The differential heating in this case would be related to the amount of solar radiation received by that surface (see Sections 7.5 and 7.6).

The point to be made above is that even when air temperatures are measured at a specific height, these values may not be generalizable to a large area. That is, if we are interested in knowing the absolute air temperature at several specific locations, it should be measured at each location if there is reason to believe that spatial heterogeneity in air temperatures exists.

It should be clear from Equation 7.3 above that an accurate estimate of air temperature is critical to being able to predict absolute leaf temperatures. If the variation in air temperatures at a given height is small (say 1–2°C), the absolute air temperature may be measured at a single location and the spatial variation in air temperatures can be overcome by then measuring the differences between leaf and air temperatures at each location. Temperature differentials can be easily measured using thermocouples, and this allows for the precise determinations necessary to evaluate the leaf energy balance equation (see Section 7.4 for a discussion of how to measure leaf/air temperature differentials).

Variation in leaf temperatures exists not only because of differences in air temperatures, but also due to differences in leaf energy balance. For a given air temperature, the maximum range of leaf temperatures (ΔT) can be calculated from Miller (1971) as

$$\Delta T = \frac{aS_d \sin(\alpha)\tan(zn)}{4\,\varepsilon\sigma\,(T_a + 273.15)^3 + h_c + (L)(d\rho/dT)/(r_l + r_a)} \qquad (7.5)$$

where S_d is the direct component of the solar radiation and zn is the zenith angle, which is equal to 90° minus the solar altitude. Since virtually 100% of all observations occur within eight standard deviations, the standard deviation can be estimated as ΔT divided by 8.

The sample size (n) then required to estimate the mean leaf temperature with a specified confidence interval (c) can be calculated as

$$n = (2ts/c)^2 \qquad (7.6)$$

where t is Student's t, usually taken in a first approximation to equal 1.96, and s is the estimated standard deviation in leaf temperatures (Miller, 1971). These equations indicate that in practice to estimate the mean leaf temperature reliably within 0.5°C under high irradiance conditions requires between four and six separate leaf measurements. The required sample size increases to six to ten leaves if a confidence interval of ±0.1°C is necessary. As the solar radiation levels decrease, so does the required sample size for any specified confidence interval. Thus, at low irradiances two to three leaves are usually sufficient to estimate the mean leaf temperature.

7.3.2 Vertical variations in temperature

Vertical gradients in air temperature are well documented (Geiger, 1966; Campbell, 1977; Lee, 1978; Rosenberg *et al.*, 1983). They are associated with surface heating and sensible heat transfer to upper air layers. Steep air temperature gradients of 3–5°C per meter can be established under moderate to dense canopy situations and gradients of 15°C per meter or greater can be established within open arid land communities (Ehleringer, 1985). As a consequence it is critical to measure leaf and air temperatures at equal heights. The magnitude of the error introduced by not measuring leaf and air temperatures at the same height will decrease with increasing height. Thus, a 5 cm difference in sensor placement may result in a 2–5°C error in air temperature estimate near the ground, but only a 0.2–1.0°C error at 1 m.

7.4 TEMPERATURE AND ITS MEASUREMENT

7.4.1 Temperature sensors

In principle there are numerous ways for measuring temperatures of different components within the environment. However, for a variety of reasons most of which will become clear in the following discussion, the commonly practised method for measuring temperatures in physiological ecology is to use thermocouples. Before discussing thermocouples though, it is instructive to present briefly alternative means of temperature measurement and the limitations associated with these approaches.

(a) Bimetallic thermometers

When strips of two different metals having dissimilar expansion coefficients are bonded together, a change in temperature will cause a distortion of the bonded strip. Bimetallic thermometers have bonded bimetallic strips which are shaped into a coil with one end fixed and the other end attached to a needle on a dial. The coiled portion is enclosed within a metal sheath. Temperature changes along the coiled region then cause a rotational motion at the free end. Bimetallic thermometers require that the entire coiled region be in contact with the substance being measured. Thus, they are most useful for measuring solution temperatures. Such sensors are usually very slow in responding, lack the resolution of other temperature measurement devices, and the temperature reading usually must be recorded manually. A common application of bimetallic thermometers is thermographs for recording air temperatures. In this instrument, one end of the metallic strip is attached to a lever at the end of which is an ink pen. Changes in rotational motion in response to temperature change are recorded on a paper trace.

(b) Liquid-in-glass thermometers

In liquid-in-glass thermometers, the liquid (usually mercury) is enclosed within glass in a bulb–capillary arrangement. Temperature is sensed in the bulb region. Thermal expansion of the liquid causes it to rise within the capillary; the calibration is etched on to the glass which encloses the mercury column providing a temperature reading. While inherently more precise than bimetallic thermometers, liquid-in-glass thermometers share many of the same drawbacks as bimetallic thermometers: temperature is measured over a relatively large region, the sensor is slow in responding, and the readout is manual. Liquid-in-glass thermometers are quite precise and are most useful when the temperatures to be measured are uniform over a relatively large distance and a slow response time is acceptable.

(c) Resistance thermometers

Resistance thermometers are perhaps the most precise method for temperature measurement. All metals change electrical resistance with temperature, but metals differ in their temperature coefficient of resistance. Temperature is measured simply by measuring the resistance of a piece of uniform metal. The most common metal used is platinum and the sensor must be enclosed in a casing to protect the metal, which increases the sensor lag time. Unlike the previous two temperature measurement devices, resistance thermometers do not require that the readout device be at the same location as the sensor. Additionally, the resistance can be measured with a meter or automatic data acquisition devices (see Chapter 2). However, since resistance is the parameter being measured, the distance between the sensor and readout device is usually not very long since resistance of the lead wires will add to the total resistance of the circuit. This lead-wire resistance can be compensated for with a four-wire-ohm measurement system, but this adds costs if the distance is long. Mostly because of their relatively large size (when both sensor and casing are included), high expense and relatively slow response time, resistance thermometers are not commonly used temperature-measurement devices within ecology. However, dew-point mirrors used to measure absolute humidities (Chapters 3 and 8) often use resistance thermometers for temperature measurement.

(d) Thermistors

A thermistor is similar to a metal resistance thermometer in that its resistance changes in response to a change in its temperature. Unlike platinum resistance thermometers, thermistors are made of a semiconductor (usually a metallic oxide). Thermistors have greater sensitivity than platinum resistance thermometers (a higher temperature resistance coefficient) and can be made with either positive or negative temperature coefficients, whereas platinum resistance thermometers have only a positive temperature resistance coefficient. Thermistors come in different sizes ranging from small beads (as small as 0.25 mm) to relatively large flat disks. The response time for a temperature change is dependent on thermistor size, and so the smaller sensors can have a relatively fast response. Thermistors can be placed in a different location from their readout device and can be easily incorporated into automated data-acquisition systems. The principal drawbacks to thermistors are that they are more expensive and more fragile than thermocouples, and that in data-acquisition systems, automated resistance reading systems are presently far less common than voltage-measuring systems (as used by thermocouples).

(e) Infrared thermometers

Since all objects at temperatures above 0 K emit radiation, it is possible to measure surface temperatures utilizing this principle. An infrared thermometer senses the amount of longwave radiation emitted by a surface (usually leaf or ground). The intensity of this signal is then converted back to a surface temperature using the Stefan–Boltzmann law of radiation emission: $E = \varepsilon\sigma T_k^4$. The utility of this approach is that the surface need not be touched in order to obtain a temperature reading and that it integrates over a larger surface area. Two drawbacks are that temperatures are usually only accurate to within 0.5°C and that slight differences in emissivity result in large systematic errors in predicted temperature. Additionally, since most hand-held instruments must be focused on the object being measured, it is not practical for experiments involving large numbers of continuous measurements where some degree of automation is required. More sophisticated infrared imaging equipment is available and provides automatic gradient processing not available in hand-held instruments. However, it does so at a much greater cost.

(f) Thermocouples

Thermocouples are small sensors formed by bringing together two dissimilar metals to form a junction. If two dissimilar metallic wires are connected to each other at their ends, a current will flow in this circuit when the two junctions are at different temperatures (Fig. 7.2(a)). The phenomenon was first observed by Johann Seebeck in 1821 and the junction of the two dissimilar metals is called a thermocouple. Both the direction of the current flow and the magnitude of the electromotive force gradient (voltage) produced depend upon the absolute temperature difference between thermocouple junctions J_1 and J_2. The Seebeck effect is the observed conversion of the thermal energy to electrical energy at the thermocouple junction. The Seebeck coefficient is the measure of the voltage gradient produced per unit change in temperature between the two thermocouple junctions. Table 7.4 lists Seebeck coefficients for various kinds of thermocouple junctions.

Shortly thereafter in 1834 James Peltier found that when a current flows across a junction of two dissimilar metals, heat is either given off or absorbed by the junction. The absorption or emission of heat at the junction depends on the direction of current flow and is called the Peltier effect. As more current is applied to the thermocouple junction, there is proportionally more change in its temperature and this is the basis for temperature control in many photosynthetic-measuring systems (Chapter 11). However, with respect to thermocouples as temperature-measuring devices, there is no measurable change in the temperature of a thermocouple when the current flowing across it is due solely to the thermal electromotive force.

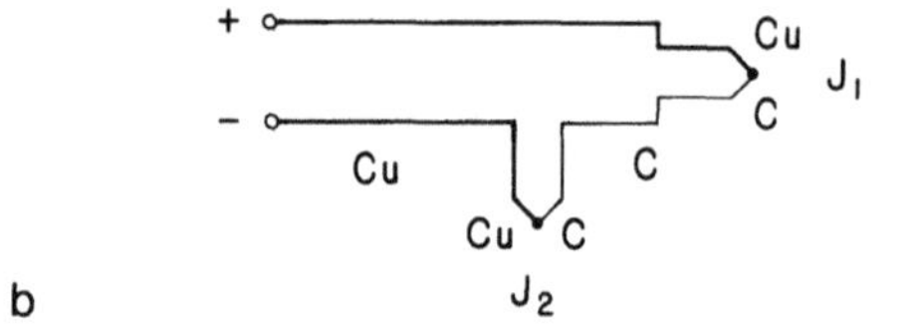

Fig. 7.2 (a) Seebeck effect – current will flow as long as two junctions are at different temperatures. (b) A system for temperature measurement using two thermocouple junctions (one for measuring and the other as reference) where J_1 is the sensing thermocouple and J_2 is the reference thermocouple and Cu and C indicate copper and constantan wire respectively.

Table 7.4 Seebeck coefficients for various kinds of thermocouples at 0°C (from *Temperature Measurement Handbook,* Omega Engineering, Stamford, Connecticut)

Type	*Metal* +	*Metal* −	*Seebeck coefficient* ($\mu V\ °C^{-1}$)
E	Chromel	Constantan	58.5
J	Iron	Constantan	50.5
K	Chromel	Alumel	39.4
T	Copper	Constantan	38.7

In 1851 William Thompson found that current flowing through a single homogeneous conductor will liberate or absorb heat if a temperature gradient exists within that conductor. The Thompson effect occurs both within wires to which a current has been applied as well as to those generated by thermocouples. The thermoelectric electromotive force (Seebeck effect) is then the sum of the Peltier electromotive forces at the junctions and the Thompson electromotive forces in the two different wires.

Thermocouples are inexpensive, come in various sizes and respond relatively fast to a change in temperature. An additional feature which makes thermocouples quite popular is that the distance between the sensor and the readout device can be variable and long without affecting the signal. The measurement signal is a voltage and the sensors are easily incorporated into data-acquisition systems. Their principal drawback was that before the availability of operational amplifiers the signal voltage was low and more difficult to resolve. That problem has been easily overcome with today's accurate and more powerful voltage amplification systems.

For most applications, we are interested in knowing the absolute temperature of one of the thermocouple junctions. In these situations the temperature of the other thermocouple junction (called the reference junction) must be known and the voltage gradient within the circuit is then measured with a voltmeter. The reference junction is most commonly at 0°C, and this is achieved by either placing the reference junction into a container (usually a thermos or Dewar flask) of ice water or utilizing an electronic compensation circuit (most common approach) which achieves the same effect. Fig. 7.2(b) illustrates a typical copper–constantan thermocouple setup where J_1 is the measuring thermocouple junction and J_2 is the reference junction; in this setup, the voltage is measured by a voltmeter across the positive/negative leads.

Copper–constantan thermocouples are the most commonly used in environmental sciences (Table 7.4). The Seebeck coefficient for this combination of two different metals is relatively high, ranging from 38.7 $\mu V\ °C^{-1}$ at 0°C to 46.8 $\mu V\ °C^{-1}$ at 100°C. The use of copper as a junction minimizes potential additional thermocouple effects when leads are connected to a measuring device. Thermocouples may be commercially purchased, but they are also easily constructed from paired thermocouple wire. The thermocouple junction is the sensing element for temperature. It may be constructed by twisting the two different wires together (in which case it is the first junction of the twist which becomes the measurement location), and then binding them together with solder (no effect on temperature measurement), or by butt-welding the two wires.

Thermocouples have the added advantage that the wire is relatively inexpensive and that it is very uniform in composition. The uniformity means that there will be little variation between different thermocouples, and thus they do not have to be individually calibrated. However, there are some minor disadvantages to thermocouples, especially to those in which copper forms one of the junctions. The disadvantages are that copper has a high thermal conductivity so that heat conduction to the thermocouple junction could be a problem (especially at low wind speeds when the wire is in full sun) and that, in the smaller gauges, copper is not very stiff and therefore thermocouple penetration of a surface (such as a leaf epidermis) might be difficult.

Thermocouples can be arranged together for two very different applications. When closely linked together in series, they provide a larger signal output (roughly equivalent to the product of the Seebeck coefficient and the number of thermocouple junctions). Many thermocouple junctions in series are called a thermopile and this is the basis of many pyranometers (see Chapter 6). Alternatively, when thermocouples are arranged in parallel, they then estimate the average temperature. One common application is to measure the average leaf temperature by placing three to five thermocouples in a parallel arrangement. The obvious advantage is that only one junction to a data-acquisition device is needed to get the average temperature reading, saving two to four input channels for additional kinds of measurements.

Thermocouple wire comes in various diameters (Table 7.5). Two of the more commonly used gauges are 36 AWG for leaves and other thin objects and 24 AWG for air and soil temperatures.

When given a step change in the temperature of the object being measured, thermocouples and other sensors respond in an exponential manner. Thermocouples respond relatively fast to changes in temperature. The time constant, which is the time

Table 7.5 Wire gauge and diameter

*AWG**	*Diameter (mm)*
16	1.29
24	0.51
36	0.13
40	0.08
44	0.05

* AWG = American wire gauge.

required for the sensor to complete 1/*e* of the step change in temperature, is a function of wire diameter. For 16 AWG wire, the time constant is 3.1 s, which means that approximately 14.3 s are required for the thermocouple to go 99% of the step change in temperature. The time constant decreases with wire diameter; for 24 AWG wire, the time constant is 0.9 s, and for 36 AWG wire it is 0.09 s.

While long time constants may not be useful if you wish to know the temperature of an object that is changing rapidly, there are occasionally applications in which an averaged temperature is preferred. In such situations, thermocouples with longer time constants are desired because they will better integrate the 'running average' temperature.

7.4.2 Thermocouple sensor errors

(a) Thermocouple size

Choosing the appropriate thermocouple size for the object to be measured is quite important. The mass of the thermocouple should be small relative to the object being measured (such as a leaf), otherwise the sensor unduly influences the temperature of the object. Thus, a 36 or 40 AWG thermocouple is appropriate for most thin leaves, but a 24 AWG is much too large as its mass and diameter are of similar magnitude to that of the leaf thickness. Virtually any gauge is appropriate for sensing air and/or soil temperatures. The ultimate choice will depend on the time constant appropriate for the object being measured.

The same thermocouple gauge need not be maintained all the way from the measuring junction to the measurement device. It is common to use a 36 AWG thermocouple of 50 cm total length to measure leaf temperature and then to use a 24 AWG wire (which is more rugged) to continue on to the meter or data-acquisition device. When this is done, it is essential to use thermocouple connectors which maintain copper and constantan continuity within the line.

(b) Attachment to leaf

Ideally, thermocouples should be inserted into the center of the leaf mesophyll tissues from the underside. This should be done with a minimum of disruption to the epidermal layer to avoid causing excessive transpirational changes. The thermocouple should be attached in such a way that it does not change the leaf orientation (angle and azimuth). Some investigators have found that thermocouples can be attached to the lower leaf surface using surgical tape with no errors in the measurement.

Vertical leaf temperature gradients do not occur in leaves (Hays, 1975), but large temperature gradients can develop across leaves depending on the magnitude of the leaf boundary layer (Drake *et al.*, 1970; Wiebe and Drake, 1980). Thus, for large leaves or in low wind speed and high irradiance environments, several thermocouples are necessary to estimate accurately the average leaf temperature (Miller, 1971).

(c) Radiation errors

It is assumed that the thermocouple is in equilibrium with the object whose temperature it is measuring, but this will not necessarily be the case if the thermocouple is exposed to solar radiation. Solar radiation loads incident on a thermocouple will cause its temperature to increase and thus deviate from the temperature of the object. The extent of the deviation can be predicted by the energy budget equation, since thermocouple temperatures are subject to the same constraints as any other object.

Consider a thermocouple which has been placed on top of a leaf to measure the leaf temperature. If the thermocouple is at the true temperature, then the second term of Equation 7.3 will be zero. The error in the thermocouple temperature reading will be a function of the total solar radiation load, wind speed at the leaf surface (usually close to zero), and thermocouple size. The largest of these error components will be the solar radiation load. Under extreme conditions, the error in leaf temperature measurement caused by exposing a leaf thermocouple to direct sunlight can approach several degrees. To reduce this error, thermocouples should be inserted into the lower surface tissue.

This same type of error can appear in air and soil surface temperature measurements. For soil surface temperature measurements, the thermocouple should be inserted several millimeters below the surface to minimize potential radiation errors, yet allowing for reliable soil surface temperature estimates. For air temperature measurements, the thermocouple should be shielded from solar radiation yet should receive full wind exposure. This is normally accomplished using two flat plates to sandwich the thermocouple, each painted white on their outer surfaces and black on their inner surfaces. These plates are placed several centimeters above and below the thermocouple to allow for adequate ventilation (Fig. 7.3).

Leaf minus air temperature differentials can be most accurately measured by using the air temperature as the reference junction instead of an ice point reference. In this way the absolute leaf temperature is not measured, but instead the difference between temperature of the air and leaf is directly sensed. This allows for the air temperature to be measured near the leaf avoiding potential spatial variation errors. It also provides a more accurate estimate of the true temperature differential, since the differential is measured directly rather than calculating it from two absolute values. When temperature differentials are measured, an additional absolute and independent reading of either air or leaf temperature is required in order to obtain absolute estimates for each reading.

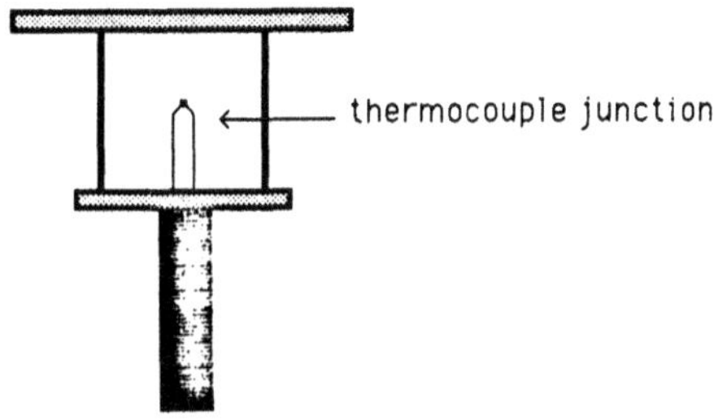

Fig. 7.3 An air temperature shield to reduce radiation error on the thermocouple.

7.4.3 Time constants

When measuring temperatures of various components within the environment, it is important to make sure that the thermocouple sensor is able to track adequately and rapidly the changes in temperature of the object being measured. In the previous section, we mentioned that time constants, the time required to complete $1/e$ (37%) of the step change, varied from 0.9 s for 24 AWG to 0.09 s for 36 AWG thermocouples. For most studies in physiological ecology these time constants are more than sufficient to track changes in air, soil and tissue temperatures. The time constant of a single exposed leaf or a leaf within a canopy to a change in solar radiation is approximately 10 s, and complete equilibration usually occurs within 40–60 s (Wiegand and Swanson, 1973). Soils respond somewhat slower and the time constant for the soil surface was estimated to be approximately 100 s by Wiegand and Swanson (1973).

7.5 ORIENTATION AND ITS MEASUREMENT

The orientation of a leaf or any other surface is important for understanding the diurnal and seasonal patterns of solar radiation incident on that surface. For leaves, these solar radiation intensities will significantly influence both photosynthetic rate (via its effect on the 400–700 nm photon flux) and leaf temperature (via its effect on total solar radiation). For ground surfaces, the amount of solar radiation will influence the heating of the surface, and therefore greatly affect soil and air temperature profiles. The amount of solar radiation that is incident on a leaf or ground surface can be specified for clear sky conditions if the direct and diffuse components of solar radiation incident on a horizontal surface and the leaf angle and its azimuth are known.

7.5.1 Cosine of angle of incidence

The cosine of the angle of incidence ($\cos(i)$) is a measure of the fraction of the direct solar radiation beam that is striking a planar surface. It varies between 0 and 1 depending on the relative geometrical positions of the surface and sun. We can describe $\cos(i)$ as

$$\cos(i) = \cos(\alpha_l)\sin(\alpha_s) + \sin(\alpha_l)\cos(\alpha_s)\cos(\beta_s - \beta_l) \quad (7.7)$$

where α_l and α_s are the angles above the horizon of the surface (e.g. leaf) and sun and β_l and β_s are the azimuthal positions of the surface and sun respectively (Gates, 1962).

7.5.2 Angle

The leaf angle (sometimes referred to as the leaf inclination) and solar angle (sometimes referred to as the solar altitude or elevation) are measured from the horizontal. Thus, a horizontal leaf has a leaf angle of 0° and when the sun is directly overhead (which never occurs if the latitude is greater than 23.5°), its solar angle is 90°.

Angles are often measured using an inclinometer, a small portable device that allows precise rapid angle determinations. Alternatively, leaf and solar angles can be measured with a protractor and plumb line.

In canopies with high leaf area indices, leaf angles often vary with canopy position, and so caution should be exercised to avoid introducing an unwanted variable by neglecting canopy position. It should also be noted that in some plants leaf angles are known to change seasonally in response to water stress (Comstock and Mahall, 1985) and in other plants leaves will move diurnally (Ehleringer and Forseth, 1980). In those leaves with diurnal movements, the leaves usually move so that they face the sun ($\cos(i) = 1.0$) throughout the day.

7.5.3 Azimuth

The azimuth represents the compass direction. It is measured with a compass (remember that a magnetic deviation must be incorporated to give the true azimuthal direction). Azimuths are recorded with respect to due south, thus south has a value of 0°, east −90°, west +90°, and north ±180°. The azimuth of the sun is relatively easy to determine when solar angles are low, but is more difficult to measure when the sun is at angles above 70°. This potential problem can be overcome by using a vertical piece of paper above the compass and orienting the compass until there is no shadow cast by the paper. For leaves, the direction that a leaf blade faces is again determined using a compass, but this direction is not necessarily the same as the petiolar–blade axis. The leaf azimuth is the direction that water would roll off a planar leaf surface.

7.6 CALCULATION OF INCIDENT SOLAR RADIATION ON DIFFERENT SURFACES

The effects of changes in latitude, season and

slope characteristics can have a pronounced impact on the intensity of solar radiation received by a surface. The diurnal changes in the intensity of solar radiation incident on any surface (leaf, ground, etc.) with a specified slope and azimuth at any latitude and at any time of the year can be reliably predicted for clear sky conditions given several parameters.

Under clear sky conditions, the intensity of solar radiation on any surface can be calculated knowing the altitude and azimuth of the sun and angle and azimuth of the surface of interest. The solar altitude (angle above the horizon) is

$$\sin(\alpha_s) = \sin(l)\sin(D) + \cos(l)\cos(D)\cos(h) \quad (7.8)$$

where α_s is the solar altitude, l is the latitude, D is the solar declination and h is the solar hour angle (List, 1968). The hour angle equals

$$h = (t-12)(\pi/12) \quad (7.9)$$

where t is the mean solar time in hours.

The azimuth of the sun (β_s) can be described as

$$\beta_s = -\arcsin[-\cos(D)\sin(h)/\cos(\alpha_s)] \quad (7.10)$$

The declination varies from −23.5° to +23.5° depending on the time of the year. In the northern hemisphere, the declination can be described by

$$D = 23.5\cos[2\pi(d-172)/365] \quad (7.11)$$

where d is the day of the year (Rosenberg *et al.*, 1983).

Equations 7.8 and 7.10 allow the path of the sun through the sky to be calculated for different locations and seasons.

Once these values are known, we can calculate the irradiance incident on a horizontal plane at the earth's surface as

$$I_o = S_c\sin(\alpha_a)A^{[l/\sin)\alpha_s)]} \quad (7.12)$$

where I_o is the irradiance incident on a horizontal surface, S_c is the solar constant and A is the atmospheric transmission coefficient. The solar constant has a value of 1350 W m^{-2} for the 400–3000 nm waveband and for photosynthetically useful wavelengths (400–700 nm) its value is 2.79 mmol m^{-2} s^{-1}. The atmospheric transmission coefficient is a function of air clarity and the thickness of the air column (which varies diurnally). Under clear skies at sea level, the atmospheric transmission coefficient is approximately 0.8. At higher elevations, it increases to approximately 0.85. Under light haze conditions, the atmospheric transmission coefficient decreases to 0.5–0.6.

For slopes with inclinations other than horizontal and with different azimuths, the intensity of solar radiation incident on those slopes (I_s) becomes

$$I_s = I_o\cos(i)/\sin(\alpha_s) \quad (7.13)$$

7.7 LEAF ABSORPTANCE AND ITS MEASUREMENT

Light is absorbed by, reflected from or transmitted through, a surface. The sum of these three fractional components is one. For leaves and other biological materials, we are often most interested in knowing what fraction of the radiation is absorbed. In the past these measurements have often been obtained by measuring leaf surface reflectance properties and assuming that transmittance through the leaf is neglible. This is incorrect and can lead to overestimates in leaf absorptance values since leaf transmittance between 400 and 700 nm can vary by 2–12% between species (Moss and Loomis, 1952).

Leaf absorptance is a measurement of the fraction of the incident photon or energy flux that is absorbed by the leaf. It can be measured at a single wavelength or over a specific waveband. When leaf absorptances are expressed for a particular waveband (such as between 400 and 700 nm), the irradiance source (such as sunlight) must be specified. Over the visible wavelengths (400–700 nm), the leaf absorptance to sunlight on either a photon or energy flux basis is nearly identical and is frequently used interchangeably (Ehleringer and Björkman, 1978; Gates, 1980). Leaf absorptances between 400 and 700 nm mentioned in this chapter have been measured on a photon basis unless otherwise noted. This is because of the rapidity and ease with which the leaf photon absorptance can be measured.

7.7.1 Measurement of leaf absorptances

Leaf absorptances (a) are measured with an integrating sphere. There are two basic types: the Ulbricht sphere (Rabideau *et al.*, 1946), in which the leaf sample is positioned inside the sphere during measurement (Fig. 7.4), and the Taylor sphere (Taylor, 1920), in which the sample is positioned at a port on the outside surface of the sphere during the measurement. A significant attribute of the Ulbricht sphere for leaf absorptance measurements is that only a single measurement needs to be made to obtain leaf absorptance. In an Ulbricht sphere, light enters the integrating sphere and strikes the leaf sample, which is suspended in the center of the sphere interior. All light not absorbed is either reflected off the leaf surface or is transmitted through the leaf. The inside walls of the integrating sphere are coated with magnesium oxide powder, which has the unusual property that its absolute reflectance is 0.99 at all wavelengths between 350 and 2000 nm (Kortüm, 1969). Occasionally, barium sulfate, which has an absolute reflectance of 0.90–0.95 between 350 and 2000 nm, is substituted for magnesium oxide (Kortüm, 1969). The intensity of the photon flux reflecting off the sphere walls, when the leaf is in the path of the incoming beam (I_{in}), is measured with either a silicon cell or a LI-COR Quantum Sensor. This value is then compared with the photon flux resulting when the sample has been pulled out of the beam (I_{out}). With the leaf out of the sample beam, light instead strikes a magnesium oxide reflectance standard. The leaf absorptance is then calculated as

$$a = 1 - I_{in}/I_{out} \tag{7.14}$$

In contrast, the Taylor integrating sphere requires two separate measurements. In the first measurement, light passes through the

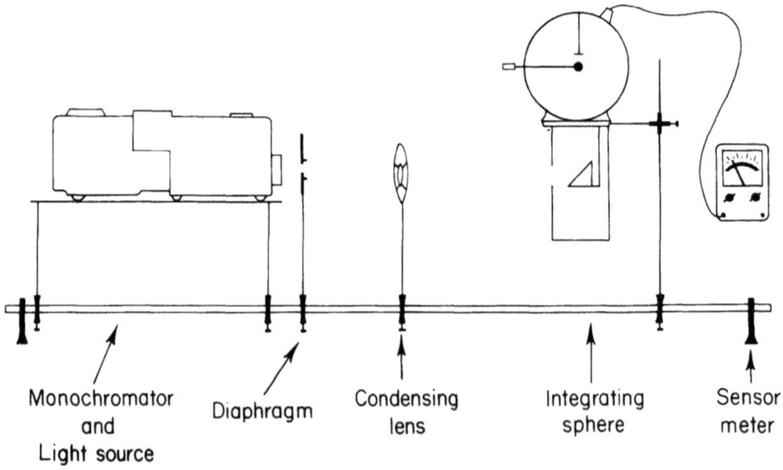

Fig. 7.4 Ulbricht integrating sphere (from Ehleringer, 1981).

sphere and strikes either the leaf sample or a reflectance standard. Light reflected back into the sphere is measured by a sensor and reflectance is then calculated as the ratio of the signal when light strikes the leaf to that when it reflects from the reference. To estimate transmittance the leaf is placed at the point where light would normally enter the sphere. The transmittance is then the ratio of the signal when the leaf is in the opening to when it is removed from the opening. The leaf absorptance is then calculated as one minus the reflectance minus the transmittance.

7.7.2 Visible versus total solar absorptance

Leaf absorptance to visible wavelengths (400–700 nm) is useful for understanding the fraction of incident photons that can be used in photosynthesis; these values are generally about 85% in most mature green leaves (Gates *et al.*, 1965; Gates, 1980; Ehleringer, 1981). In contrast, there is much less absorption of wavelengths beyond 700 nm, even though almost one half of the solar radiation at the earth's surface is in wavelengths between 700 and 3000 nm. Most of the solar radiation absorbed beyond 700 nm is due to five broad water absorption bands distributed between 900 and 3000 nm; only a small fraction of the solar radiation is absorbed by pigments such as phytochrome. As a consequence the leaf absorptance to solar radiation (400–3000 nm) is generally about 50%.

The integrated leaf absorptance to sunlight between 400 and 700 nm can be measured by integrating the leaf absorptance spectrum over these wavelengths with the solar radiation spectrum or it can be more easily measured by using sunlight as a light source into the integrating sphere and using a LI-COR Quantum Sensor (which measures only 400–700 nm photons) to determine absorptance. An even simpler approach was found by Ehleringer (1981), where it was shown that the leaf absorptance at 625 nm was equal to the integrated leaf absorptance to sunlight between 400 and 700 nm. This held true for a wide diversity of plant species with leaf absorptances ranging from 30 to 90%.

While the leaf absorptances to 400–700 nm are useful for photosynthetic studies, the leaf absorptance to total solar radiation is most useful for heat and energy balance studies. These two leaf absorptances are expected to be closely linked since leaves absorb primarily in the visible wavelengths. Ehleringer (1981) found from sampling different leaves over a large range of absorptances that the leaf absorptance to total solar radiation (%) was related to the visible waveband absorptance as

$$a_{400\text{–}3000} = 0.73a_{400\text{–}700} - 11.9 \quad (7.15)$$

7.7.3 Leaf absorptance data

The leaf absorptance will depend primarily on the epidermal characteristics and on leaf chlorophyll content. Except for developing and senescing leaves, it appears that leaf chlorophyll contents are sufficiently high that changes in leaf absorptance are largely due to

Table 7.6 Estimates of leaf absorptance to solar radiation by different species for the 400–700 nm waveband (useful for photosynthetic studies) and the 400–3000 nm waveband (useful for energy budget and heat balance studies). Data are from Birkebak and Birkebak (1964) and Ehleringer (1981, 1988).

Species	$a_{400\text{–}700}$	$a_{400\text{–}3000}$
Acer negundo	0.80	0.50
Acer saccharinum	–	0.49
Atriplex hymenelytra	0.66	0.39
Encelia farinosa	0.42	0.21
	0.63	0.35
	0.81	0.49
Fraxinus pennsylvanica	–	0.50
Geraea canescens	0.81	0.48
Malvastrum rotundifolium	0.83	0.46
Populus tremuloides	0.86	0.49
Quercus alba	–	0.49
Quercus rubra	–	0.49

variations in epidermal properties (Lin and Ehleringer, 1982, 1983). Table 7.6 lists leaf absorptances to sunlight for the 400–700 nm and 400–3000 nm wavebands for a number of species from diverse habitats. The differences in leaf absorptance are principally due to epidermal modifications such as pubescence and waxes.

7.8 BOUNDARY LAYER CONSIDERATIONS

Wind and its effects on leaf boundary layers have been discussed in Chapter 4. Suffice it to reiterate here that both wind speed and leaf size and shape will influence the magnitude of the leaf boundary layer and the resistances to heat and water vapor transfer across this layer of laminar air flow. The reader is referred to Chapter 4 for a complete description of wind and boundary layers and for values of the convection coefficients as a function of leaf size and shape.

REFERENCES

Birkebak, R.C. (1966) Heat transfer in biological systems. *Int. Rev. Gen. Exp. Zool.* **2**, 269–344.

Birkebak, R.C. and Birkebak, R. (1964) Solar radiation characteristics of tree leaves. *Ecology*, **45**, 646–9.

Campbell, G.S. (1977) *An Introduction to Environmental Biophysics*, Springer-Verlag, New York, 159 pp.

Collier, B.D., Cox, G.W., Johnson, A.W. and Miller, P.C. (1973) *Dynamic Ecology*, Prentice-Hall, Englewood Cliffs, 563 pp.

Comstock, J.P. and Mahall, B.E. (1985) Drought and changes in leaf orientation for two California chaparral shrubs: *Ceanothus megacarpus* and *Ceanothus crassifolius*. *Oecologia* **65**, 531–5.

Drake, B.G., Raschke, K. and Salisbury, F.B. (1970) Temperatures and transpiration resistances of *Xanthium* leaves as affected by air temperature, humidity, and wind speed. *Plant Physiol.*, **46**, 324–30.

Ehleringer, J.R. (1981) Leaf absorptances of Mohave and Sonoran Desert plants. *Oecologia*, **49**, 366–70.

Ehleringer, J.R. (1985) Annuals and perennials of warm deserts. In *Physiological Ecology of North American Plant Communities* (eds B.F. Chabot and H.A. Mooney), Chapman and Hall, New York, pp. 162–80.

Ehleringer, J.R. (1988) Changes in leaf characteristics in species along elevational gradients in the Wasatch Front, Utah. *Am. J. Bot.*, **75**, 680–89.

Ehleringer, J.R. and Björkman, O. (1978) Pubescence and leaf spectral characteristics in a desert shrub, *Encelia farinosa*. *Oecologia*, **36**, 151–62.

Ehleringer, J.R. and Forseth, I.N. (1980) Solar tracking by plants. *Science*, **210**, 1094–8.

Gates, D.M. (1962) *Energy Exchange in the Biosphere*, Harper and Row, New York. 151 pp.

Gates, D.M. (1980) *Biophysical Ecology*, Springer-Verlag, New York.

Gates, D.M., Keegan, H.J., Schleter, J.C. and Weidner, V.R. (1965) Spectral properties of plants. *Appl. Opt.*, **4**, 11–20.

Geiger, R. (1966) *The Climate near the Ground*, Harvard University Press, Cambridge. 611pp.

Hays, R.L. (1975) The thermal conductivity of leaves. *Planta* **125**, 281–7.

Holman, J.P. (1968) *Heat Transfer*. McGraw-Hill Book Co., New York, 401pp.

Idso, S.B., Jackson, R.D., Ehrler, W.L. and Mitchell, S.T. (1969) A method for determination of infrared emittance of leaves. *Ecology* **50**, 899–902.

Jones, H.G. (1983) *Plants and Microclimate*, Cambridge University Press, Cambridge. 323pp.

Kortüm, G. (1969) *Reflectance Spectroscopy*, Springer-Verlag, New York.

Lee, R. (1978) *Forest Microclimatology*, Columbia University Press, New York, 276pp.

Lin, Z.F. and Ehleringer, J.R. (1982) Changes in spectral properties of leaves as related to chlorophyll and age in papaya. *Photosynthetica*, **16**, 520–5.

Lin, Z.F. and Ehleringer, J.R. (1983) Epidermal effects on spectral properties of leaves of four herbaceous species. *Physiol. Plant.*, **59**, 91–4.

List, R.J. (1968) *Smithsonian Meteorological Tables*, 6th edn, Smithsonian Misc. Coll. 114., Smithsonian Press, Washington, 527pp.

Miller, P.C. (1971) Sampling to estimate mean leaf temperatures and transpiration rates in vegetation canopies. *Ecology*, **52**, 885–9.

Miller, P.C. (1972) Bioclimate. leaf temperature, and primary production in red mangrove canopies in south Florida. *Ecology*, **53**, 22–45.

Monteith, J.L. (1973) *Principles of Environmental Physics*, Edward Arnold, London, 241pp.

Moss, R.A. and Loomis, W.E. (1952) Absorption spectra of leaves. I. The visible spectrum. *Plant Physiol.*, **27**, 370–91.

Rabideau, G.S., French, C.S. and Holt, A.S. (1946) The absorption and reflection spectra of leaves, chloroplast suspensions, and chloroplast fragments as measured in an Ulbricht sphere. *Am. J. Bot.*, **33**, 769–77.

Rosenberg, N.J., Blad, B.L. and Verma, S.B. (1983) *Microclimate the Biological Environment*, 2nd edn, John Wiley and Sons, New York, 495pp.

Taylor, A.H. (1920) The measurement of diffuse reflection factors and a new absolute reflectometer. *J. Opt. Soc. Am.*, **4**, 9–23.

Wiebe, H.H. and Drake, B.G. (1980) Leaf temperature mapping with thermosensitive liquid crystal models. *BioScience* **30**, 32–3.

Weigand, C.L. and Swanson, W.A. (1973) Time constants for thermal equilibration of leaf, canopy, and soil surfaces with changes in insolation. *Agron. J.*, **65**, 722–4.

8

Measurement of transpiration and leaf conductance

Robert W. Pearcy, E.-Detlef Schulze and Reiner Zimmermann

8.1 INTRODUCTION

Measurements of leaf transpiration and calculations of leaf conductance to water vapor are important in almost all investigations of plant water relations. Transpiration is a primary determinant of leaf energy balance (Chapter 7) and plant water status (Chapter 9). Together with the exchange of CO_2 it determines the water use efficiency. The close linkage between CO_2 uptake and H_2O via the stomatal pore has allowed for separation of stomatal and biochemical limitations to photosynthesis through calculation of intercellular CO_2 concentrations. In this chapter we will cover the principles and instruments necessary for measurement of leaf transpiration and the calculation of leaf conductances to water vapor exchange. We will also consider the methodology and problems involved in determining whole-plant and canopy transpiration rates. Emphasis is placed on methods and equipment that have as their primary purpose, the direct measurement of transpiration rates or leaf conductance to water vapor loss. It should be noted that in many research problems, knowledge of both CO_2 and H_2O exchange are required. In the past, porometers that measure only leaf conductance to water vapor have sometimes been used to infer more general environmental response of gas exchange including CO_2 uptake. While a general correlation is expected, direct measurements of CO_2 exchange, which are much more feasible than even a few years ago, are clearly more appropriate. Field equipment designed for simultaneous measurements is covered primarily in Chapter 11. In this chapter we will, however, cover the water vapor sensors and the theory and procedures necessary to measure transpiration and calculate stomatal conductances in these systems.

8.2 LEAF TRANSPIRATION RATE

Measurements of transpiration and the calculation of water vapor conductances of single leaves are nearly always based on the measurement of vapor added by transpiration into the air inside a chamber enclosing the leaf or a leaf surface (Jarvis and Catsky, 1971). In an 'open' system, a flow of air passes through the chamber and the leaf transpiration rate is calculated from the difference in water vapor content of the air entering and leaving the chamber, the flow rate and the leaf area. A typical 'closed' system would have the leaf enclosed for a short period of time in a chamber along with a humidity sensor so that the rate of transpiration is a function of the rate of increase of humidity in the chamber. Closed systems, while useful for photosynthesis measurements (see Chapter 11), have limited applicability for determination of transpiration rates because the increasing air humidity causes a reduction in the vapor pressure gradient from the leaf to the air and, consequently, a reduction in the transpiration rate. This, combined with problems of water vapor adsorption to the chamber which will be discussed below, render them much less useful and reliable than open systems even though they require less instrumentation.

In an open system (Fig. 8.1), the leaf transpiration rate is equal to the additional amount of water vapor leaving the chamber above that entering. The water vapor added by the leaf via transpiration (E, mol m^{-2} s^{-1}) is given by:

$$EL = u_o w_o - u_e w_e \qquad [\text{mol s}^{-1}] \quad (8.1)$$

where w_e and w_o (mol mol^{-1}) are the mole fractions of water vapor (mol of water vapor per total mol of all gases) in the entering and outgoing air streams respectively, L (m^2) is the leaf area and u_e and u_o (mol s^{-1}) are the total molar flow rates (air plus water vapor) entering and leaving the chamber, respectively. If the flow meter is calibrated in volumetric or mass units, molar flow rate is obtained by applying the ideal gas law.

Using Dalton's law of partial pressures, which states that the mole fraction of a gas in a mixture is equal to its partial pressure, w_o and w_e can be calculated from measurements of vapor pressure (v_w, Pa) or relative humidity (R, %). If vapor pressure is measured then:

$$w = \frac{v_w}{P} \qquad [\text{Pa Pa}^{-1}] \quad (8.2)$$

where P is the total air pressure at the humidity sensor. If relative humidity is measured then:

$$w = R\left(\frac{v_{w,\,sat}}{P \times 100}\right) \qquad [\text{Pa Pa}^{-1}] \quad (8.3)$$

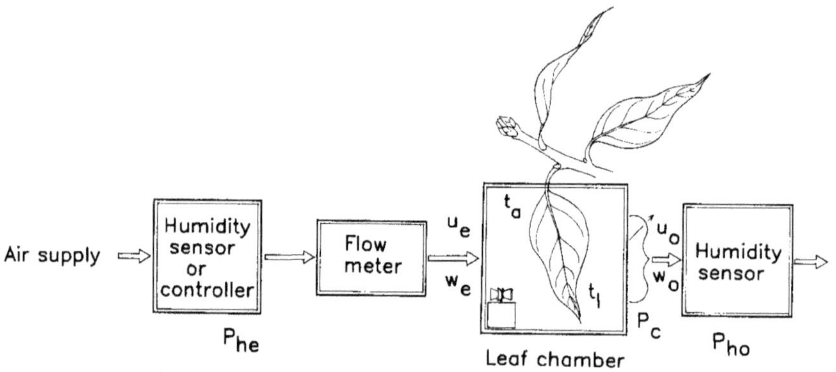

Fig. 8.1 Functional block diagram of an 'open' system for measurement of transpiration. The flow rate of air in mol s^{-1} into (u_e) and out of (u_o) the chamber, including the leaks and the mole fractions of water vapor entering (w_e) and exiting (w_o) the chamber must be known. Pressures must be known at the humidity sensors (P_{he}, P_{ho}) and in the chamber (P_c). Leaf (t_l) and air (t_a) temperatures in the chamber must be measured.

where $v_{w,sat}$ is the saturation water vapor pressure at the temperature of the humidity sensor. Air pressure can be determined by adding the pressure measured at the humidity sensors either with simple water manometers or electronic differential pressure sensors to the atmospheric pressure measured with a barometer (see Chapter 4). Alternatively, an absolute pressure transducer can be used. If the additional pressure in the system above atmospheric pressure does not exceed 0.5 kPa or so, it can be ignored, whereas atmospheric pressure varies considerably with elevation and weather.

Special consideration must be given to the determination of gas flow rates. Because of conservation of mass, the total outgoing flow from the chamber must exceed the entering flow by the amount of water vapor added by the leaf. Measurement of the flow downstream of the chamber is impractical because it is usually not possible to seal the chamber completely. Therefore the flow meter is mounted upstream of the chamber. Leaks in the chamber or after it are then usually of little consequence provided that the chamber atmosphere is well mixed so that the humidity measurement is a true sample of the air leaving the chamber by all routes, including the leaks. The total flow out of the chamber is then calculated from the entering flow plus the additional flow of water vapor. Since the difference in the flow is equal to the evaporation from the leaf, Equation 8.1 can be rewritten as

$$u_o - u_e = u_o w_o - u_e w_e \quad [\text{mol s}^{-1}] \quad (8.4)$$

and therefore with rearrangement,

$$u_o = u_e \left(\frac{1-w_e}{1-w_o} \right) \quad [\text{mol s}^{-1}] \quad (8.5)$$

Equation 8.5 can be substituted into Equation 8.1, with rearrangement, to yield the correct relationship for calculation of the transpiration rate of a leaf:

$$E = \frac{u_e(w_o - w_e)}{L(1-w_o)} \quad [\text{mol m}^{-2}\ \text{s}^{-1}] \quad (8.6)$$

Transpiration rates should be presented using SI conventions of units of mmol m^{-2} s^{-1} or by conversion to mass units of mg m^{-2} s^{-1}. Because we are frequently interested in comparing fluxes of water vapor and CO_2, molar units are generally preferable.

Leaves can have stomata on both sides and therefore transpire from either one or both surfaces. With broad leaves, however, the convention is to take the area of only one surface. With needles, stems or other surfaces there is no standard procedure and either the projected surface area, one half of the total area or the total surface area have been used. However, as long as the convention is clearly stated and followed consistently, the result will be comparable with other published work.

8.3 LEAF CONDUCTANCE TO WATER VAPOR

Transpiration rates measured in chambers are not useful parameters in themselves since they depend on properties of the particular chamber environment as well as those of the plant. It is usually very difficult to match the chamber environment to that of a leaf outside of the chamber closely enough so that the measured transpiration rates will be representative of the outside leaves. Thus a conductance of the leaf to water vapor loss is derived from the transpiration rate. Conductances are usually presented rather than their inverse, a resistance, since conductances are proportional to the flux and they express the regulatory control exerted by the stomata on transpiration rates. If transpiration rates representative of leaves in their natural environment are needed then it is necessary to determine leaf temperatures as well as the vapor pressure of the atmosphere and then recalculate a transpiration rate using the stomatal conductance determined in the chamber and a boundary layer conductance (see Section 8.3.3). Since stomata respond to the microenvironmental conditions, an important condition for the recalculation of natural transpiration rates is that the chamber conditions be sufficiently close to natural conditions that the stomatal conductances do not change when the leaf is enclosed or that the measurement be completed before the stomata have time to respond.

Conductance to water vapor loss (g_w) is derived from Ficks law of diffusion (see Nobel, 1984) and can be expressed in its simplest form as the proportionality between the rate of transpiration (E) and the driving force for evaporation, the gradient in water vapor from the intercellular spaces in the leaf to the atmosphere (ΔW)

$$g_w = \frac{E}{\Delta W} \qquad (8.7)$$

If ΔW is expressed as a mole fraction gradient (mol mol^{-1}) and E is in molar units (mol m^{-2} s^{-1}) then the units of g_w are mol m^{-2} s^{-1}. Formerly, ΔW was expressed as a concentration gradient (g m^{-2}) and E in mass units (g m^{-2} s^{-1}), giving g_w units of m s^{-1}. The relative merits of expressing g_w in molar units are discussed in Section 8.3.1.

The water vapor gradient, ΔW, can be determined from the difference in vapor pressure in the chamber and the saturation vapor pressure in the intercellular air spaces. In a well-stirred chamber, the mole fraction of water vapor in the chamber (w_a) is equal to that in the airstream exiting the chamber (w_o). If the chamber is not well stirred by a fan then it will be some average of the mole fractions in the ingoing and exiting air streams. Jarvis and Catsky (1971) discuss the various assumptions regarding mixing in the chamber and show that if the chamber is not mixed and if the change across the chamber is less than one-half of the entering mole fraction, then a linear average [$w_a = (w_o + w_e)/2$] gives a reasonable estimate. Since the air near the cell walls is assumed to be saturated with water vapor, the mole fraction in the intercellular spaces (w_i) is derived from the saturation vapor pressure at the leaf temperature divided by the total pressure (see Appendix Table A7). Thus,

$$\Delta W = w_i - w_o \qquad [\text{mol mol}^{-1}] \quad (8.8)$$

Recent experiments suggest that the assumption of water vapor saturation in the intercellular spaces near the cell walls is essentially valid for well-watered plants (Sharkey *et al.*, 1982; Mott and O'Leary, 1984). A water potential of −2 MPa at a leaf temperature of 25°C equilibrates with a relative humidity of 98.5% in the intercellular spaces and therefore results in only a very small change in w_i from that at 0 MPa. Water potential effects can become significant at a combination of high relative humidities and low water potentials.

Accurate leaf temperature measurements are critical since small errors can have more than proportional effects on the estimation of

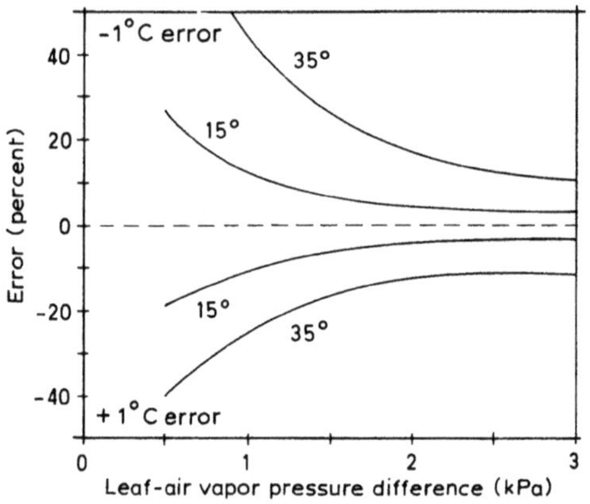

Fig. 8.2 The effect of a 1° C error in leaf temperature on the calculated conductance from measurements of transpiration in a leaf chamber. Errors are shown as a function of the true leaf–air vapor pressure difference at 15 and 35° C. The upper and lower lines are for a −1°C and +1°C error respectively.

ΔW, particularly if the measurements are performed at high humidities where ΔW is small. The problems that are frequently encountered are that temperature is not uniform across a leaf and is not sampled adequately, or that there is insufficient thermal contact between the sensor (usually a thermocouple) and the leaf. Fig. 8.2 shows the error in g_w resulting from a 1°C error in leaf temperature at different temperatures and vapor pressure deficits. Ideally, leaf temperatures should be measured to an accuracy of ±0.1°C but this is rarely achieved. Precautions for making accurate leaf temperature measurements are discussed in Chapter 7.

8.3.1 Conductance units

While conductances based on concentration differences and having units of mm s^{-1} are commonly seen, particularly in the older literature, Cowan (1977) derived an alternative formulation that has several distinct advantages and is receiving increasing use. If we apply Ficks law, expressing the concentration gradient as a finite difference and ignoring additional corrections for the time being (see Section 8.3.2), then:

$$E = D_{wa} \frac{\Delta\rho}{\Delta X} \tag{8.9}$$

where $\Delta\rho$ is the water vapor density gradient from the leaf to the air, ΔX is the effective length of the diffusion path (m) and D_{wa} is the binary diffusion coefficient of water vapor in air ($m^2\ s^{-1}$). It follows that

$$g_w = \frac{D_{wa}}{\Delta X} \tag{8.10}$$

According to Fuller *et al.* (1966) D_{wa} is dependent on temperature (T) and pressure (P) as follows

$$D_{wa} = D^0{}_{wa} \left(\frac{T}{T_0}\right)^{1.8} \left(\frac{P_0}{P}\right) \tag{8.11}$$

where $D^0{}_{wa}$ equals the diffusion coefficient at standard temperature (T_0; 273°C) and pressure (P_0; 101.3 kPa). Because of the temperature and pressure dependence of D_{wa} and also of water vapor concentration, conductance based on concentration gradients (g m^{-3}) are dependent on temperature and pressure. Thus, even if vapor pressure gradients and stomatal apertures and densities are identical for two leaves, the conductances will differ if the measurements are made at different pressures or temperatures. For example, it can be shown by substituting Equation 8.11 into Equation 8.9, and adjusting the water vapor concentrations using the gas law, that a temperature increase from 20 to 30°C will increase the calculated value of g by 6.3%. A change from sea level to 1000 m elevation will result in a 10% higher calculated value of g_w even with no change in stomatal aperture.

Mole fractions, however, are independent of temperature and pressure. By using the gas law to convert a concentration gradient to a mole fraction gradient, it can be shown that g_w is proportional to $T^{1.8}$ if concentrations are used and to $T^{0.8}$ if mole fractions are used (see Hall, 1982; Nobel, 1984). Consequently, the change from 20 to 30°C causes only a 2.8% increase in g calculated from the mole fraction gradient. Moreover, g_w expressed in molar units is independent of pressure. Molar units are easily used since E and g have the same dimension. Thus, ΔW expressed as a dimensionless mole fraction multiplied by g_w in molar units results in E having molar units. We recommend that molar units be used whenever possible. Conversion of conductances in units of cm s^{-1} to molar units is easily done, assuming isothermal conditions, if pressure and temperature are known since:

$$g_w(\text{mol m}^{-2}\ \text{s}^{-1}) = g_w(\text{cm s}^{-1})\ 0.446 \left[\frac{273}{(T+273)}\right] \left[\frac{P}{101.3}\right] \tag{8.12}$$

where temperature is in °C and pressure is in kPa. Appendix Table A9 gives conversion factors for a range of temperatures and pressures.

If conditions are not isothermal then it can be shown that the error is minimized by using leaf temperature (D. McDermitt, personal communication). However, even for a relatively large leaf–air temperature difference of 5°C within a chamber the error is less than 1–2%.

8.3.2 Calculation of total conductances to water vapor

Equation 8.7 is not strictly true because the evaporation rate through the stomatal pore is driven not only by diffusion but also by a small mass flow. When water evaporates in the intercellular spaces it displaces air causing a small mass flow through the stomatal pore. This flow carries with it a small amount of water vapor that must be added into the strictly diffusional movement of water vapor. In addition, the diffusional movements of water vapor, air and CO_2 can interact (see Jarman, 1974; von Caemmerer and Farquhar, 1981), slightly altering the diffusion of each. However, the effect of CO_2 diffusion on water vapor diffusion is negligible because of the much lower concentration of CO_2 than water vapor. But the additional transpiration occurring because of mass flow is significant and is equal to the mean water vapor mole fraction along the stomatal and boundary layer path times the evaporation rate. Thus, Equation 8.7 becomes:

$$E = g_{tw}(w_i - w_a) + E\left(\frac{w_i + w_a}{2}\right) \quad [\text{mol m}^{-2}\,\text{s}^{-1}] \quad (8.13)$$

Rearranging Equation 8.13 yields:

$$g_{tw} = \frac{E\{1+[(w_i+w_a)/2]\}}{w_i - w_a} \quad [\text{mol m}^{-2}\,\text{s}^{-1}] \quad (8.14)$$

The value of $(w_i + w_a)/2$, which is the mean water vapor concentration along the diffusion path from inside the leaf to the atmosphere, is typically 0.020 to 0.035 so that the correction is on the order of −2 to −3.5%. While it is small, the correction is easily made and should be done. These corrections are especially important for calculation of CO_2 conductances and intercellular CO_2 pressures (see Chapter 11).

8.3.3 Calculation of stomatal conductances

The g_{tw} calculated in Equation 8.14 is a total leaf conductance to water vapor and includes, in addition to the stomata, the pathway through the cuticle and the effect of the boundary layer. Thus, in order to calculate a stomatal conductance, the network of water loss pathways needs to be considered (Fig. 8.3). Conductances that are in parallel, such as those for two sides of a leaf or for stomata and the cuticle, are additive. However, for conductances that are in series, such as those for the stomata and boundary layer, the reciprocal of the total conductance is equal to the sum of the reciprocals of the conductances (= sum of the resistances).

Under most circumstances, the cuticular conductance, g_{cw}, can be ignored because it usually appears to be very small relative to stomatal conductance, g_{sw}, when the stomata are open. However, when the stomata are closed such as in low light or under conditions of water stress, a significant fraction of the transpiration may occur across the cuticle, leading to erroneous estimates of g_{sw}. These errors can be important when g_{sw} is used to calculate intercellular CO_2 pressures (see Chapter 11). Unfortunately, there is no readily apparent way to evaluate g_{cw} independent of the stomata. While g_{cw} can be measured for the abaxial surface of a hypostomatous leaf, there is no easy way of knowing if this value would apply to the abaxial cuticle as well. Similarly, g_{lw} measured in the dark may be close to g_{cw} but one cannot be sure that the stomata were fully closed.

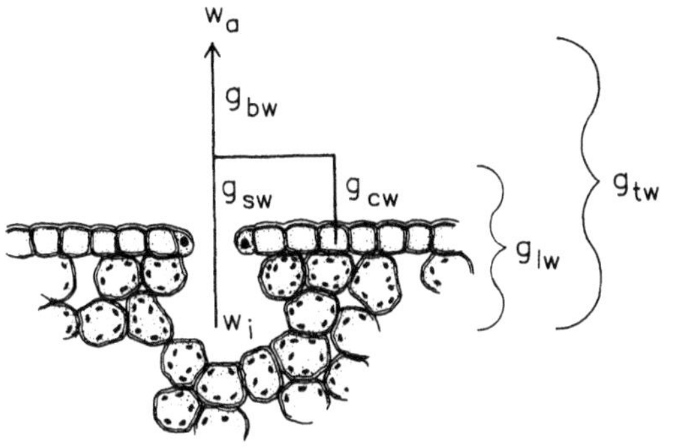

Parallel:

$$g_{lw} = g_{sw} + g_{cw}$$

$$g_{sw} = g_{lw} \text{ if } g_{sw} \gg g_{cw}$$

Series:

$$\frac{1}{g_{sw}} = \frac{1}{g_{tw}} - \frac{1}{g_{bw}} \quad \text{or:}$$

$$g_{sw} = \frac{g_{tw}\, g_{bw}}{g_{bw} - g_{tw}}$$

Fig. 8.3 Network of conductances at the leaf surface showing the rules for simplifying complex networks to calculate the stomatal conductance to water vapor.

The boundary layer conductance, g_{bw}, is an important component in the path for water vapor loss and must be included in any analysis. In a well-stirred chamber, g_{bw} is usually much larger than g_{sw} so that the primary component in the total conductance is g_{sw}. The boundary layer conductance can be evaluated using wet filter paper replicas of the leaf (see Chapter 4) and after taking the reciprocals, subtracted to yield $1/g_{sw}$. Because of the series path through the stomata and boundary layer, the gas exchange through the two sides of a leaf should ideally be measured independently in a double-sided chamber. However, if g_{bw} is large relative to g_{sw}, then little error results by calculating an average conductance from measurements where both surfaces are enclosed within a single chamber (Jones, 1983).

8.4 INSTRUMENTATION FOR TRANSPIRATION MEASUREMENTS

In order to measure transpiration and calculate conductances, accurate determinations of vapor pressures, air flow and leaf temperatures are necessary. The measurement of leaf temperatures is covered in Chapter 7 and air flow in Chapter 11. Consequently, only humidity sensors will be covered here.

There is a large variety of humidity sensors available and many types have been used with varying degrees of success in measurements of transpiration. Most humidity sensors have been developed for industrial applications of sensing humidity levels but not for flux rate measurements where accuracy and repeatability requirements are more demanding. Some other highly accurate techniques function over only narrow humidity ranges. Thus, of the many techniques available for measuring humidity, only a few are appropriate for use in measuring transpiration under field conditions. Among those developed for industrial applications are lithium chloride dew cells, Dunmore cells and carbon resistance elements. All of these provide an electrical output that is a function of relative humidity or dew point of the air but none possesses the stability and accuracy necessary for transpiration measurements. They tend to be slow, quite temperature dependent and exhibit considerable hysteresis. Psychrometric systems based on wet- and dry-bulb thermocouples have also been used under laboratory conditions (Slatyer and Bierhuizen, 1964), but these are not as reliable or as convenient as other sensors that are now available. Only infrared water vapor analyzers, dew-point mirrors and thin-film capacitance-type sensors are presently used in transpiration measurements.

8.4.1 Infrared gas analyzers

Infrared gas analyzers are typically used in ecophysiology for measurement of CO_2 but can be made with a water vapor detector. In contrast to the CO_2-filled luft-type detector found in most CO_2 analyzers, those in water vapor analyzers are filled with ammonia, which has infrared absorption bands that overlap those of water. Infrared water vapor analyzers suitable for transpiration measurements are manufactured by Leybold-Hereaus (West Germany) and Analytical Development Co. (Great Britain). Infrared water vapor analyzers provide one of the most accurate and sensitive methods for measuring either absolute water vapor pressures or the difference in vapor pressure between two gases. The disadvantages are that they are expensive and bulky and require relatively large amounts of power. Consequently, they are best suited for transportable systems but not truly portable systems. The only analyzers specifically suited for field measurements are the Binos analyzers from Leybold-Hereaus (West Germany), which can be operated from battery power, are relatively insensitive to vibration, and are available in versions for simultaneously measuring water vapor and CO_2 pressures. Analyzers for measuring absolute water vapor pressure are suitable when transpiration rates are moderate to high. With low transpiration rates, such as those occurring in many conifers or in desert plants under water stress, differential analyzers that can be used to measure the difference in water vapor pressure of the ingoing and outgoing air steam of the chamber may be preferred. Differential analyzers are more difficult to calibrate since it is necessary to establish known differences in water vapor concentration at several different reference concentrations. Leybold-Hereaus now manufactures an analyzer having split cells (model 452) that is capable of both differential and absolute water vapor and CO_2 pressure measurements. This model, although expensive, gives maximum flexibility and convenience by combining all gas measurements in a single instrument.

8.4.2 Dew-point mirrors

Dew-point mirrors operate by chilling a mirror with a small Peltier unit (a thermoelectric heat pump) until water vapor from the air above it condenses. The point of condensation is detected with a small light source (usually a light-emitting diode) and a photodetector (Fig. 8.4). The condensed water scatters the light so that less light reaches the photodetector than when no condensation is on the mirror. A feedback control circuit from the photodetector to the Peltier power supply controls the temperature of the mirror such that water vapor is just condensing. At this point the mirror temperature, which is measured with a thermistor or a platinum resistance thermometer, is equal to the dew-point temperature of the surrounding air. Dew point mirrors can resolve dew points of ±0.1°C with an accuracy of ±0.2°C. The major limitations are the tendency for the mirror to become contaminated by aerosols, which shift the condensation point to higher temperatures, and the relatively slow response. Some systems include detection and correction circuits that eliminate the effect of optical contamination by dust, etc. However, these circuits will not correct errors due to deposited salts or organic compounds that alter the vapor pressure of the condensed water. A platinum resistance thermometer (PRT) is the preferred sensor for measuring

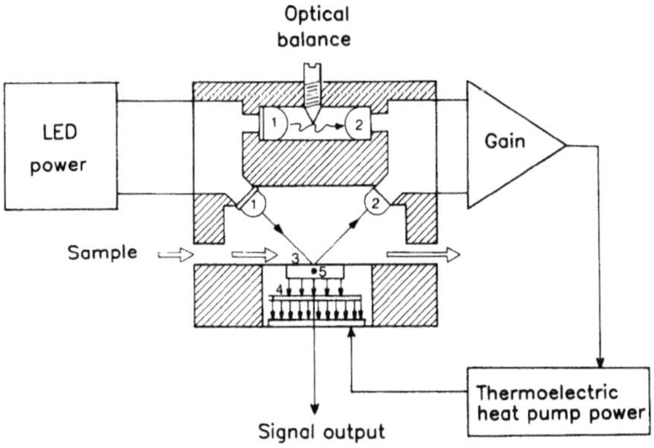

1 Light emitting diode (LED)
2 Photodiode
3 Mirror
4 Peltier heat pump
5 Platinum resistance thermometer

Fig. 8.4 Diagram showing the major components of a dew-point mirror. Adapted from literature from General Eastern.

the mirror temperature because of its accuracy and stability.

Dew-point mirror systems suitable for field use in transportable gas-exchange systems are available from H. Walz (model STR14) in West Germany and General Eastern (model Dew 10) in the United States. The Dew 10 sensor is manufactured primarily for industrial humidity control applications but has proven useful for transpiration measurements both in the laboratory and field. Both are quite compact and easily adapted for measurements in gas streams associated with leaf chambers. Both require 24 V power at a maximum of 300 mA. Both use PRT temperature sensors and for the most accurate measurements, a three- or four-wire resistance measuring circuit is required. One particular advantage is that dew-point mirrors can function at the high humidities characteristic of environments such as tropical or boreal forests where other sensors are often less reliable. But dew-point mirrors require careful maintenance if they are used to measure the dew-point differences of ingoing and outgoing air of a chamber with sufficient accuracy.

8.4.3 Thin-film capacitance sensors

These sensors consist of a thin hygroscopic polymer film separating two metal electrodes that are deposited on a thin glass chip as support. The first electrode is deposited on the glass followed by the 1 μm thick polymer film. The second electrode is then deposited on to the film (Fig. 8.5). This second electrode is sufficiently thin (approx. 0.02 μm) that water molecules can diffuse through it and enter the pores of the film, changing the electrical capacitance of the sensor (Salasmaa and Kostamo, 1975, 1986). The change in capacitance is measured by exciting the sensor and a reference capacitor with a high-frequency AC voltage. A demodulator circuit then produces a voltage proportional to the change in capacitance (humidity). By careful adjustment of the properties of the polymer and electrodes, these sensors can be made to respond nearly linearly from 10 to 80% relative humidity. The temperature coefficient is about 0.05% $°C^{-1}$ and the time constant is about 1 s at 25°C. It is important that they be excited at high frequencies, which was not done in some early circuits (Bingham *et al.*, 1980), since at lower frequencies the response is nonlinear and more temperature dependent. Above 80% relative humidity thin-film capacitance-type sensors typically exhibit hysteresis because additional water vapor is slowly absorbed and causes swelling of the film, resulting in an output that changes for hours. Upon returning the sensor to lower

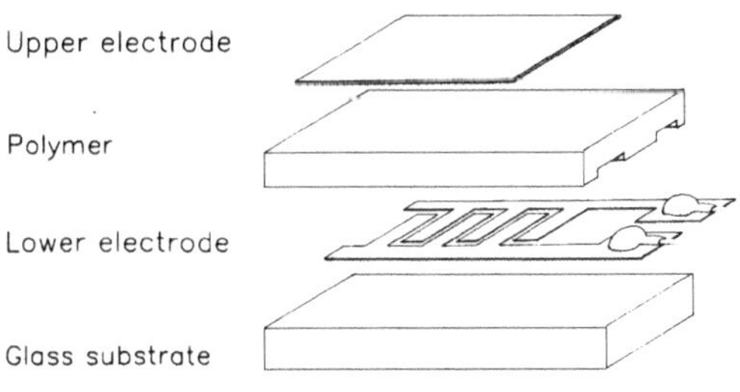

Fig. 8.5 Exploded diagram of a thin-film capacitance sensor (Vaisala Humicap).

humidities, the additional water vapor is only slowly lost. Some of the newer versions of the sensor have improved response at high humidities achieved by changing the polymer characteristics but the hysteresis has not yet been entirely eliminated. This is one of the most serious limitations to the operation of portable porometers utilizing these sensors in humid environments such as tropical forests. Thin-film sensors can be damaged by contaminants such as sulfur dioxide which oxidize the electrodes.

Thin-film capacitance sensors are inexpensive and compact, and are easily mounted within a porometer chamber. They are available from several companies. Those manufactured by Vaisala in Finland and by Coreci in France have been used as humidity sensors in transpiration measurements but others are probably suitable as well. Tables comparing the various sensors have been compiled by Schurer (1986). Thin-film capacitance sensors are probably the sensor of choice for porometers where true portability is a priority. Provided that the sensor is not exposed to high humidity, the sensitivity is very stable. There is, however, a tendency for the offset to drift but this can be corrected for by periodically comparing the measured output to a known humidity using, for example, one of the simple dew-point columns (see Section 8.5.2). Any offset drift can then be added or subtracted from the signal to give the true humidity.

With some precautions, thin-film capacitance sensors can also function well in transportable gas-exchange apparatus. In general they are satisfactory in null-balance instruments. For differential measurements of ingoing and outgoing air it is necessary to thermostat the sensors and to provide a system for conveniently checking the zero against air with a known water vapor pressure obtained, for example, from a thermostated condenser column. Thermostating can be achieved by mounting the sensors in a temperature-controlled block (Fig. 8.6). The air stream and the sensor temperature can thus be raised so that relative humidity never exceeds 60%, avoiding problems with hys-

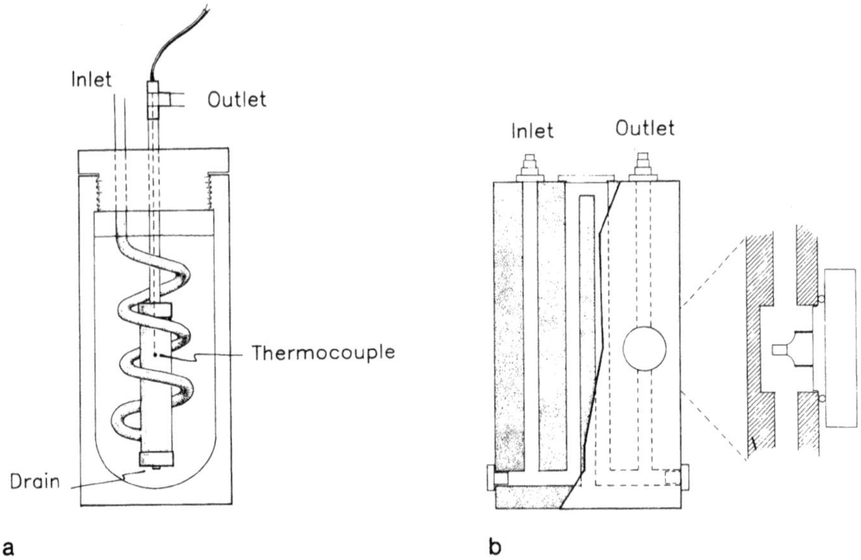

Fig. 8.6 (a) A calibration system for humidity sensors using a simple copper tubing column in an insulated flask of the type used to hold hot drinks. (b) Diagram of a thermostated block designed to hold a capacitance humidity sensor. Thermostating can be achieved with a heater pad pressed to the block and a temperature controller.

teresis. However, the bulk and cost of such additions reduce the advantages of having a small inexpensive sensor.

8.5 CALIBRATION OF WATER VAPOR SENSORS

Calibration of humidity sensors requires known humidities to be established. Two general approaches are available and adaptable for field use: (1) obtain equilibrium humidities over saturated salt solutions or hydrated salts or (2) condensation of humidified air to a known dew point. Additionally, other humidities can be obtained by mixing dry air with air of known humidity using mass flow controllers (see Chapter 11).

8.5.1 Salt solutions and hydrated salts

Saturated salt solutions, made by adding enough of the salt to distilled water so that crystals remain in the bottom of the container, can be used to obtain a wide range of relative humidities. Table A13 in the Appendix gives the relative humidities and vapor pressures as a function of temperature for a series of salts covering nearly the full range of relative humidities. The temperature coefficient of the equilibrium relative humidity is quite small in most cases but temperature gradients or changes are important because they prevent establishment of a true equilibrium. Any impurities in either the salt or the water will give an incorrect relative humidity. The major disadvantage of this technique is that the container must be closed and thermally insulated in order to establish an equilibrium. Thus, either the sensor must be removed from the instrument and sealed into the container or the container sealed on to the porometer. After sealing, at least 5–10 min are required to establish the equilibrium.

Hydrated salts may also be used to establish known vapor pressures in an air stream. The water vapor pressure in equilibrium with a salt hydrate pair is a function of the nature of the hydrate pair and the temperature, but not of the relative proportions of each hydrate. Parkinson and Day (1981) have used this principle to develop a calibration system in which the salt is ferrous sulfate heptahydrate ($FeSO_4.7H_20$). With dehydration it undergoes a transition to $FeSO_4.4H_2O$ and, if allowed to come to equilibrium, establishes a dew-point temperature given by:

$$T_d = 1.134\ T_s - 11.6 \qquad [°C] \qquad (8.15)$$

where T_d is the dew-point temperature and T_s is the temperature of the salt.

A calibration system utilizing this principle is commercially available from the Analytical Development Company (UK; model WG600) but requires line power for heating the columns, limiting its usefulness under field conditions. A calibration system can be constructed by filling glass or plastic columns with $FeSO_4.7H_20$ and then passing an air stream through the columns. The air stream entering the column should be drier than the equilibrium dew point in order to establish the correct equilibrium. Moreover, the column volume needs to be sufficiently large and the flow slow enough that both thermal and vapor pressure equilibriums are established. The temperature can be measured with a thermocouple or a thermometer embedded in the salt crystals of the last column. The main disadvantages of this method are that care must be taken to assure that a vapor pressure equilibrium is established and only one vapor pressure can readily be obtained without fairly elaborate systems for changing temperature. However, they can be used in conjunction with gas-mixing systems to obtain different vapor pressures.

8.5.2 Condensation techniques

The second approach to establishing humidities for calibration is by condensing water vapor out of a saturated air stream to a known dew point. The vapor pressure in the air stream can then be calculated with sufficient accuracy from the relationship:

$$v_w = \exp(52.57633 - 6790.4985/T - 5.02808\ln(T)) \quad [\mathrm{kPa}] \quad (8.16)$$

where T (K) is the dew-point temperature and v_w is the water vapor pressure in kPa (Campbell, 1977). If a calibration of a relative humidity sensor is desired then it is also necessary to measure the temperature at the sensor and to calculate the saturation vapor pressure at this temperature. Also, pressure measurements must be made if there is more than a 0.5 kPa or so difference in the system or if it is significantly above atmospheric pressure. The humidifier can be a gas washing bottle or a column lined with water-saturated filter paper. The temperature of the humidifier must clearly be greater than that of the column. Because of the evaporative cooling in the humidifier, it is often better to use several in series so that the last one remains warm. Condenser systems incorporating thermoelectric cooling to set the dew-point temperature are available from H. Walz (West Germany) and W. Gries (West Germany). A simple but functional design is discussed in Chapter 11. An even simpler column which is suitable for calibrations can be constructed from copper tubing immersed in an insulated flask containing water of the desired dew-point temperature (Fig. 8.6). Heat transfer between the copper and the air is quite efficient so that only a short length of tubing is adequate for low to moderate flows. By adding water of different temperatures, different dew points can be obtained. Temperatures in the flask change only very slowly so that over the time required for the calibration, the dew point is essentially constant. The water should be agitated periodically to prevent development of any thermal gradients. The dew-point temperature can be measured with a thermocouple placed in the outlet below the water level in the thermos. The flows should be kept low so that pressure drops are small, otherwise, the pressure must be measured with either a simple water manometer or an electronic pressure transducer and the humidity corrected for the change in pressure. While this condensation system is very simple, it functions remarkably well and is essentially trouble-free. The only difficulty is that if drops of water inadvertently get into the outlet tubing, which can occur if too much condensate collects, they will obviously add water vapor.

8.6 SYSTEMS FOR MEASURING TRANSPIRATION AND LEAF CONDUCTANCE

The most widely used instruments for measuring stomatal conductance are diffusion porometers. In principle, all diffusion porometers measure transpiration which is then used to derive a value for the stomatal conductance. Three general types of porometers are in wide use: (1) transient porometers which measure the rate of humidity increase within a closed chamber attached to the leaf, (2) null-balance porometers where the humidity is held constant by matching a flow of dry air into the chamber so that it balances the water vapor being lost from the leaf and (3) constant-flow porometers where the steady-state increase in vapor pressure occurring after the leaf is enclosed in the chamber is measured.

8.6.1 Transient porometers

The earliest porometers were of the transient type which was first developed by Wallihan (1964). Various modifications to the basic design have been made over the years (Kanemasu *et al.*, 1969; Byrne *et al.*, 1970; Turner and Parlange, 1970; Stiles *et al.*, 1970) but the principle of operation has remained essentially the same. A small chamber containing a humidity sensor is clamped on to the leaf surface and either the time required for an increase in humidity between two preselected levels or the change in humidity for a given time interval is determined. Unstirred porometers relying on diffusion of water vapor from the leaf to the sensor are dependent on the geometry of the chamber, sensor and leaf. They cannot be used with conifers or with leaves that do not completely cover the chamber aperture. Ventilated transient porometers such as those developed by Kaufmann and Eckard (1978) and Körner and Cernusca (1976) can be used with a wider variety of plants but are more difficult to calibrate.

While it is theoretically possible to calculate a stomatal conductance from the rate of increase in humidity in a transient porometer, the response time of the sensors and absorption of water vapor within the chamber necessitate an empirical calibration. Typically a flat plate containing fields of pores and backed with wet filter paper is used in place of the leaf. The diffusion resistance of the pore fields can be calculated from the size and number of pores. Since water vapor absorption and the sensor characteristics are dependent on temperature, calibrations must be carried out over a range of temperatures similar to that present when the porometer is to be used. During the calibration, care must be taken so that the filter paper is not so wet that water is squeezed into the pores or so dry that an additional resistance is present. Ventilated porometers cannot be calibrated in this manner because strictly diffusional conditions may not hold within the pores. Instead, a wet surface such as scintered glass plate or filter paper connected to a glass capillary can be used to mimic transpiration (Körner and Cernuska, 1976). Movement of a bubble along the capillary can be used to calculate the rate of evaporation into the chamber. Transient porometers are relatively simple and comparatively inexpensive, but the laborious calibration, temperature response and susceptibility to drift and errors associated with water vapor sorption cause many problems. Certainly, the earlier and simpler designs are not at all reliable and should be avoided. The design of Stiles *et al.* (1970) overcomes some of the limitations and an improved version is manufactured by Delta T Devices (model Mk II) in Great Britain. The Delta T Device model Mk II porometer has a small molded polypropylene chamber enclosing a capacitance-type humidity sensor which is clamped on to the leaf surface. The humidity is automatically cycled over preselected ranges via silica gel desiccant and a pump, which is turned on to lower the humidity and then off to allow the humidity to rise because of transpiration. The time for the humidity to increase between two preselected levels (the transit time) is measured and displayed. Generally four to five cycles, each lasting typically less than 5 s are required to obtain repeatable transit times. The improvements in performance of this porometer over earlier transient porometers result from the cycling until repeatable transit times are achieved, the use of polypropylene and other materials to minimize vapor sorption, and the use of a capacitance-type humidity sensor.

8.6.2 Null-balance porometers

The null-balance porometer, first developed by Beardsell *et al.* (1972), uses a flow of dry air through the chamber to balance the transpiration rate of the leaves and therefore maintain a steady-state humidity in the chamber (Fig. 8.7). Because the humidity of the entering air is known (usually 0% or very close to 0% relative humidity) it can be shown that under isothermal conditions, determination of g_w requires only a humidity measurement of the air within the chamber, the flow rate of dry air entering the chamber and the leaf area. In practice, however, conditions are not isothermal so that leaf and air temperatures must also be measured.

The null-balance porometer has several inherent advantages over transient porometers. Since a steady-state condition can be achieved at a humidity similar in most cases to ambient levels, errors due to water vapor

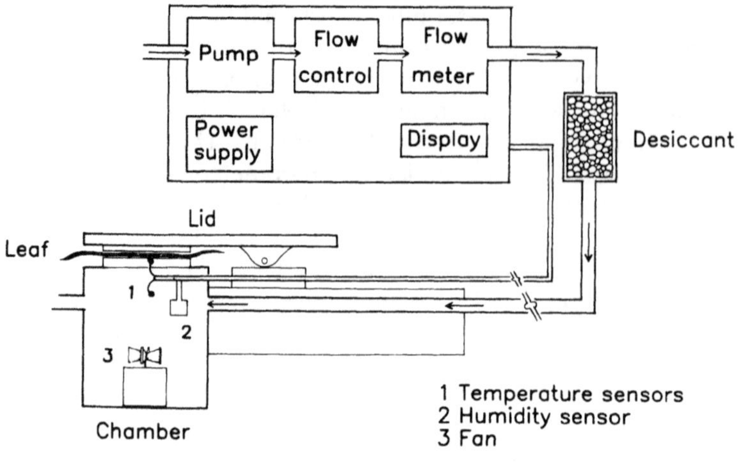

Fig. 8.7 Block diagram showing the components of a null-balance porometer.

sorption can be minimized. The calibration is also much easier since the humidity sensor can be calibrated at constant humidity levels over saturated salt solutions or with air of a known dew point. The flow meter can also be fairly easily calibrated (see Chapter 13), or at least checked for accuracy.

A commercially available null-balance porometer (model LI-1600) is available from LI-COR Inc. (USA). This instrument utilizes a capacitance-type (Vaisala) sensor mounted in a ventilated chamber. The apertures are interchangeable so a wide variety of leaf shapes and sizes can be accommodated. For broad leaves, the surface not enclosed by the porometer is uncovered, which facilitates energy exchange and avoids over temperatures (Schulze *et al.*, 1982). An external shroud with a second fan helps to maintain the chamber temperature close to ambient air temperatures. A flow of dry air is supplied to the chamber via a pump and desiccant tubes containing silica gel, and the flow rate is measured with an electronic mass flow meter. An electronic bleed valve system connected to a feedback circuit regulates the flow into the chamber so that steady-state humidity is maintained at a level preselected by the operator. In practice, the ambient humidity is usually measured with an open chamber and selected as the set point. Other humidities can also be selected if necessary provided that the instrument is conditioned for a while at this humidity. Manual adjustment is only required to bring the flow into the range where the automatic valve can control properly. A microprocessor-based data-acquisition system within the control console measures the signals from the leaf and air temperature thermocouples, the flow meter and humidity sensor and provides a readout of stomatal conductance on an LCD display. Measurements of stomatal conductance generally require between 15 s and 1 min with this porometer. A rechargeable battery allows 5–6 h of continuous use under field conditions.

Transpiration measurements in the LI-COR 6200 photosynthesis system (see Chapter 11 for a full description) are intermediate between a transient and a null-balance system. While the flow in this instrument can be set so that the relative humidity is maintained constant, in practice, it usually increases or decreases. Thus the calculations are based both on the transient rate of increase in the relative humidity and the flows out of the chamber to the CO_2 analyzer.

8.6.3 Constant-flow porometers

In contrast to the null-balance porometers where the flow is varied to achieve a compensation of the vapor added by transpiration, several porometers have been developed that maintain a constant flow of dry gas through the chamber (Parkinson and Legg, 1972; Day, 1977). After a steady state has been achieved, the resulting humidity is measured. The major disadvantage of this type of porometer is that the leaf and chamber may be exposed to quite different humidities from the ambient levels, depending on the flow through the system. If the change in vapor pressure is large, absorption or desorption of water may occur, leading to errors in the conductance measurements, or stomata may respond to the changed vapor pressure.

Recently, porometers utilizing a constant-flow principle for the simultaneous measurement of CO_2 and H_2O exchange have become available in which a constant flow rate of the ambient air is maintained through the chamber during the measurement, and, for transpiration, the difference in vapor pressure of the entering and exiting air is measured (Fig. 8.8). Flows through the chamber must be kept fairly high so that there is not an excessive depletion of CO_2 or increase in vapor pressure. Consequently, the vapor pressure difference between the reference and sample is kept small and an infrared analyzer is required in order to obtain sufficient accuracy. A CO_2/H_2O porometer of this type was developed by Schulze *et al.* (1982) and is available commercially from H. Walz (West Germany). The chamber is similar to that of the LI-COR 1600 porometer, but an air stream of constant flow is passed through the chamber and monitored with a multichannel gas analyzer (Binos). The thermo-

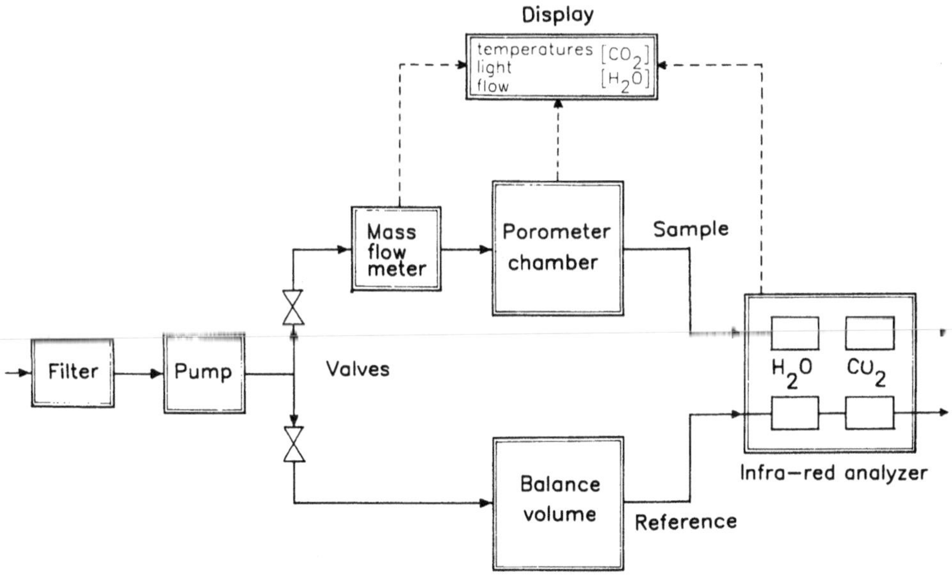

Fig. 8.8 Block diagram showing the components of a constant-flow porometer.

couple and Vaisala humidity sensor in the chamber determine the leaf–air vapor pressure gradient. A similar system manufactured by W. Gries (West Germany) uses a four-channel Binos gas analyzer to obtain the absolute and differential CO_2 and H_2O pressures and incorporates a Sharp computer for data processing. Both systems can be equipped with temperature-controlled cuvettes, CO_2-gas mixing systems and vapor condensors, and thus have considerable flexibility. They have the great advantage of providing both CO_2 and H_2O exchange measurements, which for many studies is essential. The disadvantages are that they are quite expensive and considerably less portable than, for example, a LI-COR 1600 porometer. Details of systems designed to measure both CO_2 and H_2O exchange are given in Chapter 11.

8.6.4 Comparisons of porometer types

Null-balance porometers are more complex and therefore more expensive than transient porometers. Constant-flow porometers are less complex than null-balance porometers but the requirement for an infrared analyzer for the CO_2/H_2O porometer makes it considerably more expensive. However, constant-flow porometers are also less portable than either transient or null-balance porometers. Both null-balance and constant-flow porometers are in effect equivalent to a leaf chamber and the calculated conductances in both are based on mole fraction gradients of water vapor and the measured transpiration rates. Thus, the estimates of gas fluxes and conductances are derived from first principles. Conductance estimates with the transient porometer are based on a comparison with the calculated conductance of the calibration plate. With care this can provide a standard but it may not be directly comparable with diffusion through stomata.

Null-balance porometers function best at high conductances where the balancing flow is high whereas for transient porometers, the time response of the sensor, timing errors and failure to establish uniform humidity gradients within the chamber cause them to be less accurate at high conductances. Because of the greater inherent accuracy of the infrared analyzer as compared with thin-film capacitance sensors used in null-balance porometers, continuous-flow porometers are potentially the most accurate. In null-balance and continuous flow porometers, the error in the humidity measurement affects both the estimate of the vapor pressure gradient and the transpiration rate. For conductance measurements the effect of the error enters twice in the same direction into the calculations. An error analysis of the equations used to calculate g_w for a null-balance porometer shows that a 1% humidity error gives a 4% error in g_w at a relative humidity of 50% (Campbell, 1975). The error increases dramatically at humidities below 20% because of error in the transpiration term and above 80% because of error in the gradient. If the real error in a humidity sensor is 4%, which is possible because of drift in thin-film capacitance detectors that are not frequently calibrated, then the error in g_w would be 16%. Errors in transient porometers are more difficult to specify because it depends on the calibration standard; where the calibration is based on a measured evaporation rate using for example a microcapillary pipette, a 1% humidity error has less than a 2% effect on the calculated g_w below 50% relative humidity. Above 80% relative humidity the error rises dramatically. In practical terms, however, the slight advantage of dynamic over null-balance porometers with regard to dependence on the humidity sensor accuracy is outweighed by other errors introduced in the calibration. The use of porometers at high humidities (i.e. greater than 75% as is often found in tropical or boreal forests) presents special problems. The potential errors shown above are compounded by poor performance of thin-film capacitance sensors at high humidities. In these circumstances the constant flow porometer has inherent advantages, although it should be recognized that all conductance estimates are subject to more error under these conditions because of the low rates of transpiration and low vapor pressure gradients. If a null-balance porometer such as the LI-COR 1600 is to be used in high-humidity environments, then it is best to store the head at lower humidity and use a lower humidity set point during the determinations.

One of the most critical factors in all porometers is temperature. The calibration of transient porometers assumes equal leaf and air temperatures; if they are not, the errors can be substantial (Morrow and Slatyer, 1971). Unfortunately, the leaf temperature thermistor is typically mounted in the foam gasket surrounding the aperture away from the evaporating leaf surface. Consequently, it is very difficult to tell if the leaf and air temperatures are the same. Leaf and air temperatures are measured separately in the LI-COR null-balance porometer. In all porometers, a changing temperature after closure will result in absorption or desorption of water vapor in the chamber that will be added to or subtracted from the true transpiration rate. The problem is minimized by using materials that have low water absorbances (Bloom *et al.*, 1980; Dixon and Grace, 1982; also, see Chapter 11) but it cannot be eliminated. If the porometer is moved from a cool to a warm location or into a growth chamber just prior to use, significant errors could result. Also, in a null-balance porometer such as the LI-COR 1600, an increasing air temperature causes a decreasing relative humidity and consequently causes the flow to decrease as the controlling system attempts to adjust relative humidity back to the set point. Because flow is the primary input for all rate calculations, conductances will be underestimated. These problems are most severe for low conductances. The best way to minimize temperature problems is to keep the poromoter in the shade between measurements so that it will be close to air temperature. Because of the differential water vapor measurements made in the constant-flow porometer, it should be less subject to these temperature errors.

8.7 WHOLE-PLANT MEASUREMENTS OF TRANSPIRATION

8.7.1 Extrapolation from porometry

One of the most tempting approaches to estimating whole-plant transpiration is to extrapolate from porometer measurements on single leaves to the canopy. It must be made quite clear, however, that this is not easily accomplished. Mainly because of boundary layer conditions, transpiration rates measured in a porometer will be different from that of a leaf transpiring under undisturbed natural conditions. Consequently, the only completely valid approach is to calculate transpiration from conductances measured with a porometer and vapor pressure gradients determined independently from leaf temperatures and the ambient vapor pressure of leaves in their natural position. Boundary layer conductances should be determined (see Chapter 4) and reciprocally added to the stomatal conductances, since they will be quite different for leaves in the chamber as compared to natural conditions, especially for large leaves. This requires estimates of wind speed within the canopy. In addition, stomatal conductances, leaf temperatures, vapor pressures and wind speeds must be known in different canopy layers. The problem is less acute for needle-leaved species, such as conifers, than for broad-leaved plants because of the high boundary-layer conductances and close coupling to air temperature. The extrapolation of porometer transpiration data may be justified when: (1) the leaves are small, (2) the aerodynamic (canopy) boundary layer resistance is low (Jarvis, 1985) so that vapor pressure and wind gradients are small, and (3) the leaf area distribution among the different vapor pressure regimes in the canopy are known. Under these special conditions an extrapolation was successfully made for *Picea* and *Larix* by Schulze *et al.* (1985), but this may not be possible for conditions of lower canopy roughness (Jarvis, 1985).

Even when the extrapolation is possible, the requirement for sufficient sampling dictates that measurements will be achievable for only one or a very limited set of conditions. Thus extrapolation to daily or seasonal transpiration requires simulation modeling. Models incorporating stomatal responses to environmental variables have been developed for different species (Thorpe *et al.*, 1980; Küppers and Schulze, 1985). To scale up from a conductance model of this type, additional models of radiation penetration into the canopy and leaf microenvironment are required but are generally difficult to adjust for a specific situation. Caldwell *et al.* (1986) have used this approach to predict daily and annual totals for *Quercus coccifera* canopies in a Portuguese macchia. A major difficulty is validation of the models; Caldwell *et al.* used the good agreement between soil water use and predicted transpiration as a validation. Models of this type may be most useful for examining the consequences in a hypothetical sense of changes in canopy structure such as leaf area index or leaf angles on whole-plant water use.

8.7.2 Xylem flow measurements

The most immediate way of determining the total water flow through a plant is to measure the amount of water passing through the stem between the root system and the canopy. Because the opposite flow in the phloem is so much smaller than the xylem flow, the latter can be equated with the total flow. Measurements in intact plants were first attempted by Huber and Schmidt (1932), who developed the 'heat pulse method', which measures the velocity of the water flow in the xylem. In order to calculate the total amount of water, the actual conducting area must be known, which is generally impossible, since it is not known which vessels participate in transport of water. To overcome this limitation, a 'heat balance' method was applied by Vieweg and Ziegler (1960) and further developed by Čermák *et al.* (1973), using the specific heat capacity of water for keeping a temperature gradient constant, allowing long-term and continuous observations of xylem flow in the field.

(a) Determination for xylem flow velocity

The xylem flow velocity, V, can be determined by inserting a heating-device in a certain distance, D, below a temperature sensor (commonly a thermocouple) in the hydroactive xylem. The velocity is then determined by measuring the time interval, t, between a heat-pulse applied by the heating device and the detection of a temperature increase at the temperature sensor.

$$V = \frac{D}{t} \tag{8.17}$$

If the water content of the conducting xylem vessels is known, one can then calculate the mean rate of water transport through the measured section of the stem. However, this calculation contains several unknowns. First, the determination of the amount of conducting xylem area is difficult to determine even after cutting the stem and examining its cross-section. Because of wall friction, only part of the vessel cross-section is actually conducting (Čermák *et al.*, 1973). Attempts to estimate it have been made by coloring the xylem during transpiration with dyes (Čermák *et al.*, 1984) but the pattern which emerges is very complicated. The xylem flow can vary within a great range from young to old conducting vessels. But even these calculations by Čermák *et al.* (1984) show that not all vessels contribute to the flow, although the total area may be dyed. Moreover, xylem vessels often anastomose, making it difficult to determine the true distance between the heat source and the temperature sensor.

(b) Measurement of xylem sap mass flow
To overcome the problem of determining the effective area of the xylem, a steady-state mass flow technique, using the specific heat capacity of water C_w (4186.8 J K^{-1} kg^{-1} at 15°C) can be used. There are two principal methods of measurement: constant heating and null balance.

Constant-heating method. In this method xylem water is permanently heated with a constant amount of energy. The increase of the xylem sap temperature at a selected distance above the heating point can be used to calculate the mass flow of water through the xylem vessels. Although it is simple to install a constant-heating device and to measure the temperature difference between the point of heating and the higher point, problems can occur when xylem flow is low or stops during the night, leading to low removal of heat from the heater and the potential for xylem heat injury. Problems arise also at high mass flow rates occurring when transpiration rates are high which result in small temperature increases and poor signal resolution.

Constant-heating methods are described for herbaceous species with small stem diameters by Sakuratani (1981) and for trees by Granier (1985). In the instrument of Granier (1985, 1987) the xylem is heated with constant energy supply at one point by a small cylindrical probe containing a resistance wire heater and a thermocouple which is inserted 2 cm into the stem. Approximately 10 cm downstream, a second probe with a thermocouple but no heater measures the temperature. The temperature difference between the two probes is strictly influenced by the sap flow density around the heating probe. The sap flow (F, m^3 h^{-1}) of the tree can be calculated as follows:

$$F = 0.4284 Sa \left(\frac{T_m - T_d}{T_d} \right)^{1.231} \qquad (8.18)$$

Where Sa is the sapwood area at the heating probe level (m^2), T_m is the maximum temperature difference (K) obtained when sap flow is zero (measured at the end of the night when the xylem water potential is high and there is no water vapor pressure difference between plant and atmosphere), and T_d is the temperature difference between the two probes (K). This method needs calibration of each new installed device for the value of T_m, which depends on the thermal characteristics of the wood. Additionally the whole device must be thermally insulated to minimize temperature drifts due to direct solar radiation striking the stem or wind. In contrast to the heat pulse method, the calculation is not dependent on knowledge of the conducting area, since a mass flow and not a velocity is the basis of the measurement.

Null-balance method. A steady-state null-balance method maintains a preselected temperature difference of between 1 and 2 K between the reference temperature measured downstream from the heating point and temperature measured by a sensor at the point of maximum heat input. The electric energy is controlled such that the temperature difference is held constant independent of the xylem flux. The advantage of this system is that the amount of energy input to the heaters is directly proportional to the mass flow of water in the xylem irrespective of area and that no tissue overheating occurs at low xylem flux.

The main practical problem in the steady-state system is the application of an even temperature field to the whole conductive xylem area. For herbaceous species or small woody branches, a resistance wire spiral wound around the shoot and insulated against heat loss to the atmosphere is normally sufficient. This principle is described by

Schulze and Fichtner (1988) and was applied successfully to herbs and lianas.

For large trees, it is practical to heat only small sections of the sapwood, which can be done for example on opposite sides of the tree. Moreover, the heat cannot be supplied externally because of the insulating effects of the bark. In this case, a resistance-wire-heater probe or an alternating current (AC) supplied via electrodes inserted in the xylem can be used. Current flows between the electrodes, which remain cold, but because the xylem sap through which the current passes has a resistance, it is heated up. The AC current avoids problems with ion migration in the xylem sap. This principle of heat balance xylem sap measurements was first applied by Vieweg and Ziegler (1960) and was further developed by Čermák and Kucera (1981) by adding thermal compensation for natural temperature gradients occurring at the measuring point. The measuring device, which is commercially available from S.E.P. Gröger (West Germany), consists of five vertical stainless-steel electrode plates which are inserted in parallel into the xylem at a spacing of about 20 mm. A network of four thermocouple pairs inserted to various depths between the electrodes (the point of maximum temperature) measure the difference in temperatures with reference to points 10 cm below the plates. There is a controller circuit that regulates the power applied to the electrodes so that a constant temperature gradient (usually 1–2°C) is maintained. The power applied to the electrodes is recorded. Measurements from the eight thermocouples are also used to compensate automatically for thermal gradients caused by wind or solar radiation. In addition, the whole trunk section is covered with foam insulation and aluminium foil to minimize these externally imposed gradients. The depth of insertion of the thermocouples and electrodes depends on the diameter of the sapwood area, with the objective being that heating occurs across the entire hydroactive xylem segment. The calculation of xylem sap mass flow requires an estimate of the total conductive xylem area at the electrode level and the proportion which is heated by the electrodes. This can be obtained by analyzing a stem core. The total xylem mass flow (F_m, kg h^{-1}) is given by the difference of the measured mass flow without compensation of the heat losses of the measuring point Q and the fictive mass flow Q' which can be obtained during night when actual xylem flow is supposed to be zero.

$$F_m = Q - Q' \tag{8.19}$$

Both the recorded mass flow Q and the fictive mass flow Q' can be calculated according to the following equation:

$$Q = \frac{Pk}{C_w T} \tag{8.20}$$

where P is the power input (J s^{-1}) into the heating electrodes, C_w is the specific heat capacity of water (J K^{-1} kg^{-1}), T is the temperature difference (compensated against drifts) between the heated and the unheated part of the xylem in K. The dimensional constant, k, relates the section flux to the total circumference:

$$k = \frac{\theta}{d(n-1)} \tag{8.21}$$

where θ is the circumference (m) of the woody part of the tree at the measuring device, d is the distance between the electrodes (m) and n is the number of heating electrodes.

This measuring system has been used for xylem flow measurements in a variety of conifers and deciduous trees. The device, once installed, can monitor continuously the xylem mass flow for some months without problems. Changes in water flow on a scale of minutes can readily be observed. For

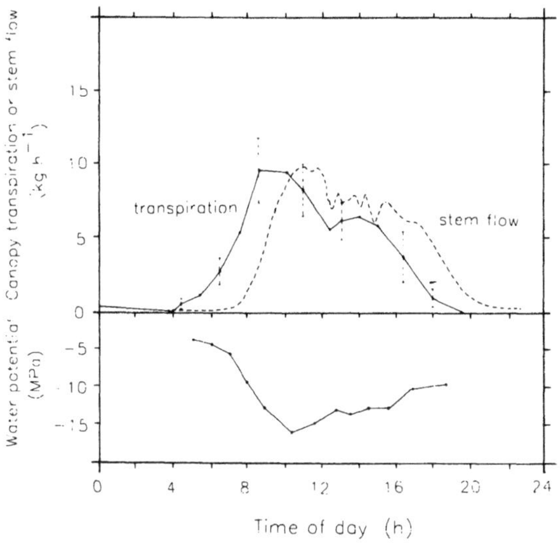

Fig. 8.9 The daily course of transpiration (solid line) of a *Larix* hybrid tree canopy, as calculated from porometer measurements and canopy leaf area, and the xylem water flow (dashed line) measured with the heat balance method of Čermàk *et al.* (1983). Stem flow lags behind because of the capacitance of the stem. In the morning the water content of the stem declines because transpiration exceeds water uptake while in the late afternoon the stem is recharged. Redrawn from Schulze *et al.* (1985).

example, Schulze *et al.* (1985) measured xylem flow of *Larix* and *Picea* trees and compared it to estimates of transpiration derived from a porometer and a controlled climate chamber (Fig. 8.9). Xylem flow can be seen to lag behind the canopy transpiration estimates from the porometer because of the capacitance of the stems and foliage, but on a daily basis there was close agreement (77 kg day^{-1} for the xylem flow and 75 kg day^{-1} for the extrapolation from the porometer). Comparisons with gravimetric water loss measurements suggest that the accuracy of the technique is around 10% (Penka *et al.*, 1979).

The major difficulties with the technique arise because of poor insulation and temperature gradients and from the accuracy of the measurement of the temperature difference. If the difference is controlled at a set point of 1 K and temperatures are measured to 0.1 K, then there is a 10% uncertainty in the flow measurement. Insulation can be applied effectively to trees but not easily to herbaceous plants or small branches. Measurement with small herbaceous plants is further complicated by temperature differences between the soil and air, which may then result in a temperature gradient in the xylem. Errors may also arise if the stem is not heated homogeneously. For large trees there is little problem since the electrodes can be spaced at equal distances. However, for small trees the electric field density between electrodes may not be constant. Also, in systems heated by wire wound around the stem, water flowing in the center may not reach the set point at high flow rates, whereas that close to the bark may rise above the set point.

8.7.3 Lysimeter measurements

For monitoring the total water loss of a potted single plant within a brief time interval the change in total weight is often used in laboratory experiments as the most direct and exact method of whole plant water loss. The same principle can also be used in the field, but it needs extensive preparations and control. After installation it is limited to monitoring at most a few individuals. Regardless of the restrictions it is a valuable possibility to measure transpiration and to compare these results with other methods for estimating, modeling or measuring transpiration of individual plants although it is much more expensive than measuring xylem flow directly. But in contrast to the xylem flow measurement the lysimeter measures the total mass changes of the root-stem-canopy system to the environment. It measures uptake and evaporative water loss which bypasses the plant. A disadvantage is that the root system and the soil structure invariably undergo some disturbance.

Commonly a lysimeter is installed by isolating a section of undisturbed soil which is rooted by a single plant in the case of trees or by a plant population in the case of crops from the circumfering soil. The soil block is enclosed in a container, sitting on a balance. For bigger plants, e.g. trees, the appropriate balance device is a ring of elastic tubes at the bottom of the soil container. The weight is measured by filling the tubes with airless water or oil and applying a hydrostatic pressure column to compensate the container weight. Pressure changes of the tubing system indicate weight changes. The hydrostatic balancing has a series of advantages. The total weight of the container can reach over 25 tons for trees with root systems. The accuracy is only dependent on the accuracy of the pressure sensor. The instrumentation with tubes also needs only a small space for installation below the lysimeter soil container.

For measuring the transpiration of the plant under field conditions, several parameters influencing the weight of the container have to be taken into account. Weight losses can occur due to transpiration water loss (via stomata, cuticle and bark), evaporation from the soil, respiration of tissue and soil and biomass loss (litter fall and loss of twigs or bark). Weight gain can occur from water uptake by rainfall, fog interception or snow deposition and from pressure changes in the surrounding soil. For a longer period of time also the carbon-uptake via photosynthesis has to be considered. The main problems are related to the water content of the soil in the container. Since the free capillary exchange of ground water is stopped by the container walls, water can accumulate during rainy periods or the soil dries much faster than under normal conditions.

By installing a separate drainage/watering system by protecting against rain and litter fall, factors influencing the total weight changes can be eliminated, leaving the net plant water losses over short periods. But in these cases the system is by no means 'natural'. Depending on the sensitivity of the installed pressure sensor, the accuracy of such lysimeters can reach values of 0.06 to 0.02 (mm water column per m^2 soil surface) (van Bavel and Meyers, 1962; Fritschen *et al.*, 1973). Using lysimeter systems in combination with photosynthesis and tissue respiration measurements on the one hand and soil water potential measurements on the other provides a complex approach to water dynamics with larger plants. However, they are clearly expensive, difficult to install and present many technical and logistical difficulties. For many purposes, the use of xylem flow devices is a satisfactory alternative.

Still lysimeters play an important role in estimating the water balance and evapotranspiration, but this may also be done by water budgets and neutron probe measurements in a natural soil profile. Furthermore, lysimeters may be used as 'pots' in order to define the root development of plants if root studies are desired, but in this case a weighting installation is not necessary.

REFERENCES

Beardsell, M.F., Jarvis, P.G. and Davidson, B. (1972) A null-balance diffusion porometer suitable for use with leaves of many shapes. *J. Appl. Ecol.*, **9**, 677–90.

Bingham, G.E., Coyne, P.I., Kennedy, R.B. and Jackson, W.L. (1980) Design and fabrication of a portable minicuvette system for measuring leaf photosynthesis and stomatal conductance under controlled conditions. Lawrence Livermore Laboratory, Livermore CA Publication no. UCRL-52895.

Bloom, A.J., Mooney, H.A., Björkman, O. and Berry, J. (1980) Materials and methods for carbon dioxide and water vapor exchange analysis. *Plant Cell Environ.*, **3**, 371–6.

Byrne, G.F., Rose, C.W. and Slatyer, R.O. (1970) An aspirated diffusion porometer suitable for use with leaves of many shapes. *J. Appl. Ecol.*, **9**, 39–44.

Caemmerer, S., von and Farquhar, G.D. (1981) Some relationships between the biochemistry of photosynthesis and the gas exchange of leaves. *Planta*, **153**, 376–87.

Caldwell, M.M., Meister, H. -P., Tenhunen, J.D.

and Lange, O.L. (1986) Canopy structure, light microclimate and leaf gas exchange of *Quercus coccifera* L. in a Portuguese macchia: measurements in different canopy layers and simulations with a canopy model. *Trees*, **1**, 25–41.

Campbell, G.S. (1975) Steady state diffusion porometers. Measurements of stomatal aperture and diffusive resistance. Washington State University Agriculture Research Center bulletin 809, Pullman, WA pp. 20–3.

Campbell, G.S. (1977) *An Introduction to Environmental Biophysics*, Springer-Verlag, New York, Heidelberg and Berlin.

Čermák, J., Deml, M. and Penka, M. (1973) A new method of sap flow rate determination in trees. *Biol. Plant.*, **15**, 171–78.

Čermák, J., Jenik, J., Kucera, J. and Zidek, V. (1984) Xylem sap flow in a crack willow tree (*Salix fragilis*) in relation to diurnal changes of the environment. *Oecologia*, **64**, 223–9.

Čermák, J. and Kucera, J. (1981) The compensation of natural temperature gradients at the measuring point during the sap flow rate determination in trees. *Biol. Plant.*, **23**, 469–71.

Cowan, I.R. (1977) Stomatal behaviour and environment. *Adv. Bot. Res.*, **4**, 117–227.

Day, W. (1977) A direct reading continuous flow porometer. *Agric. Meteorol.*, **18**, 81–9.

Dixon, M. and Grace, J. (1982) Water uptake by some chamber materials. *Plant, Cell Environ.*, **5** 323–7.

Fritschen, L.J., Cox, L. and Kinerson, R. (1973) A 28-meter Douglas fir in a weighing lysimeter. *For. Sci.*, **19**, 256–61.

Fuller, E.N., Schettler, P.D. and Geddings, J.C. (1966) A new method for the prediction of binary gas-phase diffusion coefficients. *Ind. Eng. Chem.*, **58**, 19–27.

Granier, A. (1985) Une nouvelle methode pour la mesure du flux seve brute dans le tronc des arbres. *Ann. Sci. For.*, **42**, 193–200.

Granier, A. (1987) Mesure du flux de seve brute dans le tronc du Douglas par une nouvelle methode thermique. *Ann. Sci. For.*, **44**, 1–14.

Hall, A.E. (1982) Mathematical models of plant water loss and plant water relations. In *Encyclopedia of Plant Physiology*, new series, Vol. 12b, *Physiological Plant Ecology* (eds O.L. Lange, P.S. Nobel, C.B. Osmond and H. Ziegler), Springer-Verlag, Heidelberg, Berlin and New York, pp. 231–60.

Huber, B. and Schmidt, E. (1937) Eine Kompensationsmethode zur thermoelektirschen Messung langsamer Saftstrome. *Ber. Dtsch. Bot. Ges.*, **55**, 512–29.

Jarman, P.D. (1974) The diffusion of carbon dioxide and water through stomata. *J. Exp. Bot.*, **25**, 927–36.

Jarvis, P.G. (1985) Transpiration and assimilation of tree and agricultural crops: the 'omega factor'. In *Attributes of Trees as Crop Plants* (eds M.G.R. Cannell and J.E. Jackson), Institute of Terrestrial Ecology, pp. 460–80.

Jarvis, P.G. and Catsky, J. (1971) Chamber microclimate and principles of assimilation chamber design. In *Plant Photosynthetic Production. Manual of Methods* (eds Z. Sestak, J. Catsky and P.G. Jarvis), W. Junk, The Hague, pp. 59–77.

Jones, H.G. (1983) *Plants and Microclimate*, Cambridge University Press, Cambridge.

Kanemasu, E.T., Thurtell, G.W. and Tanner, C.B. (1969) Design, calibration and field use of a stomatal diffusion porometer. *Plant Physiol.*, **44**, 881–5.

Kaufmann, M.R. and Eckard, A.N. (1978) A portable instrument for rapidly measuring conductance and transpiration of conifers and other species. *For. Sci.*, **23**, 227–37.

Körner, C. and Cernusca, A. (1976) A semi-automatic, recording diffusion porometer and its performance under alpine field conditions. *Photosynthetica*, **10**, 172–81.

Küppers, M. and Schulze, E.-D. (1985) An empirical model of net photosynthesis and leaf conductance for the simulation of diurnal courses of CO_2 and H_2O exchange. *Austr. J. Plant Physiol.*, **12**, 512–26.

Morrow, P.A. and Slatyer, R.O. (1971) Leaf temperature effects on measurements of diffusive resistance to water vapor transfer. *Plant Physiol.*, **47**, 559–61.

Mott, K.A. and O'Leary, J.W. (1984) Stomatal behavior and CO_2 exchange characteristics in amphistomatous leaves. *Plant Physiol.*, **74**, 47–51.

Nobel, P.S. (1984) *Biophysical Plant Physiology and Ecology*, W.H. Freeman, San Francisco.

Parkinson, K.J. and Day, W. (1981) Water vapor calibration using salt hydrate transitions. *J. Exp. Bot.*, **32**, 411–18.

Parkinson, K.J. and Legg, B.J. (1972) A continuous flow porometer. *J. Appl. Ecol.*, **9**, 669–75.

Penka, M., Čermák, J., Stepánek, V. and Palat, M. (1979) Diurnal courses of transpiration rate and transpiration flow rate as determined by the gravimetric and thermometric methods in a full-grown oak tree (*Quercus robur*). *Acta Universitatis Agric. (Brno) Ser. C.*, **48**, 3–30.

Sakuratani, T. (1981) A heat balance method for measuring water flux in the stem of intact plants. *J. Agric. Meteorol.*, **37**, 9–17.

Salasmaa, E. and Kostamo, P. (1975) New thin film

humidity sensor. In *Third Symposium on Meteorological Observations and Instrumentation*, American Meteorological Society, Boston, pp. 23–8.

Salasmaa, E. and Kostamo, P. (1986) HUMICAP thin film humidity sensor. In *Advanced Agricultural Instrumentation Design and Use* (ed. W. Gensler), Martinus Nijhoff, Dordrecht, pp. 135–48.

Schulze, E.-D., Čermák, J., Matyssek, R., Penka, M., Zimmermann, R., Vasicek, F., Gries, W. and Kučera, J. (1985) Canopy transpiration and water fluxes in the xylem of the trunk of *Larix* and *Picea* trees – a comparison of xylem flow, porometer and cuvette measurements. *Oecologia*, **66**, 475–83.

Schulze, E.-D. and Fichtner, K. (1988) Xylem water flow of tropical lianas. *Oecologia* (in press).

Schulze, E.-D., Hall, A.E., Lange, O.L. and Walz, H. (1982) A portable steady-state porometer for measuring the carbon dioxide and water vapor exchanges of leaves under natural conditions. *Oecologia*, **53**, 141–5.

Schurer, K. (1986) Water and plants. In *Advanced Agricultural Instrumentation Design and Use* (ed. W. Gensler), Martinus Nijhoff, Dordrecht, pp. 429–56.

Sharkey, T.D., Imai, K., Farquhar, G.D. and Cowan, I.R. (1982) A direct confirmation of the standard method of estimating intercellular partial pressure of CO_2. *Plant Physiol.*, **69**, 657–9.

Slatyer, R.O. and Bierhuizen, J.F. (1964) A differential psychrometer for continuous measurements of transpiration. *Plant Physiol.*, **39**, 1051–6.

Stiles, W. Montieth, J.L. and Bull, T.A. (1970) A diffusive resistance porometer for field use. *J. Appl. Ecol.*, **7**, 617–38.

Thorpe, M.R., Warrit, B. and Landsberg, J.J. (1980) Responses of apple leaf stomata: a model for single leaves and a whole tree. *Plant, Cell Environ.*, **3**, 23–7.

Turner, N.C. and Parlange, J.-Y. (1970) Analysis of operation and calibration of a ventilated diffusion porometer. *Plant Physiol.*, **46**, 175–7.

van Bavel, C.H.M. and Meyers, L.E. (1962) An automatic weighing lysimeter. *Agric. Eng.*, **43**, 580–8.

Vieweg, G.H. and Ziegler, H. (1960) Thermoelektrische Registrierung der Geschwindigkeit des Transpirationsstromes. *Ber. Dtsch. Bot. Ges.*, **73**, 221–6.

Wallihan, E.F. (1964) Modification and use of an electric hygrometer for estimating relative stomatal apertures. *Plant Physiol.*, **39**, 86–90.

9

Plant water status, hydraulic resistance and capacitance

Roger T. Koide, Robert H. Robichaux, Suzanne R. Morse and Celia M. Smith

9.1 INTRODUCTION

Many excellent reviews have been written about plant water status and its measurement (e.g. Slatyer, 1967; Barrs, 1968; Boyer, 1969; Brown and van Haveren, 1972; Slavik, 1974; Turner, 1981). The reader is referred to these sources for a more complete review, particularly of the older literature. In this chapter, our major goals are to introduce the reader to the concept and measurement of plant water potential and its components, and then to discuss the consequences of gradients in these components within the plant. First, we describe the most commonly used techniques for measuring the water potential of higher plants growing under field conditions, specifically the psychrometric and pressure chamber techniques. Second, we describe methods for measuring the components of water potential, particularly turgor pressure and osmotic potential, and water content. Since the transpirational path of the plant can be regarded as a hydraulic resistor, transpirational fluxes occur only when gradients in the various components of water potential exist within the plant. Thus, the third concept that we introduce is hydraulic resistance. Techniques for its calculation are described for steady-state transpiration, when root water absorption equals shoot evaporation. Nonsteady-state flux occurs when there is a net movement of water between the transpirational path and tissues adjacent to it. Such tissues are regarded as capacitors, the final topic of our discussion.

9.2 WATER POTENTIAL AND ITS COMPONENTS

The thermodynamic parameter commonly used to describe the energy status of water in plants is the water potential (Slatyer, 1967; Passioura, 1982; Nobel, 1983). This parameter, designated Ψ, is defined as:

$$\Psi = \frac{\mu_w - \mu_w^*}{\bar{V}_w} \qquad (9.1)$$

where μ_w is the chemical potential, or free energy per mole, of water at some point in the system at constant temperature and pressure, μ_w^* is the chemical potential of pure water at the same temperature and at atmospheric pressure, and $\bar{V}_w$ is the partial molal volume of water. The quantity, $\mu_w - \mu_w^*$, in Equation 9.1 represents the work involved in moving one mole of water from some point in the system to a pool of pure water at the same temperature and at atmospheric pressure. In plants, Ψ varies from zero to negative values. The units of Ψ are those of pressure, with the common units being MPa (megapascals). Alternative units are bars (1 MPa = 10 bars) and J m^{-3} (1 MPa = 10^6 J m^{-3}).

For a plant cell, Ψ may be expressed as the sum of three components:

$$\Psi = P + \pi + \tau \tag{9.2}$$

where P is the hydrostatic pressure (or turgor pressure), π is the osmotic potential and τ is the matric potential (Tyree and Jarvis, 1982). At equilibrium, Ψ is the same across the heterogeneous phases of the cell, i.e. across the vacuole, cytoplasm and cell wall. However, the components of Ψ may differ markedly across these phases. For water in the vacuole and cytoplasm (the symplasmic water), the dominant components are usually P and π, with P usually having a positive value. In a cell at equilibrium, P and π are probably uniform throughout the symplasm, since it is unlikely that significant pressure gradients can exist across the bounding membranes of the vacuole and cytoplasmic organelles (Tyree and Jarvis, 1982). However, the particular solutes contributing to the reduction in π may differ between the various symplasmic compartments. For water in the cell wall (the apoplasmic water), the dominant component is usually P, with π and τ contributing to Ψ mainly in the region immediately adjacent to the charged wall surface (Tyree and Karamanos, 1981). In this latter region, which is often less than 3 nm wide, P may be positive. For most of the water in the cell wall, however, P is negative. In some halophytes, both P and π may be important components of Ψ in the apoplasm.

Within a tissue, the symplasmic values of P and π may vary significantly from cell to cell, even at equilibrium. Given this variation, the most appropriate parameters for describing the water relations of the tissue symplasm are the bulk, weight-averaged values of P and π (Tyree and Hammel, 1972). These weight-averaged values, designated $\bar{P}$ and $\bar{\pi}$, respectively, are defined as:

$$\bar{P} = \sum_{i=1}^{n} \frac{w_s^i}{W_s} P^i \tag{9.3}$$

and

$$\bar{\pi} = \sum_{i=1}^{n} \frac{w_s^i}{W_s} \pi^i \tag{9.4}$$

where P^i, π^i, and w_s^i are the turgor pressure, osmotic potential and weight of symplasmic water respectively in the ith cell in the tissue, W_s is the total weight of symplasmic water in the tissue and n is the number of cells in the tissue (Tyree and Jarvis, 1982). Thus, at equilibrium, Ψ in the tissue symplasm may be expressed as:

$$\Psi = \bar{P} + \bar{\pi} \tag{9.5}$$

(Compared to the symplasmic values of $\bar{P}$ and $\bar{\pi}$, the symplasmic value of $\bar{\tau}$ is negligible.) Analogous weight-averaged values may also be defined for the components of Ψ in the tissue apoplasm. In addition to water in the walls of living cells, the tissue apoplasm includes water in the walls and lumina of dead cells, such as vessel elements, tracheids and fibers. The dominant component of Ψ in the tissue apoplasm is usually $\bar{P}$. At equilibrium, Ψ in the tissue apoplasm equals Ψ in the tissue symplasm.

A gravitational term is often included as a component of Ψ. As emphasized by Passioura (1982), however, the definition of Ψ in terms of μ_w ignores external force fields. Hence, it is preferable to use a separate gravitational potential, Ψ_z, in evaluating the effects of gravity. The total water potential may then be defined as $\Psi + \Psi_z$. The value of Ψ_z increases with height at a rate of 0.0098 MPa

m^{-1} (Nobel, 1983). Thus, it is usually very small, except in tall trees.

Following the development of techniques for accurately measuring Ψ and its components in higher plants growing under field conditions, numerous studies have analyzed the ecological significance of variation in these parameters for plants growing in diverse habitats and exhibiting diverse growth forms. Recent studies, for example, have examined plants growing in habitats as varied as tropical rain forest (Oberbauer, 1983; Myers *et al.*, 1987), savannah (Meinzer *et al.*, 1983) and paramo (Goldstein *et al.*, 1984), Mediterranean woodland and scrub (Hinckley *et al.*, 1980; Calkin and Pearcy, 1984; Davis and Mooney, 1986), and temperate forest (Roberts *et al.*, 1980, 1981; Parker *et al.*, 1982), prairie (Knapp, 1984; Barnes, 1985), desert (Monson and Smith, 1982; Nilsen *et al.*, 1983, 1984, 1986), salt marsh (Drake and Gallagher, 1984) and coastal dune (Pavlik, 1984). These studies have documented significant variation among species in these parameters, particularly in relation to habitat water availability. They have also shown that these parameters may vary within individuals in response to ontogenetic, diurnal and seasonal factors.

9.2.1 Techniques for measuring Ψ

Several techniques are currently available for measuring Ψ in plant tissues (Turner, 1981). The principal methods used with higher plants growing under field conditions are the psychrometric and pressure chamber techniques.

(a) Psychrometric techniques

Techniques employing thermocouple psychrometers are widely used for measuring Ψ in higher plants. For plants growing under field conditions, psychrometric techniques are useful primarily under circumstances in which excised tissue samples can be transferred to temperature-controlled laboratory facilities. Recent advances in the use of these techniques with attached organs such as leaves and stems may enhance their suitability under more remote circumstances as well. Excellent reviews of the theory, design and application of thermocouple psychrometers have been provided by Brown and van Haveren (1972) and Briscoe (1986).

When a tissue sample is placed in an enclosed psychrometer chamber held at a constant temperature, the water in the sample will equilibrate with the chamber atmosphere. By measuring the equilibrium relative humidity (h) of this atmosphere, thermocouple psychrometers enable the value of Ψ in the sample to be calculated as:

$$\Psi = \frac{\boldsymbol{R}T \ln(h)}{\bar{V}_w} \tag{9.6}$$

where $\boldsymbol{R}$ is the universal gas constant and T is the Kelvin temperature (Turner, 1981; Nobel, 1983). A major assumption is that loss of water from the sample to the chamber atmosphere during the equilibration period does not affect the value of Ψ in the sample.

Three techniques are commonly available for measuring h in the psychrometer chamber. In the first technique, pure water is allowed to evaporate into the chamber atmosphere from the surface of a thermocouple junction. The resulting depression in the junction temperature (the so-called 'wet-bulb depression') is related to h by:

$$h = \frac{e_w - ac(T - T_w)}{e} \tag{9.7}$$

where a is the atmospheric pressure, c is the psychrometric constant, T is the psychrometer temperature (or dry-bulb temperature), T_w is the wet-bulb temperature, e is the saturation vapor pressure at T, and e_w is the saturation vapor pressure at T_w (List, 1968). In practice, the psychrometer is calibrated with solutions of known concentration to provide a conversion factor relating thermocouple output (in microvolts) to Ψ. Two procedures have been developed for placing pure water on the thermocouple junction. One procedure uses a cooling current (the Peltier effect) to reduce the temperature of the junction below the dew point of the chamber atmosphere in order to condense water on it (Spanner, 1951). Another procedure involves manually placing a small amount of pure water on the junction prior to the measurement (Richards and Ogata, 1958). Because of its greater convenience, the procedure employing the Peltier effect has proven more adaptable for field use. Thermocouple psychrometers employing the Peltier effect are commercially available from JRD Merrill Specialty Equipment Corp. (Logan, Utah) and Wescor Corp. (Logan, Utah), while those employing a modified Richards and Ogata procedure are commercially available from Decagon Devices Corp. (Pullman, Washington).

The second technique, known as dew-point hygrometry, measures the dew-point temperature rather than the wet-bulb depression in order to determine *h* (Neumann and Thurtell, 1972; Campbell *et al.*, 1973). Dew-point hygrometers hold the thermocouple junction at the stable dew point by pulsed cooling currents. One advantage of this technique is that the dew-point depression is larger than the wet-bulb depression. Hence, the thermocouple signal is larger. Additionally, since no net movement of water from the thermocouple junction to the chamber atmosphere occurs at the dew point, the signal is stable for a long period, and the vapor equilibrium in the chamber is not disturbed. Calibration is the same as for the first technique. Thermocouple psychrometers capable of operating in the dew-point mode are manufactured by Wescor Corp. (Logan, Utah).

In the third technique, the change in the temperature of the thermocouple junction is measured when solutions with different Ψ values are placed on the junction (Boyer and Knipling, 1965). If a solution with a Ψ value equal to that of the sample is used, and if the sample has come to equilibrium with the chamber atmosphere, then no net vapor movement will occur and the thermocouple output will be zero. This is the isopiestic point. In practice, since the relationship between the thermocouple output and the Ψ value of the solution is linear, only two solutions are used, typically distilled water and a solution whose Ψ value is close to that of the tissue sample. The isopiestic point is then determined by extrapolation. Although this technique involves more steps than the others, it has the same advantage as the dew-point technique in that the disturbance of vapor equilibrium in the chamber is minimized near the isopiestic point. According to Boyer and Knipling (1965), this minimizes the error associated with a high resistance to vapor exchange between the sample and the chamber atmosphere. The isopiestic technique has seen wide use in laboratory studies. Since it is more complicated than the other techniques, however, it has not been widely adopted as a practical technique for field use.

Thermocouple psychrometers have been used extensively to measure Ψ in excised tissue samples. The need for continuous nondestructive monitoring of Ψ in intact attached leaves has also promoted the development of *in situ* psychrometers (Boyer, 1972; Neumann and Thurtell, 1972; Campbell and Campbell, 1974). Partial dissolution or abrasion of the leaf cuticle (Neumann and Thurtell, 1972; Brown and Tanner, 1981; Savage *et al.*, 1984) and the use of compounds such as wax and lanolin to seal the psychrometer chamber to the leaf surface (Campbell and Campbell, 1974; Brown and Tanner, 1981) have helped to insure that vapor equilibrium is reached with *in situ* psychrometers. These innovations have also helped to reduce the initial equilibration period and to reduce the response time for changes in Ψ to minutes (Turner *et al.*, 1984). As a result, *in situ* leaf psychrometers appear to be suitable for the continuous monitoring of Ψ even in plants where Ψ fluctuates rapidly. *In situ* stem (Michel, 1977; McBurney and Costigan, 1982) and root (Fiscus, 1972) psychrometers have also been used with some success.

In some cases, values of Ψ measured with *in situ* leaf psychrometers agree closely with those measured with pressure chambers (Boyer, 1967; Kikuta *et al.*, 1985). Good correlations are generally found when leaves are not transpiring and when gradients in Ψ are not present within the leaf. In transpiring leaves and in severely stressed large leaves, however, gradients in Ψ may occur across the leaf, such that measurements with *in situ* psychrometers may differ greatly from those with pressure chambers (Turner *et al.*, 1984; Shackel and Brinckmann, 1985). *In situ* psychrometers unavoidably cause some portion of the leaf to be shaded, thereby reducing transpiration in the area measured. If large resistances to water movement occur within the leaf, this may lead to overestimations of Ψ (Brown and Tanner, 1981; Savage *et al.*, 1983).

In situ leaf psychrometers are commercially available from Wescor Corp. (Logan, Utah), but often require some modifications to increase their thermal stability (Brown and Tanner, 1981; Savage *et al.*, 1983; Shackel, 1984). The precautions necessary for their use have been discussed at length by Shackel (1984).

The three most common sources of error associated with thermocouple psychrometers are: (1) vapor pressure disequilibria, (2) thermal gradients and instability, and (3) changes in Ψ in the sample due to growth and excision.

Vapor pressure disequilibria, except for those related to thermal gradients and instability or to the special problem of making a vapor seal with *in situ* psychrometers, are caused by water sinks within the psychrometer chamber and by the resistance to vapor diffusion from the sample to the chamber atmosphere. The first of these may be reduced by proper selection of chamber materials (Boyer, 1972; Dixon and Grace, 1982), scrupulous cleanliness, and maximization of sample surface area per unit chamber volume. Salts and other foreign materials on the sample surface may also absorb water vapor. Washing the sample with distilled water and drying it thoroughly prior to measurement are good precautions. Techniques for reducing the resistance to vapor exchange from the sample, such as partial dissolution or abrasion of the leaf cuticle, have been used primarily with *in situ* psychrometers.

Thermal gradients and instability are caused by respiratory heat production from the sample (Barrs, 1964), by gradients or fluctuations in environmental temperature, and by absorption of solar irradiance. Thermal gradients and instability produce errors in the measurement of Ψ by causing the chamber atmosphere to differ in temperature from the sample or by causing electrical potentials to develop within the measuring circuit. The effects of respiratory heat production may be reduced by assuring good contact between the sample and the chamber and by using chamber materials with a high thermal conductivity. Gradients or fluctuations in environmental temperature may be minimized by performing operations in a controlled-temperature water bath if laboratory facilities are available. For *in situ* psychrometers, whose use has often been limited by the need for precise temperature control, Brown and Tanner (1981), Savage *et al.* (1983), and Shackel (1984) have discussed methods for increasing thermal stability. Absorption of solar irradiance may be reduced by using chamber materials with a high reflectivity.

The high resistance to vapor diffusion from the sample to the chamber atmosphere typically results in a long equilibration period, often on the order of hours. In an excised tissue sample, growth during the equilibration period will cause a reduction in Ψ since $\bar{P}$ will decline in the absence of a water supply (Cosgrove *et al.*, 1984). Errors of this sort may be reduced by working at low temperature or with mature tissue. Changes in Ψ in an excised tissue sample will also occur if water and solutes released from damaged cells at the cut surface are taken up by intact cells, thus altering their values of π and P (Barrs and Kramer, 1969; Nelsen *et al.*, 1978). Excision errors may be minimized by using samples with a low ratio of cut surface area to sample surface area. Errors resulting from water and solute release by damaged cells are eliminated with *in situ* psychrometers, provided that the epidermal cells are not damaged by partial dissolution or abrasion of the leaf cuticle.

(b) Pressure-chamber technique

Since its rediscovery by Scholander *et al.* (1965), the pressure-chamber technique has emerged as the most widely used method for measuring Ψ in higher plants growing under field conditions. This stems in large part from its simplicity and reliability, and its lack of requirement for precise temperature control (Turner, 1981). Excellent reviews of both the theoretical and practical aspects of this technique have been provided by Ritchie and Hinckley (1975) and Turner (1981, 1987).

The pressure-chamber technique measures the apoplasmic value of $\bar{P}$ (Scholander *et al.*, 1965). If the apoplasmic values of $\bar{\pi}$ and $\bar{\tau}$ are negligible, then this technique also measures the apoplasmic value of Ψ. If the tissue is at equilibrium, then the apoplasmic value of Ψ will equal its symplasmic value.

The apoplasmic value of $\bar{\pi}$ is usually higher than -0.1 MPa and is often higher than -0.02 MPa, though it can be much lower in halophytes (Ritchie and Hinckley, 1975; Turner, 1981). Thus, the apoplasmic value of $\bar{\pi}$ is often negligible. As a precaution, however, this value should be checked for the species of interest in a given study, particularly for species growing under conditions of high salinity. (The technique for measuring the apoplasmic value of $\bar{\pi}$ is discussed in Section 9.2.3.)

The apoplasmic value of $\bar{\tau}$, which results from the interaction between the water dipole and the electric field associated with the charged cell wall surface, is negligible unless substantial intrusion of air–water interfaces occurs in the cell wall pores (Tyree and Jarvis, 1982). For most cell walls, this intrusion does not occur until Ψ drops below -14 MPa, which is well below the lowest Ψ values experienced by most higher plants.

In a transpiring leaf with a finite resistance to water flow between the apoplasm and symplasm, the apoplasmic and symplasmic values of Ψ will not be equal (Turner, 1981). Once transpiration stops, however, the apoplasmic and symplasmic Ψ values will converge (Passioura, 1982). This convergence will occur quite rapidly in most leaves, with any disequilibrium between the apoplasm and symplasm disappearing during the time period between excising a leaf and making a measurement with the pressure chamber (Turner, 1981). The equilibrium Ψ value will be influenced by the apoplasmic and symplasmic capacitances (Passioura, 1982) (see Section 9.4). Since the capacitance of the symplasm is usually much greater than that of the apoplasm, the equilibrium Ψ value will closely reflect the symplasmic Ψ value at the time of leaf excision.

Thus, in theory, the pressure-chamber technique is capable of measuring the symplasmic value of Ψ under most conditions. The technique itself is elegant in its simplicity (Scholander *et al.*, 1965; Turner, 1981, 1987). A leaf (or branch) is excised from the plant with a sharp razor blade, and is inserted into the pressure chamber with the cut surface of the petiole (or stem) protruding slightly through the rubber sealing gasket. Pressure is then increased gradually in the chamber by adding nitrogen or air from a compressed gas source. The pressure is increased until water in the xylem first appears at the cut surface of the petiole. The accurate determination of this end point is aided by the use of a dissecting microscope or high-magnification hand lens. The balance pressure in the chamber at this end point, taken as a negative value, equals the apoplasmic value of $\bar{P}$ in the leaf, which in turn equals the symplasmic value of Ψ under most conditions.

Important precautions to be taken in measuring Ψ with the pressure-chamber technique have been nicely summarized by Turner (1981, 1987). These include:

1. Immediately prior to being excised from the plant, the leaf should be enclosed in a plastic sheath (except for the protruding petiole) to prevent water loss during the measurement period. The sheath should remain on the leaf inside the pressure chamber. In rapidly transpiring leaves, Turner and Long (1980) measured significant decreases in Ψ within the first 10–30 s following excision, and recommended enclosing the leaf in a plastic sheath 1–2 s prior to excision. The attached leaf should not be enclosed in the sheath for much longer than 1–2 s, however, as Ψ may begin to increase.
2. The petiole should not be recut after the leaf is excised from the plant. When the petiole is cut, the water in the xylem, which is under tension, will recede a short distance from the cut surface (Scholander *et al.*, 1965). The water will recede until the air–water interface is unable to pass through the pores in the end walls of the xylem elements. As long as the petiole has been cut only once, water will refill the emptied xylem elements when the end point is reached in the pressure chamber, resulting in no net water movement into the tissue symplasm. If the petiole has been cut more than once, however, a net movement of water into the symplasm will occur, which will result in an erroneously high measurement of Ψ. This error will be greatest in species with large xylem elements and in species experiencing low Ψ values.
3. The length of petiole protruding from the pressure chamber should be minimized, both to prevent evaporation from the exposed petiole and to avoid exclusion errors (Millar and Hansen, 1975).
4. The rubber sealing gasket should fit tightly around the petiole such that no gas escapes from the pressure chamber. Damage to herbaceous petioles resulting from constriction by the gasket may be minimized by the use of a quick-setting epoxy resin (Nobel and Jordan, 1983).
5. The rate of pressurization in the chamber should be very slow to prevent the development of Ψ disequilibria within the tissue symplasm. Turner (1981) recommends a rate of 0.003–0.005 MPa s^{-1}. Internal Ψ disequilibria may be particularly problematical in branches, compound leaves and severely stressed large leaves, within which resistances to water transport may vary significantly from point to point (Turner *et al.*, 1984).

The extent to which these precautions need to be followed will depend on the species of interest in a given study. With some species, for example, the petiole can be recut several times without significantly affecting the measured value of Ψ. In addition, it is possible with some species to use a moderately rapid initial rate of chamber pressurization, followed by a slow rate of pressurization near the end point. The latter method enables more leaf samples, and hence a larger number of individuals within the population, to be processed in a short period of time. Exploratory measurements with each new species of interest will aid in determining the optimal procedures to be used.

Even when these precautions are stringently followed, however, the ease with which Ψ can be measured with the pressure-chamber technique may vary from species to species. In most species, the end point is very well defined, enabling Ψ to be measured with a high degree of precision. In some species, however, determination of the end point is complicated by resin exudation from ducts in the xylem or fluid exudation from the pith and cortex (Ritchie and Hinckley, 1975).

Removal of these exudates from the cut surface by repeated blotting with lintless tissue during chamber pressurization, or by constriction of the stem, may aid with the end point determination (Turner, 1981, 1987).

In the US, pressure chambers are commercially available from PMS Instrument Co. (Corvallis, Oregon) and Soil Moisture Equipment Corp. (Santa Barbara, California).

9.2.2 Techniques for measuring the symplasmic values of $\bar{\pi}$ and $\bar{P}$

Pressure-chamber and psychrometric techniques are also the principal methods for measuring the symplasmic values of $\bar{\pi}$ and $\bar{P}$ in higher plants growing under field conditions. The pressure-probe technique, which enables P to be measured directly in the cells of intact tissues, has recently been adapted successfully for use with mesophytic plants under precisely controlled laboratory conditions (Husken *et al.*, 1978; Zimmermann *et al.*, 1980; Brinckmann *et al.*, 1984; Cosgrove *et al.*, 1984; Shackel and Brinckmann, 1985). With appropriate modifications, this technique may eventually be suitable for use with plants growing under field conditions.

(a) Pressure-chamber technique

The most widely used method for measuring the symplasmic values of $\bar{\pi}$ and $\bar{P}$ in intact living tissues is the pressure-chamber technique. This technique was originally developed by Scholander *et al.* (1964, 1965) and has subsequently been refined by Tyree and coworkers (Tyree and Hammel, 1972; Tyree, 1981; Tyree and Karamanos, 1981; Tyree and Richter, 1981, 1982; Tyree and Jarvis, 1982). The simplicity of the technique enables it to be used even in remote field sites.

The theoretical aspects of the pressure-chamber technique have been discussed in detail by Tyree and Jarvis (1982). In brief, the symplasmic value of $\bar{\pi}$ is approximated by:

$$\bar{\pi} = -\frac{\phi \rho \boldsymbol{R} T N}{W_s} \tag{9.8}$$

where ϕ is an osmotic coefficient (to account for the nonideality of solute behavior), ρ is the density of water in the symplasm, and N is the number of moles of solutes in the symplasm (ionic species counted separately). As the water content of an initially saturated tissue decreases, the symplasmic value of $\bar{P}$ will decrease and eventually will reach zero. If negative values of $\bar{P}$ do not occur in the tissue symplasm, then at all lower water contents, Ψ equals $\bar{\pi}$ (see Equation 9.5). Thus, from Equation 9.8:

$$\frac{1}{\Psi} = \frac{1}{\bar{\pi}} = -\frac{W_s}{\phi \rho \boldsymbol{R} T N} \tag{9.9}$$

The tissue relative water content (R') is defined as:

$$R' = \frac{W_s + W_a}{W_s^o + W_a^o} \tag{9.10}$$

where W_a is the weight of apoplasmic water, W_s^o is the weight of symplasmic water at full hydration, and W_a^o is the weight of apoplasmic water at full hydration. Thus:

$$W_s = R'(W_s^o + W_a^o) - W_a \tag{9.11}$$

Substituting into Equation 9.9 the expression for W_s in Equation 9.11 yields:

$$\frac{1}{\Psi} = \frac{1}{\bar{\pi}} = \frac{W_a}{\phi \rho \boldsymbol{R} T N} - \frac{R'(W_s^o + W_a^o)}{\phi \rho \boldsymbol{R} T N} \tag{9.12}$$

For values of R' less than that at which $\bar{P}$ reaches zero, Equation 9.12 describes a linear relationship between $1/\Psi$ (or $1/\bar{\pi}$) and R', assuming that W_a and ϕ are constant. Hence, a plot of $1/\Psi$ versus R' yields a straight line in the region where $\bar{P} = 0$ (Fig. 9.1). For values of R' greater than that at which $\bar{P}$ reaches zero, the relationship between $1/\Psi$ and R' is no longer linear, since $\Psi = \bar{P} + \bar{\pi}$. However, the relationship between $1/\bar{\pi}$ and R' remains linear, assuming that W_a and ϕ remain constant in this region. Hence, from Equation 9.12, it is possible to calculate $1/\bar{\pi}$ for all values of R' greater than that at which $\bar{P}$ reaches zero. The latter relationship is illustrated by a dashed line in Fig. 9.1. At full hydration, where $R' = 1$, Equation 9.12 reduces to:

$$\frac{1}{\bar{\pi}} = -\frac{W_s^o}{\phi\rho \boldsymbol{R} TN} \tag{9.13}$$

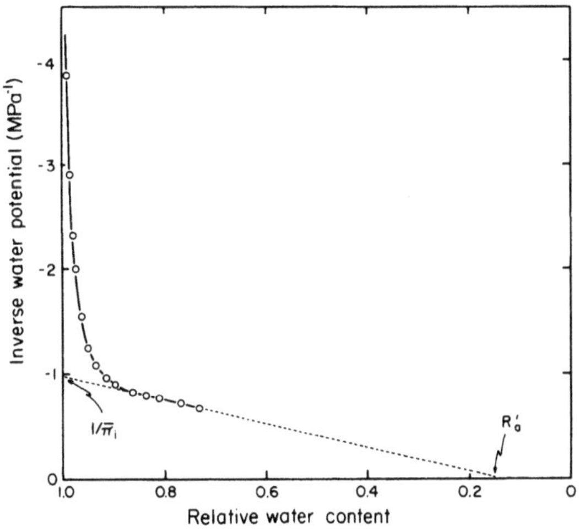

Fig. 9.1 Relationship between the reciprocal of tissue water potential ($1/\Psi$) and tissue relative water content (R') (solid line). An additional data point exists at $1/\Psi = -7.41\ \mathrm{MPa}^{-1}$ and $R' = 0.997$. The dashed line defines the calculated relationship between the reciprocal of tissue osmotic potential ($1/\bar{\pi}$) and R'. The intercept of the dashed line with the y-axis yields the reciprocal of tissue osmotic potential at full hydration ($1/\bar{\pi}_i$). The value of $\bar{\pi}_i$ in this example is -1.02 MPa. The intercept of the dashed line with the x-axis yields the relative water content of the tissue apoplasm (R'a). The value of R'a in this example is 0.14. Given the measured relationship between $1/\Psi$ and R', together with the calculated relationship between $1/\bar{\pi}$ and R', it is possible to calculate the relationship between tissue turgor pressure ($\bar{P}$) and R' from Equation 9.5, since $\bar{P} = \Psi - \bar{\pi}$ (see Fig. 9.2). It is also possible to calculate the relationship between $\bar{P}$ and Ψ in a similar manner (see Fig. 9.3). The data are for the Hawaiian bog species, *Dubautia paleata*.

The right side of Equation 9.13 thus equals the reciprocal of the tissue osmotic potential at full hydration ($\bar{\pi}_i$).

Given the calculated relationship between $1/\bar{\pi}$ and R', together with the measured relationship between $1/\Psi$ and R', it is also possible to calculate relationships between $\bar{P}$ and R' (Fig. 9.2) and $\bar{P}$ and Ψ (Fig. 9.3) using Equation 9.5. Species (or individuals) may differ markedly in their capacities for maintaining high $\bar{P}$ values as both R' and Ψ decline. These differences in turgor maintenance capacity may result from two separate phenomena (Robichaux, 1984). First, the maximal value of $\bar{P}$ may increase, reflecting a decrease in the value of $\bar{\pi}$ at full hydration ($\bar{\pi}_i$). Second, the rate at which $\bar{P}$ declines with decreasing R' (or Ψ) may decrease. The

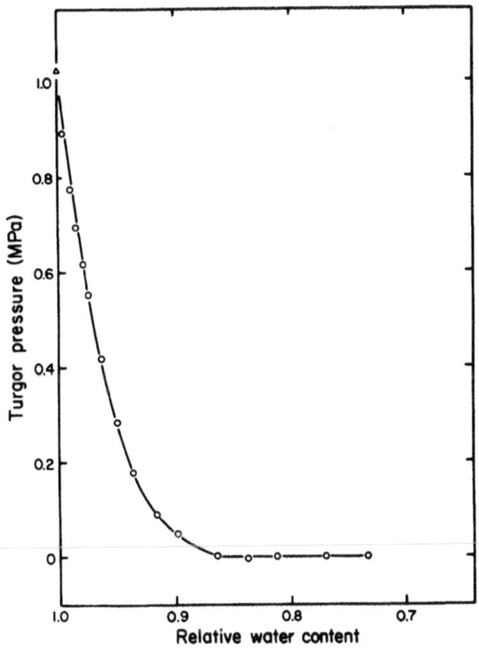

Fig. 9.2 Relationship between tissue turgor pressure and tissue relative water content. The relationship was obtained from the data in Fig. 9.1. The maximal value of $\bar{P}$, denoted by the triangle, was obtained from the calculated value of $\bar{\pi}_i$, which is equal in magnitude but opposite in sign.

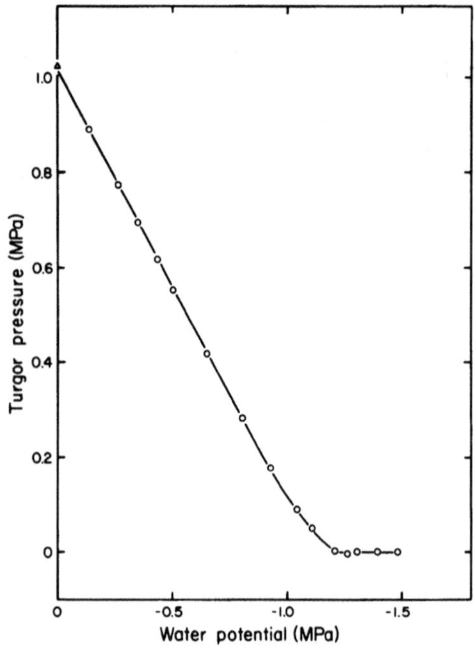

Fig. 9.3 Relationship between tissue turgor pressure and tissue water potential. The relationship was obtained from the data in Fig. 9.1. The maximal value of $\bar{P}$, denoted by the triangle, was obtained from the calculated value of $\bar{\pi}_i$.

latter change reflects an increase in the degree of tissue elasticity.

The bulk tissue elastic modulus ($\bar{E}$) is defined as the change in $\bar{P}$ for a given fractional change in the weight of symplasmic water (Tyree, 1981; Tyree and Jarvis, 1982). Hence:

$$\bar{E} = \frac{d\bar{P}}{dW_s} W_s \tag{9.14}$$

If W_a remains constant as W_s decreases, then:

$$\bar{E} = \frac{d\bar{P}}{dR'}(R' - R'_a) \tag{9.15}$$

where R'_a is the relative water content of the tissue apoplasm, or the apoplasmic fraction (Robichaux *et al.*, 1986). In practice, $\bar{E}$ may be calculated using the finite form of Equation 9.15. To calculate the value of $\bar{E}$ near full hydration ($\bar{E}_i$), for example, one obtains $\Delta\bar{P}/\Delta R'$ from the linear slope of the first four to five points of the $\bar{P}$ versus R' relationship (Fig. 9.2) and R' from the mean value for these four to five points. One obtains R'_a from the original plot of $1/\Psi$ versus R', by extrapolating Equation 9.12 to the point at which $1/\Psi = 0$ (Fig. 9.1). At this point:

$$R' = R'_a = \frac{W_a}{W_s^o + W_a^o} \tag{9.16}$$

From Equation 9.15, it follows that an increase in the degree of tissue elasticity is reflected in a decrease in $\bar{E}$. In addition to calculating $\bar{E}_i$, it is important to calculate the relationship between $\bar{E}$ and $\bar{P}$ when comparing the tissue elastic properties of different species or individuals (Tyree and Jarvis, 1982; Robichaux, 1984; Robichaux and Canfield, 1985). To

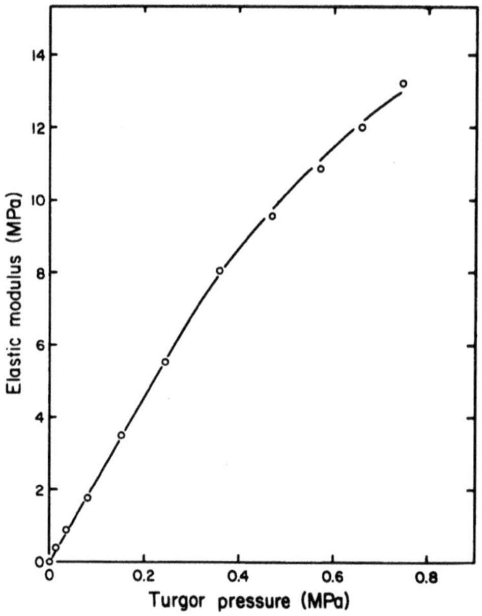

Fig. 9.4 Relationship between the tissue elastic modulus and tissue turgor pressure. The relationship was obtained from the data in Fig. 9.2. The tissue elastic modulus is defined as the change in $\bar{P}$ for a given fractional change in the weight of symplasmic water (see Equations 9.14 and 9.15).

determine this relationship, one calculates $\bar{E}$ for each successive set of four points in the $\bar{P}$ versus R' relationship (Fig. 9.2), following the procedure outlined above for $\bar{E}_i$. In other words, one calculates $\bar{E}$ for points 1–4, 2–5, 3–6, 4–7, etc., beginning near full hydration. The mean $\bar{P}$ value for each set of four points is then used in plotting $\bar{E}$ as a function of $\bar{P}$ (Fig. 9.4). (We have found four points to be optimal when the number of data points in the $\bar{P}$ versus R' relationship is moderately high. When fewer data points are available, $\bar{E}$ may be calculated for each successive set of three points.)

Data obtained with the pressure-chamber technique may be used to estimate diurnal changes in $\bar{P}$ in plants growing under field conditions. For comparable tissues, one measures the diurnal changes in Ψ according to the procedures discussed in Section 9.2.1, and the relationship between $\bar{P}$ and Ψ according to the procedures discussed in this section (Fig. 9.3). Given the assumption that the latter relationship does not change during the day, one may then calculate the diurnal changes in $\bar{P}$. We emphasize that the assumption of a diurnally constant relationship between $\bar{P}$ and Ψ may not always be valid, particularly for herbaceous species (Acevedo *et al.*, 1979) and for species growing under conditions of low water availability (Bowman and Roberts, 1985). Thus, this assumption should be checked for the species of interest in a particular study.

Three procedures are available for generating the relationship illustrated in Fig. 9.1. In all three procedures, a leaf (or branch) is collected in the field, sealed in a plastic bag and returned to the laboratory or field station, where the petiole (or stem) is recut under water to eliminate air in the xylem. (If the initial cut can be made under water when the leaf is harvested, there is no need to recut the petiole.) Care should be taken to keep the leaf blade and as much of the petiole as possible dry during this procedure. This minimizes the potential for errors associated with nonsymplasmic water loss. The leaf is allowed to rehydrate by storing it in a cool dark humid chamber with the end of its petiole submerged in distilled water. As a general rule, the rehydration period should not extend much beyond the point at which Ψ approaches zero. This minimizes the potential for errors associated with metabolic changes within the leaf. Once the leaf is fully rehydrated, its saturated weight is measured, preferably with a balance having 0.1 or 1 mg accuracy. Prior to measuring the saturated weight, it is advisable to recut the petiole above the level to which it was submerged during the rehydration. This further minimizes the potential for errors associated with nonsymplasmic water loss. Following the saturated weight determination, the initial Ψ value is measured with the pressure chamber. This initial Ψ value should be higher than -0.1 MPa if the leaf has fully rehydrated.

At this point, the three procedures diverge. In the first procedure, the leaf is allowed to dry under ambient conditions on the laboratory bench (Hinckley *et al.*, 1980; Robichaux, 1984; Kikuta *et al.*, 1985). At periodic intervals, the fresh weight and water potential of the leaf are measured. For each Ψ measurement, two fresh weight measurements are obtained, one immediately preceding and one immediately following the Ψ determination. The average of these two fresh weight measurements is used in the subsequent calculation of R'. To prevent a significant weight change during the measurement, the leaf may be enclosed in a plastic sheath while it is in the pressure chamber. The precautions discussed in Section 9.2.1 should be followed in measuring Ψ with the pressure chamber at each point. In addition, the rate at which the pressure is released from the chamber after each reading should be very slow, typically on the order of 0.003–0.005 MPa s^{-1}. A rapid release rate will result in a rapid decline in leaf temperature, which may cause condensation on the leaf surface, damage to the leaf membranes or even freezing of the leaf.

In the second procedure, the leaf loses water by overpressurization in the pressure chamber rather than by evaporation on the laboratory bench (Wilson *et al.*, 1979; Wilson and Ludlow, 1983). After the saturated weight and initial Ψ value are obtained, the leaf is exposed to a pressure 0.2–0.3 MPa in excess of the initial balance pressure. This overpressurization forces water out of the leaf more rapidly than it would be lost by evaporation. The expressed water is blotted from the petiole and discarded. After 4–10 min of overpressurization, the pressure is slowly released from the chamber, the leaf is removed and its fresh weight is measured. The leaf is then reinserted into the pressure chamber and the corresponding Ψ value measured. Following the Ψ determination, the leaf is exposed to another period of overpressurization, and the measurement cycle is repeated. Thus paired measurements of fresh weight and Ψ are obtained at periodic intervals as the leaf water content declines.

In the third procedure, the leaf also loses water by overpressurization in the pressure chamber (Cheung *et al.*, 1975; Ritchie and Roden, 1985). In this case, however, the expressed water is collected and weighed, while the leaf remains enclosed in the pressure chamber. After the saturated weight and initial Ψ value are obtained, the leaf is exposed to a pressure 0.2–0.3 MPa in excess of the initial balance pressure for 10 min, during which the expressed water is collected in a preweighed section of plastic tubing filled with dry tissue. (The tubing should fit tightly on the exposed end of the petiole to minimize evaporative water loss.) After 10 min, the tubing is removed and weighed rapidly. The weight of expressed water is subtracted from the leaf saturated weight to obtain the leaf fresh weight. The pressure in the chamber is lowered by 0.2–0.3 MPa for approximately 15 min, after which a new Ψ value is measured. The new Ψ value will depend on the amount of water lost during the previous overpressurization period. Following the new Ψ determination, the leaf is exposed to another period of overpressurization, and the measurement cycle is repeated. The leaf should be enclosed in a plastic sheath while it is in the pressure chamber, since evaporative water loss will introduce a significant error into the fresh weight calculation. In addition, a partial pressure of approximately 0.2 MPa of compressed air should be maintained in the pressure chamber throughout the measurement period, since continuous exposure of the leaf to pure nitrogen gas for a prolonged period may cause cell membrane damage (Cheung *et al.*, 1975).

In all three procedures, the measurements are carried out until the water content of the leaf has declined to 55–65% of its original saturated value. The leaf is then dried in an oven at 70–80°C for 18–24 h. For each Ψ measurement, the corresponding R' value is calculated as:

$$R' = \frac{\text{fresh weight} - \text{dry weight}}{\text{saturated weight} - \text{dry weight}} \qquad (9.17)$$

A plot of $1/\Psi$ versus R' yields the relationship illustrated in Fig. 9.1. The region over which this relationship is linear may be determined graphically with the aid of a stepwise linear regression program. When the data exhibit more scatter than that shown in Fig. 9.1, more sophisticated analytical methods may be used to determine the linear region (Powell and Blanchard, 1976; Sinclair and Venables, 1983). At high R' values, points will deviate from linearity because of positive $\bar{P}$ values. In some cases, points may also deviate from linearity at very low R' values (Wilson *et al.*, 1979). The latter deviation appears to result from cell death (Tyree *et al.*, 1973), together with the associated leakage of solutes into the apoplasm.

The three procedures offer different practical advantages. The first procedure offers the major advantage that six to seven leaf samples may be processed concurrently by one individual working with one pressure chamber. In the second and third procedures, in contrast, several pressure chambers are required for processing multiple samples, since the pressure chamber is used both for expressing water from the leaf and for measuring Ψ. The latter procedures, however, offer the advantage of being more rapid in many instances.

In principle, the three procedures should yield the same result. However, a recent analysis by Ritchie and Roden (1985) suggests that this may not be true in all cases. For those cases in which different results are obtained, there is currently no consensus as to which procedure yields the more accurate result.

Regardless of the procedure used, the pressure chamber technique has four major assumptions. These are: (1) the symplasmic value of Ψ is accurately measured by the pressure chamber (see Section 9.2.1), (2) negative values of $\bar{P}$ do not occur in the symplasm (i.e. $\bar{P} \geq 0$), (3) the net water loss from the tissue is entirely from the symplasm (i.e. W_a = constant = W_a^o), and (4) as water is lost from the symplasm, the concentration of solutes increases in an ideal manner (i.e. ϕ = constant). The consequences of violating these assumptions have recently been analyzed by Tyree and Karamanos (1981), Tyree and Richter (1981, 1982), Tyree and Jarvis (1982), Cortes and Sinclair (1985) and Robichaux *et al.* (1986).

The pressure-chamber technique is also subject to several other potential sources of error (Tyree and Karamanos, 1981; Tyree and Richter, 1981). These include: (1) gradual changes in $\bar{\pi}$ over the time course of the measurements, (2) plastic deformation of the cell walls over the time course of the measurements, (3) systematic and random errors in the measurement of Ψ, and (4) internal Ψ disequilibria. As discussed in Section 9.2.1, internal Ψ disequilibria may be particularly problematical in branches, compound leaves and severely stressed large leaves. According to Turner *et al.* (1984), allowing the leaf to lose water by evaporation rather than overpressurization may minimize the existence of internal Ψ disequilibria.

(b) Psychrometric techniques

Several techniques employing thermocouple psychrometers are also available for measuring the symplasmic values of $\bar{\pi}$ and $\bar{P}$ (Turner, 1981). The value of Ψ in a tissue sample is first measured with the psychrometer. Sap is then extracted from the sample with a small press. The sample may be pressed either in the fresh state (Markhart and Lin, 1985) or following freezing and thawing (Turner, 1981). The latter procedure aids in membrane disruption. The sap extract is then immediately placed on a filter paper disc in the psychrometer chamber and its value of $\bar{\pi}$ determined (Turner, 1981). Alternatively, the tissue sample may be frozen and thawed and its Ψ value measured a second time (Acevedo *et al.*, 1979). Since $\bar{P}$ is zero following freezing, Ψ equals $\bar{\pi}$ in the second measurement. Once Ψ and $\bar{\pi}$ are known, the value of $\bar{P}$ in the sample prior to pressing and freezing is calculated with Equation 9.5.

With these techniques, the symplasmic values of $\bar{\pi}$ and $\bar{P}$ may be measured much more rapidly than with the pressure-chamber technique. The psychrometric techniques are subject to a major source of error, however, in that the symplasmic water is diluted with apoplasmic water following membrane disruption by pressing or freezing (Tyree and Jarvis, 1982). Hence, the symplasmic value of $\bar{\pi}$ may be significantly overestimated with these techniques, resulting in erroneously low $\bar{P}$ estimates. In addition, ion exchange between symplasmic ions and exchange sites in the cell wall following membrane disruption may result in incorrect $\bar{\pi}$ and $\bar{P}$ estimates (Tyree and Jarvis, 1982).

9.2.3 Technique for measuring the apoplasmic value of $\bar{\pi}$

A major assumption of the pressure-chamber

technique for measuring Ψ is that the apoplasmic value of $\bar{\pi}$ is negligible (see Section 9.2.1). The latter value may be determined by placing a leaf in a pressure chamber, expressing water (or sap) from the xylem by overpressurization (see Section 9.2.2), and then measuring the $\bar{\pi}$ value of the expressed water (or sap) with a thermocouple psychrometer. It is best not to use the water that is first expressed, since it may be contaminated by solutes released from damaged cells at the cut surface of the petiole.

9.3 WATER CONTENT

In certain instances, decreases in tissue water content may be more important than decreases in Ψ or $\bar{P}$ in terms of influencing plant growth (Sinclair and Ludlow, 1985; Ludlow, 1987). For example, Kaiser (1982) has demonstrated a strong correlation between changes in leaf protoplast volume and changes in leaf photosynthetic activity. Thus, it is often important to measure both water content and Ψ for plants growing under field conditions.

Tissue water content may be expressed in several ways, including the amount of water per unit dry weight, per unit fresh weight and per unit weight of water at full hydration (Slatyer, 1967). The third expression, which equals R', is the most satisfactory in terms of quantifying tissue water deficits, since it is not influenced by changes in tissue dry weight (Slatyer, 1967; Turner, 1981).

Diurnal changes in R' either may be estimated using data obtained with the pressure chamber technique or may be measured directly. In the former case, the procedure is analogous to that described previously for estimating diurnal changes in $\bar{P}$. Similar precautions also hold for estimating R' in this manner.

In contrast to $\bar{P}$, R' may also be measured directly for higher plants growing under field conditions. Leaf samples (often as small discs) are collected in the field and placed immediately into hermetically sealed tared containers (Turner, 1981). (Any evaporative water loss introduces a significant error into the R' calculation.) After their fresh weights are obtained, the samples are rehydrated to obtain their saturated weights, then oven-dried to obtain their dry weights. Values of R' are then calculated with Equation 9.17.

As a modification of this method, some workers have used a β-ray absorption technique (Nakayama and Ehrler, 1964) to monitor the water content of a given leaf repeatedly, then harvested the leaf to obtain saturated and dry weights for the R' calculations (e.g. Schulze *et al.*, 1972). To be operational, however, the β-ray absorption technique has to be calibrated initially against direct measurements of leaf water content.

9.4 HYDRAULIC RESISTANCE AND CAPACITANCE

Even in the absence of soil water deficits, diurnal reductions in shoot water potential occur as a consequence of water loss in transpiration (Kozlowski, 1968; Ehrler *et al.*, 1978). The cause of these reductions in shoot water potential is the resistance to liquid water flux (hydraulic resistance) in the soil/plant transpirational path. Increased plant hydraulic resistance has been shown to be the mechanism by which some fungal pathogens cause wilting (Duniway, 1977; Olsen *et al.*, 1983). The measurement of plant hydraulic resistance has also been used to assess the functional significance of xylem type (Calkin *et al.*, 1985) and to help evaluate the factors limiting plant water uptake (Elfving *et al.*, 1972; Bates and Hall, 1982). Plant hydraulic resistance may also play a role in the determination of habitat, especially in relation to stomatal behavior. For example, Kuppers (1984) found that the combination of high hydraulic resistance and low stomatal response to humidity was possibly responsible for limiting *Ribes uva-crispa* to shaded habitats. When leaves were exposed to high light levels, desiccation occurred rapidly. Both the soil and the plant contribute to the overall resistance of the transpirational path. Only the contributions of the plant are discussed here.

9.4.1 Steady-state flux: calculation of the transpirational path hydraulic resistance

The most appropriate equation for calculating plant hydraulic resistance relates the steady-state water flux to the driving forces for liquid water movement and to the hydraulic resistance. Thus:

$$F = \frac{(\Delta\bar{P} + \bar{\sigma}\Delta\bar{\pi}) + \Delta\Psi_z}{R_h} \tag{9.18}$$

where F is the water flux through the resistor, $\Delta\bar{P}$ is the hydrostatic pressure gradient across the resistor, $\bar{\sigma}$ is the effective reflection coefficient of the resistor, $\Delta\bar{\pi}$ is the osmotic potential gradient across the resistor, $\Delta\Psi_z$ is the gravitational potential gradient across the resistor and R_h is the hydraulic resistance. The resistor may be the entire plant body or any part thereof. $\Delta\bar{P}$ and $\bar{\sigma}\Delta\bar{\pi}$ represent the hydrostatic and osmotic driving forces, respectively. Water flux through the entire plant body, from soil to atmosphere, may involve several parallel paths, some of which are dominated by hydrostatic driving forces and some of which have significant osmotic driving forces. If $\bar{\sigma}$ is 0 (as in the xylem), no osmotic driving forces are present.

It should be emphasized that Equation 9.18 is valid only for steady-state flux. For the whole plant, this would require that the water flux into the root system be equivalent to the water flux leaving the canopy. This strict steady-state condition probably rarely occurs in the field because a net movement of water between tissue water sources (capacitors) and the transpirational path can occur (see Section 9.4.2). It is probably true, however, that for the many plants for which the magnitude of the total water flux is much greater than that contributed by capacitors, the dynamics of water flux may be described by a series of steady states.

Despite the appropriateness of Equation 9.18 for describing water fluxes within the plant body, its utility is extremely limited, particularly under field conditions. This is primarily because of our current inability to measure accurately the component driving forces. If we assume $\bar{\sigma} = 1$ (most plant membranes have reflection coefficients of 1 for most pertinent solutes), then Equation 9.18 reduces to:

$$F = \frac{\Delta\Psi + \Delta\Psi_z}{R_h} \tag{9.19}$$

where $\Delta\Psi$ is the water potential gradient across the resistor. Although some theoretical situations have been described for which Equation 9.19 would not be appropriate (Fiscus *et al.*, 1983), many data sets are adequately handled by this equation. This is true because at moderate to high fluxes, osmotic forces probably play only a small role in water movement. Our inability to measure accurately $\Delta\bar{P}$, $\bar{\sigma}$ and $\Delta\bar{\pi}$ for field-grown plants makes necessary the use of Equation 9.19, at least as a first approach. Except in tall trees, Equation 9.19 may be further simplified by ignoring $\Delta\Psi_z$.

In many cases, a linear relationship between $\Delta\Psi$ and F is evident, particularly at moderate to high fluxes (Fig. 9.5). The slope of the linear region is taken to indicate R_h.

In the field, F through intact plants can be measured using a porometer or other gas-exchange apparatus (Roberts and Knoerr, 1978; Bates and Hall, 1982; Nobel and Jordan, 1983; Meinzer *et al.*, 1983; Kuppers, 1984), heat-pulse or heat-flow apparatus (Kucera *et al.*, 1977; Allaway *et al.*, 1981; Cohen *et al.*, 1983), or energy balance equations (Abdul-Jabbar *et al.*, 1984). Schulze *et al.* (1985) compared the use of the porometer and heat-

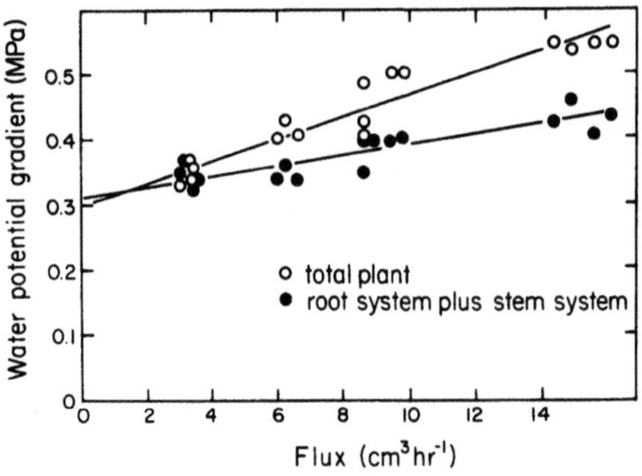

Fig. 9.5 Relationships between the water potential gradient ($\Delta\Psi$) and the steady-state water flux (F) in 8-week-old sunflower plants grown in soil in pots. Flux was measured gravimetrically. The soil was kept moist and was assumed to be at $\Psi = 0$ MPa at the time of the measurements. For the total plant, $\Delta\Psi$ was calculated from the soil to the transpiring leaf. Leaf Ψ was measured with a pressure chamber. The equation for the linear regression is $\Delta\Psi = 0.0171(F) + 0.297$. For the root system plus stem system, $\Delta\Psi$ was calculated from the soil to the stem. Stem Ψ was measured using covered (i.e. non-transpiring), fully-expanded leaves. The equation for the linear regression is: $\Delta\Psi = 0.0079(F) + 0.313$.

flow techniques. They found that the two techniques yield similar results in the estimation of water flow through the plant, though careful attention must be paid to the boundary layer resistance within the porometer cuvette. A high boundary layer resistance can lead to severely underestimated transpiration rates. For water flow through rootless plants, potometers can be used to measure F (Running, 1980).

Measurement of $\Delta\Psi$ can be accomplished using the psychrometric or pressure chamber techniques discussed in Section 9.2.1. Usually, the Ψ of transpiring leaves is measured to obtain one end of the Ψ gradient (the downstream end). Although the leaf system R_h may be large (Tyree and Cheung, 1977; Black, 1979; Koide, 1985a), in some cases it is the root system plus stem system R_h that is of most interest (Duniway, 1977; Koide 1985b). In these cases, it is sufficient to measure the Ψ of nontranspiring leaves to obtain the downstream end of the Ψ gradient. Plastic or aluminum foil wrappings are often used for this purpose (Begg and Turner, 1970; Black, 1979; Koide, 1985b). Values of root system plus stem system R_h have been calculated in this fashion for intact sunflower plants (Fig. 9.5). Alternatively, stem Ψ may be measured with stem thermocouple psychrometers (McBurney and Costigan, 1982), although problems associated with thermal gradients under field conditions may limit thermocouple psychrometer utility (see Section 9.2.1).

Measurement of Ψ at the upstream end of the Ψ gradient can be difficult. In theory, it is the average Ψ at the surface of the absorbing roots that is used in the calculation of root system or whole plant R_h. As mentioned previously, root psychrometers have been used (Fiscus, 1972), but it is usually easier to arrive at an average root surface Ψ by other means. The bulk soil Ψ measured with soil psychrometers has been used for this purpose (Allen *et al.*, 1981). There is some justification for doing so in comparative studies if it can safely be assumed that the Ψ gradients from bulk soil to roots are the same for all treatments. However, if the plants that are compared have markedly different transpiration rates or different rooting densities, and particularly if they grow in very dry light-textured soils, large differences in the Ψ gradients from bulk soil to roots might be expected to occur (Faiz and Weatherley, 1977). Jones (1983) showed that the root surface Ψ of field-grown water-stressed plants could be estimated from the leaf Ψ and leaf vapor conductances of the stressed plants and the well-watered controls. Other workers have measured pre-dawn leaf Ψ to estimate root surface Ψ, but upon initiation of transpiration, the Ψ of the rooting zone may be somewhat lower than it was before dawn.

If it can be assumed that the root surface Ψ does not change with time or transpiration rate (such as in moist soils of high hydraulic conductivity), then it may not be necessary to know the actual $\Delta\Psi$ driving the transpirational flux. Instead, the slope of the relationship between leaf Ψ and F should be R_h. The extrapolated intercept on the ordinate will equal the effective root surface Ψ, provided that $\bar{\sigma} = 1$ and steady-state flux prevails.

Richter (1973) and Jarvis (1975), among others, have pointed out that resistances, which are based on fluxes ($cm^3\ s^{-1}$), are often confused with resistivities, which are based on flux densities ($cm^3\ s^{-1}\ cm^{-2}$), the area being that of leaves, roots or membranes. Resistivities are not serially additive as are resistances, but their calculation may be worthwhile for comparative purposes, particularly for organs such as roots that differ in size (Newman, 1973; Fiscus and Markhart, 1979). Calculation of resistivities on a leaf area basis may also be justified since leaf area can be linearly related to the xylem cross-sectional area supplying the leaf (Salleo *et al.*, 1985).

9.4.2 Nonsteady-state flux: calculation of capacitance

Under many natural circumstances, plant water flux cannot be considered to be steady state; that is, the amount of water absorbed by the root system is not equal to that transpired by the canopy. In plants with meristematic and young tissue, some nonsteady-state water flux occurs owing to the expansion of growing cells (Boyer, 1985). In other cases, a net flux of water can occur out of or into nongrowing tissues not normally viewed as part of the transpirational path. Such tissue water sources (or sinks), whether living (symplasmic) or nonliving (apoplasmic), are called water capacitors.

Any tissue that is not separated from the transpirational path by an infinite hydraulic resistance can serve as a capacitor. Water exchange from capacitors occurs on a diurnal and seasonal basis (Kozlowski, 1968; Waring and Running, 1978). The shrinking and swelling of tissues are the result of symplasmic capacitors losing and gaining water (Kozlowski, 1968). Tissues such as leaves (Meidner, 1952), stems, fruits (Kozlowski, 1968) and roots (Huck *et al.*, 1970) have been shown to change physical size as a consequence of water exchange. Root shrinkage may be important to overall plant hydraulic resistance because it can reduce the contact surface area between root and soil (Faiz and Weatherley, 1982). Wood may serve as an apoplasmic capacitor, although cavitation, not shrinkage, is the result of water loss (Siau, 1971; Waring and Running, 1978).

In some cases, capacitor water can make a significant contribution to transpiration. In *Agave deserti*, stored water can sustain maximal transpiration for up to 16 h (Nobel and Jordan, 1983). Many trees also store water that can contribute significantly to transpiration (Jarvis, 1975; Waring and Running, 1978; Waring *et al.*, 1979; Schulze *et al.*, 1985). In other cases, capacitor water may make only a slight contribution to transpiration. Nevertheless, any exchange of water between plant tissues and the transpirational path may have an effect on the local tissue Ψ. The amount of change in Ψ for a given amount of water exchange depends on the capacitance of the tissue.

The tissue capacitance (C) is defined as the change in tissue water volume (V) for a given change in tissue Ψ (Nobel, 1983). Hence:

$$C = \frac{\Delta V}{\Delta\Psi} \tag{9.20}$$

Clearly, the capacitance of a tissue consisting of 100 cells would be 10 times greater than that of a tissue consisting of 10 cells, assuming cell size and all else to be equal. Thus, for comparing differences in capacitors that are not solely related to differences in total water volume, a volume-normalized capacitance ($\bar{C}$) can be defined as:

$$\bar{C} = \frac{\Delta V}{\Delta\Psi}\,\frac{1}{V_o} \tag{9.21}$$

where V_o is the tissue water volume at full hydration. Since V/V_o equals R', $\bar{C}$ can be expressed as:

$$\bar{C} = \frac{\Delta R'}{\Delta \Psi} \tag{9.22}$$

(Nobel and Jordan, 1983).

The principal method for measuring $\bar{C}$ in higher plants growing under field conditions is the pressure-chamber technique. This technique is very similar to that discussed in Section 9.2.2 for measuring the symplasmic values of $\bar{\pi}$ and $\bar{P}$. Instead of plotting the relationship between $1/\Psi$ and R' (Fig. 9.1), one simply plots the relationship between Ψ and R' (Fig. 9.6). The reciprocal of the slope of the latter relationship equals $\bar{C}$. The various procedures for generating this relationship are discussed in detail in Section 9.2.2.

The rate of water exchange between the capacitor and the transpirational path depends on the gradient in Ψ and on the hydraulic resistance to water exchange (the transfer hydraulic resistance). By analogy to electrical circuits, Nobel (1983) showed that the time constant (τ') for water exchange between the capacitor and the transpirational path equals R_h^tC, where R_h^t is the transfer hydraulic resistance.

The pressure-chamber technique for measuring τ', and thus R_h^t, has been discussed in detail by Nobel and Jordan (1983). A leaf (or other organ) is placed in the pressure chamber and an initial Ψ value (Ψ_a) is obtained. The pressure in the chamber is then increased to a value (corresponding to Ψ_b) that is 0.2 MPa in excess of the initial balance pressure. Thus, $\Psi_b = \Psi_a - 0.2$. (For some species, a value other than 0.2 MPa may be more appropriate for this overpressurization.) The water expressed from the leaf by this overpressurization is blotted from the petiole and discarded. At various time intervals, the pressure in the chamber is briefly lowered to the point at which exudation just ceases. Thus, for each time interval (t), a corresponding Ψ value (Ψ_t) is obtained, with $\Psi_a > \Psi_t > \Psi_b$. The value of τ' is then calculated with the equation:

$$\Psi_t = (\Psi_a - \Psi_b)e^{-t/\tau'} + \Psi_b \tag{9.23}$$

If C is known for this same range of Ψ values, then R_h^t may be calculated as τ'/C (Nobel and Jordan, 1983). The precautions associated with this technique are similar to those discussed for the pressure-chamber techniques of Sections 9.2.1 and 9.2.2.

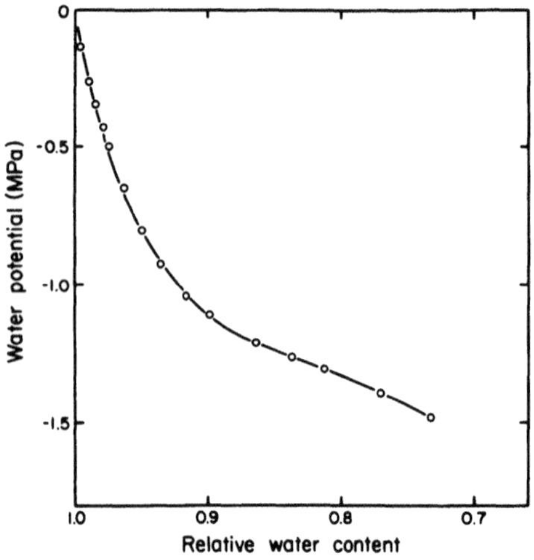

Fig. 9.6 Relationship between tissue water potential and tissue relative water content. The data are the same as in Fig. 9.1. The volume-normalized capacitance ($\bar{C}$) of the tissue equals the change in R' for a given change in Ψ (see Equation 9.22). Thus, $\bar{C}$ equals the reciprocal of the slope of this relationship.

9.5 CONCLUSION

In this chapter, we have provided a brief introduction to the theory and measurement of plant water status, hydraulic resistance and capacitance. In conclusion, we emphasize that we currently seem limited more by the accuracy and precision of our measurements than by the theoretical framework supporting them. For example, while the steady-state water flux equation (Equation 9.18) is theoretically correct, its utility for calculating hydraulic resistance is limited by our inability to measure the component driving forces accurately. Thus, while the concepts and techniques discussed in this chapter should serve as a useful framework for field research, the limitations of our current methods should always be kept in mind.

REFERENCES

Abdul-Jabbar, A.S., Lugg, D.G., Sammis, T.W. and Gay, L.W. (1984) A field study of plant resistance to water flow in alfalfa. *Agron. J.*, **76**, 765–9.

Acevedo, E., Fereres, E., Hsiao, T.C. and Henderson, D.W. (1979) Diurnal growth trends, water potential, and osmotic adjustment of maize and sorghum leaves in the field. *Plant Physiol.*, **64**, 476–80.

Allaway, W.G., Pitman, M.G., Storey, R. and Tyerman, S. (1981) Relationships between sap flow and water potential in woody or perennial plants on islands of the Great Barrier Reef. *Plant, Cell Environ.*, **4**, 329–37.

Allen, M.F., Smith, W.K., Moore Jr., T.S. and Christensen, M. (1981) Comparative water relations and photosynthesis of mycorrhizal and non-mycorrhizal *Bouteloua gracilis* H.B.K. Lag ex steud. *New Phytol.*, **88**, 683–93.

Barnes, P.W. (1985) Adaptation to water stress in the big bluestem–sand bluestem complex. *Ecology*, **66**, 1908–20.

Barrs, H.D. (1964) Heat of respiration as a possible cause of error in the estimation by psychrometric methods of water potential in plant tissue. *Nature, London* **203**, 1136–7.

Barrs, H.D. (1968) Determinations of water deficits in plant tissues. In *Water Deficits and Plant Growth* (ed. T.T. Kozlowski), Academic Press, New York, Vol. 1, pp. 235–368.

Barrs, H.D. and Kramer, P.J. (1969) Water potential increase in sliced leaf tissue as a cause of error in vapour phase determinations of water potential. *Plant Physiol.*, **44**, 959–64.

Bates, L.M. and Hall, A.E. (1982) Relationships between leaf water status and transpiration of cowpea with progressive soil drying. *Oecologia*, **53**, 285–9.

Begg, J.E. and Turner, N.C. (1970) Water potential gradients in field tobacco. *Plant Physiol.*, **46**, 343–6.

Black, C.R. (1979) The relationship between transpiration rate, water potential, and resistance to water movement in sunflower (*Helianthus annuus* L.). *J. Exp. Bot.*, **30**, 235–43.

Bowman, W.D. and Roberts, S.W. (1985) Seasonal and diurnal water relations adjustments in three evergreen chaparral shrubs. *Ecology*, **66**, 738–42.

Boyer, J.S. (1967) Leaf water potentials measured with a pressure chamber. *Plant Physiol.*, **42**, 133–7.

Boyer, J.S. (1969) Measurement of water status of plants. *Annu. Rev. Plant Physiol.*, **20**, 351–64.

Boyer, J.S. (1972) Use of isopiestic technique in thermocouple psychrometry. II. Construction. In *Psychrometry in Water Relations Research* (eds R.W. Brown and B.P. van Haveren), Utah Agricultural Experiment Station, Utah State University, pp. 98–102.

Boyer, J.S. (1985) Water transport. *Annu. Rev. Plant Physiol.*, **36**, 473–516.

Boyer, J.S. and Knipling, E.B. (1965) Isopiestic technique for measuring leaf water potentials with a thermocouple psychrometer. *Proc. Natl. Acad. Sci., USA*, **54**, 1044–51.

Brinckmann, E., Tyerman, S.D., Steudle, E. and Schulze, E.-D. (1984) The effect of different growing conditions on water relations parameters of leaf epidermal cells of *Tradescantia virginiana* L. *Oecologia*, **62**, 110–17.

Briscoe, R. (1986) Thermocouple psychrometers for water potential measurements. In *Advanced Agricultural Instrumentation: Design and Use* (ed. W.G. Gensler), Martinus Nijhoff, Dordrecht, pp. 193–209.

Brown, P.W. and Tanner, C.B. (1981) Alfalfa water potential measurement: a comparison of the pressure chamber and leaf dew-point hygrometers. *Crop Sci.*, **21**, 240–4.

Brown, R.W. and van Haveren, B.P. (eds) (1972) *Psychrometry in Water Relations Research*, Utah Agricultural Experiment Station, Utah State University.

Calkin, H.W., Gibson, A.C. and Nobel, P.S. (1985) Xylem water potentials and hydraulic conductances in eight species of ferns. *Can. J. Bot.*, **63**, 632–7.

Calkin, H.W. and Pearcy, R.W. (1984) Seasonal progressions of tissue and cell water relations parameters in evergreen and deciduous perennials. *Plant, Cell Environ.*, **7**, 347–52.

Campbell, E.C., Campbell, G.S. and Barlow, W.K. (1973) A dewpoint hygrometer for water potential measurement. *Agric. Meteorol.*, **12**, 113–21.

Campbell, G.S. and Campbell, M.D. (1974) Evaluation of a thermocouple hygrometer for measuring leaf water potential *in situ*. *Agron. J.*, **66**, 24–7.

Cheung, Y.N.S., Tyree, M.T. and Dainty, J. (1975) Water relations parameters on single leaves obtained in a pressure bomb and some ecological interpretations. *Can. J. Bot.*, **53**, 1342–6.

Cohen, Y., Fuchs, M. and Cohen, S. (1983) Resistance to water uptake in a mature citrus tree. *J. Exp. Bot.*, **34**, 451–60.

Cortes, P.M. and Sinclair, T.T. (1985) Extraction of apoplastic water during pressure-volume dehydrations. *Agron. J.*, **77**, 798–802.

Cosgrove, D.J., van Volkenburgh, E. and Cleland, R.E. (1984) Stress relaxation of cell walls and the yield threshold for growth. Demonstration and measurement for micro-pressure probe and psychrometer techniques. *Planta*, **162**, 46–54.

Davis, S.D. and Mooney, H.A. (1986) Tissue water relations of four co-occurring chaparral shrubs. *Oecologia*, **70**, 527–35.

Dixon, M. and Grace, J. (1982) Water uptake by some chamber materials: technical report. *Plant, Cell Environ.*, **5**, 323–7.

Drake, B.G. and Gallagher, J.L. (1984) Osmotic potential and turgor maintenance in *Spartina alterniflora* Loisel. *Oecologia*, **62**, 368–75.

Duniway, J.M. (1977) Changes in resistance to water transport in safflower during the development of phytophthora root rot. *Phytopathology*, **67**, 331–7.

Ehrler, W.L., Idso, S.B., Jackson, R.D. and Reginato, R.J. (1978) Diurnal changes in plant water potential and canopy temperature of wheat as affected by drought. *Agron. J.*, **70**, 999–1009.

Elfving, C.C., Kaufman, M.R. and Hall, A.E. (1972) Intepreting leaf water potential measurements with a model of the soil–plant–atmosphere continuum. *Physiol. Plant.*, **27**, 161–8.

Faiz, S.M.A. and Weatherley, P.E. (1977) The location of the resistance to water movement in the soil supplying the roots of transpiring plants. *New Phytol.*, **78**, 337–47.

Faiz, S.M.A. and Weatherley, P.E. (1982) Root contraction in transpiring plants. *New Phytol.*, **92**, 333–43.

Fiscus, E.L. (1972) *In situ* measurement of root-water potential. *Plant Physiol.*, **50**, 191–3.

Fiscus, E.L., Klute, A. and Kaufman, M.R. (1983) An interpretation of some whole plant water transport phenomena. *Plant Physiol.*, **71**, 810–17.

Fiscus, E.L. and Markhart, A.H. (1979) Relationship between root system water transport properties and plant size in *Phaseolus*. *Plant Physiol.*, **64**, 770–3.

Goldstein, G., Meinzer, F. and Monasterio, M. (1984) The role of capacitance in the water balance of Andean giant rosette species. *Plant, Cell Environ.*, **7**, 179–86.

Hinckley, T.M., Duhme, F., Hinckley, A.R. and Richter, H. (1980) Water relations of drought hardy shrubs: osmotic potential and stomatal reactivity. *Plant, Cell Environ.*, **3**, 131–40.

Huck, M.G., Klepper, B. and Taylor, H.M. (1970) Diurnal variations in root diameter. *Plant Physiol.*, **45**, 529–30.

Husken, D., Steudle, E. and Zimmermann, U. (1978) Pressure probe technique for measuring water relations of cells of higher plants. *Plant Physiol.*, **61**, 158–63.

Jarvis, P.G. (1975) Water transfer in plants. In *Heat and Mass Transfer in the Biosphere. 1. Transfer Processes in the Plant Environment* (eds D.A. de Vries and M.H. Afgan), Wiley and Sons, New York, pp. 369–94.

Jones, H.G. (1983) Estimation of an effective soil water potential at the root surface of transpiring plants. *Plant, Cell Environ.*, **6**, 671–4.

Kaiser, W.M. (1982) Correlation between changes in photosynthetic activity and changes in total protoplast volume in leaf tissue from hygro-, meso- and xerophytes under osmotic stress. *Planta*, **154**, 538–45.

Kikuta, S.B., Kyriakopoulous, E. and Richter, H. (1985) Leaf hygrometer v. pressure chamber: a comparison of pressure-volume curve data obtained on single leaves by alternating measurements. *Plant, Cell Environ.*, **8**, 363–7.

Knapp, A.K. (1984) Water relations and growth of three grasses during wet and drought years in a tallgrass prairie. *Oecologia*, **65**, 35–43.

Koide, R. (1985a) The nature and location of variable hydraulic resistance in *Helianthus annuus* L. (sunflower). *J. Exp. Bot.*, **36**, 1430–40.

Koide, R. (1985b) The effect of VA mycorrhizal infection and phosphorus status on sunflower hydraulic and stomatal properties. *J. Exp. Bot.*, **36**, 1087–98.

Kozlowski, T.T. (1968) Introduction. In *Water Deficits and Plant Growth* (ed. T.T. Kozlowski), Academic Press, New York, Vol. 1, pp. 1–21.

Kucera, J., Cermak, J. and Penka, M. (1977) Improved thermal method of continual recording of transpiration flow rate dynamics. *Biol. Plant.*, **19**, 413–20.

Küppers, M. (1984) Carbon relations and competition between woody species in a central European hedgerow. II. Stomatal responses, water use and hydraulic conductivity in the root/leaf pathway. *Oecologia*, **64**, 344–54.

List, R.J. (1968) *Smithsonian Meteorological Tables*, 6th revised edn, Smithsonian Institution Press, Washington D.C.

Ludlow, M.M. (1987) Defining shoot water status in the most meaningful way to relate to physiological processes. *Proc. Int. Conf. Measurement Soil Plant Water Status*, **2**, 47–53.

Markhart, III, A.H. and Lin, T. (1985) New hand-operated press for the extraction of tissue sap for the measurement of osmotic potential. *Agron. J.*, **77**, 182–5.

McBurney, T. and Costigan, P.A. (1982) Measurements of stem water potential of young plants using a hygrometer attached to the stem. *J. Exp. Bot.*, **33**, 426–31.

Meidner, H. (1952) An instrument for the continuous determination of leaf thickness changes in the field. *J. Exp. Bot.*, **3**, 319–25.

Meinzer, F., Seymour, V. and Goldstein, G. (1983) Water balance in developing leaves of four tropical savanna woody species. *Oecologia*, **60**, 237–43.

Michel, B.E. (1977) A miniature stem thermocouple hygrometer. *Plant Physiol.*, **60**, 645–7.

Millar, B.D. and Hansen, G.K. (1975) Exclusion errors in pressure chamber estimates of leaf water potential. *Ann. Bot.*, **39**, 915–20.

Monson, R.K. and Smith, S.D. (1982) Seasonal water potential components of Sonoran Desert plants. *Ecology*, **63**, 113–23.

Myers, B.J., Robichaux, R.H., Unwin, G.L. and Craig, I.E. (1987) Leaf water relations and anatomy of a tropical rainforest tree species vary with canopy position. *Oecologia*, **74**, 81–5.

Nakayama, F.M. and Ehrler, W.L. (1964) Beta ray gauging technique for measuring leaf water content changes and moisture status of plants. *Plant Physiol.*, **39**, 95–8.

Nelsen, C.E., Safir, F.R. and Hanson, A.D. (1978) Water potential in excised leaf tissue: comparison of a commercial dewpoint hygrometer and thermocouple psychrometer on soybean, wheat and barley. *Plant Physiol.*, **61**, 131–3.

Neumann, H.H. and Thurtell, G.W. (1972) A peltier cooled thermocouple dewpoint hygrometer for *in situ* measurement of water potentials. In *Psychrometry in Water Relations Research* (eds R.W. Brown and B.P. van Haveren), Utah Agricultural Experiment Station, Utah State University, pp. 103–12.

Newman, E.I. (1973) Permeability to water of the roots of five herbaceous species. *New Phytol.*, **72**, 547–55.

Nilsen, E.T., Sharifi, M.R. and Rundel, P.W. (1984) Comparative water relations of phreatophytes in the Sonoran Desert of California. *Ecology*, **65**, 767–78.

Nilsen, E.T., Sharifi, M.R., Rundel, P.W., Jarrell, W.M. and Virginia, R.A. (1983) Diurnal and seasonal water relations of the desert phreatophyte *Prosopis glandulosa* (honey mesquite) in the Sonoran Desert of California. *Ecology*, **64**, 1381–93.

Nilsen, E.T., Sharifi, M.R., Rundel, P.W. and Virginia, R.A. (1986) Influences of microclimatic conditions and water relations on seasonal leaf dimorphism of *Prosopis glandulosa* var. *torreyana* in the Sonoran Desert, California. *Oecologia*, **69**, 95–100.

Nobel, P.S. (1983) *Biophysical Plant Physiology and Ecology*, W.H. Freeman, San Francisco.

Nobel, P.S. and Jordan, P.W. (1983) Transpiration stream of desert species: resistances and capacitances for a C_3, a C_4, and a CAM plant. *J. Exp. Bot.*, **34**, 1379–91.

Oberbauer, S.F. (1983) The ecophysiology of *Pentaclethra macroloba*, a canopy tree species in the rainforest of Costa Rica, Ph.D. dissertation, Duke University, Durham, North Carolina.

Olsen, M.W., Misaghi, I.J., Goldstein, D. and Hine, R.B. (1983) Water relations in cotton plants infected with *Phymatotrichum*. *Phytopathology*, **73**, 213–16.

Parker. W.C., Pallardy, S.G., Hinckley, T.M. and Teskey, R.O. (1982) Seasonal changes in tissue water relations of three woody species of the *Quercus–Carya* forest type. *Ecology*, **63**, 1259–67.

Passioura, J.B. (1982) Water in the soil–plant–atmosphere continuum. In *Encyclopedia of Plant Physiology*, New Series, Vol. 12B (eds O.L. Lange, P.S. Nobel, C.B. Osmond and H. Ziegler) Springer-Verlag, Berlin, pp. 5–33.

Pavlik, B.M. (1984) Seasonal changes of osmotic pressure, symplasmic water content and tissue elasticity in the blades of dune grasses growing *in situ* along the coast of Oregon. *Plant, Cell Environ.*, **7**, 531–9.

Powell, D.B.B. and Blanchard, T.W. (1976) The quantitative analysis of water potential/symplasmic water volume curves. *J. Exp. Bot.*, **27**, 597–607.

Richards, L.A. and Ogata, G. (1958) Thermocouple for vapor pressure measurement in biological and soil systems at high humidity. *Science*, **128**, 1089–90.

Richter, H. (1973) Frictional potential losses and total water potential in plants: a re-evaluation. *J. Exp. Bot.*, **24**, 983–94.

Ritchie, G.A. and Hinckley, T.M. (1975) The pressure chamber as an instrument for ecological research. *Adv. Ecol. Res.*, **9**, 165–254.

Ritchie, G.A. and Roden, J.R. (1985) Comparison between two methods of generating pressure-volume curves. *Plant, Cell Environ.*, **8**, 49–53.

Roberts, S.W. and Knoerr, K.R. (1978) *In situ* estimates of variable plant resistance to water flux in *Ilex opaca* Ait. *Plant Physiol.*, **61**, 311–13.

Roberts, S.W., Strain, B.R. and Knoerr, K.R. (1980) Seasonal patterns of leaf water relations in four co-occurring forest tree species: parameters from pressure-volume curves. *Oecologia*, **46**, 330–7.

Roberts, S.W., Strain, B.R. and Knoerr, K.R. (1981) Seasonal variation in leaf tissue elasticity in four forest tree species. *Physiol. Plant.*, **52**, 245–50.

Robichaux, R.H. (1984) Variation in the tissue water relations of two sympatric Hawaiian *Dubautia* species and their natural hybrid. *Oecologia*, **65**, 75–81.

Robichaux, R.H. and Canfield, J.E. (1985) Tissue elastic properties of eight Hawaiian *Dubautia* species that differ in habitat and diploid chromosome number. *Oecologia*, **66**, 77–80.

Robichaux, R.H., Holsinger, K.E. and Morse, S.R. (1986) Turgor maintenance in Hawaiian *Dubautia* species: the role of variation in tissue osmotic and elastic properties. In *On the Economy of Plant Form and Function* (ed. T.J. Givnish), Cambridge University Press, Cambridge, pp. 353–80.

Running, S.W. (1980) Field estimates of root and xylem resistance in *Pinus contorta* using root excision. *J. Exp. Bot.*, **31**, 555–69.

Salleo, S., Lo Gullo, M.A. and Oliveri, F. (1985) Hydraulic parameters measured in 1-year-old twigs of some Mediterranean species with diffuse-porous wood: changes in hydraulic conductivity and their possible functional significance. *J. Exp. Bot.*, **36**, 1–11.

Savage, M.J., Wiebe, H.H. and Cass, A. (1983) *In situ* field measurement of leaf water potential using thermocouple psychrometers. *Plant Physiol.*, **73**, 609–13.

Savage, M.J., Wiebe, H.H. and Cass, A. (1984) Effect of cuticular abrasion on thermocouple psychrometric *in situ* measurement of leaf water potential. *J. Exp. Bot.*, **35**, 36–42.

Scholander, P.F., Hammel, H.T., Bradstreet, E.D. and Hemmingsen, E.A. (1965) Sap pressure in vascular plants. *Science*, **148**, 339–46.

Scholander, P.F., Hammel, H.T., Hemmingsen, E.A. and Bradstreet, E.D. (1964) Hydrostatic pressure and osmotic potential in leaves of mangroves and some other plants. *Proc. Natl. Acad. Sci., USA*, **52**, 119–25.

Schulze, E.-D., Cermak, J., Matyssek, R., Penka, M., Zimmermann, R., Vasicek, F., Gries, W. and Kucera, J. (1985) Canopy transpiration and water fluxes in the xylem of the trunk of *Larix* and *Picea* trees – a comparison of xylem flow, porometer and cuvette measurements. *Oecologia*, **66**, 475–83.

Schulze, E.-D., Lange, O.L., Buschbom, U., Kappen, L. and Evenari, M. (1972) Stomatal responses to changes in humidity in plants growing in the desert. *Planta*, **108**, 259–70.

Shackel, K.A. (1984) Theoretical and experimental errors for *in situ* measurements of plant water potential. *Plant Physiol.*, **75**, 766–72.

Shackel, K.A. and Brinckmann, E. (1985) *In situ* measurement of epidermal cell turgor, leaf water potential, and gas exchange in *Tradescantia virginiana* L. *Plant Physiol.*, **78**, 66–70.

Siau, J.F. (1971) Flow in wood. In *Syracuse Wood Science*, Series 1. Syracuse University Press, Syracuse, New York, p. 131

Sinclair, R. and Venables, W.N. (1983) An alternative method for analysing pressure–volume curves produced with the pressure chamber. *Plant, Cell Environ.*, **6**, 211–17.

Sinclair, T.R. and Ludlow, M.M. (1985) Who taught plants thermodynamics? The unfulfilled potential of water potential. *Austr. J. Plant Physiol.*, **12**, 213–18.

Slatyer, R.O. (1967) *Plant–Water Relationships*, Academic Press, London.

Slavik, B. (1974) *Methods of Studying Plant Water Relations*, Academia Publishing House, Prague and Springer-Verlag, Berlin.

Spanner, D.C. (1951) The peltier effect and its use in the measurement of suction pressure. *J. Exp. Bot.*, **2**, 145–68.

Turner, N.C. (1981) Techniques and experimental approaches for the measurement of plant water status. *Plant Soil*, **58**, 339–66.

Turner, N.C. (1987) The use of the pressure chamber in studies of plant water status. *Proc. Int. Conf. Measurement Soil Plant Water Status*, **2**, 13–24.

Turner, N.C. and Long, M.J. (1980) Errors arising from rapid water loss in the measurement of leaf water potential by the pressure chamber technique. *Austr. J. Plant Physiol.*, **7**, 527–37.

Turner, N.C., Spurway, R.A. and Schulze, E.-D. (1984) Comparison of water potentials measured by *in situ* psychrometry and pressure chamber in morphologically different species. *Plant Physiol.*, **74**, 316–19.

Tyree, M.T. (1981) The relationship between the

bulk modulus of elasticity of a complex tissue and the mean modulus of its cells. *Ann. Bot.*, **47**, 547–59.

Tyree. M.T. and Cheung, Y.N.S. (1977) Resistance to water flow in *Fagus grandifolia* leaves. *Can. J. Bot.*, **55**, 2591–9.

Tyree, M.T., Dainty, J. and Benis, M. (1973) The water relations of hemlock (*Tsuga canadensis*). I. Some equilibrium water relations as measured by the pressure bomb technique. *Can. J. Bot.*, **51**, 1471–80.

Tyree, M.T. and Hammell, H.T. (1972) The measurement of turgor pressure and the water relations of plants by the pressure-bomb technique. *J. Exp. Bot.*, **23**, 267–82.

Tyree, M.T. and Jarvis, P.G. (1982) Water in tissues and cells. In *Encyclopedia of Plant Physiology*, New Series, Vol. 12B (eds O.L. Lange, P.S. Nobel, C.B. Osmond and H. Ziegler), Springer-Verlag, Berlin, pp. 35–77.

Tyree, M.T. and Karamanos, A.J. (1981) Water stress as an ecological factor. In *Plants and their Atmospheric Environment* (eds J. Grace, E.D. Ford and P.G. Jarvis) Blackwell, Oxford, pp. 237–61.

Tyree, M.T. and Richter, H. (1981) Alternative methods of analysing water potential isotherms: some cautions and clarifications. I. The impact of non-ideality and of some experimental errors. *J. Exp. Bot.*, **32**, 643–53.

Tyree, M.T. and Richter, H. (1982) Alternative methods of analysing water potential isotherms: some cautions and clarifications. II. Curvilinearity in water potential isotherms. *Can. J. Bot.*, **60**, 911–16.

Waring, R.H. and Running, S.W. (1978) Sapwood water storage: its contribution to transpiration and its effect upon water conductance through stems of old-growth Douglas fir. *Plant, Cell Environ.*, **1**, 131–40.

Waring, R.H., Whitehead, D. and Jarvis, P.G. (1979) The contribution of stored water to transpiration in Scots pine. *Plant, Cell Environ.*, **2**, 309–17.

Wilson, J.R., Fisher, M.J., Schulze, E.D. and Dolby, G.R. (1979) Comparison between pressure-volume and dewpoint-hygrometry techniques for determining the water relations characteristics of grass and legume leaves. *Oecologia*, **41**, 77–88.

Wilson, J.R. and Ludlow, M.M. (1983) Time trends for change in osmotic adjustment and water relations of leaves of *Cenchrus ciliaris* during and after water stress. *Austr. J. Plant. Physiol.*, **10**, 15–24.

Zimmermann, U., Husken, D. and Schulze, E.-D. (1980) Direct turgor pressure measurements in individual leaf cells of *Tradescantia virginiana*. *Planta*, **149**, 445–53.

10

Approaches to studying nutrient uptake, use and loss in plants

F. Stuart Chapin, III and Keith Van Cleve

10.1 INTRODUCTION

Although most basic principles and techniques for studying mineral nutrition were developed in agriculture, they can, with some modification, be used for studying the mineral nutrition of wild plants. In this chapter we describe procedures that have proven useful in ecology and forestry and discuss the advantages, disadvantages and potential sources of error involved with each method.

The study of nutritional ecology is important, because nutrients are almost never present in the exact balance and amounts that are optimal for plant growth. Consequently, plants are always compensating for stresses imposed by their nutritional environment. Because nutrient stress is ubiquitous, it strongly influences the physiological adaptations and the distribution of plants. There are many different mineral nutrients in soil; some are seldom available in sufficient supply (e.g. nitrogen), whereas others are potentially toxic (e.g. lead). Some essential nutrients (micronutrients) are required in extremely small amounts, whereas others may comprise 1–5% of tissue dry weight (Epstein, 1972; Gauch, 1972; Table 10.1). The relative importance of different soil minerals in influencing plant growth depends upon specific soil conditions and the physiological adaptations of plants that occupy a site. Moreover, nutrient stress strongly influences and is influenced by other environmental stresses (Bloom *et al.*, 1985), so that study of plant response to any environmental stress requires understanding of plant nutrition.

10.2 NUTRIENT UPTAKE

An understanding of the controls over nutrient uptake is essential to any study of the nutrient budget of a plant or its relationship to the soil-nutrient environment. Depending upon the question, nutrient uptake might be measured in different ways. The rate of nutrient acquisition *in situ* is best measured by sequential harvests. This is of particular interest in ecosystem studies. In contrast, the physiological potential of species to absorb nutrients under specific conditions is best determined by soil labeling experiments, short-term uptake measurements or growth of plants in common gardens. This may be of more interest in ecophysiological studies.

Table 10.1 Concentrations of macronutrients and micronutrients in leaves that are considered deficient, adequate or toxic for various species of crop plants. Nonessential elements are listed at concentrations typically found in crops. From Epstein (1972) and Gauch (1972)

	Concentration ($mg\ g^{-1}$)		
	Deficient	*Adequate*	*Toxic*
Macronutrients			
Nitrogen	<15	15–20	
Potassium	<4–14	7–35	
Calcium	<2–6	4–15	
Magnesium	<0.2–3	2–9	
Phosphorus	<1–2	1–4	
Sulfur	<0.4–2.4	0.8–2.7	3–8
Micronutrients			
Chlorine	<0.00004–0.0002	0.0002–26	7–21
Iron	<0.03–0.10	0.04–0.25	
Manganese	<0.0002–0.018	0.002–0.40	0.17–11.0
Boron	<0.006–0.05	0.01–0.65	0.03–1.4
Zinc	<0.004–0.026	0.004–0.23	0.53–1.5
Copper	<0.002–0.004	0.003–0.03	0.023
Molybdenum	<0.00001–0.0003	0.00003–0.113	
Nonessential elements			
Silicon		12.0	
Aluminum		1.1	
Sodium		0.02	

10.2.1 Uptake measured by harvest

(a) General principles

The rate of net nutrient uptake can be measured as the increase in the nutrient pool in live biomass, over any time interval. It must be corrected for losses due to herbivory, tissue death, reproductive output and leaching. Nutrient uptake can be measured in this way for a single shoot or plant part (e.g. aboveground biomass; Chapin *et al.*, 1980), for an entire individual (Headley *et al.*, 1985) or for the vegetation as a whole (Van Cleve *et al.*, 1971). Results are most meaningful if nutrient pools are determined for all plant parts. However, the difficulties of harvesting roots and of measuring root production and turnover (Chapter 16) often result in omission of belowground samples. Incomplete harvest can result in erroneous conclusions, because one cannot distinguish between uptake from soil and translocation from tissues that were not quantitatively harvested.

Because harvesting is a destructive process, an individual plant or plot of vegetation cannot be followed in consecutive harvests. Consequently, a single measurement of nutrient uptake can be obtained using either of the following procedures. A single individual can be measured nondestructively at each date and these measurements converted to units of nutrient pool size using allometric equations developed on similar nearby individuals that are destructively harvested (Chiariello and Roughgarden, 1984). Alternatively, individuals (or plots) can be harvested in a well-defined sampling area on consecutive dates (Chapin *et al.*, 1980). In either case blocks or sampling units must be established initially from which all samples are taken for a given replicate (Fig. 10.1). The size and location of the plots depend upon the vegetation type and the question posed.

In any comparative study, the experimental units being compared (e.g. harvests or treatments) must be within the same replicate block, and the replicates of a given sample type must be in different blocks. Although this aspect of experimental design has been ignored by most plant ecologists (Hurlbert, 1984), its importance to adequate experimental design cannot be overemphasized. For example, a comparison of nutrient uptake by two species in a community requires that a block containing sufficient individuals of both species for all harvest dates be established. This unit constitutes the individual replicate and should be replicated sufficiently to provide the necessary sample size (Fig. 10.1). At each harvest date a quadrat or plot is randomly selected from within each block, and individuals of the two species are harvested from that plot. At the next harvest date another plot is randomly selected from each replicate block and harvested, etc. Similarly, in a study of the impact of successional stage upon seedling growth, seedlings in each site constitute one replicate, and different sites of the same age must be selected to make true replicates (Walker and Chapin, 1986). True replication of this sort (e.g. replication of burned forest sites compared to unburned sites) may be impractical. However, in this case observed differences may be due either to the treatment studied or some factor unrelated to the treatment.

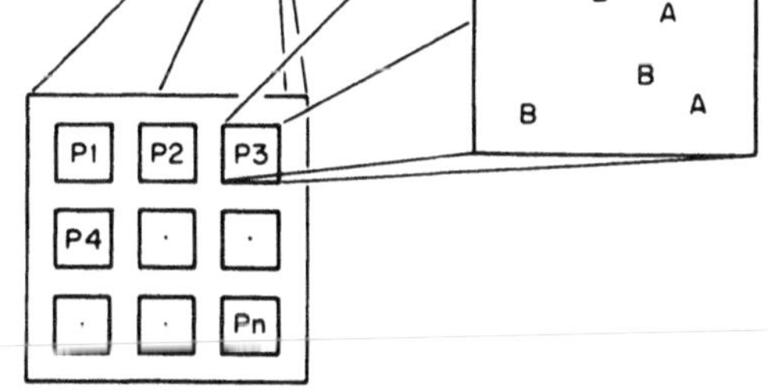

Fig. 10.1 Appropriate experimental design for comparing species (or treatments) A and B in terms of uptake by sequential harvests. The study area is subdivided into replicate blocks (R_1 to R_n). Each replicate block is subdivided into plots (P_1 to P_n); at each harvest date one plot is randomly selected. Within each plot individuals of species A and B are harvested.

(b) Methods

Nutrient uptake is measured by harvest as the increment in pool size of nutrients in the individual or plot of vegetation between consecutive harvest dates. At each date the plant should be divided into at least two parts: current year's growth and previous year's biomass (in the case of annuals, just the seed, which is generally negligible). Of course, it is usually of biological interest to divide both current and old growth into functional unit such as leaves and reproductive parts. The degree of subdivision depends upon the purpose of the study.

Nutrient uptake is the increase in nutrient pool in new growth minus any net loss of nutrient withdrawn from storage in old growth. To this estimate of nutrient uptake must be added the nutrients that have been lost between the two harvest dates to herbivores, tissue death, reproductive output and leaching (see below). The nutrient pool in any plant part is the product of biomass and concentration. Methods of measuring biomass and growth of aboveground plant parts are described by Chiariello, Mooney and Williams (Chapter 15) and of belowground parts by Caldwell and Virginia (Chapter 16). Methods for nutrient analysis are discussed in Section 10.4. Generally, biomass is much more variable among replicate samples than is nutrient concentration, so a sample size that is adequate for biomass (e.g. 10–20) should suffice for estimates of nutrient pools and uptake.

Measuring annual nutrient uptake requires harvests at least at the beginning of the

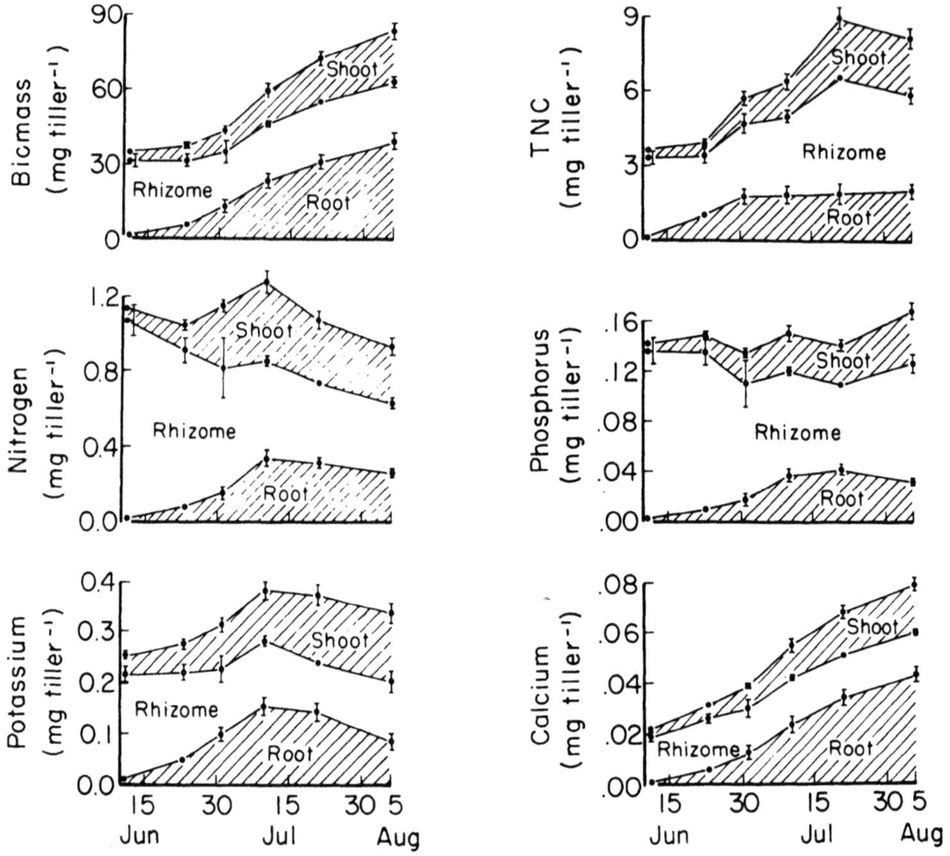

Fig. 10.2 Seasonal distribution of biomass, available carbohydrate (TNC), and various nutrients among plant parts of *Eriophorum vaginatum*. Data are means ± SE (n=4) for the pool size of the plant part immediately below the data point. From Chapin (1980b). Reproduced with permission of the Regents of the University of Colorado from *Arctic and Alpine Research*, vol. 12, 1980.

growing season and at the time of maximum plant nutrient pool. More frequent harvests can define the seasonal pattern of uptake and reduce errors due to undetected tissue loss. The time of maximum nutrient pool depends upon both the nutrient and the type of plant. Immobile nutrients (e.g. calcium) increase until the end of the growing season, unless there are large leaching losses (Fig. 10.2). For mobile nutrients (e.g. nitrogen, phosphorus and potassium) that are translocated to storage organs or reproductive parts, the timing of maximum nutrient pool depends predictably upon the type of plant (see below; Fig. 10.2). For elements of intermediate mobility (e.g. magnesium) the seasonal pattern of nutrient pool size is less predictable. For mobile nutrients in deciduous plants the nutrient pool generally plateaus during the middle of the growing season. In perennial deciduous plants the aboveground pool of mobile nutrients actually declines in autumn, as these nutrients are translocated belowground to storage organs (Fig. 10.3). In evergreen plants mobile nutrients often continue to increase until shortly before the end of the growing season (Mooney and Rundel, 1979; Chapin *et al.*, 1980). Consequently, the maximum pools

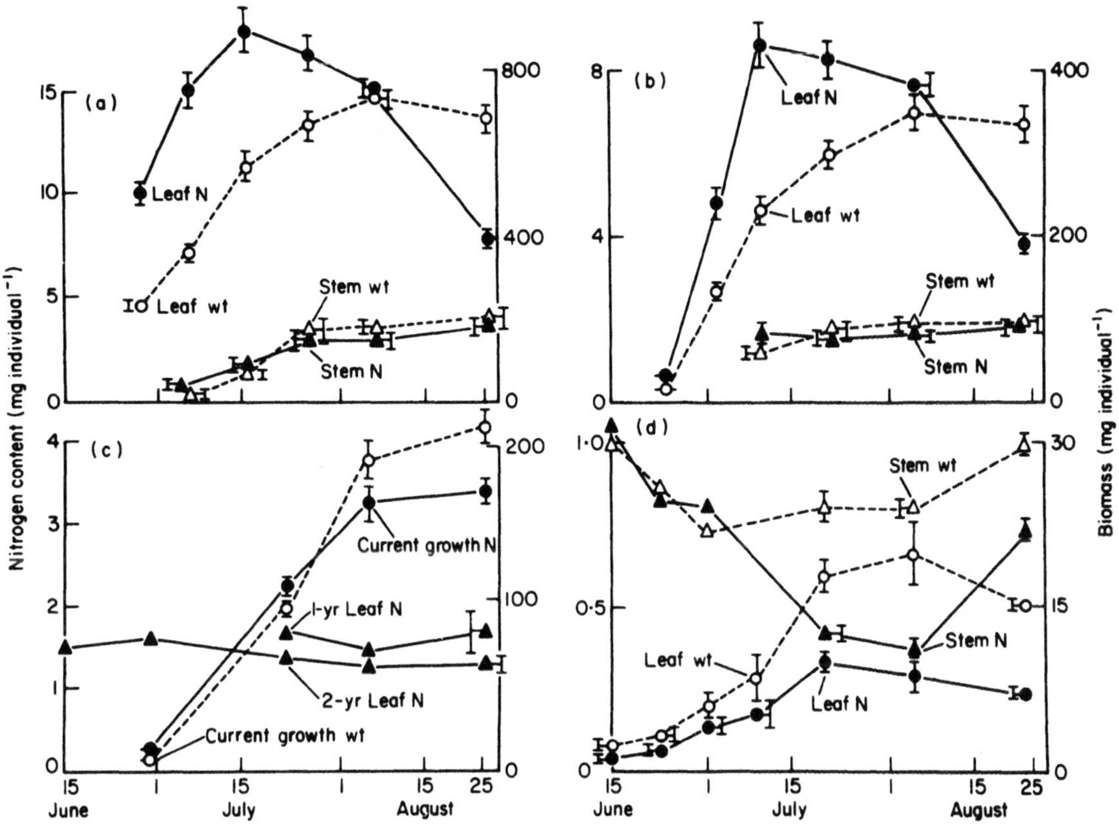

Fig. 10.3 Seasonal course of aboveground biomass (open symbols) and nitrogen (filled symbols) per individual (mean ± SE) in four species in Alaskan tundra (a) *Salix pulchra* (deciduous shrub), leaves and current year's stem, $n=20$; (b) *Betula nana* (deciduous shrub), leaves and current year's stem, $n=20$; (c) *Ledum palustre* (evergreen shrub), current year's leaves and stems combined, $n=40$, nitrogen also shown for 1-year and 2-year-old leaves; (d) *Eriophorum vaginatum* (graminord), leaves and rhizomes, $n=50$. From Chapin *et al.* (1980).

of mobile nutrients in a community containing both evergreen and deciduous species are probably best determined late in the growing season, but before visible autumn senescence. Because the actual date of maximum nutrient pool probably differs among species and communities and from year to year, it can only be determined by frequent sampling.

Foresters have used allometric relationships to estimate annual nutrient uptake from a single harvest (Ovington and Madgwick, 1959; Baskerville, 1965; Attiwill and Ovington, 1968; Barney *et al.*, 1978). From the sample stand replicate trees are selected which represent either the average tree diameter or height for the stand or the range of tree diameters and/or heights in the stand. A table for the sample stand is constructed which includes the numbers of trees in selected diameter and height classes. Trees representing the range of size classes are then harvested and sectioned in the field. Project objectives determine to what degree plant parts should be separated. For example, aboveground components might include bole, live branches, dead branches, current foliage plus twigs and older foliage. The total fresh weight of these components and their subsamples is determined separately in the field. Tree boles may be subsampled at several points from top to bottom.

Subsamples are oven-dried to a constant weight (65°C) in the laboratory. The fresh- to dry-weight ratios of the subsamples are used to calculate the dry weight of sample tree components. From this information, regression equations are developed to estimate component mass in relation to tree diameter or height. The same approach is used to calculate annual increment for individual components (current foliage, branch and trunk wood) or the total. Details of procedures are found in the above references. The stand table data and regression equations are used to estimate standing crop of total tree mass, or the mass of tree components and their annual increment.

The dry weight subsamples are chemically analyzed to estimate element concentration of the tree components. Element mass is then determined for each component, and the regression analysis described above is used to estimate element pool sizes and annual uptake for the stand. Here, nutrient uptake into aboveground parts is measured as the sum of nutrients contained in fine litter, the nutrient increment in new overwintering tissue, and the nutrients leached from the canopy by precipitation. If the community is in steady state, uptake into aboveground parts should equal nutrients in total litterfall and canopy leachates.

Nutrient uptake will be underestimated if nutrients have been lost between harvests due to herbivory, tissue death, reproductive output or leaching. Correction for these losses requires estimate of the amount of material lost (see Chapters 15 and 16) and its nutrient concentration. Forage chosen by herbivores is generally higher in nitrogen and phosphorus and lower in calcium and fiber than the average concentrations for the same plant part in the community (Westoby, 1974). Thus the researcher must sample like a herbivore to select material of similar nutrient content (e.g. choosing leaves of similar age on plants of similar phenological stage). Sources of tissue loss must be considered in any estimate of nutrient uptake by harvest. If they are not measured, this should be clearly stated in any presentation of results. Loss of nutrients by leaching is discussed in Section 10.5.

(c) Calculations

Nutrient uptake is calculated by multiplying the biomass of each sample by its nutrient concentration. All plant parts are then summed to get the total nutrient pool per individual or plot. Uptake for each plant or plot is then calculated as the difference in nutrient pool of the total plant between consecutive harvests. The uptake values for individual blocks can then be averaged to get a mean for the treatment. This calculation requires that the researcher analyze individual biomass samples from each replicate block separately for nutrients. A less satisfactory alternative is to calculate the mean and variance of the pool size from the mean and variance of the biomass and the concentration (Sokal and Rohlf, 1969).

Although uptake determined from harvest data is usually expressed as quantity of nutrient plant^{-1} or m^{-2} of ground surface, it is often useful to express it on a relative basis, i.e. g^{-1} of plant material, to account for differences in plant size. This measure, which is termed 'relative accumulation rate', is analogous to relative growth rate commonly used in growth studies (Chapter 15).

(d) Uptake by growth of plants in common gardens

Nutrient uptake can also be measured by growing plants in a common environment without competition and harvesting plants periodically. The rate of nutrient accumulation in the plant is a measure of the physiological potential of the plant to absorb nutrients under those conditions. If root biomass is determined, the rate of uptake can be expressed per plant as well as per unit root weight (Christie and Moorby, 1975). This approach has been widely used in greenhouse and growth chamber studies (e.g. White, 1972; Christie and Moorby, 1975; Shaver and Melillo, 1984), but it is equally applicable to common garden experiments in which plants of different species or ecotypes are grown in pots in the natural environment (Chapin and Chapin, 1981). By growing plants in pots it is relatively easy to recover root biomass of each individual.

10.2.2 Uptake of labeled nutrients from soil

Soils can be labeled with isotopes of important plant nutrients (e.g. ^{15}N or ^{32}P), and uptake monitored by appearance of the isotope in the plant. This procedure is most useful in determining where roots are present (or most active) in soil. For example, ^{32}P can be injected at different soil depths. By monitoring appearance of ^{32}P in leaves, one can determine maximum rooting depth of each species. The most elegant use of this technique has been through double labeling of soil by ^{32}P and ^{33}P on different sides of a plant (adjacent to different competitors; Caldwell *et al.*, 1985). By determining the relative amounts of ^{32}P

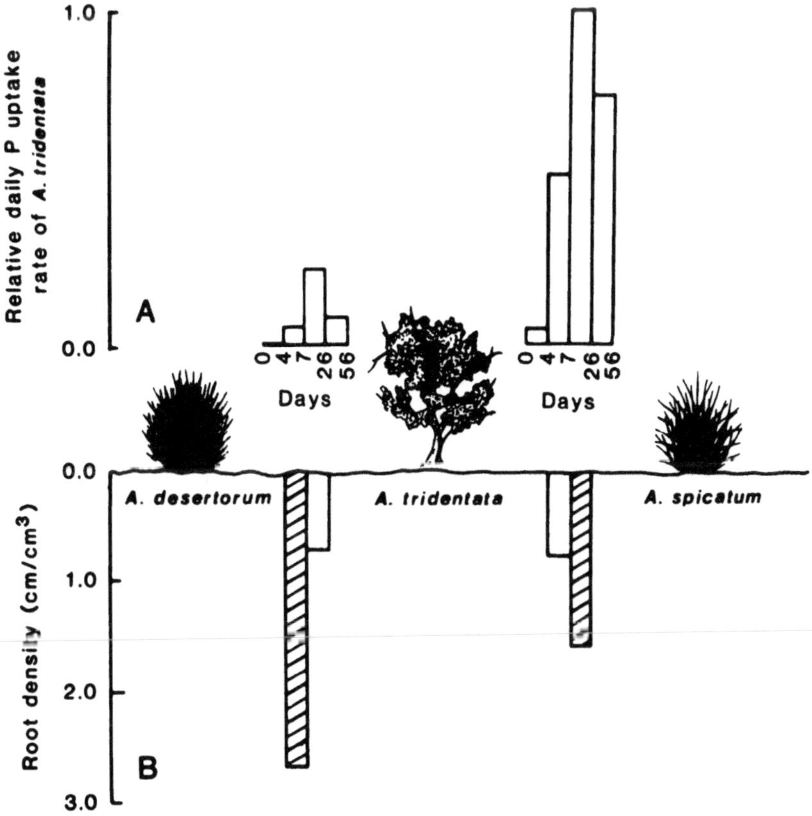

Fig. 10.4 Relative rate of phosphorus absorption by *Artemisia* from soil interspaces shared with *Agropyron desertorum* and *A. spicatum* at various times after labeling. Also shown are the rooting densities of the grasses (cross-hatched bars) and *Artemisia* (open bars) in the interspaces where the isotopes were placed. From Caldwell *et al.* (1985), © 1985 by the AAAS.

and ^{33}P in foliage it was possible to evaluate the relative potential of adjacent individuals to compete with the measured plant for phosphorus (Fig. 10.4). Such studies have tremendous potential for examining the relative competitive abilities of different species for nutrients.

The two major disadvantages in soil labeling experiments are (1) the difficulty of uniformly labeling soil and (2) predicting the amount of label that will be available to the plant in the face of the competing processes of microbial immobilization and soil adsorption reactions. Under more controlled laboratory conditions uniformly mixed soils containing radioisotopes have yielded important information on the relative importance of active uptake by roots vs diffusion to the root surface (Nye and Tinker, 1977). Such studies have not yet been attempted under field conditions.

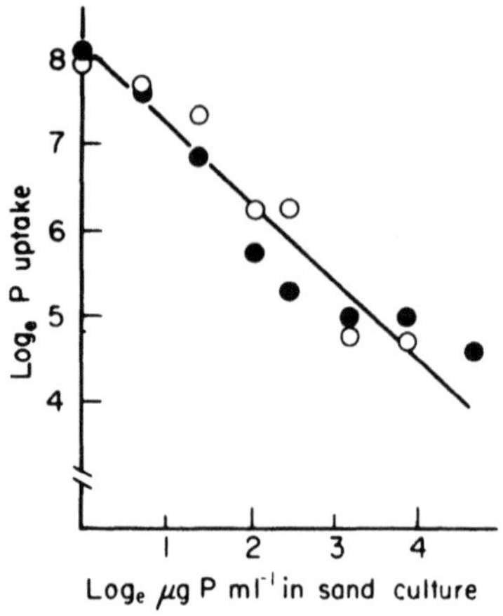

Fig. 10.5 The relationship between the uptake of ^{32}P-labeled phosphorus (pg of P mg^{-1} 15 min^{-1}) from a phosphate solution by roots of birch seedlings and the phosphorus concentration in the sand culture in which the seedlings had previously been grown. Solid symbols are 1975 experiment; open symbols are 1976 experiment. Adapted from Harrison and Helliwell (1979).

10.2.3 Short-term uptake measurements

(a) General principles

The physiological potential of roots to absorb nutrients from a solution is a sensitive measure of plant nutrient status (Fig. 10.5; Harrison and Helliwell, 1979; Lee, 1982; Glass, 1983). It differs in consistent patterns among species or ecotypes adapted to different environments (Chapin, 1974, 1980a). Most measurements of this potential have been made on plants grown in solution culture, but the technique can be adapted to certain field situations. For a given genotype, nutrient uptake potential depends upon (1) the plant's demand for nutrients, which is usually a function of growth rate (Clarkson and Hanson, 1980; Clarkson, 1985), and (2) the tissue stores of the particular nutrient. For example phosphate-deficient plants have a high capacity to absorb phosphate (Harrison and Helliwell, 1979), nitrogen-deficient plants have a high capacity to absorb nitrate and ammonium (Lee, 1982) and potassium-deficient plants have a high capacity to absorb potassium (Glass, 1983).

Because of the sensitivity of uptake potential to deficiency of a particular nutrient, Harrison and Helliwell (1979) have suggested that uptake potential be used as an assay of nutrient deficiency (Fig. 10.5) (Harrison and Helliwell, 1979). Such assays have proven very effective in laboratory growth experiments where availability of a single nutrient is varied, and other nutrients are maintained at adequate levels. However, there are two potential difficulties in interpreting such uptake measurements in field-grown roots. First, a low capacity for uptake of a particular nutrient may reflect low growth rate and low nutrient demand (Clarkson and Hanson, 1980) rather than adequate supply of the nutrient. Second, the low uptake may indicate a deficiency of some other nutrient or carbohydrate. For example, light-stressed plants show reduced uptake rates (Clement *et al.*, 1978) and nitrogen-stressed plants show reduced capacity for phosphate uptake (Cole *et al.*, 1963; Lee, 1982). Nonetheless, such measurements are useful for describing the physiological potential of the root to absorb nutrients, which is generally the question of prime interest in a field study.

There are two general questions about the validity of measuring nutrient uptake from solutions. First, because most roots are mycorrhizal under field conditions, kinetics of nutrient uptake by nonmycorrhizal roots from solutions may have little relevance to performance of mycorrhizal roots in soil. However, it can be argued (although not conclusively) that such measurements are still a useful index of the potential of the root–hyphal complex to absorb nutrients. First, the rate of nutrient absorption by root cortical cells controls the rate of transfer of phosphate (and probably other nutrients) from the fungus to the root (Cox and Tinker, 1976; Clarkson, 1985). In addition, the rate of phosphate absorption by fungal hyphae is controlled by the phosphorus status of the fungus (Beever and Burns, 1980) and, therefore, indirectly by the potential of the root to absorb phosphate. Although uptake kinetics may be useful as an *index* of the potential of mycorrhizal roots to absorb nutrients, the actual kinetic parameters are probably quite different for mycorrhizal roots in soil.

A second general question is whether kinetics of nutrient uptake have relevance for root performance in the soil–root system. Careful theoretical treatment indicates that rate of diffusion through soil to the root surface controls the rate of absorption of phosphate, nitrate, ammonium and potassium (but generally not calcium), and that kinetics of uptake by the root have minimal effect upon this process (Nye, 1977; Nye and Tinker, 1977; Clarkson, 1985). On the other hand, plants have evolved a very sensitive mechanism of adjusting uptake kinetics to compensate for plant nutrient status. This has been well documented in the laboratory (e.g. Harrison and Helliwell, 1979; Glass, 1983) and implicated in soil (Andrews and Newman, 1970). There are also consistent patterns of nutrient uptake kinetics in plants adapted to different soil fertilities and climatic regimes (Chapin, 1974, 1980a). This suggests that uptake kinetics are ecologically important and have been subject to natural selection. Root kinetics are probably most important for mobile ions (e.g. nitrate and potassium) in all soils and for immobile ions in fertile soils (Chapin, 1988). In many cases estimates of nutrient uptake calculated from harvest data are quite similar to those calculated from uptake kinetics, root biomass and soil solution concentrations (Chapin and Tryon, 1982; Headley *et al.*, 1985).

(b) Uptake by accumulation in roots

Nutrient uptake by isolated roots has most commonly been measured by rate of accumulation in roots rather than by depletion from solutions, because radioisotopes allow sensitive measurement of low rates of uptake from dilute solutions such as those normally encountered by roots in the field. There are convenient radioisotopes for phosphate (^{32}P, ^{33}P), potassium (^{42}K), sulfur, (^{35}S), etc. In addition, ^{86}Rb serves as a useful analogue of potassium in plants of high potassium status (Pettersson, 1978). Unfortunately, there are no convenient radioisotopes of nitrogen. Because ^{13}N has a 10 min half-life, it must be used in close proximity to a particle accelerator. ^{15}N is a stable isotope that must be present in much larger quantities than radioisotopes to be detected. Hence, measuring of uptake into roots using ^{15}N requires long absorption periods and/or high solution concentrations. Uptake methods using ^{15}N are currently being tested (Harrison and Van Cleve, unpublished). Although nutrient uptake methods were developed by plant physiologists, they have been used on roots sampled from soils in pot or field studies and have given results that are consistent with physiological studies (Chapin and Bloom, 1976; Harrison and Helliwell, 1979; Chapin and Tryon, 1983; Van Cleve and Harrison, 1985; Headley *et al.*, 1985).

Measuring uptake by roots from solutions

(Epstein *et al.*, 1963) involves the following steps. First, soil should be rinsed gently from roots (in the case of field and pot studies) and roots equilibrated to the temperature and pH at which uptake will be measured in a solution of 0.5 mM $CaSO_4$. Calcium maintains membrane integrity. This equilibration period also serves to rinse any unabsorbed ions out of the free space between root cells. Next, roots are transferred for a short time (e.g. 10–20 min) to a radioactively labeled solution (containing calcium) of the nutrient being studied, then rinsed in a cold (5°C) solution (containing calcium) that has the nonradioactive form of the ion being studied at a concentration much higher (50–100-fold) than in the uptake solutions. These solutions exchange off any isotope that may be in the root free space but not actively absorbed. The major exception to this rinse procedure is with nonradioactive isotopes (e.g. ^{15}N), where tissue concentrations must be analyzed chemically. In this case, roots should be rinsed in KCl. After rinsing, roots are blotted to remove excess water, dried in an oven, weighed and counted in a scintillation counter. In the case of isotopes that produce a hard beta emission, roots can be counted undigested in an aqueous solution with a wavelength shifter (7-aminonaphthalene-1,3-disulfonic acid, ANDA) by means of Cerenkov radiation (Lauchli, 1969; Chapin and Holleman, 1974). Analytical procedures for ^{15}N are described by Ehleringer and Osmond (Chapter 13). Detailed procedures for measuring nutrient uptake are given elsewhere (Epstein *et al.*, 1963; Chapin and Bloom, 1976; Harrison and Helliwell, 1979; Veerkamp and Kuiper, 1982a).

The following possible experimental artifacts should be examined each time a new set of species or conditions is used. First, does removing roots from soil damage them? The procedure has given reasonable results with roots taken from soil (e.g. Chapin and Bloom, 1976; Harrison and Helliwell, 1979). However, in some cases, roots appear damaged and have very low rates of uptake (Chapin, unpublished). Second, does excision affect uptake? Excision has a negligible effect upon uptake in some studies (Chapin, 1974) but a substantial effect in others (Nye and Tinker, 1977). In general, treatments (e.g. plant nutrient status) cause the same kind of effects upon uptake by excised versus intact roots, although the actual values may differ between the two root types (Chapin and Clarkson, unpublished). Excision is most likely to reduce uptake rate in plants of low carbohydrate status, for example those that are heavily fertilized or grown under low light (Chapin, 1980a), as in many growth-chamber studies. If uptake is measured on roots attached to shoots, it is important to make sure that isotopes have not been translocated from roots to shoots during the experiment. This is a potential problem for uptake periods longer than 10 min. Third, is uptake linear with time during the measurement period? Over long time intervals, efflux of the isotope from roots to the solution becomes appreciable, so that the rate of accumulation of isotope declines, resulting in an underestimate of the rate of nutrient influx. Fourth, are the roots rinsed enough to exchange off any isotope not actively absorbed (i.e. on the root surface or between the cells)? This can be a problem if the rinse solution is too dilute or if roots are not rinsed long enough. It is most likely to be a problem with divalent cations.

If uptake rates are measured at a series of solution concentrations, kinetics of uptake can be calculated. Although uptake is thought to be mediated by enzyme-like carriers in the plasma membrane of root cortical cells, the kinetics of uptake by a morphologically complex tissue such as a root probably do not directly correspond to the kinetics of carriers. Root uptake kinetics are nonetheless valuable for describing the maximum capacity of a root to absorb nutrients (V_{max}) and the potential of the root to function at low nutrient concentrations (the inverse of K_m). Kinetic parameters are not rigidly fixed for any species.

Both V_{max} and K_m increase substantially as uptake is measured at higher concentrations (Nissen, 1974). Therefore it is essential that kinetics be compared only among roots measured under similar conditions. Reviews that compare uptake kinetics from different studies often ignore this point.

Kinetics are most commonly calculated using the Lineweaver–Burk plot. However, this method gives undue weight to measurements made at low concentrations (generally those most prone to measurement error) and can lead to substantial errors. Other plotting procedures have different biases (Wilkinson, 1961). Methods are available, however, that weigh all sample points more evenly and provide an estimate of the variance for the kinetic parameters (Wilkinson, 1961; Cleland, 1967; Conway *et al.*, 1970; Li, 1983).

(c) Uptake by disappearance from solution

Recording nutrient depletion from solutions is a sensitive technique for measuring net uptake over relatively long time intervals (Longeragan and Asher, 1967; Clement *et al.*, 1978; Bloom and Chapin, 1981). Because of the time involved, this approach is unsuitable for excised roots and, therefore, has not been adapted to field studies. However, this approach has proven extremely useful for laboratory studies of nitrate and ammonium uptake, because measurements can be made with specific ion electrodes at low, ecologically realistic concentrations of inorganic nitrogen. Net nutrient absorption, as measured by solution depletion, is the net result of nutrient influx and nutrient efflux, whereas uptake studies with radioisotopes measure only nutrient influx.

10.3 NUTRIENT USE AND NUTRIENT STATUS

10.3.1 Tissue concentration

(a) General principles

The total concentration of a nutrient in tissue is the most commonly used index of nutrient status. This approach was developed in agriculture where for each crop the concentration of a given nutrient in a specific plant part (e.g. petiole of the third leaf) provided an accurate prediction of the degree to which that species would respond to addition of that nutrient (Bould *et al.*, 1983; Bouma, 1983). This approach has also been used in forestry with some success (van den Driessche, 1974). However, there are several potential causes of low concentrations of tissue nutrient: (1) high fiber, lignin or resin (e.g. sclerophyllous versus mesophytic leaf); (2) high carbohydrate accumulation (e.g. fruit versus leaf); (3) low concentrations of compounds containing that nutrient even under nutrient-sufficient conditions (e.g. root nitrogen versus leaf nitrogen); (4) seasonally programmed reduction in concentration (e.g. leaves in autumn or storage organs in midsummer); (5) low concentrations of nutrient reserves in vacuolar stores. Only the last of these factors is a true indicator of plant nutrient status. The first four causes commonly differ among parts of the same plant and among individuals or species for the same plant part without reflecting a difference in nutrient status. For example, sun leaves have a lower nitrogen concentration than shade leaves on the same individual due to higher fiber and carbohydrate content (Boardman, 1977). The low nitrogen concentration of old leaves may reflect lesser demand for nitrogen rather than age-related nitrogen stress (Mooney and Gulmon, 1982; Field, 1983). If nutrient concentration is to be a useful indicator of nutrient status, it is essential to control for as many of these other factors as possible. Factors that are particularly important to control for are species or ecotypes (Chapin *et al.*, 1980; Saric, 1981), plant part (van den Driessche, 1974), tissue year class (Chapin *et al.*, 1980; Hom and Oechel, 1983; Jonasson and Chapin, 1985), phenological date (as contrasted with calendar date: van den Driessche, 1974; Jonasson *et al.*, 1986), position in the canopy (van den Driessche, 1974) and proximity to nutrient sinks (e.g. leaves on vegetative versus reproductive shoots: Ernst, 1983; Gray, 1983).

Choice of the best plant part to determine plant nutrient status depends upon the nutrient and the time of year. Nitrogen, phosphorus and potassium are mobile in the phloem and are translocated from old to young leaves in response to nutrient stress. Consequently, old leaves are a more sensitive indicator of deficiency of these elements than are young leaves. In contrast, deficiency symptoms for phloem-immobile elements, such as calcium, appear first in young leaves (Epstein, 1972). During the growing season, leaves may be a more sensitive indicator of nutrient stress than are storage tissues, but at other times storage tissues (e.g. bark of some trees, young stems of other species) may be a better indicator of nutrient status (van den Driessche, 1974; van den Driessche and Webber, 1977; Chapin *et al.*, 1980).

10.3.2 Nutrient use efficiency

Nutrient use efficiency (NUE) is most commonly defined as the amount of biomass produced per unit of nutrient, and, thus, is simply the inverse of nutrient concentration. Consequently, all of the precautions discussed elsewhere in relation to nutrient concentration apply here as well. Although NUE has the same information content as nutrient concentration, it has many interesting physiological and ecological implications (White, 1972; Gerloff, 1976; Vitousek, 1982; Shaver and Melillo, 1984).

A high NUE indicates that more biomass has been produced per unit nutrient (i.e.

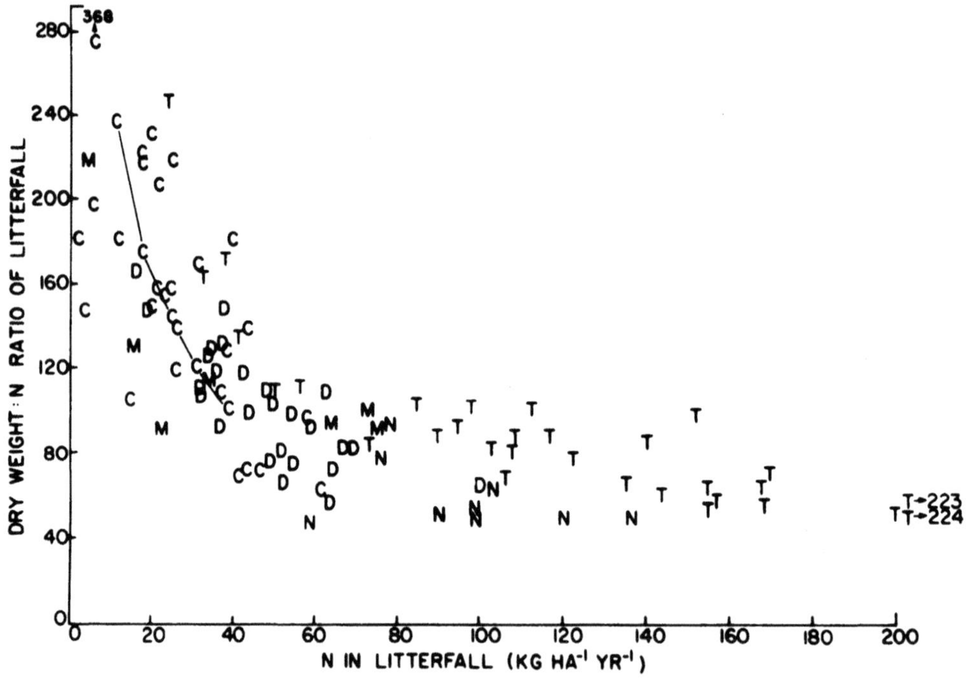

Fig. 10.6 The relationship between the amount of nitrogen in fine litterfall and the dry mass to nitrogen ratio of that litterfall. Forest types are coniferous (C), temperate deciduous (D), tropical evergreen (T), Mediterranean (M) and temperate N-fixing (N). From Vitousek (1982).

there is a low tissue nutrient concentration). This is a common plant response to low nutrient availability (White, 1972; Chapin, 1980a; Vitousek, 1982; Shaver and Melillo, 1984). However, the processes responsible for differences in NUE depend strongly upon the ages and types of tissues measured. The concept was originally applied to nutrient concentrations of mature leaves (White, 1972; Gerloff, 1976). In this case the major factor influencing NUE is nutrient status, which is a function of nutrient availability and uptake rates (Shaver and Melillo, 1984). NUE can also be calculated on a whole plant basis (i.e. total plant mass divided by total nutrient mass), in which case NUE reflects both tissue concentration and biomass allocation to high-nutrient versus low-nutrient tissue. On a whole-plant basis NUE increases in response to nutrient stress because (1) tissue nutrient concentrations decline and (2) the ratio of root biomass (with low tissue nutrient concentrations) to leaf biomass (with high tissue nutrient concentrations) increases. Finally, NUE can be calculated on leaf litter (Fig. 10.6; Vitousek, 1982), in which case it reflects initial leaf status and efficiency of nutrient resorption (and/or leaching loss).

The most appropriate measure of NUE depends upon the question posed. NUE of mature leaves may be the best indicator of plant physiological responses to nutrient stress. NUE on a whole-plant basis confounds several processes in the same measure but gives a more global response of plants to nutrient stress. NUE measured on leaf litter is the end result of many physiological processes and may be the most meaningful measure for ecosystem-level questions or for estimating how much carbon can be fixed per unit nutrient expended (and lost) by the plant. Regardless of the measure of NUE used, its method of calculation must be clearly defined.

NUE has also been discussed in a more physiological sense as the amount of carbon gain per unit of nutrient; this can be considered either in terms of the instantaneous rate of photosynthesis per unit nitrogen (Veerkamp and Kuiper, 1982b; Field and Mooney, 1983, 1986) or as the cumulative carbon gain over the life of the leaf (Small, 1972; Chapin *et al.*, 1980).

10.3.3 Nutrient ratios

Because all vascular plants apparently require a very similar balance of nutrient supply for optimal growth (Ingestad, 1971, 1982), they should show this same ratio of nutrient concentrations in tissues, unless some element is limiting growth. Redfield (1958) suggested that the ratio of nitrogen to phosphorus should be a sensitive indicator of the relative limitation of phytoplankton growth by these two elements. The same ratio can indicate nutrient limitation in terrestrial plants (Garten, 1976; Miller, 1980; Shaver and Melillo, 1984). However, nutrient ratios can also reflect temporary storage of one nutrient in excess of demand (luxury consumption) rather than deficiency of the other nutrient. For example, temporary storage of phosphorus and a low N : P ratio could result from high phosphate availability under conditions where growth is limited by carbohydrates, water or micronutrients, and may not reflect nitrogen deficiency. Calcium, in particular, accumulates as a function of cumulative transpiration, so that nutrient ratios involving calcium may not indicate deficiency of the other element in the ratio. Nutrient ratios may, however, be a good indicator of certain expected imbalances. For example, calcium : magnesium ratio in serpentine is higher in serpentine plants that exclude magnesium than in those that do not.

10.3.4 Stable isotope ratios

The ratio of ^{15}N to ^{14}N varies in plant tissue depending upon the number of biological transformations it has undergone (see Chapter 13). Soil is enriched in ^{15}N, particularly if soil N derives in part from salt spray (Virginia and Delwiche, 1982). Biological transformations discriminate in favor of ^{14}N, so that every time a portion of the total N pool is transformed, the product becomes enriched in ^{14}N and the remaining original substrate becomes enriched in ^{15}N. Consequently, a low $^{15}N : ^{14}N$ ratio can be used to recognize the following situations: a high dependence upon symbiotically fixed N and greater dependence upon N that is recycled (through internal stores or mineralization/nitrification) than upon recently arrived inorganic N. The major advantage of using stable isotope ratios is that they reflect the net transformations of N in the ecosystem, e.g. the relative degree to which two co-occurring species depend upon symbiotically fixed N. The major disadvantage of $^{15}N : ^{14}N$ ratios is that many N transformations occur in plants and soil, all of which affect this ratio in the same way. Consequently, one cannot distinguish the relative importance of these processes by simply examining isotope ratios of plant material. The uses and interpretations of stable isotope ratios are discussed more thoroughly in Chapter 13.

10.3.5 Nutrient chemical fractions

Nitrogen and phosphorus are present in plants in a variety of chemical forms that

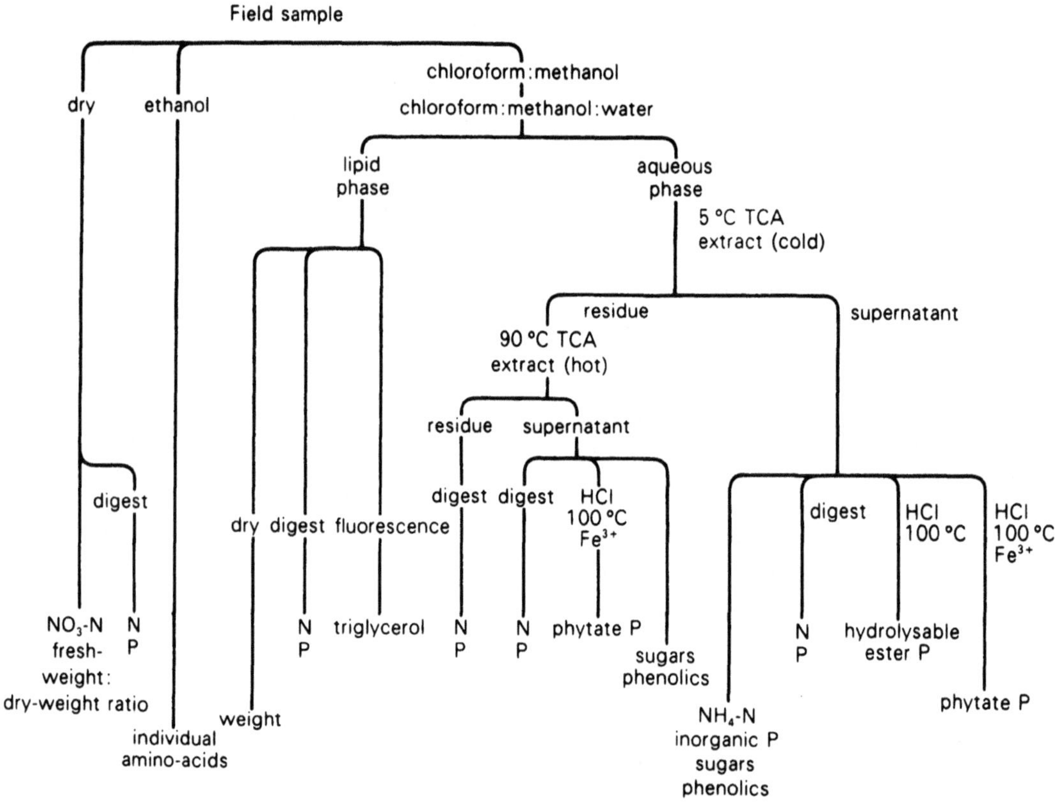

Fig. 10.7 Outline of fractionation procedure to identify carbon, nitrogen and phosphorus fractions in plant material. From Chapin *et al.* (1986).

have quite different chemical roles. Information about the chemical form of nitrogen and phosphorus in plants can be extremely useful in understanding the nutrient requirements of different species, the nature of nutrient storage in seeds or vegetative plant parts and the level of specific tissue nutrient reserves. The appropriate chemical procedures for studying nutrient fractions depend upon the fractions to be measured. In many species arginine is the primary nitrogen store (van den Driessche and Webber, 1977; Pate, 1983; Chapin *et al.*, 1986), so that in these species measuring arginine gives a more sensitive indication of nitrogen status than measuring total nitrogen. If the objective were to measure nitrogen status of arginine-storing plants, it would be appropriate to extract the plants in an aqueous solvent, such as 80% methanol. If the chemical form of the nutrient store were unknown, or all nutrient-containing fractions were important, chloroform–methanol would be appropriate, because this extracts both lipophilic and aqueous components. Many methods have been developed for studying specific chemical components. Fig. 10.7 outlines one scheme that has proven useful for characterizing most major nitrogen- and phosphorus-containing chemical fractions, as well as carbohydrates (Kedrowski, 1983; Chapin *et al.*, 1986).

There are several potential sources of error in choosing an extraction procedure and in evaluating results. The components of interest must be soluble in the extractant selected. Methanol (80%) or acetone (80%) will extract most water-soluble (polar) and moderately polar substances, such as inorganic ions, sugars, amino acids, many proteins, chlorophyll etc. Addition of chloroform to the medium results in extraction of even the most nonpolar lipids.

The second major consideration in an extraction procedure is to inactivate hydrolytic enzymes that would otherwise modify the chemical composition of the extract. Ideally plant material should be sampled into liquid nitrogen, freeze-dried, ground and stored frozen for later extraction. Alternatively, fresh material can be subsampled for fresh- to dry-weight ratio, placed in liquid nitrogen and then homogenized in the extraction medium. In this case care should be taken to insure that the fresh–dry weight subsample is representative of the subsample that is extracted. The extraction medium should be chilled, and, where phosphorus fractions are to be measured, formic acid should be included to inactivate phosphatases (Bieleski, 1968).

10.3.6 Xylem exudates

Xylem sap from a small stem can be sampled by peeling back the bark and phloem, placing the stem in a Scholander pressure bomb and collecting the sap it exudes with a micropipette. Concentrations of constituents in the xylem sap are readily altered by evaporation, and care must be taken to minimize microbial transformations after collection. The concentrations of nutrients may be affected by time of day, time of year, crown height, tree age, drought stress and amount of pressure applied in extracting the sap, in addition to the factors directly related to plant nutrient status (Stark and Spitzner, 1985; Stark *et al.*, 1985; Osonubi *et al.*, 1988). Consequently, it is important to standardize these factors. The pressure applied to extract sap should be no more than 1.0 MPa greater than the pressure required to balance plant water potential.

Under the uncommon conditions where root pressure exceeds evaporative demand, the top of the plant can be cut off and rubber tubing sealed with vaseline placed over the stump of the shoot to collect xylem exudate (Glinka, 1980). Xylem sap does not give a direct measure of plant nutrient status but indicates the balance of nutrients that is transported to shoots. This method is particularly useful in determining whether nitrogen is transported to the shoot primarily as nitrate or as amino acids.

10.4 CHEMICAL ANALYSIS

10.4.1 Sample preparation

The objective of tissue analysis is to provide a clear indication of the natural variation in selected tissue properties, while avoiding artifacts or induced variation due to sampling or analysis (Jones and Steyn, 1973; Leaf, 1973; Allen, 1974). Methods for collecting and handling materials between the field and laboratory depend upon the analytical requirements and the need to prevent contamination of, or continued biochemical activity in, the tissue. For example, sodium or amino acid analysis of plant materials requires the use of disposable gloves during sample collection to prevent contamination from human perspiration. Analysis of plant tissues for enzymes, organic acids, or plant secondary chemicals (terpenes, phenolic substances, tannins) requires rapid preparation of materials following sampling to prevent change in concentration or activity. In these cases, tissue should be extracted in the field or quick frozen with liquid nitrogen to prevent continued biochemical activity.

In general, every effort should be made to minimize the biological activity of the tissue following collection. Material can be placed in a portable cooler for transport to the laboratory. If the time between collection in the field and transport to the laboratory is short, samples can be placed in aerated containers, such as paper or muslin bags. At the laboratory, materials should be placed in a forced-draft oven and dried at 65°C to constant weight. The time required for drying depends on the bulk and type of tissue. Leaf material may dry to constant weight in 24 h, whereas a sample of tree trunk wood may require several days to reach a constant dry weight. If a delay in drying is unavoidable, samples should be stored frozen in polyethylene bags to minimize respiratory activity. A drying temperature of 65°C is sufficient to stop enzymatic activity even after 2 months of dry sample storage (Jones and Steyn, 1973). This temperature has minimal effect on tissue mineral element and nitrogen content (Allen, 1974). Drying at a lower temperature (50°C) does not stop enzymatic activity (Jones and Steyn, 1973), whereas higher temperatures cause carbohydrates and protein to complex into lignin-like materials (Robbins, 1983) and confound fiber analyses.

Where contamination from dust or other aerosols occurs, a preliminary washing with a weak (0.1–0.3%) detergent solution, followed by rinsing in pure water, may be necessary prior to oven drying and grinding (Jones and Steyn, 1973). This should be done quickly to avoid long contact between tissues and solution and to prevent leaching of soluble elements such as K.

Some type of mechanical tissue grinding is necessary in order to prepare the material in a uniform particle size for analysis. A variety of mechanical grinders are available which can rapidly reduce plant tissue to a uniform powder. A size 60 mesh produces an acceptable particle size. Wiley, hammer, jar and ball mills may be used in grinding. A large Wiley mill can quickly handle large individual samples. Since mill components are comprised of various metals, trace element contamination may occur, requiring the use of an all-porcelain ball mill (Jones and Steyn, 1973).

Ground material should be stored in air-tight moisture-proof containers, such as snap-cap glass vials or small heavy-gauge polyethylene bags (Whirlpak type). Some investigators redry ground material prior to analysis to remove water absorbed during grinding and storage (Jones and Steyn, 1973). However, because drying characteristics of the powdered material will differ from those of the unground material, drying temperatures greater than 65°C must be avoided to prevent thermal decomposition.

(a) Chemical analysis

The first step in chemical analysis is to

destroy the organic component, leaving the various elements for analysis. In these procedures, sample losses and/or contamination should be avoided. Laboratory procedures should be checked by frequent inclusion of blanks and occasional analysis of an organic standard such as the National Bureau of Standards Standard Reference Material 1571, Orchard Leaves. The weight of material digested should be held fairly constant, because small samples are digested more completely (giving higher values of tissue concentration) than are large samples.

Wet or dry ashing are the two major methods used to destroy organic material. The former uses various combinations of HNO_3, H_2SO_4, $HClO_4$, and H_2O_2 and specially designed acid fume hoods. Current techniques include the use of block digesters with digestion times up to 2 h, depending on the combination of reagents employed (Allen, 1974; Isaac and Johnson, 1976; Nelson and Sommers, 1980). Some wet-ashing procedures are suitable for analyzing nitrogen as well as alkali-, alkaline earth- and transition-metal elements. Advantages of acid digestion are that: (1) there are no volatilization losses of metallic elements or phosphorus, (2) the procedure is fairly rapid and can be adapted for routine use, and (3) large numbers of nutrient elements can be determined in one digest solution (Allen, 1974). Proper safety precautions are essential, because these methods use concentrated acids as oxidizing agents and produce corrosive and potentially explosive fumes.

Dry ashing is the least involved method, requiring a muffle furnace and ashing temperatures not exceeding 500°C for a period of 2–8 h, depending on the material and sample size. High-walled crucibles aid in preventing loss of sample ash. The residue is dissolved with 1 : 1 HCl : H_2O, filtered, and diluted to volume in preparation for analysis (Allen, 1974). Nitrogen is volatilized and cannot be analyzed by dry ashing.

Most commonly nitrogen is analyzed by the Kjeldahl method (Bremner and Mulvaney, 1982). A variety of modifications to this procedure exist, and current techniques include use of autoanalyzers for rapid colorimetric determination of NH_4 (Isaac and Johnson, 1976; Nelson and Sommers, 1980). Total nitrogen (organic and inorganic forms) can be determined using special combustion apparatus (Bremner and Tabatabai, 1971). The normal Kjeldahl procedure analyzes only organic nitrogen plus ammonium. However, nitrate and nitrite may be included in the Kjeldahl analysis by using the permanganate-reduced iron modification (Bremner and Mulvaney, 1982). Regardless of ashing method, care should be taken to include controls (reagent blanks and organic matter standards) as checks on procedures and techniques.

Following ashing, solutions of the residue are diluted to volume in preparation for element analysis. Colorimetric analysis is commonly used for determining concentration of ammonium, nitrate, nitrite and phosphate (Jones and Steyn, 1973; Allen, 1974). Autoanalyzers with computerized data-acquisition systems greatly speed the analysis. The salicylate/hypochlorite/nitroprusside system of reagents is commonly used for NH_4 analysis of Kjeldahl digests. A salicylate-based Berthelot method using dichloroisocyanurate as the chlorine-donating agent was first used to analyze Kjeldahl digests of plant material by Crooke and Simpson (1971) and then modified for microKjeldahl by Bietz (1974). Rowland (1983) further modified the Technicon (1976) method and used it for analyzing soil extracts, natural waters and acid digests of plant tissue.

For phosphorus, colorimetric procedures include the Molybdenum Blue procedure using hydrazine sulfate and stannous chloride as reductants and the ascorbic acid procedure (Burton, 1973). The former allows a wider range of phosphorus concentrations to be considered and has a lower minimum detection limit than does the ascorbic acid method.

Other elements (K, Na, Ca, Mg, Cu, Fe, Mn, Zn) contained in the digest solutions can be determined using flame emission or absorption spectrophotometry (Jones and Steyn, 1973; Allen, 1974). Inductively coupled plasma (ICP) and direct current plasma (DCP) spectrometry allow simultaneous or sequential analysis of these and other elements such as sulfur, phosphorus and boron (Soltanpour *et al.*, 1982).

10.5 NUTRIENT LOSS

10.5.1 Leaching loss

Leaching loss of nutrients in throughfall and stemflow is frequently 5–10% of the annual nutrient flux in forests (Cole and Rapp, 1981; Van Cleve and Alexander, 1981). Throughfall can be measured by comparing the nutrient pool size in rain that falls from the canopy with that of rain in an open area. This is a common practice in forestry and can be adapted to grassland or other small canopies (Tarrant *et al.*, 1968; Eaton *et al.*, 1973; Olson *et al.*, 1981; Seastedt, 1985). Stemflow is measured by placing collars around stems of plants (generally used with trees) and collecting the water and nutrients that flow down the stem into the collar (Eaton *et al.*, 1973; Seastedt, 1985).

The potential of leaves to lose nutrients via leaching has been measured by soaking leaves either in distilled water (Tukey, 1970; Chapin and Kedrowski, 1983) or in simulated rain water (i.e. water with a chemical composition typical of rain water; Boerner, 1984). The latter has the advantage of permitting measurements of the potential of leaves to either absorb nutrients from rain or lose them by exchange or leaching (Boerner, 1984).

10.5.2 Senescence and resorption

On average, plants resorb about half of the nitrogen and phosphorus from a leaf before it is shed (Table 10.2; Chapin and Kedrowski, 1983). Resorption is extremely important to the nutrient economy of a plant, because it allows nutrient redistribution either to reproductive structures and new leaves or to storage over winter. Nutrient resorption is probably best expressed as the proportion of the leaf nutrient pool that is withdrawn prior to leaf abscission. This is simply the maximum leaf nutrient pool minus the pool at abscission divided by the maximum pool. These calculations must be based on the quantity of nutrient per unit leaf area (or some other unit of leaf size: e.g. leaf length in graminoids), rather than per unit weight, because leaves lose 0–30% of their weight during autumn senescence (Table 10.2; Staaf, 1982; Chapin and Kedrowski, 1983). Consequently, estimates of resorption based on changes in *percentage* nutrient composition could substantially underestimate the degree of nutrient resorption.

Table 10.2 Resorption of biomass and nutrients from leaves during autumn senescence in different ecosystems. Data are means ± SE, calculated from Chapin and Kedrowski (1983) and Chapin and Shaver (unpublished)

	Resorption (% of maximum)		
	Dry weight	*N*	*P*
Arctic	20±5	58±4	55±6
Boreal forest	17±5	61±5	28±6
Temperate forest	18±4	50±3	50±6
Tropical forest	–	45±4	55±7

There are several potential sources of error in measuring nutrient resorption. Leaves should be sampled from the same part of the plant at times of maximum and minimum pool size, because nutrient concentration varies in different parts of the plant. Thus, if senescent leaves are collected as they fall from the plant, leaves at time of maximal pool size must be sampled throughout the canopy in approximate proportion to their abundance. A leaf generally reaches maximum pool size shortly after complete expansion and then remains at this pool size for about half of the growing season. However, this pattern varies considerably among species and may depend upon the seasonal patterns of nutrient availability and other environmental factors affecting growth rate. Consequently, it is desirable to measure pool size of leaf nutrients several times during the growing season to record the true maximum pool size. Leaf nutrient concentration declines most rapidly just prior to abscission. Therefore, leaves should be collected as they fall from the plant rather than a few days beforehand. However, at the time of abscission, leaves have high concentrations of inorganic phosphorus and amino acids (Chapin and Kedrowski, 1983) that are readily leached. Therefore leaves must be collected immediately after falling before being leached by rain, which diminishes leaf N and P, and before being invaded by fungi, which often cause an increase in the nitrogen content of the leaf (Melillo *et al.*, 1982).

Measurements of nutrient resorption rely on the assumption that all nutrients lost from the leaf are resorbed, but, in fact, some are leached and appear in throughfall. The estimates of resorption should probably be corrected for nutrients lost in throughfall during the resorption period. (Throughfall occurring before the period of resorption does not affect the resorption calculation.) Leaching of N and P during resorption is minimal except under conditions of extremely high N and P concentration (Tukey, 1970; Chapin and Kedrowski, 1983; Jonasson and Chapin, 1985; Boerner, 1984) and can probably be ignored. However, loss of K from leaves due to leaching is probably greater than losses due to resorption (Baker *et al.*, 1985), so resorption of this element should be corrected for leaching loss. Calcium is also prone to loss by leaching (Tukey, 1970) and is not resorbed (Epstein, 1972).

Resorption has not yet been measured in roots, but the same principles apply as in leaves. Nutrient pools at times of peak nutrients (shortly after the root is produced) and at times of senescence should be expressed per unit root length (or area) rather than as percentage concentration. Root concentrations are highly variable among sample dates, reflecting daily variation in nutrient availability and uptake (Chapin *et al.*, 1986). Consequently, resorption in roots will be difficult to distinguish from the normal pattern of nutrient movement from roots to shoots.

REFERENCES

Allen, S.E. (ed.) (1974) *Chemical Analysis of Ecological Materials*, Blackwell, Oxford.

Andrews, R.E. and Newman, E.I. (1970) Root density and competition for nutrients. *Oecol. Plant.*, **5**, 319–34.

Attiwill, P.M. and Ovington, J.D. (1968) Determination of forest biomass. *For. Sci.*, **14**, 13–15.

Baker, T.G., Hodgkiss, P.D. and Oliver, G.R. (1985) Accession and cycling of elements in a coastal stand of *Pinus radiata*. D. Don in New Zealand. *Plant Soil*, **86**, 303–7.

Barney, R.J., Van Cleve, K. and Schlentner, R. (1978) Biomass distribution and crown characteristics in two Alaskan *Picea mariana* ecosystems. *Can. J. For. Res.*, **8**, 36–41.

Baskerville, G.L. (1965) Estimation of dry weight of tree components and total standing crop in conifer stands. *Ecology*, **46**, 867–9.

Beever, R.E. and Burns, D.J.W. (1980) Phosphorus uptake, storage and utilization by fungi. *Adv. Bot. Res.*, **8**, 127–219.

Bieleski, R.L. (1968) Levels of phosphate esters in *Spirodela*. *Plant Physiol.*, **43**, 1297–308.

Bietz, J.A. (1974) Micro-Kjeldahl analysis by an improved automated ammonia determination following manual digestion. *Anal. Chem.*, **46**, 1617–18.

Bloom, A.J. and Chapin, III, F.S. (1981) Differences in steady-state net ammonium and nitrate influx by cold and warm-adapted barley varieties. *Plant Physiol.*, **68**, 1064–7.

Bloom, A.J., Chapin, III, F.S. and Mooney, H.A., (1985) Resource limitation in plants – an economic analogy. *Annu. Rev. Ecol. System.*, **16**, 363–92.

Boardman, N.K. (1977) Comparative photosynthesis of sun and shade plants. *Annu. Rev. Plant Physiol.*, **28**, 355–77.

Boerner, R.E.J. (1984) Foliar nutrient dynamics and nutrient use efficiency of four deciduous tree species in relation to site fertility. *J. Appl. Ecol.*, **21**, 1029–40.

Bould, C., Hewitt, E.J. and Needham, P. (1983) *Diagnosis of Mineral Disorders in Plants*, Vol. 1, *Principles*, Her Majesty's Stationery Office, London, 170 pp.

Bouma, D. (1983) Diagnosis of mineral deficiencies using plant tests. In *Inorganic Plant Nutrition, Encyclopedia of Plant Physiology* (eds A. Lauchli and R.L. Bieleski), Springer-Verlag, Berlin, Vol. 15A, pp. 120–46.

Bremner, J.M. and Mulvaney, C.S. (1982) Nitrogen-total. In *Methods of Soil Analysis. Part 2 Chemical and Microbiological Properties* 2nd edn (ed. A.L. Page), American Society of Agronomy, Madison, pp. 595–624.

Bremner, J.M. and Tabatabai, M.A. (1971) Use of automated combustion techniques for total carbon, total nitrogen, and total sulfur analysis of soils. In *Instrumental Methods for Analysis of Soils and Plant Tissue* (ed. L.M. Walsh), Soil Science Society of America, Madison, pp. 1–15.

Burton, J.D. (1973) Problems in the analysis of phosphorus compounds. *Water Res.*, **7**, 291–307.

Caldwell, M.M., Eissenstat, D.M., Richards, J.H. and Allen, M.F. (1985) Competition for phosphorus: differential uptake from dual-isotope-labeled soil interspaces between shrub and grass. *Science*, **229**, 384–6.

Chapin, F.S., III. (1974) Morphological and physiological mechanisms of temperature compensation in phosphate absorption along a latitudinal gradient. *Ecology*, **55**, 1180–98.

Chapin, F.S., III. (1980a) The mineral nutrition of wild plants. *Annu. Rev. Ecol. System.*, **11**, 233–60.

Chapin, F.S., III. (1980b) Nutrient allocation and response to defoliation in tundra plants. *Arct. Alp. Res.*, **12**, 553–63.

Chapin, F.S., III. (1988) Ecological aspects of plant mineral nutrition. *Adv. Min. Nutr.*, **3**, 161–91.

Chapin, F.S., III and Bloom, A.J. (1976) Phosphate absorption: adaptation of tundra graminoids to a low temperature, low phosphorus environment. *Oikos*, **27**, 111–21.

Chapin, F.S., III and Chapin, M.C. (1981) Ecotypic differentiation of growth processes in *Carex aquatilis* along latitudinal and local gradients. *Ecology*, **62**, 1000–9.

Chapin, F.S., III and Holleman, D.F. (1974) Radioassay of ^{32}P in intact plant roots using Cerenkov radiation detection. *Int. J. Appl. Radiat. Isot.*, **25**, 568–70.

Chapin, F.S., III, Johnson, D.A. and McKendrick, J.D. (1980) Seasonal movement of nutrients in plants of differing growth form in an Alaskan trundra ecosystem: implications for herbivory. *J. Ecol.*, **68**, 189–209.

Chapin, F.S., III and Kedrowski, R.A. (1983) Seasonal changes in nitrogen and phosphorus fractions and autumn retranslocation in evergreen and deciduous taiga trees. *Ecology*, **64**, 376–91.

Chapin, F.S., III, Shaver, G.R. and Kedrowski, R.A. (1986) Environmental controls over carbon, nitrogen, and phosphorus chemical fractions in *Eriophorum vaginatum* L. in Alaskan tussock tundra. *J. Ecol.*, **74**, 167–95.

Chapin, F.S., III and Tryon, P.R. (1982) Phosphate absorption and root respiration of different plant growth forms from northern Alaska. *Holarct. Ecol.*, **5**, 164–71.

Chapin, F.S. III and Tryon, P.R. (1983) Habitat and leaf habitat as determinants of growth, nutrient absorption, and nutrient use by Alaskan taiga forest species. *Can. J. For. Res.*, **13**, 818–26.

Chiariello, N. and Roughgarden, J. (1984) Storage allocation in seasonal races of an annual plant: optimal versus actual allocation. *Ecology*, **65**, 1290–301.

Christie, E.K. and Moorby, J. (1975) Physiological responses of semi-arid grasses. I. The influence of phosphorus supply on growth and phosphorus absorption. *Austr. J. Agric. Res.*, **26**, 423–36.

Clarkson, D.T. (1985) Factors affecting mineral nutrient acquisition by plants. *Annu. Rev. Plant Physiol.*, **36**, 77–115.

Clarkson, D.T. and Hanson, J.B. (1980) The mineral nutrition of higher plants. *Annu. Rev. Plant Physiol.*, **31**, 239–98.

Cleland, W.W. (1967) The statistical analysis of enzyme kinetic data. *Adv. Enzymol.*, **29**, 1–32.

Clement, C.R., Hopper, M.J., Jones, L.H.P. and Leafe, E.L. (1978) The uptake of nitrate by *Lolium perenne* from flowing nutrient solution. II. Effects of light, defoliation and relationship to CO_2 flux. *J. Exp. Bot.*, **29**, 1173–83.

Cole, C.V., Grunes, D.L., Porter, L.K. and Olsen, S.R. (1963) The effects of nitrogen on short-term phosphorus absorption and translocation in corn (*Zea mays*). *Soil Sci. Soc. Am. Proc.*, **27**, 671–4.

Cole, D.W. and Rapp, M. (1981) Elemental cycling in forest ecosystems. In *Dynamic Properties of Forest Ecosystems* (ed. D.E. Reichle), Cambridge University Press, Cambridge, pp. 341–409.

Conway, G.R., Glass, N.R. and Wilcox, J.C. (1970)

Fitting nonlinear models to biological data by Marquandt's algorithm. *Ecology*, **51**, 503–7.

Cox, G. and Tinker, P.B. (1976) Translocation and transfer of nutrients in vesicular-arbuscular mycorrhizae I. The arbuscule and phosphorus transfer: a quantitative ultrastructural study. *New Phytol.*, **77**, 371–8.

Crooke, W.M. and Simpson, W.E. (1971) Determination of ammonium on Kjeldahl digests of crops by an automated procedure. *J. Sci. Food Agric.*, **22**, 9–10

Eaton, J.S., Likens, G.W. and Borman, F.H. (1973) Throughfall and stemflow chemistry in a northern hardwood forest. *J. Ecol.*, **61**, 495–508.

Epstein, E. (1972) *Mineral Nutrition of Plants: Principles and Perspectives*, Wiley, New York.

Epstein, E., Schmid, W.E. and Rains, D.W. (1963) Significance and technique of short-term experiments on solute absorption by plant tissue. *Plant Cell Physiol.*, **4**, 79–84.

Ernst, W.H.O. (1983) Element nutrition of two contrasted dune annuals. *J. Ecol.*, **71**, 197–209.

Field, C. (1983) Allocating leaf nitrogen for the maximization of carbon gain: leaf age as a control on the allocation program. *Oecologia (Berlin)*, **56**, 341–7.

Field, C. and Mooney, H.A. (1983) Leaf age and seasonal effects on light, water and nitrogen use efficiency in a California shrub. *Oecologia (Berlin)*, **56**, 348–55.

Field, C. and Mooney, H.A. (1986) The photosynthesis–nitrogen relationship in wild plants. In *On the Economy of Plant Form and Function* (ed. T.J. Givnish), Cambridge University Press, Cambridge, pp. 25–55.

Garten, C.T., Jr. (1976) Correlations between concentrations of elements in plants. *Nature, London*, **261**, 686–8.

Gauch, H.G. (1972) *Inorganic Plant Nutrition*, Dowden, Hutchinson & Ross, Stroudsburg. 488 pp.

Gerloff, G.C. (1976) Plant efficiencies in the use of nitrogen, phosphorus and potassium. In *Plant Adaptation to Mineral Stress in Problem Soils* (ed. M.J. Wright), Cornell University Agricultural Experiment Station, Ithaca, New York, pp. 161–9.

Glass, A.D.M. (1983) Regulation of ion transport. *Ann. Rev. Plant Physiol.*, **34**, 311–26.

Glinka, Z. (1980) Abscisic acid promotes both volume flow and ion release to the xylem in sunflower roots. *Plant Physiol.*, **65**, 537–40.

Gray, J.T. (1983) Nutrient use by evergreen and deciduous shrubs in southern California. I. Community nutrient cycling and nutrient-use efficiency. *J. Ecol.*, **71**, 21–41.

Harrison, A.F. and Helliwell, D.R. (1979) A bioassay for comparing phosphorus availability in soils. *J. Appl. Ecol.*, **16**, 497–505.

Headley, A.D., Callaghan, T.V. and Lee, J.A. (1985) The phosphorus economy of the evergreen tundra plant, *Lycopodium annotinum*. *Oikos*, **45**, 235–45.

Hom, J. and Oechel, W.C. (1983) The photosynthetic capacity, nutrient content, and nutrient use efficiency of different needle age-classes of black spruce (*Picea mariana*) found in interior Alaska. *Can. J. For. Res.*, **13**, 834–9.

Hurlbert, S.H. (1984) Pseudoreplication and the design of ecological field experiments. *Ecol. Monogr.*, **54**, 182–211.

Ingestad, T. (1971) A definition of optimum nutrient requirements in birch seedlings. II. *Physiol. Plant.*, **24**, 118–25.

Ingestad, T. (1982) Relative addition rate and external concentration; driving variables used in plant nutrition research. *Plant Cell Environ.*, **5**, 443–53.

Isaac, R.A. and Johnson, W.C. (1976) Determination of total nitrogen in plant tissue, using a block digestor. *J. Assoc. Offic. Anal. Chem.*, **59**, 98–100.

Jonasson, S., Bryant, J.P., Chapin, F.S. III, and Anderson, M. (1986) Plant phenols and nutrients in relation to variations in climate and rodent grazing. *Am. Nat.*, **128**, 394–408.

Jonasson, S. and Chapin, F.S. III, (1985) Significance of sequential leaf development for nutrient balance of the cotton sedge, *Eriophorum vaginatum* L. *Oecologia (Berlin)*, **67**, 511–18.

Jones, J.B. and Steyn, W.J.A. (1973) Sampling, handling and analyzing plant tissue samples. In *Soil Testing and Plant Analysis* (eds L.M. Walsh and J.D. Beaton), Soil Science Society of America, Madison, pp. 249–70.

Kedrowski, R.A. (1983) Extraction and analysis of nitrogen, phosphorus and carbon fractions in plant material. *J. Plant Nutr.*, **6**, 989–1011.

Lauchli, A. (1969) Radioassay for β-emitters in biological materials using Cerenkov reaction. *Int. J. Appl. Radiat. Isot.*, **20**, 265–70.

Leaf, A.L. (1973) Plant analysis as an aid in fertilizing forests. In *Soil Testing and Plant Analysis* (eds L.M. Walsh and J.D. Beaton), Soil Science Society of America, Madison, pp. 427–54.

Lee, R.B. (1982) Selectivity and kinetics of ion uptake by barley plants following nutrient deficiency. *Ann. Bot.*, **50**, 429–49.

Li, W.K.W. (1983) Consideration of errors in estimating kinetic parameters based on Michaelis–Menten formalism in microbial ecology. *Limnol. Oceanogr.*, **28**, 185–90.

Loneragan, J.F. and Asher, C.J. (1967) Response of plants to phosphate concentration in solution culture: II. Rate of phosphate absorption and its relation to growth. *Soil Sci.*, **103**, 311–18.

Melillo, J.M., Aber, J.D. and Muratore, J.F. (1982) Nitrogen and lignin control of hardwood leaf litter decomposition dynamics. *Ecology*, **63**, 621–6.

Miller, P.C. (1980) Quantitative plant ecology. In *Analysis of Ecosystems* (eds D. Horn, G.R. Stairs and R.D. Mitchell), Ohio State University Press, Columbus, pp. 179–231.

Mooney, H.A. and Gulmon, S.L. (1982) Environmental and biological constraints on leaf structure and function in reference to herbivory. *BioSci.*, **32**, 198–206.

Mooney, H.A. and Rundel, P.W. (1979) Nutrient relations of the evergreen shrub, *Adenostoma fasciculatum*, in the California chaparral. *Bot. Gaz.*, **140**, 109–13.

Nelson, D.W. and Sommers, L.E. (1980) Total nitrogen analysis of soil and plant tissue. *J. Assoc. Offic. Anal. Chem.*, **63**, 770–8.

Nissen, P. (1974) Uptake mechanisms: inorganic and organic. *Annu. Rev. Plant Physiol.*, **25**, 53–79.

Nye, P.H. (1977) The rate-limiting step in plant nutrient absorption from soil. *Soil Sci.*, **123**, 292–7.

Nye, P.H. and Tinker, P.B. (1977) *Solute Movement in the Soil–Root System*. University of California Press, Berkeley, California.

Olson, K.O., Reiners, W.A., Cronan, C.S. and Lang, G.E. (1981) The chemistry and flux of throughfall and stemflow in subalpine balsam fir forests. *Holarct. Ecol.*, **4**, 291–300.

Osonubi, O., Oren, R., Werk, K.S., Schulze, E.-D. and Heilmeier, H. (1988) Performance of *Picea abies* (L.) Karst. at different stages of decline. IV. Correlates with xylem sap concentrations of magnesium, calcium and potassium but not nitrogen. *Oecologia (Berlin)*, **77**, 1–6.

Ovington, J.D. and Madgwick, H.A.I. (1959) Distribution of organic matter and plant nutrients in a plantation of Scots Pine. *For. Sci.*, **5**, 344–55.

Pate, J.S. (1983) Patterns of nitrogen metabolism in higher plants and their ecological significance. In *Nitrogen as an Ecological Factor* (eds J.A. Lee, S. McNeill and I.H. Rorison), Blackwell, Oxford, pp. 225–55.

Pettersson, S. (1978) Varietal differences in rubidium uptake efficiency of barley roots. *Physiol. Plant.*, **44**, 1–6.

Redfield, A.C. (1958) The biological control of chemical factors in the environment. *Am. Sci.*, **46**, 205–21.

Robbins, C.T. (1983) *Wildlife Feeding and Nutrition*, Academic Press, New York.

Rowland, A.P. (1983) An automated method for the determination of ammonium-N in ecological materials. *Commun. Soil Sci. Plant Anal.*, **14**, 49–63.

Saric, M.R. (1981) Genetic specificity in relation to plant mineral nutrition. *J. Plant Nutr.*, **3**, 743–66.

Seastedt, T.R. (1985) Canopy interception of nitrogen in bulk precipitation by annually burned and unburned tallgrass prairies. *Oecologia (Berlin)*, **66**, 88–92.

Shaver, G.R. and Melillo, J.M. (1984) Nutrient budgets of marsh plants: efficiency concepts and relation to availability. *Ecology*, **65**, 1491–510.

Small, E. (1972) Photosynthetic rates in relation to nitrogen recycling as an adaptation to nutrient deficiency in peat bog plants. *Can. J. Bot.*, **50**, 2227–33.

Sokal, R.R. and Rohlf, F.J. (1969) *Biometry*, W.H. Freeman, San Francisco.

Soltanpour, P.N., Jones, J.B. Jr., and Workman, S.M. (1982) Optical emission spectrometry. In *Methods of Soil Analysis. Part 2 Chemical and Microbiological Properties*, 2nd edn (ed. A.L. Page), American Society of Agronomy, Madison, pp. 29–65.

Staaf, H. (1982) Plant nutrient changes in beech leaves during senescence as influenced by site characteristics. *Acta Oecol.*, **3**, 161–70.

Stark, N. and Spitzner, C. (1985) Xylem sap analysis for determining the nutrient status and growth of *Pinus ponderosa*. *Can. J. For. Res.*, **15**, 783–90.

Stark, N., Spitzner, C. and Essig, D. (1985) Xylem sap analysis for determining nutritional status of trees: *Pseudotsuga menziesii*. *Can. J. For. Res.*, **15**, 429–37.

Tarrant, R.F., Lu, K.C., Bollen, W.B. and Chen, C.S. (1968) Nutrient cycling by throughfall and stemflow precipitation in three coastal Oregon forest types. USDA Forest Service Research Paper, PNW-54.

Technicon Industrial Systems (1976) Individual/simultaneous determination of nitrogen and/or phosphorus in BD acid digests. Industrial Method No. 334-74W/B. 7 pp.

Tukey, H.B., Jr. (1970) The leaching of substances from plants. *Annu. Rev. Plant Physiol.*, **21**, 305–24.

Van Cleve, K. and Alexander, V. (1981) Nitrogen cycling in tundra and boreal ecosystems. In

Terrestrial Nitrogen Cycles (eds F.E. Clark and T. Rosswall), *Ecol. Bull., Stockholm*, **33**, 375–404.

Van Cleve, K. and Harrison, A.F. (1985) Bioassay of forest floor phosphorus supply for plant growth. *Can. J. For. Res.*, **15**, 156–62.

Van Cleve, K., Viereck, L.A. and Schlentner, R.L. (1971) Accumulation of nitrogen in alder (*Alnus*) ecosystems near Fairbanks, Alaska. *Arct. Alp. Res.*, **3**, 101–14.

van den Driessche, R. (1974) Prediction of mineral nutrient status of trees by foliar analysis. *Bot. Rev.*, **40**, 347–94.

van den Driessche, R. and Webber, J.E. (1977) Seasonal variations in a Douglas fir stand in total and soluble nitrogen in inner bark and root and in total and mineralizable nitrogen in soil. *Can. J. For. Res.*, **7**, 641–7.

Veerkamp, M.T. and Kuiper, P.J.C. (1982a) The uptake of potassium by *Carex* species from swamp habitats varying from oligotrophic to eutrophic. *Physiol. Plant.*, **55**, 237–41.

Veerkamp, M.T. and Kuiper, P.J.C. (1982b) The effect of potassium on growth of *Carex* species from swamp habitats varying from ologotrophic to eutrophic and comparison of physiological reactions of *Carex* species to P stress and K stress. *Physiol. Plant.*, **55**, 242–6.

Vitousek, P.M. (1982) Nutrient cycling and nutrient use efficiency. *Am. Nat.*, **119**, 553–72.

Virginia, R.A. and Delwiche, C.C. (1982) Natural ^{15}N abundance of presumed N_2-fixing and non-N_2-fixing plants from ecosystems. *Oecologia (Berlin)*, **54**, 317–25.

Walker, L.R. and Chapin, F.S. III, (1986) Physiological controls over seedling growth in primary succession on an Alaskan flood plain. *Ecology*, **67**, 1508–23.

Westoby, M. (1974) An analysis of diet selection by large generalist herbivores. *Am. Nat.*, **108**, 290–304.

White, R.E. (1972) Studies on mineral ion absorption by plants. I. The absorption and utilization of phosphate by *Stylosanthes humilis*, *Phaseolus atropurpureus* and *Desmodium intortum*. *Plant Soil*, **36**, 427–47.

Wilkinson, G.N. (1961) Statistical estimations in enzyme kinetics. *Biochem. J.*, **80**, 324–32.

11

Photosynthesis: principles and field techniques

Christopher B. Field, J. Timothy Ball and Joseph A. Berry

'What a great economy of means. But then, what an economy of ends.'
Henry Adams after seeing a puppet show.

11.1 THE SYSTEM CONCEPT

Among the people who do photosynthesis research at the leaf, plant or canopy level, the devices used to measure photosynthesis are usually referred to as gas-exchange systems or simply as 'systems'. The concept that photosynthesis is measured with a system, rather than a single instrument, is an important place to start, for two reasons. First, the system concept emphasizes the fact that we have nothing like a discrete photosynthesis sensor. Photosynthesis is always a calculated parameter, determined from measurements of CO_2 concentrations, gas flows and sometimes other parameters, depending on the measurement philosophy. Second, the system concept reminds us that gas-exchange systems typically measure more than just photosynthesis, for the reason that photosynthesis data are greatly enhanced by the simultaneous acquisition of other kinds of information. These two aspects of the system concept also frame the material to be presented in this chapter. Developing the idea of a measurement system, we address the design and construction of gas-exchange systems, considering both measurement principles and the devices that form system components. Emphasizing the measurement and control of parameters other than photosynthesis, we discuss some of the trade-offs that influence the choice of gas-exchange systems for different kinds of research questions.

Throughout, we deal only with systems in which photosynthesis is calculated from some measure of CO_2 exchange. Chapter 12 describes techniques for measuring photosynthesis on the basis of O_2 exchange. Chapter 13 covers aspects of stable-isotope analysis relevant to photosynthesis. We also focus primarily on systems designed to measure CO_2 and H_2O exchange simultaneously.

Several publications treat various aspects of field photosynthesis measurements in more detail than is possible here. Šesták, Čatský and Jarvis' (1971) *Plant Photosynthetic Production: Manual of Methods* is the major reference in the field and a rich source of background material. Chapters in Coombs *et al.* (1985), Marshall and Woodward (1985) and Kramer *et al.* (1989) consider many aspects of the design and operation of more modern instruments. von Caemmerer and Farquhar (1981) and Ball (1987) provide complete accounts of gas-exchange calculations for particular systems.

11.1.1 Motivation for the simultaneous measurement of CO_2 and H_2O exchange

It is possible to measure photosynthesis without transpiration and vice versa, but the power of a photosynthesis measurement is greatly increased by a simultaneous measurement of transpiration for two reasons. First, the two measurements are synergistic because the two processes are interdependent. CO_2 diffuses into leaves and water vapor diffuses out along a pathway that is largely shared. Stomatal responses that decrease water loss also decrease photosynthesis, and responses that increase photosynthesis also increase water loss. Thus, the diffusion limitation to photosynthesis must be viewed as the product of a compromise between responses that tend to increase photosynthesis and those that tend to decrease water loss. Without measuring H_2O exchange, that compromise is very difficult to evaluate. Second, leaf conductance to water vapor, calculated from measurements of H_2O exchange, is a critical parameter for determining c_i, the concentration of carbon dioxide inside the leaf. c_i can be measured directly (Sharkey *et al.*, 1982), but the measurement is technically difficult. A much simpler solution is to calculate c_i, using leaf conductance to water vapor. Because the diffusion pathways for CO_2 and H_2O are mostly the same, we can calculate leaf conductance to CO_2 from the conductance to water vapor (Section 11.7). Given the leaf conductance to CO_2, rearranging the defining equation for a conductance (Nobel, 1983) yields

$$c_i = c_a - \frac{A_n}{g_{tc}} \tag{11.1}$$

where c_i is the concentration of CO_2 in the intercellular spaces, c_a is the concentration of CO_2 in the bulk air, A_n is net photosynthesis and g_{tc} is the combined leaf and boundary layer conductance to CO_2. Section 11.7 gives a more accurate expression that corrects for complications introduced when multiple gases are diffusing in opposite directions. To the extent that the diffusion pathways for CO_2 and H_2O differ, Equation 11.1 should yield incorrect values. In plants with very thick leaves, Equation 11.1 probably overestimates c_i (Parkhurst, 1986), but the error is typically less than a few per cent in leaves of normal thickness (Sharkey *et al.*, 1982).

Both g_{tc} and c_i, calculated from H_2O exchange parameters, are of tremendous value in interpreting photosynthesis data. c_i is a measure of the availability of the primary substrate for photosynthesis and a necessary parameter in partitioning the limitation of photosynthesis between biochemical and diffusion factors (cf. Farquhar and Sharkey, 1982; Woodrow *et al.*, 1988). For studies focused on the biochemistry of photosynthesis, c_i is crucial in relating CO_2 availability between whole-leaf, cell, chloroplast, and *in vitro* experiments. For comparative analyses of photosynthetic characteristics across taxa or habitats, conductance effects are key components of the comparison. And for studies oriented toward predicting photosynthesis, conductance plays a major role in rate control, through setting the availability of CO_2.

11.2 PRINCIPLES OF PHOTOSYNTHESIS MEASUREMENT

In concept, measuring photosynthesis is straightforward. Though many photosynthesis systems have been developed, almost all utilize a closed, open or isotope system philosophy. For an introduction to these techniques, we will consider idealized devices for measuring only CO_2 exchange. Real instruments can be made to measure either CO_2 or H_2O exchange or both. We present the concepts for a system making measurements on a single leaf, but the principles are the same for whole branch, plant or even community measurements. The equations in this section are intended to establish concepts and ignore some of the details discussed in Section 11.7.

11.2.1 Closed or transient systems

If a photosynthesizing and transpiring leaf, shoot or plant is placed in a closed chamber, the CO_2 concentration in the chamber decreases, and the water content of the air increases (Fig. 11.1). The rate of CO_2 depletion is the photosynthesis rate and the rate of water vapor addition is the transpiration rate. To determine photosynthesis, one measures the CO_2 concentration in the chamber at a starting point, waits some interval and measures the CO_2 concentration again. Since the actual amount of CO_2 depletion is the product of concentration and system volume,

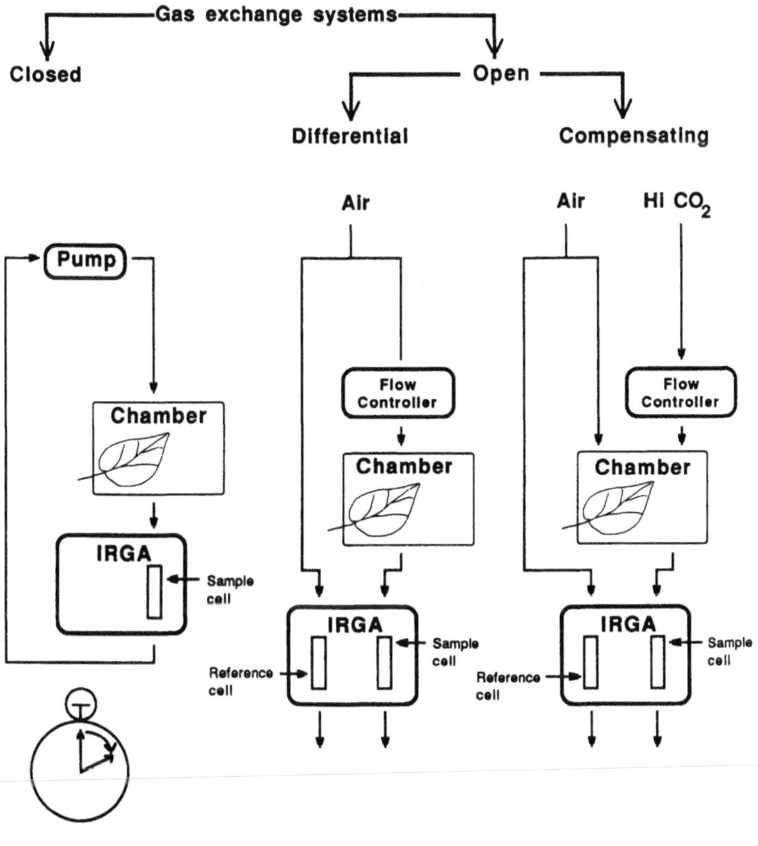

Fig. 11.1 Simplified schematic illustrating how photosynthesis is measured by three basic types of gas-exchange systems. Rates of gas exchange in all systems are determined by mass balance. The calculation of photosynthesis in a closed system is based on the rate of change of CO_2 concentration. Note that a closed system utilizes an absolute, rather than a differential, infrared gas analyzer (IRGA) and requires no flow-measuring device. A differential system calculates photosynthesis from the CO_2 depletion that occurs as air flows at a known rate past a photosynthesizing leaf. In a compensating system, CO_2 depletion by photosynthesis is compensated by CO_2 injection so that the CO_2 concentration in the air exiting the chamber is the same as that in the air stream entering the chamber.

the basic equation for total net photosynthesis (A'_n) is

$$A'_n = \frac{(c_b - c_f)V}{\Delta t} \quad (11.2)$$

where c_b is the initial concentration of CO_2, c_f is the final concentration of CO_2, V is system volume and Δt is the time between the start and the conclusion of the measurement. To express photosynthesis per unit of leaf area or weight, simply divide total photosynthesis by the area or weight (L) of the photosynthesizing tissue.

$$A_n = \frac{(c_b - c_f)V}{\Delta t L} \quad (11.3)$$

If c_b and c_f are in mol of CO_2/mol of air, V is in mol of air, elapsed time is in seconds and L is in m^2, then A_n is in mol of $CO_2\ m^{-2}\ s^{-1}$.

11.2.2 Open or steady-state systems

Consider a chamber containing a photosynthesizing and transpiring leaf. If an air stream is passed continuously through the chamber, the air leaving the chamber will be depleted in CO_2 and enriched in water vapor, relative to the air entering the chamber. If the rates of photosynthesis, transpiration and air flow through the chamber are constant, chamber conditions reach a steady state. Devices which measure photosynthesis on the basis of this equilibrium CO_2 depletion are called differential systems (Fig. 11.1). In such systems, the basic equation for net photosynthesis is

$$A_n = \frac{u_e c_e - u_o c_o}{L} \quad (11.4)$$

where u_e is the air flow entering the chamber, c_e is the concentration of CO_2 in the air entering the chamber, u_o is the air flow leaving the chamber [which differs from air flow entering the chamber because the leaf removes CO_2 and adds H_2O (Section 11.7)], c_o is the CO_2 concentration in the air leaving the chamber and L is the surface area or mass of the plant material in the chamber. If c_e and c_o are in mol of CO_2/mol of air, u_e and u_o are in mol of air s^{-1} and L is in m^2, then A_n is in mol of $CO_2\ m^{-2}\ s^{-1}$.

Under conditions where the CO_2 depletion encountered in a differential system is undesirable, compensating systems, which are also open and operate at steady state, offer practical alternatives. Compensating systems for CO_2 are particularly suited to instruments in which small air flow through the chamber causes large CO_2 depletions in differential mode. Systems in which transpiration is the only source of chamber humidity, 'null-balance systems', typically operate at low air flow and commonly utilize a compensating or combined compensating–differential approach to measuring photosynthesis. In a compensating system, the CO_2 removed by photosynthesis is replaced by injecting CO_2 into the chamber (Fig. 11.1). When the rate of CO_2 injection is adjusted such that the CO_2 concentration in the air entering and leaving the chamber is the same [with appropriate corrections (Section 11.7)], photosynthesis is equal to the rate of CO_2 injection. Photosynthesis in a 100 cm^2 leaf typically consumes less than 10 μl of $CO_2\ s^{-1}$. Because such low flows are difficult to measure and control, the gas injected to compensate photosynthesis is often not pure CO_2 but a mixture of nitrogen and CO_2, chosen to yield flow rates in a more manageable range.

11.2.3 Meteorological methods

Photosynthesis removes CO_2 from the atmosphere. But as CO_2 is depleted around photosynthesizing tissue, it tends to be replenished by atmospheric processes that transport CO_2 from regions of higher to regions of lower partial pressure. Just as a product of a CO_2 differential and a forced flow yields photosynthesis in a differential system (Equation 11.3), a combination of a CO_2 gradient and a natural transport can yield photosynthesis for an unenclosed outdoor community. The complexity of natural air movements insures that the combination is more detailed than a simple product of a concentration difference and a flow, but the basic idea is simple: the rate of CO_2 uptake by the vegetation must equal the rate of CO_2 supply from the ground below and the atmosphere above.

Two kinds of meteorological measurements are widely used. Both focus on turbulent flow as the dominant process transporting gases into and out of the vegetation canopy, but they differ in the time scale over which the turbulence is considered. In the aerodynamic or mean value method, the objective is quantifying, over a period of several minutes, (1) the time-average CO_2 profile over the vegetation and (2) the atmosphere's tendency for vertical transport, a quantity called the turbulent or eddy diffusivity. With the aerodynamic method, net photosynthesis of a canopy (A_c) is given by

$$A_c = -K_c \left(\frac{\partial \rho_c}{\partial z} \right) \tag{11.5}$$

where K_c is the eddy diffusivity and $\partial\rho_c/\partial z$ is the rate of change of CO_2 with height. If $\partial\rho_c/\partial z$ is in g m^{-3} m^{-1} and K_c is in m^2 s^{-1}, then A_c has the units g of CO_2 m^{-2} s^{-1} and is expressed per unit of ground area. The negative sign sets the direction of the flux to be toward regions of lower concentration.

Equation 11.4 is simple, but determining K_c can be quite involved. K_c varies with wind velocity, with the aerodynamic roughness of the vegetation surface and with the stability of the air layer (the tendency to suppress vertical transport). It is usually impossible to determine the eddy diffusivity for CO_2 directly, but an effective alternative is to infer the value from the eddy diffusivity calculated for a tracer, usually momentum, heat or water vapor. Because turbulent transfer is equally effective at transporting each of these quantities, the eddy diffusivity should be the same for each. The theory and techniques for determining eddy diffusivities are beyond the scope of this chapter. See Thom (1975) and Kanemasu *et al.* (1979) for background information and several papers in Monteith (1976) and Hutchison and Hicks (1985) for applications.

The eddy correlation method eliminates the need for time averaging the CO_2 gradient because it focuses on quantifying the transport associated with each pulse of turbulent flow. Wind directly above a plant canopy may move in any direction, including up and down. Upward pulses transport substances out of the canopy and downward pulses transport substances into it. Over a very short period the net flux can be in any direction, but over a period of minutes, this turbulent transport must carry into the canopy a quantity of CO_2 sufficient to replenish that removed by net photosynthesis. Thus, during periods of positive net photosynthesis, either the CO_2 concentration is, on average, greater during downward than upward pulses, or the total quantity of downward is greater than the quantity of upward air movement. Specifically, net flux of any substance is equal to the mean covariance between fluctuations in vertical wind velocity and in the quantity of the substance of interest (CO_2 for photosynthesis measurements). Since the covariance is a measure of the total extent to which variables change in concert (it is like a correlation coefficient, but without the scaling that restricts a correlation coefficient between 1 and −1), it is intuitively reasonable that it should describe the direction and magnitude of net transport.

One of the major empirical difficulties with eddy correlation is that natural pulses of turbulent air movement may be quite rapid. The frequency depends on wind conditions and on the type of vegetation surface. The most common vertical pulses over a deciduous forest occur with a frequency of one pulse per 6–10 s (Anderson *et al.*, 1986). However, sonic anemometers (Chapter 4), open path CO_2 analyzers (Jarvis and Sandford, 1985) and digital recording all make eddy correlation an increasingly powerful technique for addressing questions at the level of entire canopies. For examples of recent applications see Baldocchi *et al.* (1987) and several papers in Hutchison and Hicks (1985).

11.2.4 Isotope systems

A leaf exposed to an atmosphere containing $^{14}CO_2$ will incorporate ^{14}C into the products of photosynthesis. Calvin and colleagues elegantly exploited the opportunity of substituting a radioactive substrate for the normal substrate of photosynthesis in their classic experiments on the fate of carbon in photosynthesis (Calvin and Benson, 1948). Because ^{14}C is an inexpensive and relatively safe radioisotope, and because the technologies for measuring its decay are commonly available, many researchers have used ^{14}C uptake to measure photosynthesis.

Systems that measure photosynthesis on the basis of ^{14}C exchange can be either open or closed. The basic concepts for open and closed systems apply completely to isotope systems. However, the technological differences between measuring CO_2 concentration and radioactive disintegrations make it useful to consider isotope systems separately.

Most of the open systems for measuring ^{14}C exchange are laboratory instruments (e.g. Ludwig and Canvin, 1971). The majority of the field systems are closed and function in one of two ways. Fixation of ^{14}C into photosynthate is measured either by quantifying the radioactivity incorporated into tissue (e.g. Shimshi, 1969) or by quantifying the loss of radioactivity from the gas in the closed system (e.g. Blacklow and Maybury, 1980). In a typical tissue-sample isotope measurement, the protocol is (1) enclose the plant material in the chamber, (2) inject a known amount of $^{14}CO_2$, (3) wait a specified time (Δt), (4) remove the plant material from the chamber and (5) harvest the material for later analysis, usually with liquid scintillation counting. The basic equation for photosynthesis is

$$A_i = \frac{B}{\Delta t L \varepsilon S} \tag{11.6}$$

where A_i is photosynthesis, B is the specific activity of the plant material, L is its area or mass, ε is the counting efficiency and S is the specific activity of the gas passing over the leaf. If B is in Bq or disintegrations per second, S is in Bq per mol of CO_2 and L is in m^2, then A_i is in mol of CO_2 m^{-2} s^{-1}. As long as the $^{14}CO_2$ exposure is short, little of the ^{14}C fixed in photosynthesis is released through either photorespiration of mitochondrial respiration. Thus, A_i should be gross, rather than net, photosynthesis. However, the precise relationship between A_i and gross photosynthesis is unclear (Karlsson and Sveinbjörnsson, 1981).

The protocol for a gas-sampling system is similar, except that samples of the gas in the chamber are either withdrawn or counted *in situ* at two or more times during the measurement. The basic equation is

$$A_i = \frac{\Delta x}{\Delta t L S} \tag{11.7}$$

which differs from Equation 11.6 only in that the efficiency term is no longer necessary and the numerator becomes Δx, the change in the total activity of the gas in the system.

11.2.5 Relative advantages of each philosophy

Each of the system philosophies can yield highly accurate photosynthesis measurements. Particular devices based on each philosophy span a range of portability, ease of use and calibration, sample-processing capacity, level of control, accuracy and ruggedness. However, underlying differences in operating principles constrain the utility of every instrument. For example, closed systems never provide true steady-state measurements, because the analysis requires a concentration change. Thus, a closed-system philosophy is generally inappropriate for an instrument designed to impose environmental control and explore the responses of photosynthesis to a range of environmental conditions. Semi-closed systems like those of Musgrave and Moss (1961) and Bazzaz and Boyer (1972) are exceptions to this rule. Closed systems are inherently simpler than open systems, because the former need neither flow measurements nor differential CO_2 analyzers. This simplicity advantage makes closed systems well suited for highly portable devices intended for sampling large numbers of leaves under ambient conditions.

Meteorological techniques are really the most open of open systems, producing values for the photosynthesis of entire communities under natural conditions. These techniques obviously provide no environmental control or manipulation, and results from them are sometimes difficult to relate to the performance of individual plants. However, they provide direct access to research questions at the community level or landscape level, questions likely to increase in importance as biologists attempt to understand processes at higher levels of organization, especially in relation to global change.

Isotope systems were once quite popular because they can be very portable. However, their utility has faded as the portability and accuracy of open and closed systems has increased. Most isotope systems require destructive sampling, provide no immediate results in the field, involve a substantial amount of laboratory analysis and cannot provide simultaneous measurements of CO_2 and water-vapor exchange. Some isotope systems overcome one or more of these problems. For example, the double-isotope porometer of Johnson *et al.* (1979) does measure CO_2 and water-vapor exchanges simultaneously, and the gas-sampling system of Blacklow and Maybury (1980) does not require destructive sampling. Still, isotope systems are seldom the best option for photosynthesis studies. For further information on isotope systems, see the reviews by Voznesenskiï *et al.* (1971), Incoll (1977) and Long and Hallgren (1985).

11.3 COMPONENTS OF GAS-EXCHANGE SYSTEMS

Gas-exchange systems vary tremendously in the range of parameters measured and controlled. Here, we discuss only those parameters that are critical for a substantial proportion of field photosynthesis studies and that are not discussed elsewhere in this volume. Important parameters considered in other chapters include temperature (Chapter 7), light (Chapter 6), transpiration and conductance (Chapter 8) and field data acquisition (Chapter 2).

11.3.1 Measurement components

(a) CO_2 analyzers

All closed and open photosynthesis systems depend on CO_2 measurements (Section 11.2). CO_2 is present in the air in very small quantities. Though the level has been rising steadily, typical concentrations are still only about 350 μmol of CO_2/mol of air or 0.035% (Keeling *et al.*, 1982). The resolution requirements for accurate photosynthesis measurements are flexible, such that compensations in other aspects of system design can balance differences in CO_2 analyzer resolution. For example, extending the sampling time in a closed system or decreasing the flow in a differential system each tends to increase the CO_2 signal. However, these compensations usually entail a diminution in aspects of the system's capabilities distinct from the accuracy of the photosynthesis measurements. As a general rule, CO_2 analyzers useful in gas-exchange systems need accuracy and resölution of approximately 1 μmol mol^{-1} (1 ppm). The accuracy requirements tend to be greatest for differential systems and less stringent for closed systems. Nearly all modern gas-exchange systems measure CO_2 concentration with a nondispersive infrared gas analyzer (NDIR), also called an infrared gas analyzer or IRGA.

CO_2 is a strong absorber of infrared radiation in three bandwidths (Fig. 11.2). This infrared or thermal absorption underlies the widely discussed 'greenhouse effect' predicted to result from rising atmospheric CO_2 (Revelle and Suess, 1957). It is also the signal processed by an NDIR. The basic idea is simple. Shine an infrared source (usually a nichrome wire heated to 600–800°C) through a sample chamber and on to a detector. The energy at the detector is the total entering the system minus that absorbed in the sample chamber. The major conceptual problem in IRGA design is discriminating between CO_2 and other gases that absorb in the IR.

Some early CO_2 analyzers solved the discrimination problem by making the source wavelength specific. These dispersive instruments or spectrometers spread the infrared radiation into a spectrum and expose the sample chamber to only a narrow waveband where CO_2 absorbs strongly (McAlister, 1937). Because the dispersive instruments were costly, complex, and bulky, most modern instruments utilize a broad-band (nondispersive) source and a detector sensitive only to the CO_2 absorption bands. Until recently nearly all CO_2 IRGAs incorporated either a parallel or series Luft detector (Luft, 1943; Luft *et al.*, 1967). Both versions consist of two CO_2-filled chambers separated by a flexible membrane (Fig. 11.3). Differential IR absorption in the two chambers of the detector causes differential expansion and displacement of the flexible membrane. This displacement can be quantified by making the flexible membrane one plate of a capacitor mounted near another, fixed, plate and measuring capacitance. In a parallel Luft detector, one of the two detector chambers is exposed to IR passing through the sample cell and the other is exposed to IR coming through the reference cell. In the series design, both chambers of the detector are exposed alternately to IR coming through the sample and reference cells. Among current IRGA models, most still incorporate Luft detectors, but several manu-

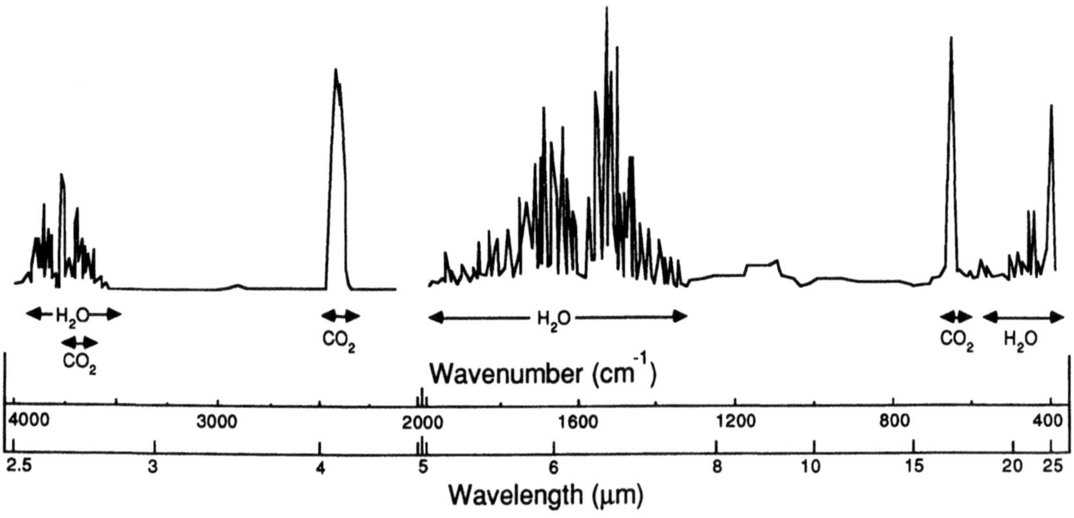

Fig. 11.2 Infrared absorption spectra of CO_2 and H_2O (modified from Sesták *et al.* (1971).

facturers are switching to other types, especially for portable instruments.

While the Luft detector has high sensitivity and relatively high selectivity for CO_2, it suffers from vibration sensitivity. Several solid-state detectors overcome the vibration sensitivity of the Luft detector, but until recently, their inherently low sensitivity limited their application to instruments with either very long sample cells or öw sensitivity requirements. New detectors that solve both problems include the mass flow detector (see below) in the Leybold Hereaus Binos analyzers, pyroelectric detectors in some analyzers from ADC, Anatek and Miran, and lead–selenium detectors in instruments from Anarad, IRI and LI-COR.

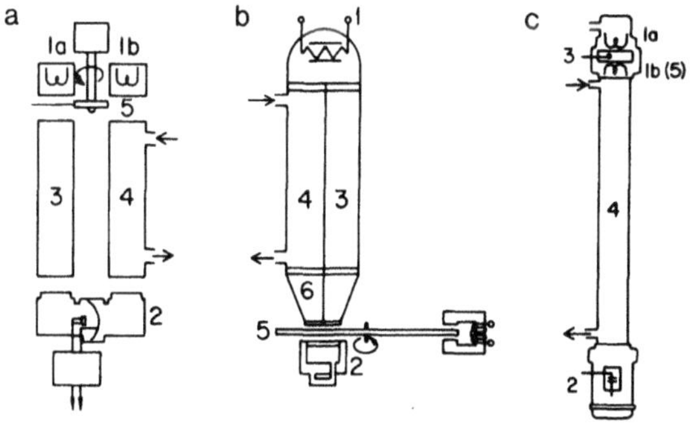

Fig. 11.3 Three types of IRGA designs. Absolute IRGAs (with sealed reference cells) are shown. The only difference in a differential instrument is that the reference gas may have some CO_2 concentration other than zero and typically flows through the reference cell. (a) An IRGA with a 'Luft' detector and chambers in parallel. (b) An IRGA with a mass flow detector (e.g. Binos). (c) An IRGA that substitutes multiple IR sources, energized at different times, for a mechanical radiation chopper (e.g. Liston-Edwards). In each instrument, the numbers indicate these components: (1) infrared radiation sources (1a and 1b indicate multiple sources), (2) detector, (3) reference cell, (4) sample cell, (5) radiation chopper [in (c), the two sources act as a chopper], (6) gas-filled filters. In (a), CO_2 concentration is calculated from the distension of the flexible membrane in the detector. In (b), the signal is the mass flow of gas between the two chambers of the detector. The detector in (c) is functionally similar to that in (a).

A second aspect of the gas discrimination problem is that both CO_2 and H_2O vapor absorb IR at wavelengths between 2.5 and 3 μm (Fig. 11.2). As a result of this, many IRGAs respond to water vapor as if it were CO_2, a problem made more serious by the fact that, in normal air, water vapor is typically 100 times more abundant than CO_2. Three techniques can be used to prevent this potential H_2O sensitivity from introducing errors in photosynthesis measurements. (1) The sample gas can be dried to a constant water content before it reaches the IRGA, but the desiccant must not influence the CO_2 concentration. Magnesium perchlorate, zinc chloride and sulfuric acid are acceptable desiccants (Janác *et al.*, 1971). Magnesium perchlorate is probably the most commonly used, though it must be used with caution (Samish, 1978). The air can also be dried by cooling to condense the water vapor. (2) The IRGA can be equipped with filters to eliminate the 2.5–3 μm band. (3) The effect of water vapor can be compensated in the calculations, but this requires independent measurements of water vapor concentration. In our laboratories, we usually dry the sample gas before it enters the IRGA.

IRGAs are manufactured in many physical configurations (Fig. 11.3). Most instruments measure the difference between the CO_2 concentrations in two cells. The reference cell is filled with CO_2-free gas in an absolute instrument and with a flowing CO_2-containing gas in a differential instrument. Because total IR absorption increases with absorbing pathlength, IRGA sensitivity generally increases with cell length. To minimize drift and noise, most IRGAs utilize some kind of a 'chopper' to interrupt the IR, referencing the measurement to a zero differential. Traditional choppers are propeller-shaped disks that rotate in the optical path, but some newer instruments have replaced mechanical choppers with multiple IR sources (the Liston-Edwards IRGAs) (Fig. 11.3) or have abandoned choppers in favor of alternately routing sample and reference gases through a single cell (the Analytical Development Co. (ADC) model RFA). The Binos IRGAs incorporate special openings in the choppers that provide a compensation for source drift. Chopper stability, a critical component of IRGA accuracy, is regulated by an internal oscillator in some instruments and by the AC-power line frequency in others. Field instruments, whether operated from a battery, a generator or a remote power grid, should incorporate IRGAs with an internally regulated chopper. For more details of the design and operation of particular IRGAs, see Jarvis and Sandford (1985) and Long and Hallgren (1985).

(b) Flow meters

All open gas-exchange systems require flow measurements. The number of flows to be

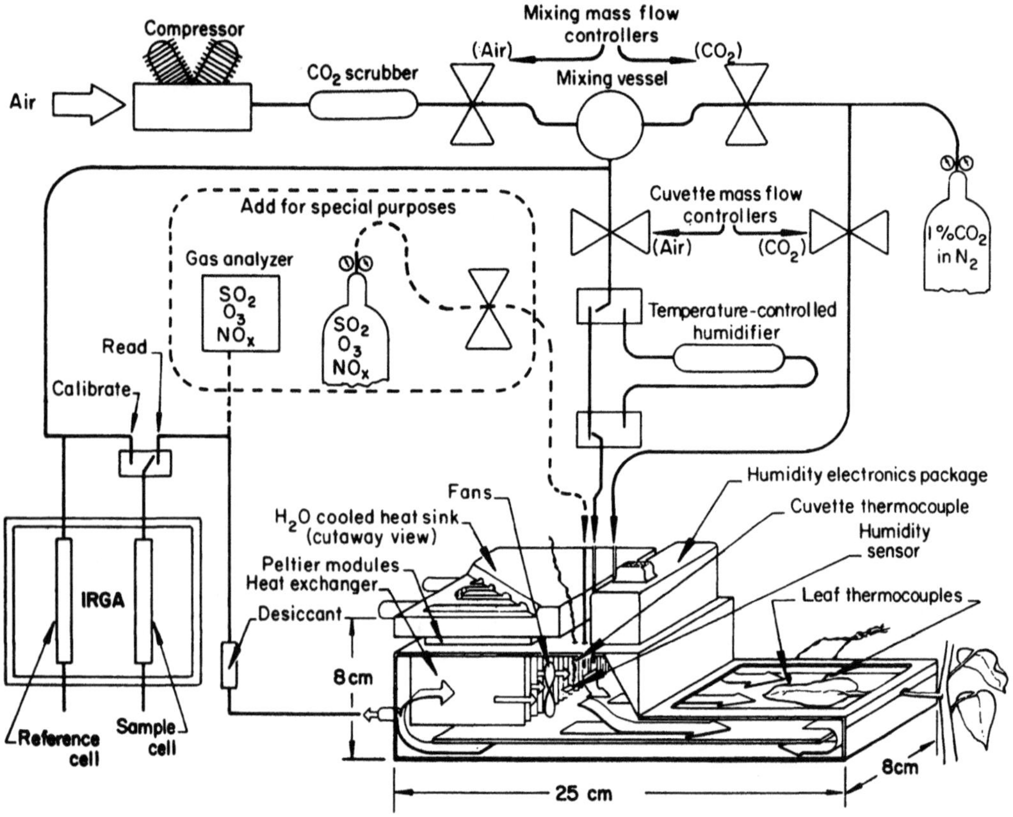

Fig. 11.4 The major components of an open gas-exchange system for controlled-environment studies. The system can operate as a compensating system (with flow through the cuvette controller for CO_2) or as a differential system (with no flow through the cuvette controller for CO_2). Note that a humidification subsystem (Fig. 11.8) can be switched into the air stream leading to the leaf chamber. In the chamber or cuvette shown here, fans stir the air to maintain a high boundary-layer conductance around the leaf. The air circulation also decreases concentration gradients of CO_2 and H_2O because each 'parcel of air' passes over the leaf several times (modified from Field and Mooney, 1989).

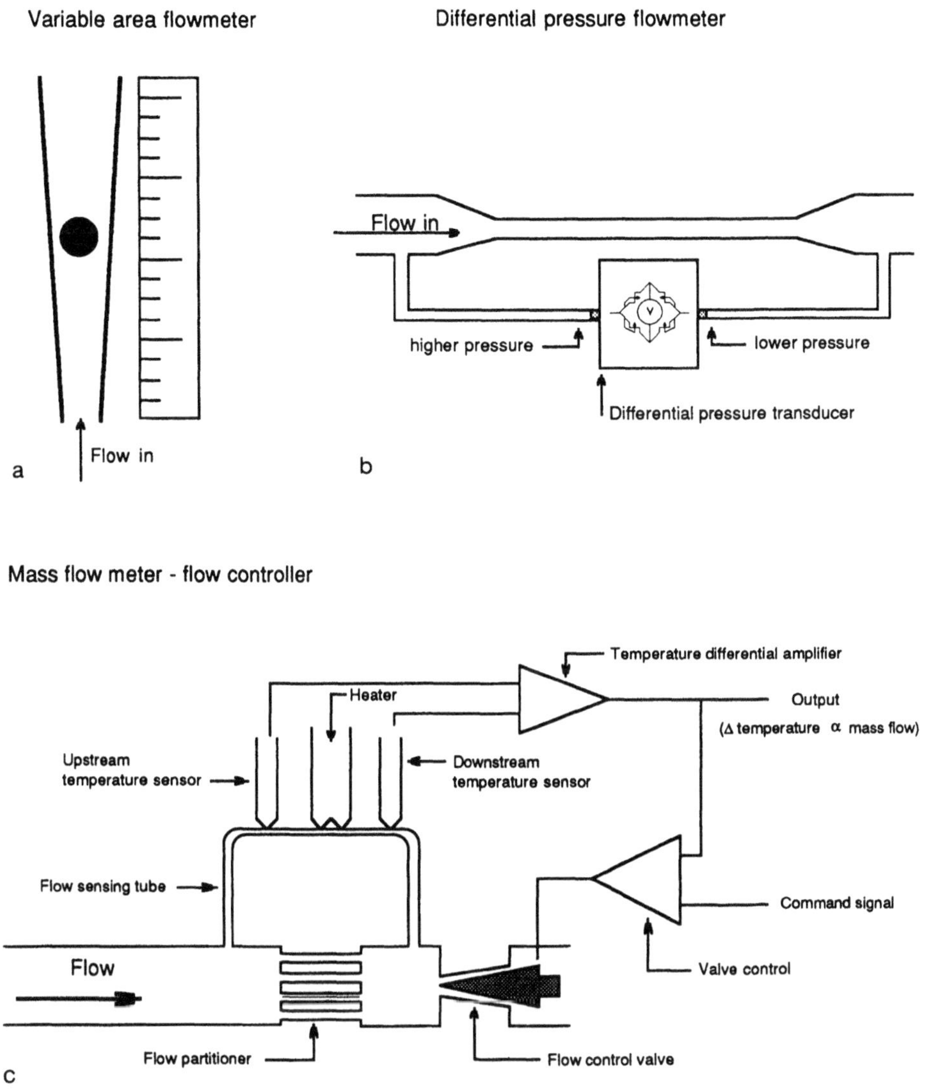

Fig. 11.5 Three types of flow metering devices. (a) In a variable area flow meter or rotameter, flowing gas pushes a float up a tapered tube to the point where the pressure difference between the upper and lower float surfaces equals the float weight. The flow is read from float height. (b) Differential pressure flow meters utilize the fact that, within limits, flow through a restriction increases with the pressure difference across the restriction. This pressure difference can be measured with a differential pressure transducer. Changing the capillary diameter and/or length allows scaling the pressure–flow relationship to make the range appropriate for different applications. (c) Mass flow meters/controllers determine flow from the heat transported by the flowing gas. Most instruments measure the heat transported by only a fraction of the total flow and use a sophisticated flow partitioner to route a constant fraction of the flow through the sensing tube. Changing the partitioner changes the effective range. Mass flow controllers compare the measured flow with the desired flow and use the difference between them to adjust an electronically operated valve.

monitored may be as low as one in a simple differential system to ten or more in elaborate systems that manufacture multicomponent gas mixtures (often CO_2 and CO_2-free air or CO_2, N_2 and O_2, occasionally with one or more pollutants under study), mix moist and dry air, and process one or more compensating gases (Fig. 11.4). The simplest closed systems require no flow measurements, but sophisticated closed systems (e.g. the LI-COR LI-6200) may include humidity or CO_2 compensation, which requires flow measurements.

Three types of flow meters are widely used in gas-exchange systems. These are variable-area flow meters (also called rotameters), differential-pressure flow meters and mass flow meters. Each type can operate with total errors of 1% or less of full scale, but the instruments differ in sensitivity to environmental conditions and ease of data acquisition.

Variable-area flow meters or rotameters are simple mechanical devices in which a flowing gas pushes one or more floats (usually spherical) up a tapered tube (Fig. 11.5). The tube diameter increases with height, so that as the float rises, it occupies a decreasing proportion of the tube cross-section. The float reaches equilibrium when the pressure difference between the float's upper and lower surface equals its weight. These flow meters are attractive for their low cost and freedom from power supplies, but they are rarely the best choice for field gas-exchange systems, because they are sensitive to attitude, temperature and pressure, and do not interface easily with automated data loggers. Only units with quite long tubes approach the accuracy of other types of flow meters.

Differential-pressure flow meters operate on the principle that, within appropriate ranges, flow through a tube or an orifice increases monotonically with the pressure differential between the two ends (Fig. 11.5). With the proper orifice configuration and upstream pressure, pressure measurements do not need to be differential because flow is supersonic and therefore largely independent of downstream pressure (Parkinson and Day, 1979). Electronic pressure transducers make differential-pressure flow meters easy to interface with data loggers and effective solutions for many applications in field gas-exchange studies. However, both variable-area and differential-pressure flow meters operate best when temperature controlled. Many of the critical parameters, including the dimensions of the tube, the viscosity of the gas, the conditions for transition from laminar to turbulent flow, and the conversion from volume to molar flow are temperature dependent.

Mass flow meters measure flow on the basis of heat transport. Because the heat-transporting capacity of a gas depends on molar rather than volume flow, mass flow meters yield mass or molar flows directly, without temperature or pressure corrections. The parameter measured in most mass flow meters is the temperature difference between sensors upstream and downstream from a heater (Fig. 11.5). As flow increases, more heat is carried from the heater to the downstream sensor and less is carried to the upstream sensor. To obtain reasonable temperature differentials, most mass flow meters direct only a small portion of the total gas through a very narrow measurement tube, and divert the majority through a larger bypass. When mass flow meters are exposed to flows of condensation-prone gases or gases carrying particulates, performance quickly degrades as the contaminants affect flow partitioning. However, some new mass flow meters (e.g. Sierra Instruments accu-mass® meters) operate without a bypass, generating a heat transport signal from a single heated temperature sensor inserted in a large-bore tube [making them similar, in principle, to hot-wire anemometers (Chapter 4)]. In addition to much reduced sensitivity to contamination, these meters have significantly shorter time constants than the bypass models, making them easier to interface with feedback circuits. Some mass flow meters are inclination sensitive because free convection in a non-horizontal flowtube appears as upward flow. However, several new instruments overcome the inclination sensitivity with sophisticated flowtube configurations. Mass flow meters are well suited to interface with automated data loggers and can be purchased as flow controllers, in which the flow-meter output is connected to an electronic valve. Tylan mass flow controllers incorporate a particularly ingenious valve in which thermal expansion and contraction of the valve stem opens and closes the valve. The flow controller configuration opens several options for electronic control, including analog and digital control of CO_2 concentration and humidity. For most gas-exchange systems, mass flow meters and flow controllers are the best option for flow measurement, combining high accuracy, good long-term stability, excellent control and output proportional to molar flow.

11.3.2 Gas-exchange chambers

The architecture, aerodynamics and materials used in a gas-exchange chamber are critical determinants of a system's functionality. Although no single chamber design is optimal for all systems and all experiments, the basic requirements for acceptable performance are very general. These requirements are: (1) to define and, in many cases, control the chamber environment, (2) to stir the air, providing thorough mixing and a high boundary layer conductance, and (3) to seal the chamber effectively. Though these requirements are conceptually simple, meeting each of them entails subtle and sometimes difficult compromises.

(a) Controlling the chamber environment

For both ambient-sampling and controlled-environment systems, effects of the chamber on the leaf environment are critical concerns. Whether the goal is to minimize or maximize effects on the leaf environment, interactions among chamber, environment and leaf place important constraints on the range of possibilities.

Temperature control. Temperature control of a gas-exchange chamber may be either passive (based on natural energy exchange) or active (based on energy supplied specifically to wrest temperature control from natural energy exchange). Both control modes have limits. Well-designed passive control can often maintain leaf temperatures inside a chamber similar to those outside a chamber, but the match is rarely perfect. Active control can fix leaf temperature far from ambient but at the cost of potentially large power demands and temperature gradients within the chamber.

The temperature of a leaf in a gas-exchange chamber can be held near the ambient temperature with passive control if the chamber is shielded from some of the incoming radiation, if it is made from materials that do not absorb a large proportion of the incoming radiation, if it allows the infrared radiation emitted by the leaf to escape, or if it has an effective heat exchanger for transferring heat to the environment. Each of these solutions provides useful options for keeping a chamber near ambient temperature under shady conditions or short periods in bright light, but none confer long-term maintenance of ambient temperature under high light. Many ambient-sampling systems with passive temperature control utilize more than one approach to minimize overheating.

Shielding in the near-infrared is very useful, especially for chambers illuminated with artificial light, where the ratio of infrared to visible radiation may be greater than with sunlight. One example of shielding is the acrylic (Perspex, Plexiglass, Lucite®) plate above the Parkinson chambers (available from ADC, Hoddesdon, England). Because acrylic absorbs strongly in the near-infrared, a simple filter above the chamber can substantially decrease the incoming energy impinging on chamber and leaf. Water is also a strong

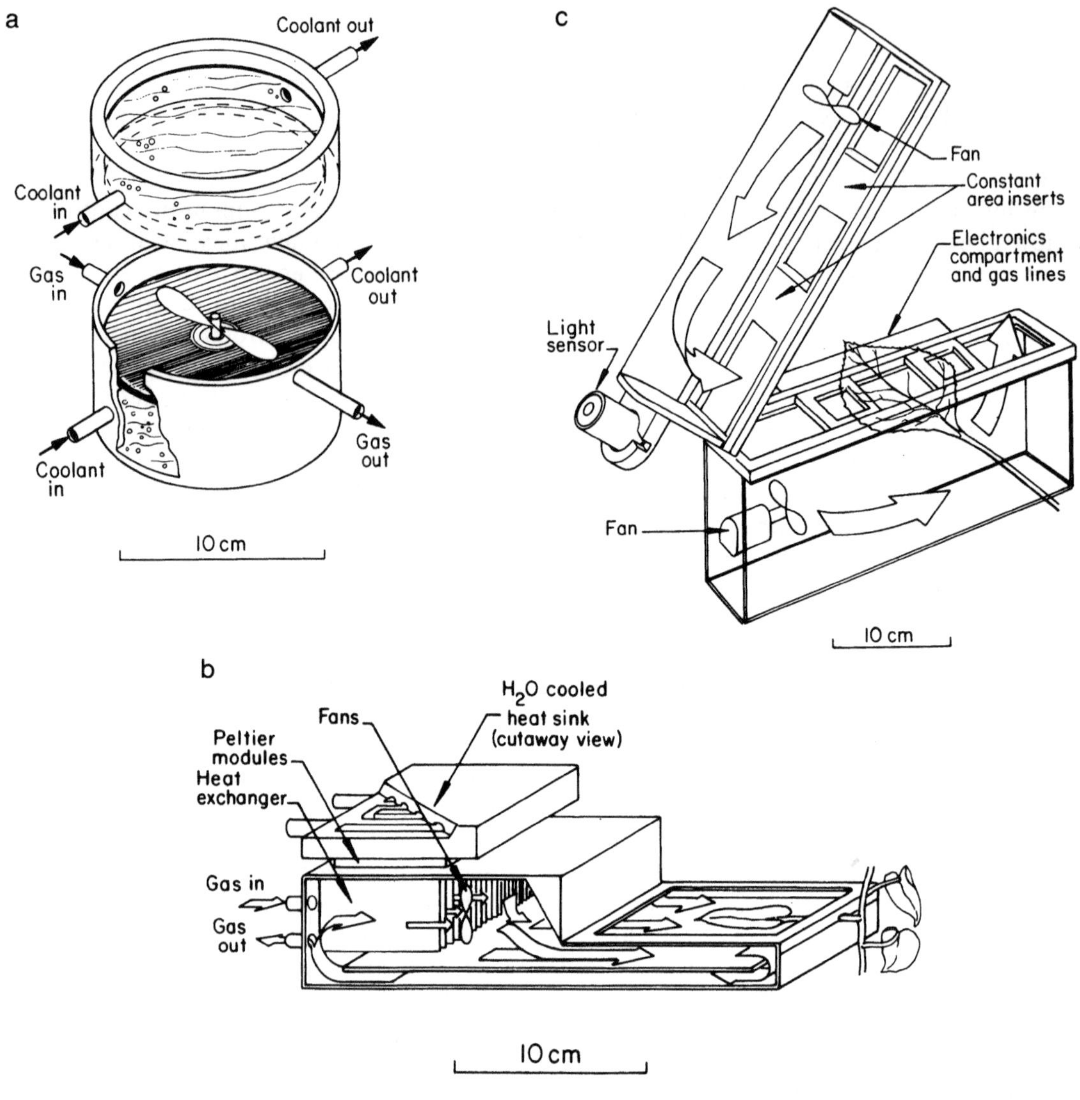

Fig. 11.6 Schematic diagrams of three types of leaf chambers. (a) A controlled-environment chamber similar to that used with the mobile laboratory of Mooney *et al.* (1971). Circulating water provides the temperature control. Most chambers of this design are made from nickel-plated brass and incorporate a glass window. Even though a large fan blows air directly against the leaf, boundary layer conductances in chambers like this are usually lower than in chambers like those shown in (b) and (c). (b) A controlled-environment chamber similar to that used in the portable system of Field *et al.* (1982). Temperature in this chamber is Peltier controlled, and circulating water acts as a heat sink or source for the Peltier modules. Most implementations of this design are nickel-plated brass or aluminum with a glass window. (c) One of the chambers (this is the 1 liter version) for use with the ambient-sampling LI-COR LI-6200. This chamber is not actively temperature controlled, but it is made from polycarbonate, a material that transmits much of the thermal infrared radiation emitted by the leaf and other chamber parts. The chamber is lined with a Teflon film to minimize CO_2 and H_2O exchange by chamber materials. This chamber includes constant-area inserts that automatically restrict sampling to a fixed leaf area, eliminating the need for measuring leaf area.

absorber in the near-infrared. An open-top water filter can be even more effective than acrylic, because evaporation cools the water, limiting its contribution to the thermal (longwave infrared) radiation on the chamber and leaf. Commercially available infrared filters ('hot mirrors') provide a third option for shielding. They have the desirable characteristic of reflecting rather than absorbing in the near-infrared.

Minimizing radiation absorption by chamber materials has two components. Low absorption in the visible and the near-infrared reduces direct heating of the chamber parts. Transparency in the thermal infrared allows the longwave infrared emitted by the leaf and other surfaces in the chamber to escape, preventing a greenhouse effect. Polished metal surfaces can effectively reflect visible and near-infrared radiation, while several

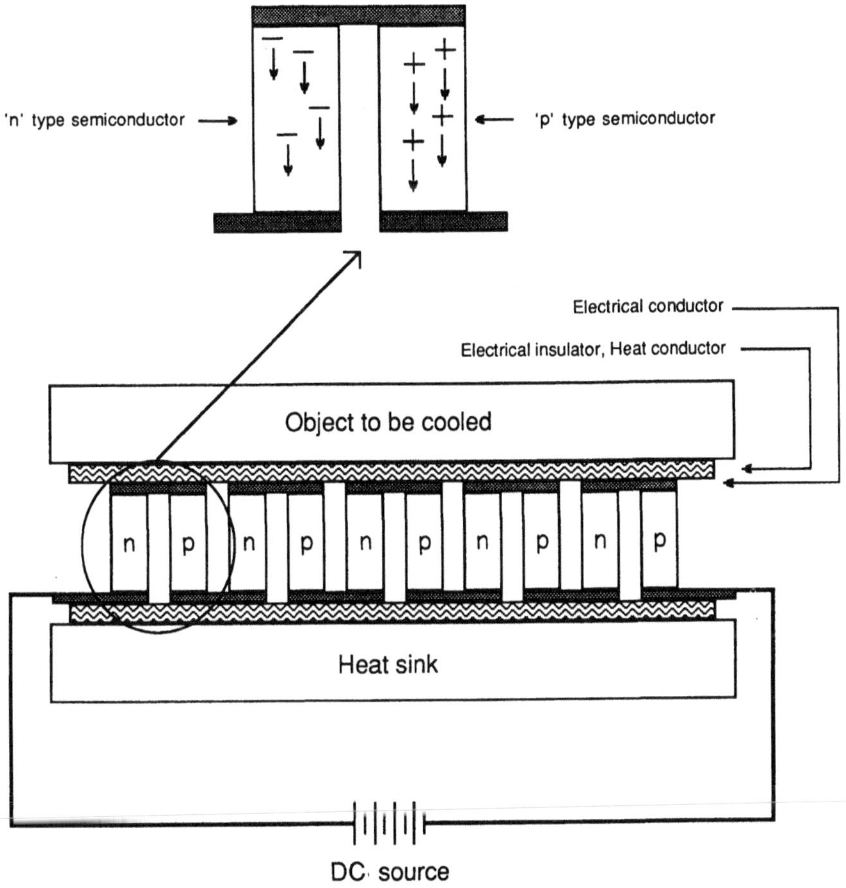

Fig. 11.7 Schematic diagram of a thermoelectric heat pump or Peltier module. The active part of the module is an array of semiconductors, usually bismuth telluride, heavily doped to create alternating 'n' type and 'p' type elements. The 'n' type and 'p' type elements are electrically in series but thermally in parallel. When a current is passed through the device, the 'n' type material carries electrons and the 'p' type material carries 'holes' to the hot surface (enlarged view). Reversing the polarity of the current reverses the direction of the heat pumping. These devices are useful in gas-exchange systems, because both heating and cooling capability are present in one small device with negligible thermal inertia.

plastics are quite transparent through the waveband where solar radiation contains significant amounts of energy. Acrylic and glass are quite transparent in the visible, but begin to absorb significantly at wavelengths longer than 1.5 and 2.5 μm respectively. Polycarbonate (Lexan®, General Electric, or Mangard®, General Electric) and FEP Teflon® (E.I. du Pont de Nemours & Co.) are quite transparent to near-infrared and relatively transparent to thermal IR. The LI-COR chambers (e.g. Fig. 11.6c) are constructed either from polycarbonate or part from polycarbonate and part from Propafilm® (ICI Americas, Inc., a polypropylene film coated with polyvinylidene chloride) to minimize absorption of thermal IR. An external acrylic skeleton supporting a skin of FEP Teflon® film makes a lightweight ambient chamber usable even under desert conditions (J.R. Ehleringer, personal communication).

Passive heat exchangers can also assist with energy dissipation. Some of the Parkinson chambers (available from ADC) have a metal heat exchanger, cooled by natural convection, on their bottom surface. With slightly more active control, the chambers of the LI-COR porometer and the gas-exchange system of Schulze *et al.* (1982) use a fan to pull ambient air past a metal chamber, employing the chamber body as a heat exchanger.

Three kinds of active temperature controllers are widely used in field gas-exchange systems. These are compressor-based refrigerators for cooling, electrical resistance heaters for heating and thermoelectric heat pumps for heating and cooling. With active temperature controllers, the most important consideration in the choice of chamber materials is high thermal conductivity, allowing efficient heat transfer through and out of the chamber. Copper, brass and aluminum are good in this respect but stainless steel is poor. Even with a good chamber material, efficient temperature control often requires a large surface area of heat exchanger for transferring heat between the air in the chamber and the heat exchanger (Fig. 11.6b).

In most laboratory gas-exchange systems, leaf chambers are water jacketed, and temperature is controlled via the temperature of the circulating water (Fig. 11.6a). The water temperature is adjusted with resistance heaters and compressor-based refrigerators. This technology is employed in some large field systems (e.g. Mooney *et al.*, 1971) but the mass and power requirements of refrigerated circulators make it generally unsuitable for portable systems.

For portable field systems, the best available solution for active temperature control is based on the thermoelectric heat pump or Peltier module (Fig. 11.7). A Peltier module is a semiconductor device that functions like an array of thermocouples, each with one junction in thermal contact with the chamber and the other in thermal contact with a heat sink (Fig. 11.6b). The amount of heat transferred through the module increases with the current driven through it. The direction of heat transfer reverses with the polarity of the current. Though these devices are still power hungry, they are very small and can be built directly into chambers, minimizing the mass that requires temperature control. Temperature control for a leaf chamber enclosing up to 40 cm^2 of leaf area consumes from 5 to 60 W of electrical power, depending on the control temperature, the ambient temperature and the radiation input.

CO_2 concentration control. The simplest technique for controlling CO_2 concentration is to supply the chamber with air from a tank of known CO_2 concentration. Photosynthesis will decrease that concentration, but if the flow rate is high, the depletion due to photosynthesis is small. Controlling CO_2 concentration near the level in a storage cylinder is impractical when an experimental design requires measurements at many CO_2 concentrations. For these measurements, it is useful to employ a gas-mixing system based

on a precision mixing pump (e.g. Type 2 G26/3-F Wösthof), supersonic orifices (e.g. ADC gas-mixing system) or mass flow controllers. Mass flow controllers do not yield the accuracy of precision mixing pumps, but they provide an excellent combination of accuracy and portability for field systems. CO_2 concentrations in the physiologically relevant range can be prepared in a single mixing stage from pure CO_2 and CO_2-free air (Küppers *et al.*, 1987) or from CO_2-free air and a high-CO_2 gas containing 0.5–10% (5000–100 000 ppm) CO_2 in nitrogen.

Humidity control. Field gas-exchange systems have employed three approaches to control the humidity of the chamber. The null-balance approach to humidity control is the simplest in terms of hardware requirements. In null-balance instruments, dry air enters the chamber and is humidified to a steady-state level by transpiration (see Chapter 8). Null-balance instruments provide accurate humidity control, but at the cost that system responses may be slow, especially when the combination of low leaf conductances and high ambient humidities necessitates low flows through the chamber. These slow responses can be overcome by humidifying the air to a known dew point before it enters the chamber. The standard technique for humidifying the air is to pass it first through water (acidified to prevent CO_2 from going into solution – phosphoric acid does not introduce gaseous pollutants) that is slightly warmer than the desired dew point and then through a condenser held at the desired dew point. Though effective, this approach adds to the system at least one component requiring temperature control plus the possibility of errors due to condensation in the gas lines. At times it may be necessary to warm the water in the humidifier (to generate a dew point greater than the desired one) and the gas line between the condenser and the chamber (to prevent condensation). We (JTB and JAB) have recently developed miniature humidifier–condenser units for field systems. These units use Peltier-controlled condensers to minimize size and power requirements (Fig. 11.8).

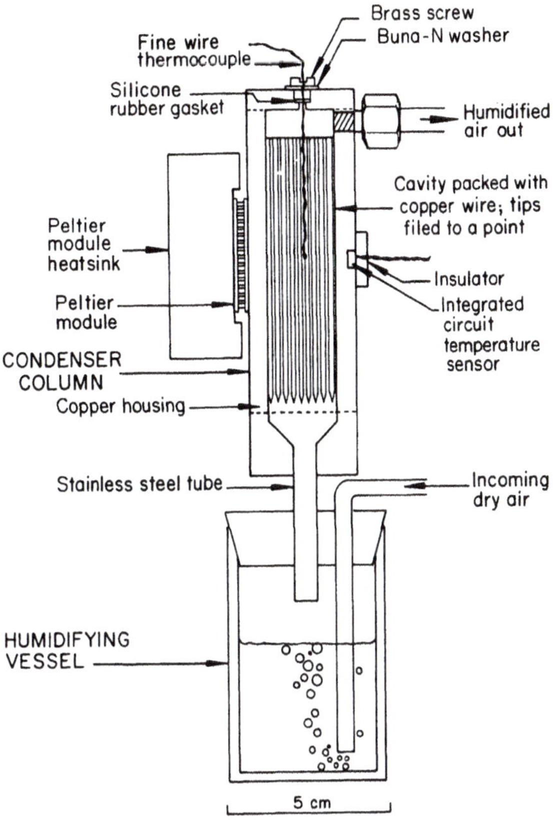

Fig. 11.8 Longitudinal section through the center of a humidifier–condenser designed for field and laboratory use. The condenser column is a copper block, hollow in the center and packed with copper rods (approximately 2 mm diameter). The temperature of the condenser is controlled by one or more Peltier modules, mounted on the copper block. The vapor pressure of the exiting air is determined by the temperature at the site of water condensation within the column, measured with a fine thermocouple between the copper rods. In this design, water from the condenser drains back into the humidifier.

A third approach to humidity control has been used extensively in German mobile laboratories. Air is humidified to a known dew point before entering the chamber, and the water added by transpiration is removed in a condensing loop through which the chamber air is rapidly circulated (Koch *et al.*, 1971). This approach yields accurate humidity control but adds enough additional hardware that it is unlikely to be the best choice for a portable system.

Light control. In sunny settings, sunlight is an excellent light source for outdoor experiments. However, sunlight is often unpredictable. It varies through the day, and can be obscured by clouds or other objects. Sunlight is the obvious light source of choice for ambient-sampling measurements, but artificial light sources are typically more useful than sunlight for controlled-environment experiments.

When sunlight is the light source, chamber materials and architecture combine to make light levels inside the chamber less than those outside. A 3 mm thickness of glass, polycarbonate or acrylic reduces the flux of photosynthetically active radiation by approximately 10%. If the material is not normal to the solar beam, the effect is even larger. A seam or sharp corner in the chamber may further reduce transmission to the leaf.

Among the artificial light sources used with field gas-exchange systems, three are broadly used with good success. When power and size are not limitations, multivapor high-intensity discharge (HID) lamps of 400–1000 W provide high intensities and good uniformity over relatively large areas. When power and size are limitations, quartz–halogen projection lamps or tungsten flood lamps of 80–250 W can give good performance, but only when the illuminated area is relatively small and the optics are well aligned. Cooling requirements for the leaf chamber can be substantially reduced by filtering the IR radiation from any light source. Tungsten flood lamps with IR-reflecting front lenses and IR-transmitting rear lenses are available [Cool-Beam® (General Electric) and Cool-Lux® (Sylvania)]. In our field systems, the standard field lights are quartz–halogen projection lamps with built-in parabolic reflectors. These are aimed at 45° cold mirrors that transmit IR but reflect visible radiation. Operating the lamps from regulated power supplies increases lamp life and decreases the effects of fluctuations in generator output.

Boundary layer. The boundary layer conductance of naturally growing leaves is low in still air and increases with wind speed (Chapter 4). It may seem reasonable that the boundary layer conductance in a gas-exchange system, especially an ambient-sampling system, should be tailored to match the ambient boundary layer conductance, and some systems incorporate provisions for matching wind speed inside and outside the cuvette (e.g. Bosian, 1960). However, a high wind speed, high enough to yield boundary layer conductances much higher than those typical under natural conditions, is important for three reasons. First, a high wind speed can effectively mix the chamber contents, insuring that the gas sampled at the chamber outlet has the same composition as the gas near the leaf. Second, a high boundary layer conductance increases the convective coupling between leaf and chamber. Strong coupling is essential for effective temperature control in controlled-environment systems and for preventing undesirable temperature rises in ambient-sampling systems. Third, a high boundary layer conductance minimizes the limitation to leaf gas exchange imposed by the chamber. The effects of boundary layer conductance on CO_2 and H_2O transport are roughly parallel to the effects of leaf conductance (Section 11.7). The absolute effects of boundary layer and the consequences of errors in estimating it increase as boundary layer conductance decreases (Moreshet *et al.*, 1968).

Boundary layer conductance is determined by interactions among chamber architecture, gas flow through the chamber, the position and size of the leaf or other plant material, and the efficiency of the fans or other stirring devices. In general, boundary layer conductance increases with the capacity of the fans, but the routing of the gas flow can be as important as the fan capacity. Some chambers use relatively elaborate architectures to obtain high boundary layer conductances (e.g. Fig. 11.6b) while others couple simple architectures with high-capacity fans (e.g. Fig. 11.6a). The boundary layer conductance is often more sensitive to leaf size and location in architecturally simple than in more elaborate chambers. Several ambient-sampling systems (e.g. Fig. 11.6c) use ingenious architectures to restrict the sampling to a fraction of the leaf area while providing rapid mixing between the gas volumes in contact with the upper and lower leaf surfaces.

Measuring boundary layer conductance is quite simple. The traditional method involves simply measuring 'leaf' conductance on a piece of wet filter paper, using the calculations appropriate for a real leaf. Two modifications of this procedure we find useful are (1) continuously irrigating the filter paper with water supplied by a syringe pump or another low-flow pump and (2) making separate measurements for upper and lower surfaces using artificial leaves made from filter paper glued to an impermeable substrate. Parkinson's (1985) method, which is less sensitive than the traditional method to temperature and humidity errors, is based on the energy balance of the filter paper. For most purposes, a chamber will not perform well unless the boundary layer conductance is greater than or equal to 1 mol m^{-2} s^{-1}. Measuring the boundary layer conductance for a range of leaf sizes in a range of positions within the chamber often indicates a surprisingly large variability.

Sealing the chamber. Sealing a chamber to prevent outward and inward leaks is simple in principle but is also a source of many problems. A tight seal is critical for the accurate operation of a closed system and much less important, sometimes even undesirable, in an open system. The seal is critical for a closed system because the calculations explicitly assume that the leaf is exchanging gases with a fixed volume (Sections 11.2 and 11.7). A leak invalidates that assumption.

In an open system operating at a slight positive pressure, any leaks should be outward and will not alter the composition of the gas measured by sensors in the chamber or downstream. To insure that the chamber operates at positive pressure, you can either (1) measure pressure in the cuvette with a liquid-level manometer or a pressure transducer (we find that 10 mm of water is a reasonable overpressure) or (2) use a flow pattern in which gas is pushed from upstream of the chamber to an outlet at atmospheric pressure downstream of the chamber. Since flow is a critical parameter in open-system calculations (Section 11.2 and 11.7), it is important to measure flow upstream of the chamber, so leaks at the chamber do not imply incorrectly low flows. The total flow leaving the chamber is always more than the total flow entering (Penning de Vries *et al.*, 1984), but the equations in Section 11.7 correct for this effect. Often, in open systems, the flow to the IRGA should be less than the flow through the chamber, and the performance of the system is improved by engineering a leak between the chamber and the IRGA. If this leak is an adjustable bleed valve, it can be used to compensate for other uncontrolled leaks.

Many of the commonest problems associated with sealing the chamber result from undesirable gas exchange by the gasket materials. Some of those problems and ways to avoid them are discussed below.

11.3.3 Construction materials in gas-exchange systems

The gas- and energy-exchange characteristics of construction materials play a major role in the performance of a gas-exchange system

(Bloom *et al.*, 1980). Poorly chosen materials can result in temperature control problems as well as temperature- and humidity-dependent CO_2 and water vapor fluxes from an empty chamber.

(a) Gas lines

For lines carrying only dry gas, the most important characteristic to avoid is excessive gas permeability, especially differential permeability. Many rubber and plastic compounds are more permeable to CO_2 than to N_2 or O_2, meaning that gas composition can change in the lines. For lines carrying moist gas, water adsorption is a dominant concern. Because CO_2 is soluble in water, any material that adsorbs water will also exchange CO_2.

Rubber tubing is unsatisfactory for gas-exchange systems as a result of high permeability and water adsorption. Polyvinyl chloride [Tygon® (TM Norton Co.)] has much lower permeability but unacceptably high water adsorption. The utility of this tubing for gas exchange is further reduced by the volatilization of the plasticizers added to increase flexibility and transparency. These plasticizers have IR absorption bands that overlap with those of CO_2 (Bloom *et al.*, 1980).

Several metals and some plastics are suitable for gas-exchange lines. Many laboratory systems use copper tubing for dry gas and stainless steel for moist gas. Since metal tubing is usually impractical for portable instruments, many field systems use tubing made from one of the fluorinated hydrocarbons [Teflon® (E.I. du Pont de Nemours & Co.), especially FEP® or Kel-F®]. We have had excellent success with a much less expensive two-layer tubing consisting of a polyethylene liner and a shell of ethyl vinyl acetate (Bev-A-Line®, Ryan Herco Industrial Plastics).

(b) Chamber materials

We have addressed the interaction of chamber materials with temperature control and light penetration in preceding sections. Interactions with water and CO_2 exchange can also be very important.

Water adsorption is an even more serious problem with chambers than with gas lines because the surface area of chambers is so large. Acrylic plastic is definitely unsatisfactory, and polycarbonate is only a little better. Both materials can be used, however, if they are coated with a material that adsorbs much less water. Plastic films made from polypropylene (e.g. Propafilm®) or FEP Teflon® are suitable for lining chambers and can dramatically decrease water adsorption. Polished copper and aluminum adsorb little water, but both quickly oxidize, and the oxides are very hygroscopic. Nickel-plated brass, stainless steel and glass adsorb extremely little water. For controlled-environment chambers, we generally use nickel-plated brass (for its combination of high thermal conductivity and low water adsorption) with glass windows. Some pollutants, especially SO_2, adsorb strongly to nickel-plated brass, and studies with these gases require chambers of stainless steel, glass or FEP Teflon®.

Other chamber components can also create problems with water adsorption. Foam gaskets are particularly problematic (Hack, 1980) because the surface area of even a small piece of foam is very large. Lange (1962) solved the gasket problem by saturating a foam gasket with liquid paraffin, but we can usually tolerate the water exchange from a small closed-cell foam gasket, especially if the edges are sealed with petroleum jelly or silicon rubber. The plastics used in chamber fans, light sensors and leaf supports may all adsorb water, and only a careful series of empty-chamber experiments can pinpoint the offenders. Some fans are also annoyingly persistent CO_2 sources (Schulze *et al.*, 1982).

11.3.4 Data-acquisition and control

The principles of field data acquisition presented in Chapter 2 generally apply to the specific needs for data acquisition in gas-exchange experiments. Automated data acquisition and reduction in field gas-exchange systems has four advantages. Automated recording (1) insures immediate access to reduced data, in the field. It also (2) increases the maximum sampling frequency, (3) increases the quantity of data that can be processed and (4) decreases transcription errors. While each of the advantages is important, we consider the first to be the strongest motivation for automating data acquisition. Immediate data reduction and, if possible, plotting has great value both because it can be a guide to detecting and solving problems with the instruments and because it encourages scientific insight in the field under conditions where the insights and their consequences can be explored.

Output signals from the sensors in gas-exchange system are usually voltages, but some of the sensors are current sources (e.g. photovoltaic light sensors), while others provide a variable resistance (e.g. thermistors, platinum resistance thermometers and some pressure transducers). Some data loggers are configured to handle all three kinds of inputs. For others capable of logging only voltages, Chapter 1 discusses circuits appropriate for converting current and resistance signals into voltages. Some of the sensors in gas-exchange systems produce relatively large signals, typically 0–5 V (e.g. mass flow meters and IRGAs), but others produce signals on the order of 1 mV (e.g. thermocouples). One important constraint in selecting a data logger for use with a gas-exchange system is the need for accurate logging over multiple-voltage ranges.

The number of channels required for logging data from a field gas-exchange system depends on the complexity of the system. A simple closed system for measuring photosynthesis and transpiration could have as few as four outputs, for CO_2, the water content of the air, leaf temperature and photon flux density. A data logger for a complex open system requires 10–20 channels per chamber.

Sampling frequency should be consistent with the general principles discussed in Chapter 2. For experiments on the responses of photosynthesis to short sunflecks (Chazdon and Pearcy, 1986), it is appropriate to sample all sensors several times per second, though other characteristics of the system may degrade the time resolution below the sampling frequency. With an ambient-sampling system like the LI-COR LI-6200, it is important to tailor the time between samples to the leaf area and photosynthetic rate. When using controlled-environment systems to generate response curves, we generally sample a few times a minute and compare groups of samples to check for equilibration in the system and the leaf.

Few field gas-exchange systems use signals from a computer to control the chamber environment, but the technology for digital to analog conversion is available, and computer control opens several options. For example, a computer-based controller can regulate the value of a calculated parameter. Using inputs from sensors that read in units of relative humidity or dew point, a computer-based controller can adjust the moisture content of the air to achieve a target vapor pressure deficit. A controller that adjusts the CO_2 and water vapor content of the air can even maintain photosynthesis or conductance at target levels. With extensive computer control, a field system can maintain chamber conditions at ambient over long periods or can automatically generate response curves. We find that as few as four channels of digital-to-analog conversion (typically setting CO_2 concentration, the injection of compensating CO_2, chamber temperature and the temperature of the humidifier–condenser) is sufficient to generate comprehensive and flexible control.

	Li - Cor Li - 6200	Ehleringer & Cook 1980	Schulze et al. 1982	Field et al. 1982	Mooney et al. 1971
	Closed system w/ integral IRGA	Closed system w/syringe sampling	Differential system	Compensating system	Differential system mobile lab
Single leaf responses					
Ps 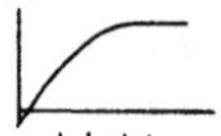 Light	+	+	+	+ + +	+ + +
Ps 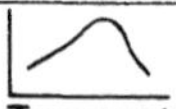 Temperature	+	+	+	+ + +	+++
Ps 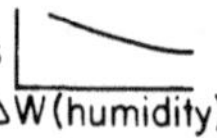 ΔW (humidity)	+	+	+	+ + +	+++
Ps 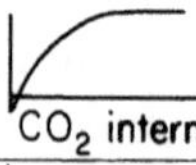 CO_2 internal	+ +		+	+ +	+ + +
Ps 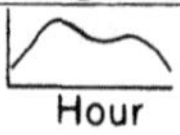 Hour	+ + +	+ +	+ + +	+ +	+++
Ps 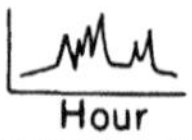 Hour	+ +	+ +	+ + +	+	+ +
Ps 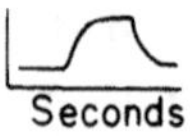 Seconds			+ +	+	+++
Leaf population studies					
Maximum Ps	+ + +	+	+ +	+ +	+
Daily course (many leaves)	+ + +	+ +	+ +		
Ps 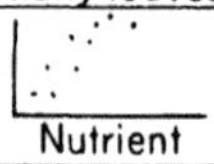 Nutrient	+ +	+	+ +	+ +	+
Ps 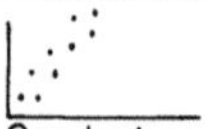 Conductance	+ +		+ +	+ +	+
Performance characteristics					
Sample handling (leaves/day)	200	200	200	20	2
Set up (hours)	0.1	0.5	0.5	1	5
System mass (kg)	5	40	20	100	4000
Power requirement (W)	10	100	50	400	5000
Initial investment ($)	15,000	10,000	20,000	20,000	50,000

Fig. 11.9 Comparative uses and attributes of five real gas-exchange systems. Suitability for a given objective increases with the number of plusses. Limitations and capabilities represent a combination of intrinsic features of design philosophies and configurations of particular devices. Both the suitability for various objectives and the performance characteristics can improve with new technology (modified from Field and Mooney, 1989).

11.4 REAL PHOTOSYNTHESIS SYSTEMS

Designing a field photosynthesis system always involves compromises among several, often incompatible, objectives. Increasing portability often means sacrificing environmental control. Increasing measuring speed sometimes entails decreasing accuracy. Increasing accuracy typically involves increased expense. Because the optimum combination of portability, speed and accuracy depends on the design of the research, no single system or measurement philosophy is the best choice for all studies. Here, we discuss a selection of the systems described in the literature or manufactured commercially. We selected examples that illustrate major lines of instrument development and are not recommending particular products. These systems are ordered by measurement philosophy and not on the basis of suitability for any particular experiment. Fig. 11.9 summarizes the performance and suitability of several systems for a range of experimental objectives.

11.4.1 Closed systems

(a) Closed systems with a remote IRGA

When an experimental design places a high premium on portability and economy, it is appealing to confine the field work to collecting gas samples for later laboratory analysis. The appeal was especially strong when all IRGAs were delicate laboratory instruments and when field gas-exchange systems were big laboratories on wheels. While recent advances in IRGA design make it increasingly unlikely that remote IRGA systems are instruments of choice for many experiments, these devices are useful for illustrating the extreme of hardware simplicity sufficient for obtaining a photosynthesis measurement.

Ehleringer and Cook (1980) evaluated a closed system consisting of only a chamber and syringes, in addition to the remote IRGA. To operate this system, one clamps a leaf in the chamber, draws an initial gas sample into one syringe, waits a specified interval and draws another sample into a second syringe. Ehleringer and Cook (1980) analyzed the CO_2 content of the air samples with the technique of Clegg *et al.* (1978) in which the sample is injected into a flowing carrier gas directed through an absolute IRGA. This technique is not necessarily more accurate than simply filling the IRGA sample cell with the unknown, but it allows measurements on smaller samples, samples less than the volume of the IRGA sample cell.

The system of Ehleringer and Cook (1980) can, if used with care, yield useful photosynthesis data. It offers simplicity and economy but at the cost of limited accuracy, the total absence of environmental control, and measurements of no parameters other than photosynthesis. The absence of immediate access to the data is perhaps this technique's biggest weakness. Without immediate feedback, it is very difficult to know which measurements are spoiled by, for example, breathing into the chamber, a poor chamber seal, leakage from the syringes, too large or too small CO_2 depletions, or unacceptable changes in temperature or humidity. These intrinsic limitations in the instrument and the sampling protocol tend to make the data noisy enough that the approach cannot yield the resolution necessary for answering many ecological questions (Fig. 11.9).

(b) Self-contained closed systems

Highly portable closed systems are a recent development, made possible primarily by advances in data loggers and by the availability of light-weight battery-powered IRGAs (e.g. Williams *et al.*, 1982). Combining these technologies, it is possible to produce an instrument that frequently samples, stores and displays the CO_2 concentration in the closed system. With a frequently updated display, it is easy to avoid the exhalation errors so common with remote-IRGA systems.

Fig. 11.10 Photographs of four gas-exchange systems operating under field conditions. (a) LI-6200 manufactured by LI-COR (photograph courtesy of LI-COR). (b) The differential system of Schulze *et al.* (1982) in ambient-sampling mode (photograph by Dr W. Beyschlag).

Fig. 11.10 (cont) (c) The controlled-environment, compensating system of Field *et al.* (1982). (d) The controlled-environment mobile laboratory of Mooney *et al.* (1971). Each of the systems makes measurements on approximately the same amount of leaf area.

With multiple CO_2 measurements during a single depletion cycle, photosynthesis can be calculated several times and stored or displayed as a mean or some other statistical parameter, thus decreasing the influence of single bad measurements.

One of the most popular self-contained closed systems is the LI-6000, Portable Photosynthesis System (LI-COR, Lincoln, NE, USA), introduced in 1982. The LI-6000 and its successor, the LI-6200, are the current 'state of the art' in portable closed systems (Fig. 11.9). The total system weighs only 9 kg, operates from internal batteries, and can easily be operated by one person. It incorporates a flexible computerized data logger that the user programs to customize measurements for a particular experiment. In addition to calculating photosynthesis and leaf conductance, the system measures light, temperature, humidity and CO_2 concentration, and the built-in computer stores a large number of independent samples and handles corrections and linearizations.

The LI-6200 is available with a number of leaf chambers, an especially useful feature for a closed system (Figs 11.6 and 11.10). Since chamber volume in a closed system influences the rate of CO_2 depletion by photosynthesis (Equation 11.4), chambers of different volume can be used to tune the system to the photosynthetic characteristics of the plants under study.

The LI-6200 provides no environmental control. Photosynthesis and transpiration are always measured under ambient conditions as modified by the system. With a closed system these modifications usually include (1) CO_2 depletion by photosynthesis, (2) humidity increase as a result of transpiration and (3) temperature increase. The LI-6200 minimizes effects of each of these perturbations away from ambient conditions. Consequences of CO_2 perturbations are minimized through two mechanisms. First, the low noise and high accuracy of the LI-6200's IRGA (LI-6250, also made by LI-COR) make it possible to obtain a photosynthesis measurement from a small CO_2 depletion (2–5 μmol mol^{-1} per sample or 10–30 μmol mol^{-1} per complete measurement). Software enabling sampling at specified depletions rather than times further tunes the system to operate efficiently with small CO_2 depletions. Second, the system statistically projects photosynthesis and intercellular CO_2 concentration at the time the system was closed. To minimize humidity changes, the LI-6200 allows the operator to adjust the air flow through the IRGA loop containing a desiccant. When the flow is adjusted such that water removal by the desiccant exactly matches transpiration, the humidity is constant during a measurement, and transpiration is calculated entirely from a compensating-system equation (Section 11.7). If the drying does not exactly balance transpiration, the humidity may increase or decrease during a measurement, and transpiration is calculated with a combination of closed- and compensating-system equations (Section 11.7). Temperature increases during measurements are always problems in systems without tmperature control. Many metals heat to well above ambient in bright light, and many plastics that transmit visible light absorb infrared reradiated by a leaf, resulting in pronounced 'greenhouse effects'. The LI-6200 minimizes temperature increases by keeping sampling times short, and by incorporating chambers constructed from plastics that transmit much of the thermal radiation emitted by the leaf and other absorbing substances within the chamber. Even modest temperature elevations may have dramatic effects on photosynthesis and, especially, transpiration, so it is wise to always sample quickly and to shade the chamber whenever possible.

The LI-6200 is basically designed for sampling large numbers of leaves under ambient conditions, but it offers some prospects for measuring response curves (Fig. 11.9). For example, one can generate a rapid CO_2 curve over a limited CO_2 range, making successive measurements while photosynthesis depletes the CO_2. Alternately, one can use the built-in CO_2 scrubber to change the ambient CO_2 rapidly. While the transient technique cannot enable all the analyses practical with response curves generated by steady-state measurements, it does increase the feasibility of collecting large numbers of response curves.

11.4.2 Open systems

(a) Differential systems without environmental control

Differential systems can be very simple. Griffiths and Jarvis (1981) and Schulze *et al.* (1982) describe sophisticated, yet conceptually simple, differential systems without environmental control (Fig. 11.10). Both systems are built around a two-channel IRGA (Binos I, Leybold Hereaus, Hanau, Federal Republic of Germany). The major difference between the systems is in the measurement protocol. Griffiths and Jarvis (1981) adjust the rate of air flow through the chamber to achieve a predetermined CO_2 depletion. Schulze *et al.* (1982) fix the flow rate and allow the CO_2 depletion to equilibrate. In the system of Schulze *et al.* (1982), a pump delivers ambient air through two pathways. One passes first through a mass flow meter, then to a chamber containing a leaf, and finally through the sample cells of both IRGA channels. The other pathway leads to the IRGA reference cells but includes a buffer volume to insure that gas in the sample and reference cells entered the system at the same time. To minimize fluctuations in the humidity and CO_2 concentration of the incoming air, systems of this type often draw air from well above the ground and mix air into a large volume before pumping it into the measurement system. Using a waterbed mattress as a mixing volume, Pearcy and Calkin (1983) developed a system that operates from ambient air but with a composition that remains constant over an entire day.

The Schulze *et al.* (1982) system equilibrates to steady state and yields a measurement under ambient conditions in times ranging from 30 s to 2 min. The equilibration time increases with the size of the CO_2 or water-vapor differential, meaning that sampling times tend to be longer for leaves with higher photosynthetic rates, a situation opposite that typically encountered with closed systems. As in all systems without temperature-controlled leaf chambers, the chamber usually heats to slightly above ambient temperature. To minimize this temperature rise, the Schulze *et al.* (1982) system incorporates an actively ventilated heat exchanger. A fan pulls ambient air over the outside of a nickel-plated aluminum chamber, keeping temperature elevation to less than 3.5°C. The system does not, in the ambient-sampling mode, provide a measure of absolute CO_2 concentration. Though the Schulze *et al.* (1982) system is portable and battery powered, it is typically operated with the chamber hand held but the IRGA set at one or a few locations during a day. Without a data logger, its operation requires two people.

The ADC portable CO_2 assimilation-transpiration measurement system (Analytical Development Co., Hoddesdon, UK) is another differential system without environmental control. It is functionally similar to the system of Schulze *et al.* (1982) but differs in important details. First, the ADC system incorporates very portable sensors for CO_2 and humidity. These sensors make the system very portable but they lack the resolution of the two-channel IRGA in the Schulze *et al.* (1982) system. Therefore, the ADC system functions best with much larger CO_2 depletions and humidity increases than the system of Schulze *et al.* (1982). Second, the ADC system is designed to operate with dry air entering the chamber while the system of Schulze *et al.* (1982) uses unmodified ambient air. Under some conditions, the equilibrium humidity in the chamber of the ADC system may be close to ambient, but under others, it is far from ambient.

In the simple configurations described here, the system of Schulze *et al.* (1982) and the ADC system are suited primarily for large samples under ambient conditions (Fig. 11.9). Both systems can, however, be modified for other purposes. Successors of the system of Schulze *et al.* (1982), available commercially from H. Walz, Mess- und Regeltechnik, D-8521 Effeltrich, Federal Republic of Germany, can be configured for quick sampling, for continuous measurements under ambient or controlled conditions, or for response curves. In the more complicated configurations, these instruments are differential systems with extensive environmental control. As is true for many of the systems discussed here, elaborations on an initial design can dramatically alter a device's capabilities. It is often incorrect to assume that characteristics discussed here apply to later versions of a given instrument.

(b) Differential systems with environmental control

From the earliest days of field gas exchange and until the last few years, large mobile laboratories set the standards for accuracy and environmental control. Bosian (1955), who developed one of the first mobile laboratories, was an early advocate of controlled-environment studies. By the early 1960s, his instruments controlled temperature, humidity and wind speed in the gas-exchange chamber (Bosian, 1965).

Most of the large mobile laboratories for photosynthesis research (Eckardt, 1966; Koch *et al.*, 1971; Mooney *et al.*, 1971; Björkman *et al.*, 1973) are either pure differential or mixed differential–compensating systems with extensive environmental control. Several of the mobile laboratories and a few portable systems (Oechel and Lawrence, 1979; Küppers *et al.*, 1987) simultaneously measure and control multiple cuvettes. The mobile laboratories made measurements comparable in accuracy and complexity to those obtained from the best laboratory gas-exchange systems, but made them on plants growing in nature, in sites as hostile to plants and equipment as Death Valley, California (Mooney *et al.*, 1976), Israel's Negev Desert (Lange *et al.*, 1969) and alpine tundra at Niwot Ridge, Colorado (Moore *et al.*, 1973).

The mobile laboratory of Björkman *et al.* (1973) effectively illustrates the convergence between the most competent laboratory and field systems (Figs 11.9 and 11.10). For many years this system, housed in a mobile home, alternated between field and laboratory use. In the field it was a self-contained laboratory. At the Carnegie Institution's Department of Plant Biology, it became, when docked at a special port, another room in the laboratory complex. There, it was the primary laboratory system. The mobile laboratory of Björkman *et al.* (1973) made few compromises to increase portability. Cooling was provided by refrigerated circulating baths (Fig. 11.6). A precision mixing pump established the gas mixtures. All the tubing carrying moist gas was heated stainless steel, and a mini-computer controlled the data-acquisition unit. Fully configured for operation in Death Valley, power consumption in this system approached 10 kW.

Recently, miniaturization has been one of the most important trends in system development, and modern differential systems reflect this trend. Pearcy and Calkin (1983) developed a controlled-environment differential system, including automated data acquisition and an artificial light, which they carried by hand to remote rain forest sites. A 500 W generator powered the system. By storing a large quantity of ambient air in a waterbed mattress, they could operate for a day or more from air of a constant CO_2 and water-vapor content. Pearcy and Calkin (1983) controlled the water content of the air entering the chamber by mixing dried air with the moist air stored in the mattress. Pearcy and Calkin's (1983) system did not have the capability of generating CO_2 response curves. However, more recent instruments, based on the same philosophy but including several mass flow controllers, do have that capability (Sharkey, 1985; Atkinson *et al.*, 1986; Küppers *et al.*, 1987),

The systems of Atkinson *et al.* (1986) and Küppers *et al.* (1987) are complex portable devices configured for measurements beyond the competence of most mobile laboratories. Both these systems use mass flow controllers for gas mixing and can humidify the gas stream to any desired level. The system of Atkinson *et al.* (1986) was optimized for field SO_2 fumigation. The system of Küppers *et al.* (1987) includes O_2 as a controlled parameter. This system independently measures and controls two leaf chambers, one for response curves and one for tracking gas exchange under ambient conditions.

We have now entered an era when truly portable (movable by hand, but often in several loads) instruments can deliver all of the gas-exchange flexibility and accuracy of a mobile laboratory, at a much lower cost, size and power consumption.

(c) Compensating systems with environmental control

Controlled-environment compensating systems for the simultaneous measurement of CO_2 and water-vapor exchange are now very similar to the most powerful portable differential system (Fig. 11.9). Systems typically weigh approximately 100 kg and require about an hour for set up in the field. Power requirements are 100–300 W without artificial light but approximately 400 W with artificial light, a necessity for response curves in all but the clearest conditions. Systems with these power requirements can be battery operated, but we find that portable generators provide a more satisfactory power source. If a portable generator is the power source, the IRGA must be frequency insensitive and the DC power supplies for individual components must be well regulated. To date, all of the leaf chambers used with controlled-environment systems have been tripod mounted and cannot be moved between leaves in less than about 5 min. System equilibration at altered chamber conditions seldom requires more than a few minutes, but the plant response, especially of stomatal conductance, may be much slower. Sample handling capacity in these systems depends on the type of measurements, but typical outputs are replicated measurements at a single set of conditions for 10–20 leaves per day or two to four response curves per day.

Bingham *et al.* (1980), Field *et al.* (1982) and Sharkey (1985) describe systems in which transpiration is the only source of water vapor for humidifying the cuvette and in which CO_2 depletion by photosynthesis is partly or fully compensated by injecting CO_2 into the leaf chamber. The major advantages of the compensating over the differential system are hardware simplicity and decreased dependence on the IRGA gain for an accurate photosynthesis measurement. The compensating system requires no components for humidifying the gas stream, for measuring the humidity of the gas entering the leaf chamber, or for preventing condensation in the gas lines. These simplifications are important, because they eliminate not only mass but also the power demands of two or more temperature-controlled components. The major disadvantages of the compensating systems described by Bingham *et al.* (1980) and Field *et al.* (1982) are: (1) Measuring CO_2 responses of photosynthesis is not convenient and typically requires several air cylinders for a complete response curve. (2) Because the leaf is the only source of water vapor, system responses can be very slow, especially when chamber humidity is high, transpiration is low, or leaf area is small. (3) Since leaf gas exchange has a large effect on the environment within the chamber, maintaining a constant chamber environment requires that the operator adjusts flows in response to changes in photosynthesis or transpiration. Recent technological improvements have eliminated or substantially decreased these problems, but at the cost of increased weight and power requirements.

Mass flow controllers can be used to make compact accurate gas-mixing devices necessary for CO_2 curves in portable systems (Sharkey, 1985) and can be used to control O_2 (Küppers *et al.*, 1987), SO_2 (Atkinson *et al.*, 1986), or other gases. The problem of the intrinsic relationship between low transpiration rates and long time lags caused by low air flow in compensating systems has been overcome with several designs for humidifying the air entering the chamber. These include Pearcy and Calkin's (1983) use of the humidity in ambient air, a Peltier-driven humidifier–condenser (Fig. 11.8) and temperature-dependent changes in the hydration state of some salts (Parkinson and Day, 1981).

Systems with humidifiers for establishing the desired dew point and mass flow controllers for adjusting the ambient CO_2 concentration and compensating leaf CO_2 uptake can control CO_2 and humidity completely independently. This capability simplifies response-curve measurements in which the goal is to vary a single factor, holding all others constant. Humidifiers and mass flow controllers also decrease the necessity for operator intervention in response to changing photosynthesis or transpiration. With humidification, flow rates are much higher and the effects of the leaf on the chamber environment are much lower. Since mass flow controllers can be set with electrical signals, it is straightforward to implement a feedback loop such that air flow is automatically adjusted to provide the desired chamber humidity.

With the addition of air stream humidification, compensating systems can be operated at high flow rates and as fully differential systems. Portable systems capable of operating in either compensating or differential mode give the investigator a combination of accuracy, control and flexibility matched in relatively few laboratory settings.

11.5 MATCHING INSTRUMENT TO OBJECTIVE

To date, no photosynthesis system is better than all others, for addressing the broad range of questions involving gas-exchange research. Increasing portability and sampling speed typically entail sacrificing environmental control and sometimes compromising accuracy. The portability–control trade-off may be ameliorated with improved technology, but sampling speed in controlled-environment experiments is unlikely to be dramatically increased. The constraint on sampling speed in experiments designed to measure steady-state responses is usually the time required for equilibration by the plant. Given that the equilibration of stomatal conductance in response to a small change in humidity requires on the order of 15 min (Field *et al.*, 1982) and that the induction of photosynthesis in response to the transition from low to high light requires several minutes (Pearcy *et al.*, 1985), we cannot hope to measure maximum photosynthetic capacity during a 1 min sample or to execute steady-state response curves at 30 s per point. For the future, we expect parallel improvements in ambient-sampling and controlled-environment systems, but we do not expect the rapid development of an outstanding all-purpose system.

The best gas-exchange approach for a particular research program depends on that program's goals. The goals of the research incorporating gas exchange are broad and dynamic, but some generalities are useful.

Response curves can be generated with both ambient-sampling and controlled-environment systems (Fig. 11.9), but the curves from ambient-sampling systems are typically amalgamations of points presented together because they span the appropriate range of some environmental variable. In response curves from ambient-sampling systems, adjacent points on a plot are rarely measured sequentially, parameters other than the parameter of interest may be changing, and the data may not represent equilibrium responses. Transient measurements of the response of photosynthesis to CO_2 concentration are possible with closed systems like the LI-COR LI-6200. It is also possible to modify ambient-sampling differential systems for measuring CO_2 responses (e.g. Schulze *et al.*, 1982), but any measurement shortly after enclosing a leaf may not represent an equilibrium.

All systems function reasonably well for tracking changes in photosynthesis over periods of hours, but an ambient-sampling system might track 10 leaves while a controlled-environment system tracks one (Fig. 11.9). However, controlled-environment systems provide the opportunity to fix some environmental variables while one or several others vary naturally, a powerful asset for experiments designed to separate effects of multiple factors that change over a day or a season. Remember, though, that even the most powerful gas-exchange systems do not control all environmental variables. Manipulating factors like soil moisture, transpiration by leaves other than those in the chamber and the plant carbohydrate status requires technologies beyond those provided by traditional gas-exchange systems.

Relatively few systems are competent to track responses of gas exchange to rapid environmental changes. Closed systems are the worst suited for following rapid changes while differential systems are the best, sometimes shaving response times to about 1 s (Pearcy *et al.*, 1985).

Studies requiring large sample sizes pose some of the most difficult choices for gas-exchange research (Fig. 11.9). Ambient-sampling systems yield by far the largest number of measurements, but the nature of the ambient sampling imposes several limitations. First, it is difficult to separate effects due to a parameter of interest from effects of environmental variation. This problem is especially severe when the parameter of interest is a continuous biological variable (e.g. leaf age or nutrient status) in contrast to a discrete experimental treatment (e.g. fertilization or irrigation). Second, it is, with ambient-sampling techniques, very difficult to assess photosynthesis and conductance under conditions that occur only rarely, for example, extremes of temperature or humidity. Similarly, ambient-sampling techniques generally prevent access to environmental conditions that never occur naturally but that may be very useful for revealing differences largely masked under normal conditions. For example, differences in the total activity of the carboxylating enzyme, ribulose-1,5-bisphosphate carboxylase/oxygenase are much more obvious at reduced than at normal CO_2 levels (von Caemmerer and Farquhar, 1981). Finally, results from ambient-sampling techniques may be misleading as a result of differences between natural conditions and conditions in the chamber. Maintaining chamber conditions at ambient sometimes requires the full sophistication of an advanced controlled-environment system.

Individual researchers must evaluate the trade-off between the large sample sizes obtainable with ambient-sampling techniques and the potentially richer data set per sample obtainable with controlled-environment systems. Some of the best solutions to the sample-size versus control dilemma involve combining multiple systems and multiple system concepts in a single research program. For example, use a controlled-environment system to identify responses that appear to differ among populations and then use an ambient-sampling system to survey populations exposed to an environment where the differences will be manifest. Or, once differences are identified with a controlled-environment system, use an ambient-sampling system to quantify consequences of those differences. For some questions, the approach might be reversed; initial data coming from ambient-sampling measurements and later probing from controlled-environment experiments.

11.5.1 Measurement scale

Thus far, we have discussed primarily measurements at the leaf level, the focus for most gas-exchange studies. Our emphasis on the leaf level is for convenience and not because experiments at other levels are less productive or important. The decision about measurement level should be, like the choice of system, resolved on the basis of the goals of the research program.

The leaf is the obvious level for most experiments probing mechanisms of photosynthesis for two reasons. The study material is relatively homogeneous at this level, and resource levels, especially light, can be made very uniform. Physiological and environmental homogeneity at the leaf level make it the natural choice for comparative studies searching for physiological differences. At higher levels, it is difficult to distinguish differences in photosynthesis from differences in canopy structure that alter light interception. In combination with measurements or models predicting microhabitats within a whole plant or a canopy, leaf level measurements can be critical components of studies attempting to integrate gas exchange to higher levels.

Measurements at levels higher than the leaf may be motivated by several factors. Among the most important are quantifying gas exchange and the components of productivity at the canopy or ecosystem level, testing predictions based on the integration of leaf-level measurements, quantifying photosynthesis in plants with small leaves or complex leaf arrays, testing hypotheses concerning resource interception and the efficiency of canopy structure, and studying gas exchange by organs other than leaves.

Twig or branch measurements, well suited for measurements on plants with small leaves, have been extensively employed in studies of conifers (Leverence and Öquist, 1987) and chaparral shrubs (Oechel and Lawrence, 1979). Whole plant chambers provide options for measuring gas exchange in plants too small for leaf or twig level studies but are especially important for studies aimed at understanding the consequences of canopy architecture, leaf aging, biomass allocation between leaves and stems, and nutrient allocation among leaves, stems and reproductive tissues (Caldwell *et al.*, 1983; McKree, 1986). Canopy chambers and meteorological methods are invaluable for assessing effects of species or species mixtures on production, especially when fluxes into and out of the soil are potentially important (Hilbert *et al.*, 1987).

For the future, we predict an increasing emphasis on the integration of photosynthesis with research on global change, ecological success and agricultural yield. Research with these emphases will need to take advantage of the opportunities intrinsic in measurements at the branch, plant and canopy level. Breakthroughs in the integration of photosynthesis with yield, competition, species distribution and ecosystem or landscape productivity will almost certainly require programs utilizing measurements at multiple levels.

11.6 CALIBRATING PHOTOSYNTHESIS SYSTEMS

Calibration is an issue of critical importance for field instruments because the bumps, dirt, moisture and temperature changes of field work tend to make calibrations change faster than under laboratory conditions. Calibrations should be performed on a regular schedule that reflects the differences in stability among instruments. Often, components of a single device require calibration on different schedules. For example, some IRGAs should be checked every few minutes for zero drift, but semiannually for sensitivity of the gain to changes in the background CO_2 concentration.

The rigors of field conditions place an extra emphasis on internal checks and 'bioassays' for instrument errors in field systems. The possibilities for internal checks vary from system to system, but most devices present several options. For example, except for small effects due to changes in the ambient CO_2 concentration, a compensating system should yield the same photosynthesis rate at a range of CO_2 injection rates. If it does not, either the IRGA calibration or the calibration for the CO_2-injection flow meter is incorrect. A differential system should yield a zero photosynthesis rate for an empty chamber (this occurs less frequently than one might guess). Bioassays for instrument errors use consistent plant characteristics to assess system performance. One useful bioassay for photosynthesis systems utilizes the constancy of the photon yield of photosynthesis (the initial slope of the light response curve, expressed on the basis of absorbed light) in leaves of C_3 plants not damaged by photoinhibition (Ehleringer and Pearcy, 1983). When an experiment indicates a photon yield substantially different from 0.06 mol of CO_2 per mol of light absorbed (typically 0.04–0.055 mol of CO_2 per mol of light incident), it is wise to look very hard for system errors before proceeding much further with the gas-exchange studies. Similarly, if an experiment fails to indicate a strong correlation between photosynthesis and leaf conductance through a light-response curve (cf. Wong *et al.*, 1979), something may need calibration. An awareness of the consistencies in gas-exchange characteristics is an important tool for evaluating instrument performance, but is of limited use once the need for calibration is identified.

11.6.1 Calibrating flow meters

(a) The bubble tower method

The basic idea behind most flow meter calibrations is to measure the time necessary for an unknown flow to fill a known volume.

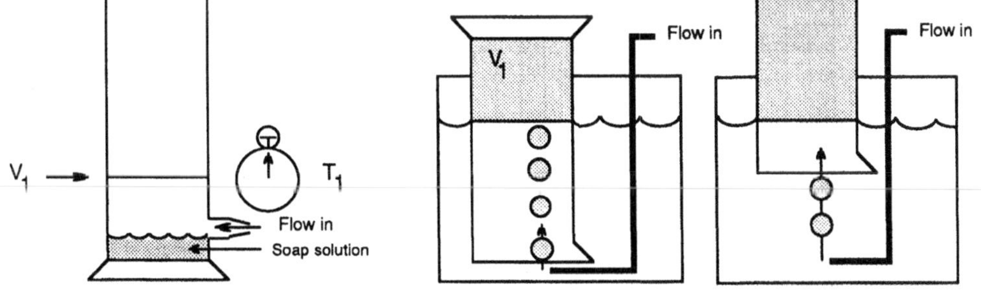

Fig. 11.11 Two methods of flow calibration. The bubble tower method (a) involves timing the displacement of a soap film as it rises through a known volume. In this method the entire system is always close to atmospheric pressure. The inverted cylinder technique (b) involves measuring the time required for the flowing gas to displace a known volume of water. To avoid pressure-related errors, the water level inside and outside the cylinder should always be the same. Both methods of flow calibration require precautions and corrections related to temperature and humidity of the gas in the cylinder. Neither method is appropriate for calibrating low flows of gases that are highly water soluble.

One of the most accurate calibrations for flows in the range from 1 to 100 ml s^{-1} requires only a stopwatch, a graduated cylinder or a buret and a soap solution. A flow bubbled through a soap solution can be coaxed to make soap films that float up a cylinder at the gas flow rate (Fig. 11.11). Ideal gas laws (Section 11.7) are used to correct the flows to other temperatures or pressures or to calculate molar flows. Bubbling a gas through a soap solution results in significant humidification. To avoid an unknown volume increase from this humidification, a humidifier is placed before the bubble tower, using a water column near ambient temperature to saturate the gas with water vapor at a known temperature. If the water column is before the flow meter, the only new calculations address the flow meter response to water vapor versus the other gas. If it is after the flow meter (as is necessary when the flow meter requires a significant pressure drop), humidification results in a known volume increase (Section 11.7) allowing a calculation of the pre-humidification flow. Since gases cool on expansion, it is necessary to correct for gas temperature on the basis of measurements on the gas flowing through the cylinder and not from a measurement on the cylinder wall.

Several flow calibrators based on the bubble tower method are now commercially available. Since most flow meters require relatively infrequent calibrations, we find that homemade calibrators are sufficient to support a large number of flow meters or controllers.

(b) The inverted cylinder method

Though usually more cumbersome than the bubble method, an alternative flow meter calibration involves collecting the flow in a cylinder inverted in a tub of water (Fig. 11.11). By carefully raising the cylinder as its gas content increases, one can avoid the pressurization that would result from pushing water out of the cylinder. Temperature and humidity precautions should be similar to those for the bubble calibration.

(c) The mixing method

For flows below 1 ml s^{-1}, measuring volume can be quite difficult, but it is often quite straightforward to calculate flow by mixing a small volume of one gas at an unknown flow with a larger volume of a second gas at a known flow, and measuring the effect on the composition of the gas. For example, to calibrate a flow meter used for CO_2 injection in a compensating system, pass a reference gas through the IRGA reference cell and a mixture of the reference and CO_2-injection gases through the sample cell. (Note: Some older IRGAs operate properly only if the CO_2 concentration is higher in the reference cell than in the sample cell. If you have one of these, switch what you call the reference and sample cells.) The flow rate through the CO_2-injection flow meter is given by:

$$u_c = \frac{u_r \Delta c}{c_c - c_r - \Delta c} \quad [\text{mol s}^{-1}] \qquad (11.8)$$

where u_c is flow through injection meter (mol s^{-1}), u_r is flow of reference gas into mixing system (mol s^{-1}), c_c is mole fraction of CO_2 in injection gas (mol of CO_2/mol of air), c_r is mole fraction of CO_2 in reference gas and Δc is CO_2 differential from IRGA (mol mol^{-1}). Δc should be calculated with the corrections for CO_2 background described in Section 11.6.2.

11.6.2 Calibrating IRGAs

Obtaining satisfactory performance from an IRGA may require a number of adjustments which differ from instrument to instrument and fall outside of the range of normal calibrations. Adjustments such as source balancing (equalizing the intensity of the IR through the two cells) and phasing (adjusting the relative times the two cells are exposed to the IR source) are explained in the manufacturers' instruction manuals.

IRGA responses are intrinsically nonlinear, but many instruments include or are available with electronic linearizers. Electronic linearizers typically work quite well when the measurements are confined to a limited range of CO_2 concentrations but are rarely satisfactory in differential instruments operated at a broad range of background CO_2 concentrations. Because IRGA nonlinearities can be accurately described with a simple equation, we find that the flexibility of a nonlinearized instrument generally outweighs the convenience of a linearized one. In theory, IRGA output (D) in a nonlinearized instrument varies as

$$D = \frac{\Delta P_c}{P_{mc}y+z} \quad \text{[Volts]} \qquad (11.9)$$

where ΔP_c is the partial pressure of CO_2 (sample cell) minus the partial pressure of CO_2 (reference cell) (Pa), P_{mc} is the mean CO_2 partial pressure in the two cells (Pa), y is the sensitivity of the IRGA gain to background CO_2 and z is the zero offset of the shift in gain with background. Equation 11.9 also applies for nonlinearized absolute IRGAs, in which the CO_2 partial pressure in the reference channel is effectively zero.

For a linearized IRGA or a nonlinearized one used over a small range of CO_2 concentrations, calibration requires only setting the gain and zero. Nonlinearized instruments used over a wide range of CO_2 concentrations can be calibrated in either of two ways. One, the zero and gain can be adjusted every time the background CO_2 concentration is changed. Simple in concept, this approach is generally time-consuming and awkward to integrate into experiments, unless an IRGA with a split cell is used (Parkinson and Legg, 1978), which makes the procedure much less complex. The alternative we use in our laboratories is a calibration that allows us to determine y and z in Equation 11.9. Knowing y and z, we always set the IRGA zero at one CO_2 concentration and the gain at a second and calculate D from Equation 11.9. In general, y and z are very stable and need to be checked once or twice a year. The gain should be checked daily or weekly in a field instrument, and the zero should be checked several times a day and whenever the reference CO_2 concentration is changed.

Setting the gain and zero requires two gases of known composition. The zero is set with one gas (usually CO_2-free air in an absolute IRGA and a gas near the background CO_2 concentration in a differential instrument) in both the sample and reference cells. The gain is set with the zero gas (the gas used to zero the instrument, which may or may not be a gas with zero CO_2) in the reference cell and a span gas, which should be different from the zero gas by approximately the full-scale IRGA reading, in the sample cell. Sources for these calibrations standards may be cylinders of certified composition or gas mixed on site using one of several accurate mixing systems, including precision mixing pumps (e.g. Type 2 G26/3-F, Wösthoff, West Germany), supersonic-orifice based mixers (e.g. CO_2 calibrator, Analytical Development Co., UK), volume-based mixers (e.g. Gas Calibration Cylinder, LI-COR, USA) or combinations of mass-flow controllers (e.g. FC 360, Tylan, USA). Instruments incorporating split cells (Parkinson and Legg, 1978) can be calibrated using only one gas of known composition.

Determining y and z requires a full-range calibration (Bloom *et al.*, 1980). Rearranging Equation 11.9 to

$$\frac{\Delta P_c}{D} = P_{mc}y+z \qquad (11.10)$$

it is clear that there should be a linear relationship between ($\Delta P_c/D$) and P_{mc}, with a slope y and y-intercept z. For nonlinearized differential analyzers, we use the following procedure to determine ($\Delta P_c/D$) and P_{mc} over the entire CO_2 range at which the instrument will be operated, and then use linear regression to obtain values for y and z.

1. Adjust the IRGA zero and gain under standard conditions as described above;

2. Run CO_2-free air through both cells and note the reading (D_s);
3. Trap the CO_2-free air in the reference cell and fill the sample cell with CO_2 in air at the full-scale deflection [2.5, 5 or 10 Pa, (25, 50 or 100 ppm) depending on the IRGA], and note the IRGA reading (D_d);
4. Repeat step 2 but instead of CO_2-free gas, use the same gas as in the sample cell in step 3. Note the IRGA reading (D_s);
5. Repeat step 3, but trapping the gas from step 4. Then, increase the concentration in the sample cell by one more full-scale deflection. Note the reading (D_d);
6. Continue repeating steps 4 and 5, increasing the partial pressure of CO_2 by the full scale of the IRGA with each repeat.

For each pass through the repeated sequence,

$$D = D_d - D_s$$

$$P_{mc} = \frac{(P_{sc} + P_{rc})}{2}$$

where P_{sc} is partial pressure of the CO_2 in the sample cell (Pa) in step 5, P_{rc} is partial pressure of CO_2 in the reference cell (Pa) in step 5 and ΔP_c equals P_{sc} minus P_{rc} in step 5.

To get from CO_2 measurements as mole fractions to partial pressures, use the relation

$$P_c = \frac{c_m}{P} \quad \text{[Pa]} \qquad (11.11)$$

where P_c is CO_2 as a partial pressure (Pa), c_m is CO_2 as a mol fraction (μmol mol^{-1}) and P is the ambient total pressure (Pa).

Once we know y and z, we calculate the CO_2 partial pressure in the sample cell (the usual goal of an IRGA measurement) as

$$P_{sc} = \frac{P_{rc}\left(1 + \frac{yD}{2}\right) + zD}{1 - \left(\frac{yD}{2}\right)} \quad \text{[Pa]} \qquad (11.12)$$

a simple rearrangement of Equation 11.9.

IRGAs are not intrinsically flow sensitive, but because they measure the concentration of molecules in the cells, they are pressure sensitive. To minimize pressure effects on IRGA readings, it is useful to keep flows low and the same in both cells. Liquid level manometers can provide a simple check for equal pressure in the two cells. For field systems used at a range of altitudes, the ideal gas law (Section 11.7) allows a simple pressure correction. Note that the full-range calibration must be in terms of partial pressures to enable appropriate shifts in IRGA sensitivity with changing ambient pressure.

11.7 CALCULATING GAS-EXCHANGE PARAMETERS

The basic outline of the gas-exchange calculations is presented in Sections 11.1 and 11.2. Here, we add most of the detail necessary for writing a computer program to calculate photosynthesis, transpiration, intercellular CO_2 and leaf conductance. Throughout, we use the units proposed by Cowan (1977). For simplicity, all calculations are presented without scaling prefixes (i.e. milli-, micro-, etc.). For the area-based calculations, it is common to express results per unit of projected leaf area (the area of one surface) rather than per unit of total area (the area of both surfaces). Other conventions are often more useful for conifer needles and other photosynthesizing tissue not organized into flat sheets. Ball (1987) gives a more complete discussion of the material in this section. Section 11.8 gives a list of the symbols used here.

11.7.1 Molar flows

Flow is a critical parameter in open-system calculations, and flow calculations are much simpler if the flows are expressed in mol s^{-1}. Regardless of whether flow meter outputs are proportional to molar flows, flow meters are almost always calibrated in volume units, and we need the volume of one mole of a gas. The ideal gas law does an excellent job of describing the pressure–temperature–volume relationships of gases present as minor components of mixtures (such as CO_2 and H_2O) near normal atmospheric conditions of temperature and pressure. The ideal gas law states

$$PV = n\boldsymbol{R}T \qquad (11.13)$$

where P is pressure (Pa) (1 bar = 10^5 Pa), V is volume (m^3), n is number of moles, $\boldsymbol{R}$ is universal gas constant (8.311 m^3 Pa mol^{-1} K^{-1} or 8.311×10^{-5} m^3 bar mol^{-1} K^{-1}) and T is temperature (K = °C + 273.16). At a pressure of 1 atmosphere (1.013×10^5 Pa or 29.92 inches of mercury), and a temperature of 20°C, the volume of one mole of an ideal gas is 2.405×10^{-2} m^3 or 24.04 liters.

11.7.2 Water content of air

(a) Saturation vapor pressure

The World Meteorological Association accepts the Wexler equation as the standard for calculating saturation vapor pressure. However, Richards' (1971) equation is much simpler and closely matches the Wexler equation over the temperature range relevant to gas-exchange studies. Richards' (1971) equation for saturation vapor pressure (v_{sat}) in Pa is

$$v_{sat} = 101325 \exp(13.3185t - 1.976t^2 - 0.6445t^3 - 0.1229t^4) \qquad (11.4)$$

where

$$t = 1 - (T_s/T) \qquad (11.15)$$

T_s is steam temperature at standard pressure (373.16 K) and T is the air temperature of interest (K). To convert this partial pressure to a mole fraction (mol of H_2O/mol of air) (w_{sat}), use

$$w_{sat} = \frac{v_{sat}}{P} \quad [\text{mol mol}^{-1}] \qquad (11.16)$$

where P is ambient pressure (Pa).

(b) Relative humidity

Relative humidity (RH) is given by

$$\text{RH} = \frac{v_a}{v_{sat}} \qquad (11.17)$$

where v_a is ambient vapor pressure (Pa). Equation 11.17 can be rearranged to yield v_a, given v_{sat} and RH. The ambient vapor pressure as a mole fraction (w_a, mol of H_2O/mol of air) is

$$w_a = \frac{v_a}{P} \quad [\text{mol mol}^{-1}] \qquad (11.18)$$

where P is ambient pressure (Pa).

(b) Dew-point temperature

The dew-point temperature (T_d) is the temperature at which the ambient vapor pressure equals the saturation vapor pressure. Given the dew-point temperature, you can calculate the ambient vapor pressure directly from Equation 11.14.

11.7.3 Transpiration

In an open system, transpiration (E) (mol m^{-2} s^{-1}), is given by

$$E = \frac{u_o w_o - u_e w_e}{L} \quad [\text{mol m}^{-2}\ \text{s}^{-1}] \qquad (11.19)$$

where u_o is flow leaving chamber (mol s^{-1}), w_o is the mole fraction of water in air leaving the chamber (mol mol^{-1}), u_e is the flow entering chamber (mol s^{-1}), w_e is the mole fraction of water in the air entering the chamber (mol mol^{-1}) and L is leaf area (m^2).

u_o is usually greater than the flow entering the chamber because transpiration increases the total volume. u_o is given by

$$u_o = u_e \left(\frac{1 - w_e}{1 - w_o}\right) \quad [\text{mol s}^{-1}] \qquad (11.20)$$

Photosynthesis decreases u_o relative to u_e, but the effect is so small that we ignore it. In a compensating system, u_e represents the sum of the air flow and the flow of the CO_2-enriched gas used to compensate photosynthesis.

In a closed system, transpiration is given by

$$E = \frac{[w_f(P_f/T_f) - w_b(P_b/T_b)]V}{\Delta t L P_v/T_v} \quad [\text{mol m}^{-2}\text{ s}^{-1}] \tag{11.21}$$

where w_f is the mole fraction of water in air at the end of the measurement (mol mol^{-1}), P_f is system pressure at the end of the measurement (Pa), T_f is system temperature at the end of the measurement (K), w_b is the mole fraction of water in the air at the beginning of the measurement (mol mol^{-1}), P_b is the system pressure at the beginning of the measurement (Pa), T_b is the system temperature at the beginning of the measurement (K), V is the system volume (mole, at P_v and T_v). Δt is the time between beginning and end of the measurement (s), L is leaf area (m^2), P_v is system pressure at the time of volume determination (Pa) and T_v is the system temperature at the time of volume determination (K). The pressure terms are included because, in a closed system, the quantity of gas in the system increases during a measurement as a result of transpiration. The pressure terms prevent this increase from effectively diluting the water content of the air. The temperature terms adjust the molar volume.

(a) Conductances

Leaf conductance to water vapor g_{lw} (mol m^{-2} s^{-1}) is defined as the proportionality constant between transpiration and the vapor concentration gradient between the leaf interior and the leaf surface. We generally assume that the leaf interior is saturated with water vapor at the leaf temperature. This assumption is very robust, though the vapor pressure in equilibrium with the leaf water decreases slightly as leaf water potential decreases (Nobel, 1983). Since it is impossible to measure the water content of the air at the leaf surface, we calculate a total conductance g_{tw}, the proportionality constant between transpiration and the water concentration gradient between the leaf interior and the bulk air in the chamber. g_{tw} is the series combination of g_{lw} and the boundary layer conductance to water g_{bw}.

The above definition of a conductance applies, strictly, only to cases of binary diffusion. The situation between the sites of evaporation inside a leaf and the bulk atmosphere outside is more complex, because water vapor is diffusing out at the same time CO_2 is diffusing in. In addition, the fluxes in opposite directions are not equal, because the inside is a source of water and a sink for CO_2. These complexities, first quantified by Jarman (1974), add a few additional terms to the following equations, based on the work of Jarman (1974) as simplified by von Caemmerer and Farquhar (1981). Leuning's (1983) more extensive analysis yields similar corrections.

Total conductance to water vapor g_{tw} (mol m^{-2} s^{-1}) is given by

$$g_{tw} = \frac{E(1-w_m)}{(w_i - w_a)} \quad [\text{mol m}^{-2}\text{ s}^{-1}] \tag{11.22}$$

where

$$w_m = \frac{(w_i + w_a)}{2}$$

w_i is the water content of the air inside the leaf (mol of H_2O/mol of air) and w_a is water in the ambient air (mol of H_2O/mol of air).

It is impossible to separate boundary-layer conductance g_{bw} (Section 11.3) precisely from leaf conductance g_{lw} unless one knows the distribution of both parameters between the leaf's upper and lower surface. Given the spatial distributions, the complete expression for g_{tw} is

$$g_{tw} = \frac{g_{ll}g_{bl}}{g_{bl} + g_{ll}} - \frac{g_{lu}g_{bu}}{g_{bu} + g_{lu}} \quad [\text{mol m}^{-2}\text{ s}^{-1}] \tag{11.23}$$

where g_{ll} is leaf conductance of the lower surface (mol m^{-2} s^{-1}), g_{lu} is leaf conductance of the upper surface (mol m^{-2} s^{-1}), g_{bl} is boundary layer conductance of the lower surface (mol m^{-2} s^{-1}), g_{bu} is boundary layer conductance of the upper surface (mol m^{-2} s^{-1}) and g_{lw} is the sum of g_{ll} and g_{lu}). It is not possible to solve Equation 11.23 for g_{ll} and g_{lu} (or their sum) without additional information. If $g_{bu} = g_{bl}$ $(= 0.5g_{bw})$ and if $g_{ll} = g_{lu}$, then

$$g_{lw} = \frac{1}{\left(\frac{1}{g_{tw}} - \frac{1}{g_{bw}}\right)} \quad [\text{mol m}^{-2}\text{ s}^{-1}] \quad (11.24)$$

At the other extreme where the leaf has stomata on only one side (e.g. $g_{lu} = 0$), then

$$g_{lw} = \frac{1}{\left(\frac{1}{g_{tw}} - \frac{2}{g_{bw}}\right)} \quad [\text{mol m}^{-2}\text{ s}^{-1}] \quad (11.25)$$

For most amphistomous leaves, the correct value for g_{lw} is between those given by Equations 11.24 and 11.25 (Moreshet *et al.*, 1968). If g_{bw} is large, the difference between the results of the two equations is small. Calculating an exact value for g_{lw} requires the ratio of g_{ll} and g_{lu}, in addition to g_{tw} and g_{bw}.

It is very difficult to calculate CO_2 conductances from CO_2 fluxes, but conductances to water and carbon dioxide are related by the diffusion coefficients of the two gases in air. Under pure diffusion (i.e. through still air), as through the stomatal pores, the diffusion coefficient for water vapor is approximately 1.6 times that of CO_2. Thus,

$$g_{lc} = \frac{g_{lw}}{1.6} \quad [\text{mol m}^{-2}\text{ s}^{-1}] \quad (11.26)$$

where g_{lc} is the leaf conductance to CO_2 (mol m^{-2} s^{-1}). The situation in the boundary layer is complicated, because the actual transport is a combination of diffusion and bulk flow. Assuming that leaf boundary layers approximate laminar flow over isothermal flat plates

$$g_{bc} = \frac{g_{bw}}{1.37} \quad [\text{mol m}^{-2}\text{ s}^{-1}] \quad (11.27)$$

where g_{bc} is the boundary layer conductance to CO_2. The assumption of laminar flow may not be entirely correct, but as long as the boundary layer conductance is high, the precise value of the coefficient has little effect on the subsequent calculations.

11.7.4 Ambient CO_2

If the air is dried between the chamber and the IRGA, the CO_2 concentration in the IRGA sample cell is greater than the concentration in the chamber by an amount related to the dilution of the air in the chamber by water vapor. The ambient CO_2 in the chamber (c_a) (mol mol^{-1}) is given by

$$c_a = \frac{c_s}{(1 + w_a - w_s)} \quad [\text{mol mol}^{-1}] \quad (11.28)$$

where c_s is the CO_2 mole fraction in the IRGA sample cell, w_a is the mole fraction of water in the chamber and w_s is the mole fraction of water in the IRGA sample cell. If the IRGA is calibrated to read CO_2 levels as partial pressures, a wise practice for instruments operated over significant altitude ranges, the conversion from partial pressure to mole fraction is

$$c_s = \frac{P_{sc}}{P} \quad [\text{mol mol}^{-1}] \quad (11.29)$$

where P_{sc} is partial pressure of CO_2 in the IRGA sample cell (Pa) and P is ambient pressure (Pa).

11.7.5 Photosynthesis

In an open system, net photosynthesis (A_n, mol m^{-2} s^{-1}) is given by

$$A_n = \frac{(u_{pe}c_e - u_{po}c_o) + u_c(c_c - c_o)}{L} \quad (11.30)$$

where u_{pe} is air flow entering the chamber, corrected for photosynthesis (mol s^{-1}), c_e is the mole fraction of CO_2 in the air entering the chamber, with air at the water content at which it passes through the IRGA (mol of CO_2/mol of air), u_{po} is the air flow leaving the chamber, corrected for photosynthesis (mol s^{-1}), c_o is the mole fraction of CO_2 in the air leaving the chamber, with air at water content at which it passes through the IRGA (mol of CO_2/mol of air), u_c is the flow rate of compensating gas in a compensating system (mol s^{-1}), c_c is the mole fraction of CO_2 in the compensating gas (mol of CO_2/mol of air or %) and L is leaf area (m^2). u_{pe}, the entering flow corrected for photosynthesis, is simply the air flow entering the chamber, at the water content at which it passes through the IRGA. u_{po} is the sum of the air flow plus the compensating flow, at the water content at which they pass through the IRGA.

In a closed system, net photosynthesis is given by

$$A_n = \frac{[c_b(P_b/T_b) - c_f(P_f/T_f)]V}{\Delta t L P_v/T_v} \quad [\text{mol m}^{-2}\ \text{s}^{-1}] \quad (11.31)$$

where c_b is the mole fraction of CO_2 in air at the start of the measurement, P_b is pressure in the system at the start of the measurement (Pa), T_b is temperature in the system at the start of the measurement (K), c_f is the mole fraction of CO_2 in the air at end of the measurement, P_f is the pressure in the system at end of the measurement (Pa), T_f is the temperature in system at the end of the measurement (K), V is the system volume under conditions of P_v and T_v (mol), Δt is time between the start and end of the measurement (s), L is the leaf area (m^2), P_v is pressure at time of volume determination (Pa) and T_v is temperature at the time of volume determination (K). If the air is not dried before passing through the IRGA, the IRGA readings must be corrected for water-vapor sensitivity.

11.7.6 Intercellular CO_2

Intercellular CO_2 (c_i) (μmol mol^{-1}) is given by

$$c_i = \frac{\left(\left(g_{tc} - \frac{E}{2}\right)c_a\right) - A_n}{\left(g_{tc} + \frac{E}{2}\right)} \quad [\text{mol mol}^{-1}] \quad (11.32)$$

where g_{tc} is total conductance to CO_2 (mol m^{-2} s^{-1}), E is transpiration (mol m^{-2} s^{-1}), c_a is the mole fraction of CO_2 in the ambient air (mol of CO_2/mol of air) and A_n is net photosynthesis (mol m^{-2} s^{-1}).

11.8 LIST OF SYMBOLS

A_c	net photosynthesis per unit of ground area	g m^{-2} s^{-1}
A_i	net photosynthesis per unit of leaf area in an isotope system	mol m^{-2} s^{-1}
A_n	net photosynthesis per unit of leaf area	mol m^{-2} s^{-1}
A'_n	net photosynthesis	mol s^{-1}
B	specific activity of plant material exposed to $^{14}CO_2$	Bq m^{-2}
c_a	CO_2 concentration in the bulk air	mol mol^{-1}
c_b	CO_2 concentration at the start of a closed system measurement	mol mol^{-1}
c_c	CO_2 concentration in injection gas in a compensating system	mol mol^{-1}

c_e	CO_2 concentration in the air entering a gas exchange chamber	mol mol^{-1}
c_f	CO_2 concentration at the end of a closed system measurement	mol mol^{-1}
c_i	CO_2 concentration in the leaf intercellular spaces	mol mol^{-1}
c_m	CO_2 concentration as a mole fraction	mol mol^{-1}
c_r	CO_2 concentration in a reference gas	mol mol^{-1}
c_s	CO_2 concentration in the sample cell of an IRGA	mol mol^{-1}
D	IRGA output due to the difference between the gases in the two cells	
D_d	IRGA output when the two cells contain the two different gases	
D_s	IRGA output when the two cells contain gas	
E	transpiration	
g_{bc}	boundary layer conductance to CO_2	mol m^{-2} s^{-1}
g_{bl}	boundary layer conductance of leaf lower surface	mol m^{-2} s^{-1}
g_{bu}	boundary layer conductance of leaf upper surface	mol m^{-2} s^{-1}
g_{bw}	boundary layer conductance to water vapor	mol m^{-2} s^{-1}
g_{lc}	leaf conductance to CO_2	mol m^{-2} s^{-1}
g_{ll}	leaf conductance of lower leaf surface	mol m^{-2} s^{-1}
g_{lu}	leaf conductance of upper leaf surface	mol m^{-2} s^{-1}
g_{lw}	leaf conductance to water vapor	mol m^{-2} s^{-1}
g_{tc}	total (leaf and boundary layer) conductance to CO_2	mol m^{-2} s^{-1}
g_{tw}	total (leaf and boundary layer) conductance to water vapor	mol m^{-2} s^{-1}
K_c	eddy diffusivity for CO_2	m^2 s^{-1}
L	leaf area	m^2
n	number of moles	
P	ambient total pressure	Pa
P_b	system pressure at the start of a closed system measurement	Pa
P_f	system pressure at the end of a closed system measurement	Pa
P_{mc}	mean CO_2 partial pressure in the two IRGA cells	Pa
P_{rc}	CO_2 partial pressure in an IRGA reference cell	Pa
P_{sc}	CO_2 partial pressure in an IRGA sample cell	Pa
P_v	system pressure at the time of volume determination	
$\boldsymbol{R}$	universal gas constant (8.311 m^3 Pa mol^{-1} K^{-1})	
RH	relative humidity	
S	specific activity of the gas in an isotope system	Bq mol^{-1}
T	absolute temperature	K
T_b	system temperature at the start of a closed system measurement	K
T_d	dew-point temperature	K
T_f	system temperature at the end of a closed system measurement	K
T_s	steam temperature at standard temperature (373.16 K)	
T_v	system temperature at the time of volume determination	K
u_c	flow through injection meter in a compensating system	mol s^{-1}
u_e	total flow entering a gas-exchange chamber	mol s^{-1}
u_o	total flow leaving a gas-exchange chamber	mol s^{-1}
u_{pe}	air flow entering chamber, corrected for photosynthesis	mol s^{-1}
u_{po}	air flow leaving chamber, corrected for photosynthesis	mol s^{-1}
u_r	flow of a reference gas into a mixing system	mol s^{-1}
V	volume of 1 mole (m^3) or of a closed gas-exchange system (moles)	
v_a	ambient water vapor partial pressure	Pa

v_{sat}	saturation water vapor partial pressure	Pa
w_a	ambient water vapor concentration	mol mol^{-1}
w_b	water vapor concentration at start of a closed system measurement	mol mol^{-1}
w_e	water vapor concentration in air entering a gas-exchange chamber	mol mol^{-1}
w_f	water vapor concentration at end of a closed system measurement	mol mol^{-1}
w_i	water vapor concentration in leaf intercellular air spaces	mol mol^{-1}
w_m	mean of water vapor concentration in ambient air and inside leaf	mol mol^{-1}
w_o	water vapor concentration in air leaving a gas-exchange chamber	mol mol^{-1}
w_s	water vapor concentration in IRGA sample cell	mol mol^{-1}
w_{sat}	water vapor concentration of air at saturation	mol mol^{-1}
y	sensitivity of IRGA gain to background CO_2	
z	zero offset of the IRGA shift in gain with background	
Δc	CO_2 differential at the IRGA as a mole fraction	mol mol^{-1}
ΔP_c	CO_2 differential at the IRGA as a partial pressure	Pa
Δt	time between start and conclusion of a closed system measurement	s
Δx	change in total activity of the gas in an isotope system	Bq
ε	counting efficiency in an isotope system	
ρ_c	atmospheric density of CO_2	g m^{-3}

REFERENCES

Anderson, D.E., Verma, S.B., Clement, R.J., Baldocchi, D.D. and Matt, D.R. (1986) Turbulence spectra of CO_2, water vapor, temperature and velocity over a deciduous forest. *Agric. For. Meteor.*, **38**, 81–99.

Atkinson, C.J. Winner, W.E. and Mooney, H.A. (1986) A field portable gas-exchange system for measuring carbon dioxide and water vapour exchange rates of leaves during fumigation with SO_2. *Plant Cell. Environ.*, **9**, 711–19.

Baldocchi, D.D., Verma, S.B. and Anderson, D.E. (1987) Canopy photosynthesis and water-use efficiency in a deciduous forest. *J. Appl. Ecol.*, **24**, 251–60.

Ball, J.T. (1987) Calculations related to gas exchange. In *Stomatal Function* (eds E. Zeiger, G.D. Farquhar and I. Cowan), Stanford University Press, Stanford, pp. 445–76.

Bazzaz, F.A. and Boyer, J.S. (1972) A compensating method for measuring carbon dioxide exchange, transpiration, and diffusive resistances of plants under controlled environmental conditions. *Ecology*, **53**, 343–9.

Bingham, G.E. Coyne, P.I., Kennedy, R.B. and Jackson, W.L. (1980) Design and fabrication of a portable minicuvette system for measuring leaf photosynthesis and stomatal conductance under controlled conditions. Lawrence Livermore National Laboratory, Livermore CA, UCRL-52895.

Björkman, O., Nobs, M., Berry, J., Mooney, H., Nicholson, F. and Catanzaro, B. (1973) Physiological adaptation to diverse environments: Approaches and facilities to study plant responses to contrasting thermal and water regimes. *Carnegie Inst. Wash. Ybk.*, **72**, 393–403.

Blacklow, W.M. and Maybury, K.G. (1980) A battery-operated instrument for non-destructive measurements of photosynthesis and transpiration of ears and leaves of cereals using $^{14}CO_2$ and a lithium chloride hygrometer. *J. Exp. Bot.*, **31**, 1119–29.

Bloom, A.J., Mooney, H.A., Björkman, O. and Berry, J.A. (1980) Materials and methods for carbon dioxide and water exchange analysis. *Plant Cell Environ.*, **3**, 371–6.

Bosian, G. (1955) Über die Vollautomatisierung der CO_2-Assimilations-bestimmung und zur methodik des küvettenklemas. *Planta*, **45**, 470–92.

Bosian, G. (1960) Zum Kuvettenklimaproblem: Beweisführung für die Nichtexistenz 2-gipfeliger Assimilationskurven bei Verwendung von Klimatisierten Küvetten. *Flora*, **149**, 167–88.

Bosian, G. (1965) Control of conditions in the plant chamber: Fully automatic regulation of wind velocity, temperature and relative humidity to conform to microclimatic field conditions. In *Methodology of Plant Eco-physiology* (ed. F.E. Eckardt), UNESCO, Paris, pp. 233–8.

Caemmerer, S., von and Farquhar, G.D. (1981) Some relationships between the biochemistry of photosynthesis and the gas exchange of leaves. *Planta*, **153**, 376–87.

Caldwell, M.M., Dean, T.J., Novak, R.S., Dzurec, R.S. and Richards, J.H. (1983) Bunchgrass architecture, light interception, and water-use efficiency: Assessment by fiber optic point quadrats and gas exchange. *Oecologia*, **59**, 178–84.

Calvin, M. and Benson, A.A. (1948) The path of carbon in photosynthesis. *Science*, **107**, 476–80.

Chazdon, R.L. and Pearcy, R.W. (1986) Photosynthetic responses to light variation in rainforest species. I. Induction under constant and fluctuating light conditions. *Oecologia*, **69**, 517–23.

Clegg, M.D., Sullivan, C.Y. and Eastin, J.D. (1978) A sensitive technique for the rapid measurement of carbon dioxide concentrations. *Plant Physiol.*, **62**, 924–6.

Coombs, J., Hall, D.O., Long S.P. and Scurlock, J.M.O. (1985) *Techniques in Bioproductivity and Photosynthesis*, 2nd edn, Pergamon Press, Oxford, 298 pp.

Cowan, I.R. (1977) Stomatal behaviour and environment. *Adv. Bot. Res.*, **4**, 117–228.

Dixon, M. and Grace, J. (1982) Water uptake by some chamber materials. *Plant Cell Environ.*, **5**, 323–7.

Eckardt, F.E. (1966) Le principe de la soufflerie aerodynamique climatisee appliqué a l'etude des echanges gazeux de la couverture vegetale. *Oecol. Plant.*, **1**, 369–99.

Ehleringer, J. and Cook, C.S. (1980) Measurements of photosynthesis in the field: Utility of the CO_2 depletion technique. *Plant Cell Environ.*, **3**, 479–82.

Ehleringer, J. and Pearcy, R.W. (1983) Variation in quantum yield for CO_2 uptake among C_3 and C_4 plants. *Plant Physiol.*, **73**, 555–9.

Farquhar, G.D. and Sharkey, T.D. (1982) Stomatal conductance and photosynthesis. *Annu. Rev. Plant Physiol.*, **33**, 317–45.

Field, C. Berry, J.A. and Mooney, H.A. (1982) A portable system for measuring carbon dioxide and water vapour exchanges of leaves. *Plant Cell Environ.*, **5**, 179–86.

Field, C.B. and Mooney, H.A. (1989) Measuring photosynthesis under field conditions – Past and present approaches. In *Instruments in Physiological Plant Ecology* (eds P.J. Kramer, B.R. Strain, S. Funada and Y. Hashimoto), Academic Press, London (in press).

Griffiths, J.H. and Jarvis, P.G. (1981) A null balance carbon dioxide and water vapour porometer. *J. Exp. Bot.*, **32**, 1157–68.

Hack, H.R.B. (1980) The uptake and release of water vapour by the foam seal of a diffusion porometer as a source of bias. *Plant Cell Environ.*, **5**, 53–7.

Hilbert, D.W., Prudhomme, T.I. and Oechel, W.C. (1987) Response of tussock tundra to elevated carbon dioxide regimes: Analysis of ecosystem CO_2 flux through modelling. *Oecologia*, **72**, 446–72.

Hutchison, B.A. and Hicks, B.B. (eds) (1985) *The Forest–Atmosphere Interaction*, Reidel, Dordrecht, 684 pp.

Incoll, L.D. (1977) Field studies of photosynthesis. Monitoring with $^{14}CO_2$. In *Environmental Effects on Crop Physiology* (eds J.J. Landsberg and C.V. Cutting), Academic Press, London.

Janác, J., Catsky, J., Brown, K.J. and Jarvis, P.G. (1971) Gas handling system. In *Plant Photosynthetic Production: Manual of Methods*, (eds Z. Šesták, J. Čatský and P.G. Jarvis), Junk, The Hague, pp. 132–48.

Jarman, P.D. (1974) The diffusion of carbon dioxide and water vapour through stomata. *J. Exp. Bot.*, **25**, 927–36.

Jarvis, P.G. and Sandford, A.P. (1985) The measurement of carbon dioxide in air. In *Instrumentation for Environmental Physiology*, (eds B. Marshall and F.I. Woodward), Cambridge University Press, Cambridge, pp. 29–57.

Johnson, H.B., Rowlands, P.G. and Ting, I.P. (1979) Tritium and carbon-14 double isotope porometer for simultaneous measurements of transpiration and photosynthesis. *Photosynthetica*, **13**, 409–18.

Kanemasu, E.T., Wesley, M.L., Hicks, B.B. and Heilman, J.L. (1979) Techniques for calculating energy and mass fluxes. In *Modification of the Aerial Environment of Plants* (eds B.J. Barfield and J.F. Gerber), American Society of Agricultural Engineering, St Joseph, Michigan, pp. 156–82.

Karlsson, S. and Sveinbjörnsson, B. (1981) Methodological comparison of photosynthetic rates measured by the $^{14}CO_2$ technique and infrared gas analysis. *Photosynthetica*, **15**, 447–52.

Keeling, C.D., Bacastow, R.B. and Whorf, T.P. (1982) Measurements of the concentration of carbon dioxide at Mauna Loa Observatory, Hawaii. In *Carbon Dioxide Review: 1982* (ed. W.C. Clark), Clarendon Press, Oxford, pp. 377–85.

Koch, W., Lange, O.L. and Schulze, E.-D. (1971) Ecophysiological investigations on wild and cultivated plants in the Negev Desert. I. Methods: A mobile laboratory for measuring carbon dioxide and water vapour exchange. *Oecologia*, **8**, 296–309.

Kramer, P.J., Strain, B.R., Funada, S. and Hashimoto, Y. (eds) (1989) *Scientific Instruments in Physiological Plant Ecology*, Academic Press, Orlando, in press.

Küppers, M., Swan, A.G., Tomkins, D., Gabriel, W.C.L., Küppers, B.I.L. and Linder, S. (1987) A field portable system for the measurement of gas exchange of leaves under natural and controlled conditions: examples with field-grown *Eucalyptus pauciflora* Sieb. ex Spreng. ssp. *pauciflora*, *E. behriana* F. Muell. and *Pinus radiata* R. Don. *Plant Cell Environ.*, **10**, 425–35.

Lange, O.L. (1962) Eine 'Klapp-Küvette' zur CO_2-Gaswechselregistrierung an Blättern von Freilandpflanzen mit dem URAS. *Ber. Dtsch. Bot. Ges.*, **75**, 41–50.

Lange, O.L., Koch, W. and Schulze, E.-D. (1969) CO_2-Gaswechsel und Wasserhaushalt von Pflanzen in der Negev-Wüste am ende der Trockenzeit. *Ber. Dtsch. Bot. Ges.*, **82**, 39–61.

Leuning, R. (1983) Transport of gases into leaves. *Plant Cell Environ.*, **6**, 181–94.

Leverence, J.W. and Öquist, G. (1987) Quantum yields of photosynthesis at temperatures between −2° C and 35° C in a cold-tolerant C_3 plant (*Pinus sylvestris*) during the course of one year. *Plant Cell Environ.*, **10**, 287–95.

Long, S.P. and Hallgren, J.-E. (1985) Measurement of CO_2 assimilation by plants in the field and the laboratory. In *Techniques in Bioproductivity and Photosynthesis* (eds J. Coombs, D.O. Hall, S.P. Long and J.M.O. Scurlock), 2nd edn, Pergamon Press, Oxford, pp. 62–94.

Ludwig, L.J. and Canvin, D.T. (1971) An open gas-exchange system for the simultaneous measurement of the CO_2 and $^{14}CO_2$ fluxes from leaves. *Can. J. Bot.*, **49**, 1299–133.

Luft, K.F. (1943) Über eine neues Methode der registrierenden Gasanalyse mit Hilfe der Absorption ultrarot Strahlen ohne spectrale Zerlegung. *Zeitschr. Tech. Phys.*, **24**, 97–104.

Luft, K.F., Kesseler, G. and Zorner, K.H. (1967) Nicht dispersive Ultrarot-Gasanalyse mit dem UNOR. *Chemie Ingenieur Technik*, **39**, 937–45.

Marshall, B. and Woodward, F.I. (eds) (1985) *Instrumentation for Environmental Physiology*, Cambridge University Press, Cambridge, 238 pp.

McAlister, E.D. (1937) Spectrographic method for determining the carbon dioxide exchange between an organism and its surroundings. *Plant Physiol.*, **12**, 213–15.

McKree, K.J. (1986) Measuring whole-plant daily carbon balance. *Photosynthetica*, **20**, 82–93.

Monteith, J.L. (1976) *Vegetation and the Atmosphere. Vol. 2 Case Studies*, Academic Press, London, p. 437.

Mooney, H.A., Björkman, O., Ehleringer, J. and Berry, J. (1976) Photosynthetic capacity of *in situ* Death Valley plants. *Carnegie Inst. Wash. Ybk.*, **75** 410–13.

Mooney, H.A., Dunn, E.L., Harrison, A.T., Morrow, P.A., Bartholomew, B. and Hays, R.L. (1971) A mobile laboratory for gas exchange measurements. *Photosynthetica*, **5**, 128–32.

Moore, R.T., Ehleringer, J. Miller, P.C., Caldwell, M.M. and Tieszen, L.L. (1973) Gas exchange studies of four alpine tundra species at Niwot Ridge, Colorado, In *Primary Production and Population Processes, Tundra Biome* (eds L.C. Bliss and F. Wielgolaski), University of Alberta, Edmonton, pp. 211–17.

Moreshet, S., Koller, D. and Stanhill, G. (1968) The partitioning of resistances to gaseous diffusion in the leaf epidermis and the boundary layer. *Ann. Bot.*, **32**, 695–701.

Musgrave, R.B. and Moss, D.N. (1961) Photosynthesis under field conditions. I. A portable, closed system for determining net assimilation and respiration of corn. *Crop Sci.*, **1**, 37–41.

Nobel, P.S. (1983) *Biophysical Plant Physiology and Ecology*, W.H. Freeman, San Francisco, 608 pp.

Oechel, W.C. and Lawrence, W.T. (1979) Energy utilization and carbon metabolism in Mediterranean scrub vegetation of Chile and California. I. Methods: A transportable cuvette field photosynthesis and data acquisition system and representative results for *Ceanothus gregii*. *Oecologia*, **39**, 321–36.

Parkhurst, D.F. (1986) Internal leaf structure: A three dimensional perspective. In *On the Economy of Plant Form and Function* (ed. T.J. Givnish), Cambridge University Press, Cambridge, pp. 215–49.

Parkinson, K.J. (1985) A simple method for determining the boundary layer resistance in leaf cuvettes. *Plant Cell Environ.*, **8**, 223–6.

Parkinson, K.J. and Day, W. (1979) The use of orifices to control the flow rate of gases. *J. Appl. Ecol.*, **16**, 623–32.

Parkinson, K.J. and Day, W. (1981) Water vapour calibration using salt hydrate transitions. *J. Exp. Bot.*, **32**, 411–18.

Parkinson, K.J. and Legg, B.J. (1978) Calibrations of infrared gas analysers for carbon dioxide. *Photosynthetica*, **12**, 65–7.

Pearcy, R.W. and Calkin, H.W. (1983) Carbon dioxide exchange of C_3 and C_4 tree species in the understory of a Hawaiian forest. *Oecologia*, **58**, 26–32.

Pearcy, R.W., Osteryoung, K. and Calkin, H.W. (1985) Photosynthetic responses to dynamic light environments by Hawaiian trees: Time course of CO_2 uptake and carbon gain during sunflecks. *Plant Physiol.*, **79**, 896–902.

Penning de Vries, F.W.T., Akkersdijk, J.W.J. and van Oorschot, J.L.P. (1984) An error in measuring respiration and photosynthesis due to transpiration. *Photosynthetica*, **18**, 146–9.

Revelle, R. and Suess, H.E. (1957) Carbon dioxide exchange between atmosphere and ocean and the question of an increase in atmospheric CO_2 during the past decades. *Tellus*, **9**, 18.

Richards, J.M. (1971) Simple expression for the saturation vapor pressure of water in the range −50 degrees to 140 degrees. *Br. J. Appl. Phys.*, **4**, L15–L18.

Samish, Y.B. (1978) Measurement and control of CO_2 concentration in air is influenced by the desiccant. *Photosynthetica*, **12**, 73–5.

Schulze, E.-D., Hall, A.E., Lange, O.L. and Walz, H. (1982) A portable steady-state porometer for measuring the carbon dioxide and water vapour exchanges of leaves under natural conditions. *Oecologia*, **53**, 141–5.

Šesták, Z., Čatský, J. and Jarvis, P.G. (1971) *Plant Photosynthetic Production: Manual of Methods*, Junk, The Hague, 800 pp.

Sharkey, T.D. (1985) O_2-insensitive photosynthesis in C_3 plants: Its occurrence and a possible explanation. *Plant Physiol.*, **78**, 71–5.

Sharkey, T.D., Imai, K., Farquhar, G.D. and Cowan, I.R. (1982) A direct confirmation of the standard method of estimating intercellular partial pressure of CO_2. *Plant Physiol.*, **69**, 657–9.

Shimshi, D. (1969) A rapid field method for measuring photosynthesis with labelled carbon dioxide. *J. Exp. Bot.*, **20**, 381–401.

Thom, A.S. (1975) Momentum, mass, and heat exchange of plant canopies. In *Vegetation and the Atmosphere. Vol. 1 Principles* (ed. J.L. Monteith), Academic Press, London, pp. 57–110.

Voznesenskiĭ, V.L., Zalenskiĭ, O.V. and Austin, R.B. (1971) Methods of measuring rates of photosynthesis using carbon-14 dioxide. In *Plant Photosynthetic Production: Manual of Methods* (eds Z. Sesták, J. Catsky and P.G. Jarvis), Junk, The Hague, pp. 276–93.

Williams, B.A., Gurner, P.J. and Austin, R.B. (1982) A new infrared gas analyzer and portable photosynthesis meter. *Photosyn. Res.*, **3**, 141–51.

Wong, S.C., Cowan, I.R. and Farquhar, G.D. (1979) Stomatal conductance correlates with photosynthetic capacity. *Nature, London*, **282**, 424–6.

Woodrow, I.E., Ball, J.T. and Berry, J.A. (1987) A general expression for the control of the rate of photosynthetic CO_2 fixation by stomata, the boundary layer and radiation exchange. In *Progress in Photosynthesis Research* (ed. J. Biggins), Martinus Nijhoff, the Netherlands, pp. 225–8.

12
Crassulacean acid metabolism

C. Barry Osmond, William W. Adams III and Stanley D. Smith

12.1 INTRODUCTION

Plants with crassulacean acid metabolism (CAM) are rarely the most abundant in plant communities, and rarely attain high biomass, but they are capable of an extraordinary array of physiological activities in a wide range of environments. The peculiar morphology and nocturnal physiology of CAM plants have attracted the curiosity of plant biologists for many years. The most comprehensive syntheses of the environmental biology of desert CAM plants are those of Gibson and Nobel (1986) and Nobel (1988). Epiphytic CAM plants are dealt with by Smith *et al.* (1986a) and aquatic CAM plants by Boston and Adams (1986). Recent reviews of the metabolic activities of CAM plants have delineated phases of the complex nocturnal diurnal metabolic cycle and component biochemical and physiological events (Kluge and Ting, 1978; Osmond, 1978; Osmond and Holtum, 1981). However, a decade or so of investigation of these plants in many habitats has uncovered a bewildering array of exceptions and variations. In response to this we now have a new set of terminologies (Cockburn, 1985), pleas that 'more representative studies are needed from diverse taxa before a generalized theory of CAM will be forthcoming' (Ting, 1985) and for the need to integrate ecological, physiological and biochemical studies (Lüttge, 1987). Obviously, further studies of the physiological ecology of CAM plants are likely to be most rewarding and this chapter seeks to outline appropriate methodologies for this research. For the most part we will be concerned with modifications of techniques to meet the special problems posed by CAM plant morphology and physiology.

CAM plants are best defined as those in which CO_2 fixation into malic acid in the dark contributes, after decarboxylation in the light, to net CO_2 fixation in the light. The detailed biochemistry of CAM plants has been reviewed recently (Osmond and Holtum, 1981; Winter, 1985) and need not concern us unduly here, but comparative aspects are relevant, (Osmond *et al.*, 1982). Briefly, the carboxylation processes which result in nocturnal accumulation of malic acid in CAM plants are analogous to those responsible for primary CO_2 fixation in mesophyll cells of C_4 plants during photosynthesis. The decarboxylation processes in CAM plants which make CO_2 available from malic acid in the light are also analogous to those which mediate the secondary CO_2 fixation in bundle sheath cells during C_4 photosynthesis. In CAM plants, as in C_3 and C_4 plants, net photosynthesis ultimately depends on the photosynthetic carbon reduction cycle. In CAM, as in C_4 photosynthesis, the concerted action of primary carboxylation and subsequent decarboxylation represents a CO_2-concentrating mechanism which overcomes the O_2 inhibition of photosynthesis observed in C_3 plants. The physiological bases of the CO_2 concentrating mechanisms in CAM plants depend on temporal separation of carboxylation and decarboxylation, tonoplast acid transport and stomatal closure in the light. This suite of properties is distinct from that in C_4 plants, which depends on spatial separation of carboxylation and decarboxylation, symplastic transport and relatively slow diffusion of gases in solution.

The extent of malic acid contribution to photosynthetic CO_2 fixation in CAM plants varies enormously with stage of development, environmental conditions and between species. Although it is a relatively straightforward matter to estimate the amount of malic acid accumulated at night, it is not always evident that the subsequent transfer of CO_2 to photosynthesis is quantitative, and it is often difficult to quantify other CO_2 exchanges. From an ecophysiological viewpoint, the ancillary physiological consequences of CAM, such as stomatal closure and exaggerated energy budgets, are at least as important as carbon budgets. It is fair to say that we have only a rudimentary appreciation of the adaptive significance of CAM in these terms. This is not surprising, for although CAM is found in massive desert succulents which endure temperatures from below freezing to above 60°C, high light and low water availability, it is also common in tropical forest epiphytes exposed to benign temperatures and low light (McWilliams, 1970; Winter *et al.*, 1983), and CAM has also been described in submerged aquatic plants (Keeley and Morton 1982).

12.2 MEASUREMENT OF SUCCULENCE

Gibson (1982) has reviewed many aspects of the anatomy of stem and leaf succulents, giving particular attention to the hypodermal organization of stem succulents. For many purposes, simple measurements of leaf or frond thickness are a useful index of succulence in CAM plants. Teeri *et al.* (1981) found a good correlation between the expression of CAM and leaf thickness among *Sedum* species and hybrids, and Winter *et al.* (1983) reported similar relationships among epiphytic orchids and ferns. However, the thickness of a leaf can vary due to conditions during growth such as light and water status, as can the expression of CAM. Not all CAM plants are succulent in the general sense, and the term is so imprecise that it needs particular definition if it is to be applied to CAM plants.

Kluge and Ting (1978) discuss several criteria for assessing succulence and favor an index of mesophyll succulence S_m, where

$$S_m = \frac{\text{water content (g)}}{\text{chlorophyll content (mg)}}$$

This index emphasizes that it is the succulence of chloroplast-containing tissues which is important in CAM. Quantitative use of this method depends on accurate estimation of chlorophyll which can be difficult in tissues containing high concentrations of free acid. Thus samples for chlorophyll estimation should be collected in the afternoon, or if this is impracticable, a method of chlorophyll estimation based on total conversion to pheophytin (Vernon, 1960) should be used.

A better index of succulence is the ratio of mesophyll cell surface area (A_{mes}) to leaf surface area (A), measured as described by Nobel *et al.* (1975). Nobel and Hartsock (1978) found $A_{mes}/A = 82$ for *Agave deserti*, compared with an average of about 15 for leaves of a C_3 plant. The index A_{mes}/A is more useful because it is directly applicable to calculations of internal resistances to CO_2 diffusion. Nobel and Hartsock suggest that the high value of A_{mes}/A is important in maintaining high liquid phase conductance (low resistance) in these plants with low stomatal conductances but high biochemical conductances to CO_2 exchange (depending on the malic acid status of the tissue).

12.3 NOCTURNAL ACIDIFICATION

A fundamental criterion for characterization of a CAM plant is evidence that chloroplast containing tissues show a nocturnal increase in free malic acid concentration with a stoichiometry of malate^{2-} to $2H^+$ of unity. Although many early observers relied on their sense of taste to distinguish CAM plants (Heyne, 1815), the presence of acids other than malic acid, interference from tannins and the presence of toxins make this a potentially hazardous and unreliable method. By and large, leaf tissues with high levels of free acids other than malic acid (e.g. oxalic acid) tend not to accumulate acid at night or to show other day–night variations in acid (Crombie, 1952). In some CAM plants organic acids other than malic acid have been observed to accumulate at night (Milburn *et al.*, 1968; Pucher *et al.*, 1949; von Willert *et al.*, 1977), but it is not clear whether these are present as the free acid and contribute to titratable acidity. High background malate, which is found in leaves of many CAM and other plants (Table 12.1), is usually associated with ionic balance (Osmond, 1976a). Although the stoichiometry of malate^{2-} and $2H^+$ is rarely measured, studies of leaf and stem succulents (Lüttge and Ball, 1974; Medina and Osmond, 1981; Nobel and Hartsock, 1983; Osmond *et al.*, 1979a) confirm that it is unity in active CAM plants. This stoichiometry is difficult to establish in plants which display only small nocturnal increases in acidity, especially when background malate levels are high (e.g. Keeley *et al.*, 1984).

For most purposes, cold water extraction of tissues by grinding in a glass homogenizer, or with sand in a mortar, is a satisfactory method of malic acid extraction. If immediate titration of extracted tissue is impracticable, tissue samples can be collected in large numbers and stored in 80% ethanol, on dry ice or in liquid N_2. Tissue collected in the field may be killed by heating at 90°C (von Willert *et al.*, 1985), but because malic acid is not very

Table 12.1 Stoichiometry of acidity and malate during CAM, measured in boiling 80% ethanol extracts as described in the text. In both species the nocturnal increase in acid (equivalents) shows a stoichiometry of 2 compared to nocturnal increase in malate (mol). However, in *Cereus*, residual malate equals about 50% of the nocturnal increase, and this does not show the same stoichiometry to residual acid, indicating pools of malate which are not involved in CAM. In *Sedum* the residual is very small, and shows the same stoichiometry as the nocturnal increase (unpublished data, C.B. Osmond)

Tissue and time	*Acid* ($\mu Eq\ g^{-1}$ *fresh wt*)	*Malate* ($\mu mol\ g^{-1}$ *fresh wt*)	$\Delta Acid/\Delta malate$
Cereus sp., am	92.0	55.9	
pm	22.1	18.9	1.17
Δ=am−pm	69.9	37.0	1.89
Sedum sp., am	201.0	101.7	
pm	10.0	5.1	1.96
Δ=am−pm	191.0	96.6	1.98

stable at higher temperatures, especially if tissues are dried, this method is not recommended. If stored in 80% ethanol the sample should be boiled and further extracted in boiling water (malic acid is not very soluble in ethanol), and frozen or dried samples should be extracted in boiling water, taking care to avoid evaporation to dryness. The cooled combined extracts should be titrated with freshly prepared standardized alkali (0.01–0.05 N) using phenolphthalein as an indicator, or a pH meter. The presence of free malic acid in extracts converts chlorophyll (green) to pheophytin (green–brown), which serves as an internal indicator of acidification to some extent. Malic acid has a pK_1 of 3.4 and a pK_2 of 5.1 at 25°C, so titration to pH 6.5 or 7.0 is adequate.

The most convenient method for specific determination of malate is by enzymatic analysis using malate dehydrogenase, and the following methods are modified after Hohorst (1965). Extraction of tissues by any of the above methods is satisfactory, but care should be taken that samples are not excessively diluted. A final concentration of malate in the extract ranging from 0.5 to 5 mM is desirable and this can usually be obtained with the extraction of about 1 g of tissue in 25 ml of H_2O. The assay buffer and malate standard should be prepared freshly. Glycine–hydrazine sulfate buffer containing 7.5 g of glycine, 5.2 g hydrazine sulfate, and 0.2 g of EDTA is dissolved in water, titrated to pH 9.5 with NaOH and made to 100 ml. The malate standard (10 mM) contains 13.41 mg of L-malic acid in 10 ml of water. Pig heart malate dehydrogenase [EC 1.1.1.37: e.g. Sigma no. M9004 or Boehringer no. 127 256 in about 3 M $(NH_4)_2SO_4$] is diluted with 3 M $(NH_4)_2SO_4$ to yield a solution containing approximately 500 units ml^{-1}. This solution, and the NAD solution (60 mM; 40 mg ml^{-1}), are stable for some weeks if stored at 4°C. A standard curve is prepared by mixing 2.8 ml of buffer, 5–50 μl (0.05–0.5 μmol) of malic acid standard, 30 μl (1.8 μmol) of NAD and 30 μl of malate dehydrogenase (15 units) with water to make 3.0 ml. The blank contains 140 μl of water. These assays are incubated at 25–28°C for 1 h and the absorptance change due to NADH production is measured at 340 nm with a UV spectrophotometer. The malate content of tissue extracts containing 0.5–5 mM malate is measured using 10–100 μl of extract in place of the standard malate. The volume added should give an absorptance change of about 0.2 unit. Optical interference is controlled by preparing a blank for each sample, containing the same volume of tissue extract but no malate dehydrogenase. If chemical or enzymatic interference from some component of the tissue extract is suspected, an internal standard, of 10 μl (0.1 μmol) of malate, should be used. Data obtained by these methods are shown in Table 12.1.

Should these latter steps prove insufficient to overcome interfering compounds, preliminary purification of the organic acid fraction may be necessary. An aliquot of the extract is passed through a 5 mm diameter (2 ml) column of freshly prepared Dowex 1 × 10 formate ion exchange resin (100–200 mesh), the column washed with 10 ml of H_2O, and malic acid eluted with 20 ml of 2 N formic acid. The eluate is then neutralized, made to volume and the malate content estimated as above. For some purposes, estimation of malate by quantitative gas chromatography may be preferred (Medina *et al.*, 1977), and the methods of Nierhaus and Kinzel (1971) are recommended.

12.4 NOCTURNAL CO_2 FIXATION

Medina and Troughton (1974) surveyed tropical bromeliads for nocturnal CO_2 fixation by suspending tissue segments (1 g fresh weight) over a bicarbonate indicator solution (10 mg l^{-1} Cresol Red) in a closed glass tube. Net CO_2 fixation in darkness was indicated by alkalinization of the solution, and net CO_2 evolution by acidification. This qualitative method is more convenient than field methods based on $^{14}CO_2$ fixation (Austin and Longdon, 1967; Szarek *et al.*, 1973) which Osmond *et al.* (1979a) concluded were, at best, qualitative indicators of stomatal opening. The principal difficulties associated with the $^{14}CO_2$ methods, which measure gross $^{14}CO_2$ fixation rather than net CO_2 exchange, include continued ^{14}C incorporation after the short-term exposure to calibrated $^{14}CO_2$, and the internal dilution of $^{14}CO_2$ by respiratory CO_2. With *Opuntia stricta* (formerly *inermis*) the $^{14}CO_2$ method overestimated the rate of dark CO_2 fixation by about 2-fold compared with that estimated from the rate of malic acid accumulation, presumably because $^{14}CO_2$ which entered the tissue in the 20 s exposure continued to be fixed during excision and slow freezing of the bulky tissue. More reliable results were obtained with less succulent *Tillandsia usneoides*. Martin *et al.* (1981) found good agreement between $^{14}CO_2$ fixation and acidification in spring and summer, but $^{14}CO_2$ fixation underestimated nocturnal CO_2 fixation in winter, when leaf tissues were continually wet. This may reflect CO_2 diffusion limitations in water films, high internal CO_2 concentrations, or both.

As discussed above malic acid accumulation is the best measure of nocturnal CO_2 fixation, irrespective of CO_2 source. The assumption of a stoichiometry of 1 mol of CO_2 fixed per mol of malic acid accumulated is justified by the known biochemistry of malic acid synthesis via the β-carboxylation of phosphoenolpyruvate (Cockburn and McAuley, 1975). These stoichiometries have rarely been established in practice. In leaf succulents such as *Kalanchoë daigremontiana* (Medina and Osmond, 1981) or well-watered stem succulents such as *Opuntia ficus-indica* (Nobel and Hartsock, 1983), the stoichiometry of mol of CO_2 fixed per mol of malic acid accumulated is 1 (or 1 : 2 per mol of H^+ accumulated) under a range of environmental conditions. Under water stress, however, the molar stoichiometry of CO_2 fixation and malic acid accumulation is less than 1, presumably because acid synthesis now depends to some extent on internal refixation of respiratory CO_2 (Fig. 12.1; Nobel and Hartsock, 1983; Osmond *et al.*, 1979a).

The magnitude of this refixation can be estimated by measuring changes in acidity of tissues kept in an airtight container in the dark (Medina and Osmond, 1981). It should be possible to assess the relative significance of internal refixation of CO_2 and the net fixation of external CO_2 by any CAM plant, under field conditions, by titration of tissues kept in air and in a small closed container. To our knowledge, this simple comparison is not yet available for any species or environment. Winter *et al.* (1986b) used flowing CO_2-free air which, depending on stomatal conductance, will tend to flush respiratory CO_2 from the system and underestimate refixation. Griffiths *et al.* (1986) recognized the importance of CO_2 recycling in epiphytic bromeliads and by comparing total noctural CO_2 uptake and malic acid content estimated 60–90% of net acidification was due to this source.

12.5 ANALYSIS OF DAY–NIGHT AND SEASONAL PATTERNS OF CO_2 AND H_2O VAPOR EXCHANGE

Nishida (1963) used a simple viscous flow porometer to demonstrate that stomata of CAM plants showed an inverted rhythm, being open in the dark and closed in the light. Neales (1975) identified four patterns of CO_2/H_2O vapor exchange which he described as nonCAM (no nocturnal CO_2 uptake, as found in well-watered C_3 or C_4 plants), weak-CAM (some nocturnal CO_2 uptake and a small morning depression of CO_2 uptake), full-CAM (predominantly nocturnal CO_2 uptake, pronounced morning depression of CO_2 uptake but some CO_2 uptake in the late afternoon), and super-CAM (negligible CO_2 uptake or water loss in the light). During prolonged drought under natural conditions

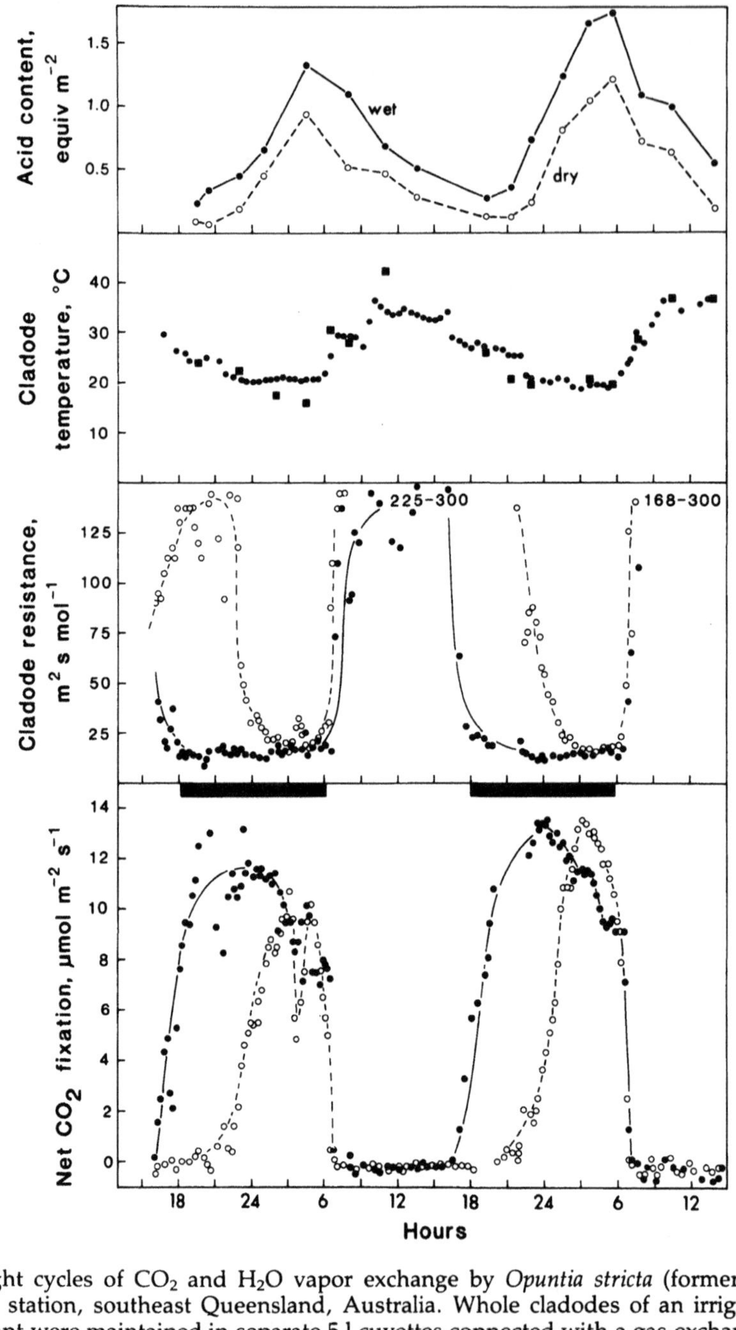

Fig. 12.1 Day–night cycles of CO_2 and H_2O vapor exchange by *Opuntia stricta* (formerly *inermis*) at Narayen research station, southeast Queensland, Australia. Whole cladodes of an irrigated (●) or a droughted (○) plant were maintained in separate 5 l cuvettes connected with a gas-exchange system in an adjacent mobile laboratory, designed by M.M. Ludlow and described in Osmond *et al.* (1979b). Temperature of adjacent unenclosed cladodes (■) was monitored and that of enclosed cladodes (●) was controlled manually by the temperature of cuvette air. Integration of curves for net CO_2 exchange shows that fixation of external CO_2 accounted for 70–100% of acid accumulated in irrigated plants, but only 41–53% of the acid accumulated in droughted plants.

stem succulents exhibit negligible CO_2 or H_2O vapor exchange with the environment yet continue to display a damped nocturnal accumulation of malic acid due to the internal refixation of respiratory CO_2 (Szarek *et al.*, 1973). Now termed CAM-idling, this pattern is distinguished from another, termed CAM-cycling, in which plants with normal C_3 or C_4 patterns of photosynthetic CO_2 exchange show nocturnal acidification in the absence of net CO_2 uptake in the dark (Ting, 1985). Osmond (1976b, 1978) delineated four phases of CO_2 and H_2O vapor exchange in CAM plants (full-CAM sensu Neales, 1975) for purposes of biochemical and physiological analysis. All of these patterns are susceptible to modification by genotype, plant development and environment.

In early laboratory studies whole plants were enclosed in simple chambers, and the integrated patterns of CO_2 and H_2O vapor exchange followed on plants with leaves of different age and light orientation. It is now evident that young cladodes on stem succulents have patterns of H_2O vapor exchange which differ from those of mature cladodes (Acevedo *et al.*, 1983), that some stem succulents have leaves which display C_3 photosynthesis rather than CAM (Lange and Zuber, 1977), and that CAM activity of leaf succulents depends very much on leaf age and environmental conditions (Jones, 1975; Winter *et al.*, 1978). Single-leaf chambers are thus to be preferred and were subsequently used to define the different phases of day–night CO_2–H_2O vapor exchange for physiological and biochemical analysis (Osmond and Björkman, 1975). Generally, flow-through systems have been employed with infrared gas analysers and dew-point hygrometers as sensors. Nobel and Hartsock (1978) used a null-point closed system, the principles of which are discussed in Chapter 11.

Quantitative estimates of day–night CO_2 fixation in CAM plants under field conditions require special attention to materials used for chambers and tubing. Nobel (1976) used a small leucite chamber affixed to *Agave* leaves by means of a nonhardening putty (Terostat; Teroson GmbH., Heidelberg) and connected by long Tygon tubing, both of which could present problems (see Chapter 11). More complex open systems, which permit tracking of leaf or cladode temperature to match external environment or adjacent unenclosed tissues, have been applied in studies of CAM (Lange *et al.*, 1975; Osmond *et al.*, 1979a). Fig. 12.1 shows two 24 h cycles of CO_2 and H_2O vapor exchange in *Opuntia* under field conditions. The patterns were followed simultaneously on irrigated and droughted plants (Osmond *et al.*, 1979a). In the latter studies the large chamber and long tubing made it difficult to provide flow rates sufficiently high to maintain realistically low relative humidities when large amounts of tissue were enclosed in such a system. The problems of gas-exchange measurements in humid situations, reviewed by Field and Mooney (1984), are likely to be amplified in the case of tropical CAM plants, in which rates of CO_2 and H_2O vapor exchange may be exceedingly slow. A detailed discussion of these difficulties, in the context of gas-exchange studies of epiphytes using the Walz portable steady-state porometer separated by 25 meters tubing between porometer and analyzer, is given by Lüttge *et al.* (1986a).

However, so far as we are aware, there have been few long-term studies to establish seasonal patterns of CO_2–H_2O vapor exchange in CAM plants under field conditions using gas-exchange systems. Lange *et al.* (1975) constructed spring through summer carbon budgets for *Caralluma* in the Negev based on daily CO_2-exchange profiles, and highlighted the daily carbon deficit in summer. Sale and Neales (1980) used large canopy chambers to measure 24 h patterns of CO_2 exchange in small plots of *Ananas* and thereby reconciled discrepancies between previous measurements of leaf CO_2 assimilation and crop growth rates. In stem succulents which do not engage in net CO_2 fixation in the light, nocturnal acidification is a reliable method for assessing net CO_2 exchange over a long period of time (Osmond *et al.*, 1979a; Szarek and Ting, 1974), and for establishing morphological and environmental influences on assimilation for subsequent photosynthetic processes and productivity modeling (Nobel, 1983). This is also the method of choice, so far, for long-term ecophysiological studies of submerged CAM plants (Boston and Adams, 1985, 1986).

It seems likely that commercially available field portable systems are also applicable to gas-exchange studies of CAM plants under some field conditions. Instantaneous measurements of CO_2 fixation, calculated from the rate of drawdown of CO_2 concentration in a small cuvette, can be obtained using a LI-COR 6000 CO_2 porometer (LI-COR, Lincoln, Nebraska 68504, USA), even at extremely low rates of net fixation in the dark (Winter *et al.*, 1986b). Lüttge *et al.* (1986a) describe the application of a Walz porometer (H. Walz, Mess- und Regeltechnik, D-8521 Effeltrich, BRD) to gas exchange of C_3 and CAM bromeliads in the tropics. It has yet to be established if these instruments are useful for CO_2 and H_2O flux measurements in CAM plants which have low stomatal conductances under high irradiance in desert conditions, when tissue temperature measurement and cuvette temperature changes are likely to introduce substantial errors (for further details see Chapters 8 and 11).

A double-isotope porometer which measures the diffusion and retention of ^{14}C and ^{3}H in tissues (Johnson *et al.*, 1979) has been applied to analyses of CO_2–H_2O vapor exchange in CAM plants (Rayder and Ting, 1983; Ting and Hanscom, 1977). Tissues are exposed for 20 s to a flowing stream of dry air labeled with $^{14}CO_2$ which is humidified by passage through $^{3}H_2O$. A disc of exposed tissue is removed and immediately placed in 80% methanol at dry ice temperature, and then subsequently counted. Although conductance calculated from $^{3}H_2O$ vapor diffusion shows reasonable agreement with conductance measured gravimetrically (Johnson *et al.*, 1979), other calibrations have not been made. This method potentially suffers from the drawback of both ^{14}C methods of photosynthetic measurement and of diffusion porometry, as discussed in Chapters 8 and 11.

Estimates of stomatal conductance based on the use of transit-time diffusion porometers or steady-state porometers may also be problematic. In some semiarid climates the energy budgets of CAM plants at night lead to dew formation during the more interesting phases of gas exchange (Osmond *et al.*, 1979a), making it impossible to use porometry. A principal limitation of transit-time diffusion porometers is that at low stomatal conductance, long periods of time are required for measurement. During high light conditions, this leads to rapid heating of the sensing systems, resulting in potentially large errors (Morrow and Slatyer, 1971). Shading of the sensor cup does not help much with massive succulents such as *Opuntia*, which may be 10° C or more above air temperature. Moreover, conductances can be lower than water desorption from the measurement system. Bartholomew and Kadzimin (1976) designed a Teflon porometer cup for use with *Ananas* leaves, and concluded that water desorption from Plexiglass cups was evidently responsible for a fourfold lower resistance measured with Plexiglass instruments. Many observations based on commercially available porometers, and the extensive extrapolations as to water loss made from these data, need to be viewed with caution. These problems may be less significant in arid than in humid environments. Steady-state porometers, with rapid response times, are susceptible to errors associated with low flow rate and humidity sensing at low humidity, and may not be much better. Moreover the configuration of commercially available instruments such as the LI-COR 1600 steady-state porometer is unsuited for most succulents.

Neales *et al.* (1968) were among the first to

draw attention to the lower water cost of CO_2 uptake at night (26–80 g of H_2O lost g^{-1} of CO_2 gained) compared to that in the light (110–210 g of H_2O lost g^{-1} of CO_2 gained) in succulents with CAM. These data were obtained from whole-plant cumulative CO_2 and H_2O vapor exchange analyses under controlled environmental conditions, and though widely touted, this attribute of CAM has not been widely assessed in ecophysiological studies, especially under field conditions. Meinzer and Rundel (1973) found reasonable correlations in glasshouse experiments between average transpiration ratio measured during mid-period dark CO_2 exchange and that measured by dry weight increase in growth, and water loss measured by pot weight. This must have been a fortuitous correlation because *Echeveria* used in these experiments lost considerable amounts of CO_2 during the day. Not surprisingly, growth was negatively correlated with transpiration ratio. Dinger and Patten (1974) compared cumulative CO_2–H_2O vapor exchange measurements and pot weighing methods for three *Echinocereus* species from different elevations. They showed the improved water use efficiency (reciprocal of transpiration ratio) at lower night temperature was a result of reduced transpiration in *E. fendleri* and to increased CO_2 fixation in *E. triglochidiatus*.

In *Opuntia* cladodes, measured in the field under temperature and atmospheric conditions which tracked conditions outside the cuvette, dark CO_2 fixation was about twice as water use efficient as CO_2 fixation in the light (Osmond *et al.*, 1979b). However, this intrinsic difference is further amplified by the thermal conditions, atmospheric water vapor concentration and stomatal behavior of plants in the dark and light. Stomatal conductance of *Opuntia* in the dark, but not in the light, shows a feed-forward response to changes in atmospheric humidity which reduces transpiration (Osmond *et al.*, 1979b; see also Lange and Medina, 1979). Average vapor pressure difference between cladode and air at night was only 1 kPa, but was about 3 kPa in the afternoon, and this produced a 10-fold difference in transpiration ratio (20 g of H_2O lost g^{-1} of CO_2 gained in dark, and 180 g of H_2O lost g^{-1} of CO_2 gained in the light). Because *Opuntia* assimilates most of its CO_2 at night the overall daily transpiration ratio for these plants in summer, estimated from 24 h patterns of CO_2 and H_2O vapor exchange, was about 35 g of H_2O lost g^{-1} of CO_2 gained.

Although detailed analyses of the complex CO_2/H_2O vapor exchange patterns are necessary to provide the key parameters for physiological analysis of water use efficiency in CAM plants, it is dubious whether such studies are the best way to pursue ecological analyses of water relations under desert conditions. For these purposes, much more pragmatic observations seem satisfactory. For example, Nobel (1976, 1977) made long-term biweekly observations of day–night stomatal conductance, measured leaf or cladode temperature of *Agave* and *Ferocactus* and recorded air water vapor concentration. From these he derived estimates of average annual transpiration ratios of 25 g of H_2O lost g^{-1} of CO_2 gained for *Agave* and 70 g of H_2O lost g^{-1} of CO_2 gained for *Ferocactus*. Diffusion porometry was used, and the methodology is suspect for reasons discussed above. However, stomata were so tightly closed, that errors in conductance measurements during the day become irrelevant to overall water balance considerations. Nobel (1976) estimated that 60% of the annual water loss by *Agave* occurred during only 19% of the year when conditions favored nocturnal stomatal opening. The good agreement between short-term laboratory studies and long-term field estimates of transpiration ratio is principally due to the fact that most of the species investigated in the field do not normally exchange CO_2 or H_2O vapor during the day. There have been no comparable comparisons of the water costs of CAM under humid atmospheric conditions in the tropics, where one might expect daunting problems associated with measurement of transpiration, or in leaf succulents under field conditions in which there is significant diurnal CO_2 and H_2O vapor exchange.

There have been few long-term studies of productivity in CAM plants, but these too depend on acidification and porometry, photosynthetic area and dry weight measurements (Osmond *et al.*, 1979a; Acevedo *et al.*, 1983). Nobel (1984) has formulated an 'environmental productivity index' which is based on laboratory-derived responses of net CO_2 uptake to photosynthetically active radiation, day–night air temperatures and soil water availability. An index (maximum value of unity) is assigned to each environmental variable from monthly measurements of these environmental parameters in the field combined with the responses measured in the laboratory. Their product is termed the environmental productivity index, which indicates the fraction of maximal CO_2 uptake expected under field conditions. The model has been successfully validated for *Agave deserti* using independently measured dry weight changes in the field (Nobel, 1984), and has since been used to predict productivity of *A. deserti* over its elevational range (Nobel and Hartsock, 1986) as well as for other species of cultivated and native agaves (Nobel, 1985; Nobel and Meyer, 1985; Nobel and Quero, 1986). This model has yet to be applied to CAM plants which also exhibit significant diurnal uptake of atmospheric CO_2.

12.6 MEASUREMENT OF PHOTOSYNTHESIS AND RESPIRATION BY O_2 EXCHANGE

Although conventional infrared gas analyzer (IRGA)-based techniques can be used for day–night measurement of CO_2 and H_2O vapor exchange in CAM plants, and to characterize the CO_2 response curves of photosynthesis, light responses based on CO_2 exchange are complicated because of the onset of dark CO_2 fixation at low light (Osmond *et al.*, 1979b; Winter, 1980). Moreover, one of the more interesting phases of CAM, photosynthesis during the deacidification process, cannot be studied by CO_2 exchange because stomata are closed and internal CO_2 concentrations are high (Cockburn *et al.*, 1979). Measurements of photosynthetic O_2 exchange provide a simple and convenient way around these difficulties, and allow accurate measurements of quantum yield, light response characteristics and maximum rates of photosynthesis. The method depends on adaptation of the leaf disc oxygen electrode system (Delieu and Walker, 1981, 1983) for steady-state measurements with a calibrated light source (Adams *et al.*, 1986a; Björkman and Demmig, 1987; Walker and Osmond, 1986). The system, illustrated in Fig. 12.2 and manufactured by Hansatech, Kings Lynn, Norfolk, UK (or also available from Decagon Devices, Pullman, Washington 99163, USA), is compact, portable and relatively inexpensive. It is primarily designed for laboratory studies, but can be readily set up and operated adjacent to field sites (Winter *et al.*, 1986b). A key requirement is constant-temperature cooling water, obtained from a thermoregulated circulator, or a large reservoir flow-through system.

Leaf pieces of known area or discs (10 cm^2) are placed in a small thermostated chamber fitted with a gas-phase O_2 electrode. When using particularly thick tissues, such as *Opuntia* cladodes, the chamber must be enlarged by machining the upper chamber Plexiglass plug as required. The leaf disc is provided with CO_2 at H_2O saturation by using a carbonate buffer (0.9 M Na_2CO_3/0.1 M $NaHCO_3$)-impregnated pad, or is filled with air containing 5% CO_2 bubbled through water. A simple method of generating H_2O-vapor-saturated CO_2 to 5 or 10% requires the operator to breathe gently through the chamber until the O_2 electrode signal indicates

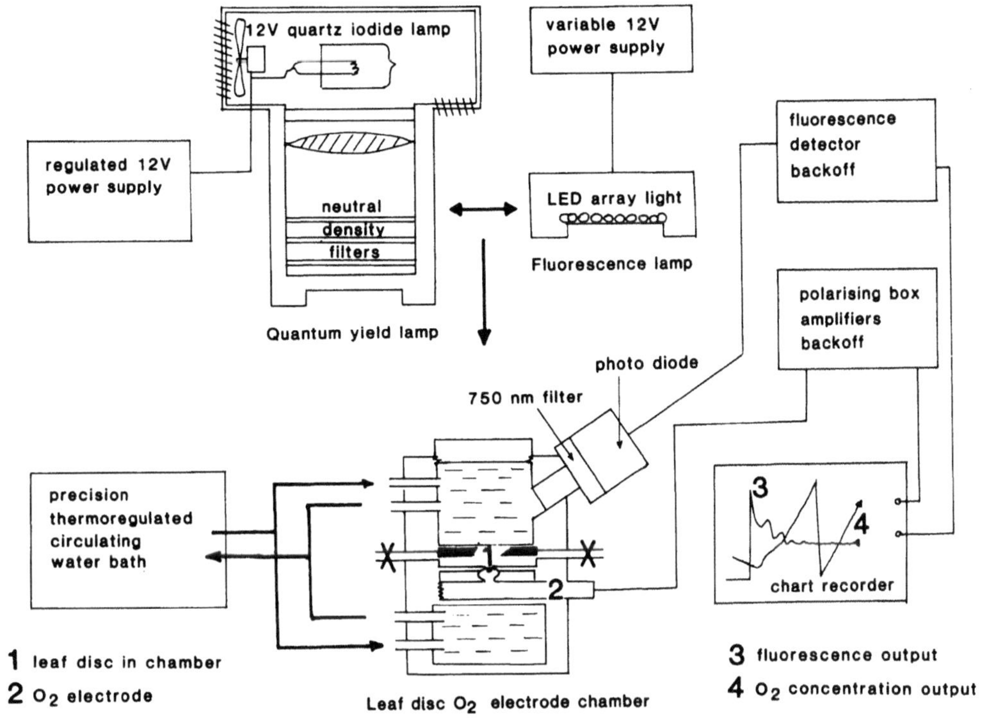

Fig. 12.2 Leaf disc electrode chamber for steady-state measurement of photosynthetic O_2 exchange and room temperature fluorescence (diagramatic only; after Delieu and Walker, 1983).

the desired decline from 21% O_2 due to CO_2 in the breath. The chamber is closed, and after thermal adjustment a steady rate of O_2 uptake in the dark due to respiration is established. The leaf disc is then illuminated by a previously calibrated 12 V quartz–iodide light source and maximum photosynthetic rate established.

The quartz–iodide light source fitted with a 50 W lamp is adequate for saturation of photosynthesis in most CAM plant tissues but light output can be doubled by using a 100 W lamp if required. It provides very uniform illumination and, when used with a regulated power supply and neutral density filters, gives very reproducible photon fluxes between experiments. Care should be taken to use lower photon fluxes to saturate photosynthesis if leaves of shade-grown plants are used, so as to avoid photoinhibition. The rate of change in O_2 concentration is easily measured at high sensitivity with amplification and back-off provided by the electrode control box, using a multirange recorder. Photosynthetic rate is calculated from the slope of this line, using the O_2 content of the chamber volume, corrected for O_2 partial pressure and temperature. The volume of the chamber with the tissue disc in place is calculated from the change in O_2 concentration recorded after pressurizing the closed chamber by injection of 1 ml of air from a syringe (Delieu and Walker, 1981). A computer-operated system, which uses an ultra-bright LED light source and which dispenses with chart recording, has been developed by D.A. Walker (Walker, 1988).

If gas-phase CO_2 sources are used when

generating a light response curve with deacidified CAM plants it is necessary to periodically recharge the system with CO_2 as it is utilized in photosynthesis. Adequate time must be allowed for temperature reequilibration after charging or the rates of O_2 evolution will be underestimated. This problem is most acute for tissues with low rates of photosynthesis. If acidified CAM tissues are used, then the internal CO_2 concentrations resulting from the decarboxylation of malic acid are sufficient for photosynthetic saturation. Charging with CO_2 is still recommended, however, to saturate the chamber with water vapor, and to alleviate subsequent pressure changes within the chamber. We have also found that CAM tissues taken from the dark or low light require a long period of exposure to light (30 min or more) before maximum rates of photosynthetic O_2 evolution can be obtained.

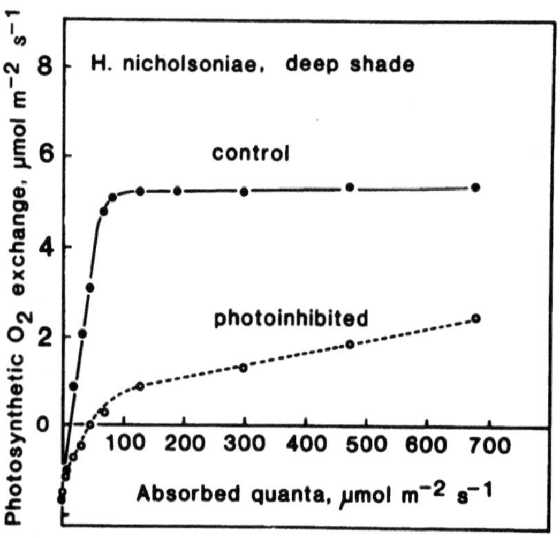

Fig. 12.3 Light response curve and quantum yield of photosynthetic O_2 exchange in leaves of *Hoya nicholsoniae*, a CAM leaf succulent grown in deep shade (max 80 µmol of photons m^{-2} s^{-1}) in a North Australian rain forest, before and after transfer to bright light (1800 µmol photons m^{-2} s^{-1}) for 2 h. Measurements were made using a Hansatech leaf disc O_2 electrode set up in an adjacent motel room (after Winter *et al.*, 1986a).

Accurate estimates of the temperature dependency of photosynthesis, respiration, light compensation and light- and CO_2-saturated photosynthesis, as well as quantum yield, may be obtained from the leaf disc electrode system. Fig. 12.3 shows some of these parameters obtained with leaf discs of *Hoya nicholsoniae*, an epilithic climbing CAM plant from a deeply shaded habitat in Eastern Australia. The light-saturated rate of photosynthesis, light requirement for saturation and light compensation point are all quite low compared with other CAM plants from bright habitats. When transferred to bright light for 2 h, *H. nicholsoniae* suffers substantial photoinhibition, which is indicated by an increase in light compensation point, a decrease in quantum yield (from 0.107 to 0.029 mol of O_2 mol^{-1} of absorbed photons) and a decrease in light- and CO_2-saturated photosynthetic rate. The Kok effect appears to become especially pronounced following photoinhibition, possibly because the respiratory rate is now about equal to the photosynthetic rate (Sharp *et al.*, 1984; Adams *et al.*, 1986a). In a study of glasshouse-grown epiphytic CAM plants Lüttge *et al.* (1986b) found higher apparent quantum yields using the leaf disc O_2 electrode than by CO_2 exchange. However, many of the values reported by these authors were very low indeed and may reflect chronic photoinhibition.

Quantum yields and maximum rates of photosynthesis obtained in this way are useful indicators of the potential of CAM plants and their responses to environmental stress. However, if CO_2 fixation is confined to the dark, as is often the case with stem succulents, then an effective, overall quantum yield can be derived simply from the dependency of nocturnal acidification on the sum of photosynthetically active radiation intercepted the previous day. Nobel (1977) first used this technique with *Ferocactus* in the field. He mounted CO_2-exchange chambers at eight locations on the surface of a barrel cactus, each of which received different illumination. His estimate of quantum yield for overall CO_2 exchange, 0.017 mol of CO_2 mol^{-1} of absorbed photons (correcting for absorptance of 85%), is about 20% of that measured in other CAM plants using the leaf disc O_2 electrode. Subsequent estimates by Nobel, based on the nocturnal acidification technique with other cacti and *Agave* yielded higher values, from 0.022 to 0.045 mol of CO_2 mol^{-1} of absorbed photons (data of Nobel, cited by Nobel and Hartsock, 1983). These values suggest that CO_2 losses in the carboxylation/decarboxylation cycles of CAM, or noneffective absorption of light which is used to recycle CO_2 via photorespiration behind closed stomata after deacidification, reduce overall photosynthetic efficiency by up to 80%. All of the quantum yields measured in this way are based on studies of plants from high light habitats – it will be instructive to obtain similar data for rain forest epiphytic CAM plants and aquatic CAM plants.

12.7 WATER RELATIONS

Generally speaking, CAM plants represent the archetype of water stress avoidance at high water potential. Low surface to volume ratios, low stomatal frequencies and low stomatal and cuticular conductances have long been held to facilitate water retention under conditions of high evaporative demand. MacDougal and Spalding (1910) kept a large detached stem of *Ibervillia sonorae* in air for 8 years in Tucson, during which time it lost only 50% of its water content and made some apical growth each year. Only recently, however, have quantitative analyses of plant water relations, using modern techniques, given some insight into these conventional wisdoms. Water conservation due to stomatal behavior and water capacitance serve to maintain high plant water potential, after soil water potential falls below that of the plant, and to prolong the period of net CO_2 exchange at night (by up to 40 days in the case of *Ferocactus*, Nobel, 1977). The fundamental importance of water relations parameters in overall performance of CAM plants in desert conditions has been highlighted recently by Nobel (1984). He found that production of new leaves in *Agave deserti*, which is the 'driving parameter' for his environmental productivity index (see above), was highly correlated with the water status of the tissue.

Water relations of CAM plants have not been extensively researched because it is often difficult to measure water relations parameters of these succulents, especially under field conditions. Gravimetric measurement of water loss from detached tissues, with cut surfaces sealed by vaseline, is often the most accurate and convenient way to estimate transpiration under field conditions. Pot weighing (Meinzer and Rundel, 1973) and lysimetry (Ekern, 1965) have been employed at different scales of enquiry.

Measurements of plant water potentials in CAM succulents have been made by pressure bomb and psychrometric techniques. A simple direct approach using ceramic tip Wescor PT 51–05 soil thermocouple psychrometers inserted in bulky tissues was first described by Szarek and Ting (1974). A core of tissue about the same diameter as the psychrometer cup is removed from the leaf or cladode and the cup fully inserted into the tissue, then sealed into place with viscous grease. The cups should be inserted so that grease does not melt and cover the cup, and grease should be removed carefully before attempting to remove the cup. In spite of the best precautions, plant mucilages and saps, and the sealants, tend to clog the pores of the ceramic cups. Nevertheless, Nobel (1976) found good agreement between this method in leaf blades and the pressure chamber method applied to leaf tips of *Agave*, and has continued to use the inserted soil probes (Acevedo *et al.*, 1983). Osmond *et al.* (1979a) correlated changes in water potential measured with these probes and tissue water content (g of H_2O g^{-1} dry wt) of *Opuntia* following irrigation after drought, but only a small proportion of the probes gave reliable results throughout these studies. The small clip on leaf psychrometer chamber, Wescor Model L-51A, is difficult to use with CAM plants, even under controlled environmental conditions, requiring unacceptably long equilibration periods and being subject to large temperature variation between tissue and the sensor block (Osmond *et al.*, 1976).

The use of Wescor C-51 and C-52 sample chambers in conjunction with the Wescor HR-33T microvoltmeter for plant water potential measurement by psychrometric or dew-point methods is described in Chapter 9. Although suitable for measurement of osmotic pressure of cell saps (by placing macerated tissue samples in small equilibration chambers), water potential estimates with CAM tissues require lengthy equilibration times, often up to 6 h (Gerwick and Williams, 1978). However, Earnshaw *et al.* (1985) reported that 60 min was sufficient time for measuring total leaf water potential of *Sedum* and *Sempervivum* leaf discs. The potential benefits of a 10-place sample chamber for psychrometric measurements (Decagon Devices) seem to be offset by large sample chamber size and longer equilibration times. This system requires the walls of each sample chamber to be covered with tissue, and so it appears to be useful for obtaining bulk leaf water potential from leafy samples but not for osmotic potential measurements. Thus our measurements of water potentials of *Opuntia basilaris* in Death Valley using this system were about 1.0 MPa more negative than those reported elsewhere (Adams *et al.*, 1986b) and indeed, the most negative water potentials for cacti (−3.4 MPa) have been obtained using psychrometric methods (Green and Williams, 1982).

A primary source of error in psychrometric determination of either total plant water potential or osmotic potential of the cell sap is the mixing of apoplastic and symplastic water in either leaf discs or macerated samples. How these potential errors in CAM plant samples relate to similar errors in C_3 and C_4 plants, and how the day–night cycle of malic acid metabolism relates to these errors, needs to be experimentally assessed. Older notions that the mucilages and gums in some CAM plants conferred high matric potentials and retained cell water have been dispelled by Wiebe and Al Saadi (1976). In some CAM plants, such as the atmospheric tillandsias, epidermal trichomes may absorb water directly from humid air and complicate the interpretation of tissue averaged water relations parameters. Barcikowski and Nobel (1984) showed that the chlorenchyma and pith water storage tissue of cacti have very different relative water contents during desiccation cycles. Clearly, in these plants, tissue-averaged water relations parameters need to be interpreted cautiously. Cellular water relations parameters in CAM plants have been measured directly on single mesophyll cells in *Kalanchoë daigremontiana* using the pressure probe technique (Steudle *et al.*, 1980). Their findings of a generally low value of turgor pressure (<0.2 MPa) and extremely short half-times for water exchange measured at the cell level match those inferred from tissue averaged measurements in the field and laboratory (Nobel and Jordan, 1983; Sinclair, 1983a,b).

Leaf succulents are not often amenable to the insertion of soil probes but leaf water potential can be measured with a pressure chamber. Smith *et al.* (1985) removed flanges from the leaf blade of epiphytic CAM plants so that the mid-rib could be inserted in the port of a pressure chamber. This presumably overcomes problems of exudation from pithy tissues around the veins, often found when leaf succulents are pressurized. Using precautions described by these authors, it was possible to distinguish CAM epiphytes from C_3 epiphytes on the basis of day–night changes in the leaf water potential of about 0.2 MPa using the pressure chamber. Sinclair

(1983a) also used the pressure chamber to establish pressure–volume curves for epiphytic CAM plants. Starting with tissues which were fully hydrated, he found an average discrepancy of about 12% between tissue weight loss and sap collected, which he ascribed to transpiration in the chamber, in spite of the usual precaution of including water saturated paper in the chamber. Nobel and Jordan (1983) used pressure–volume curves to estimate the water capacitance of *Agave* and other species. Fully hydrated tissues were sealed into the pressure chamber port using a quick-setting expoxy resin (Devcon S-209). These studies establish that CAM plants from both habitats show a large change in relative water content for a small change in plant water potential.

None of the above methods was used in perhaps the most important study of CAM plant water relations. Instead, Lüttge and Ball (1977) used the liquid exchange method (Shardakov method) for estimation of leaf water potential throughout the day–night cycle in *Kalanchoë daigremontiana*. In spite of the shortcomings of this method, they demonstrated that plant water potential was lowest pre-dawn, in marked contrast to the pre-dawn maximum measured in most other tissues. This method has recently been calibrated against the pressure chamber (Smith and Lüttge, 1985), providing more detail in relation to the roles of the nocturnal accumulation of osmotica (malic acid) and transpiration as factors determining the changes in tissue water relations of CAM plants. Similar trends have been clearly established in tropical epiphytic CAM plants under field conditions, and used to distinguish CAM epiphytes from the C_3 epiphytes (Smith *et al.*, 1986). Laboratory studies of CO_2–H_2O vapor exchange with simultaneous measurements of water uptake by potometry have established that malic acid accumulation enhances water uptake from dusk to dawn (Ruess and Eller, 1985). In the field this rapid rehydration depends upon resumption of water uptake by surviving roots, and the initiation of new roots (Kausch, 1965; Nobel and Sanderson, 1984). As in ecophysiology generally, these important functional relationships in roots and soils have received scant attention.

12.8 STRESS PHYSIOLOGY

CAM plants have long attracted attention from ecophysiologists because they show extremes of stress tolerance (high temperature stress) and stress avoidance (water stress). With the emerging realization that light may exaggerate responses to stress in the photosynthetic system, and the growing appreciation of epiphytic CAM plants, light stress provides an additional perspective in these studies (Osmond *et al.*, 1987). Many of the techniques developed for study of stress physiology in nonsucculent C_3 and C_4 plants can be applied to CAM plants, but require some creative modification, as discussed below.

Thermal tolerance in CAM plants has been assessed by examination of intrinsic indicators of chloroplast membrane function (chlorophyll fluorescence as a function of temperature) and by examination of cell membrane integrity by vital staining and other techniques (Didden-Zopfy and Nobel, 1982). Chlorophyll fluorescence is the least labor intensive of these methods, and more applicable to field use. Downton *et al.* (1984) showed that the temperature threshold of fluorescence rise corresponds to the threshold for a time-unstable decline in CO_2- and light-saturated photosynthesis. The measurements were made with leaf discs or tissue segments mounted on a brass block, fitted with a thermocouple appressed to the tissue as described by Schreiber and Berry (1977). The block is heated at about 1°C min^{-1} and temperature and fluorescence recorded on an X/Y recorder. Fluorescence is excited by weak 480 nm light (0.1 μmol of photons m^{-2} s^{-1}) and monitored at 690 nm using a photomulti-

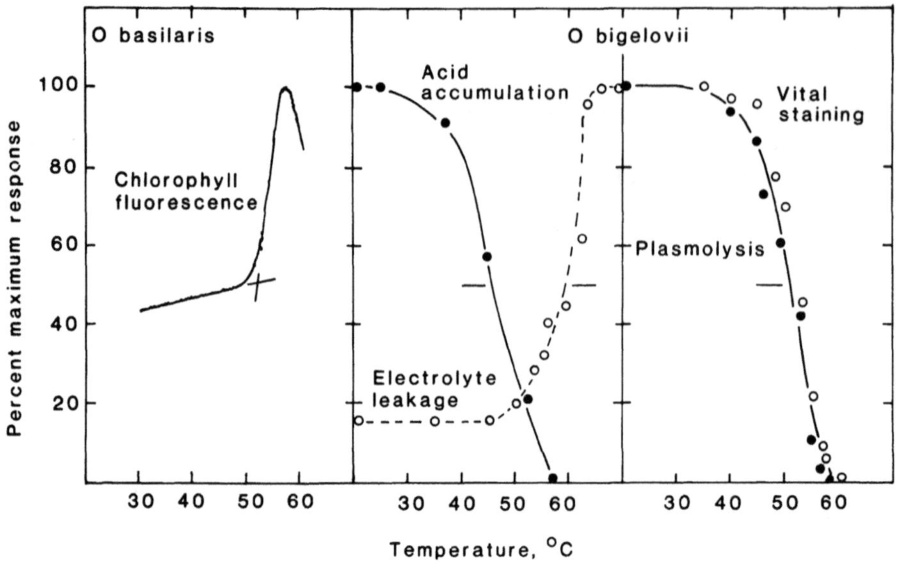

Fig. 12.4 Methods for estimating the upper limits of thermal tolerance in CAM plants. (a) Fluorescence break point (52°C) in *Opuntia basilaris* from Death Valley, California, obtained using the method of Schreiber and Berry (1977) (data provided by W.J.S. Downton). (b) Response of nocturnal acidification (at 20°C) and of electrolyte leakage into water for 30 min (conductivity relative to tissue boiled in water for 1 h) in *Opuntia bigelovii* from Deep Canyon, California following 1 h treatment at different temperatures. The critical temperature (for 50% change) was 46°C for acidification and 61°C for electrolyte leakage (redrawn from Didden-Zopfy and Nobel, 1982). (c) Response of vital staining with Neutral Red, and plasmolysis (induced with 0.5 M KNO_3 or 1 M mannitol) in *Opuntia bigelovii* from Deep Canyon, California after 1 h heat treatment at different temperatures. The critical temperature (50% decrease in uptake or plasmolysis) was 52°C in both cases (redrawn from Didden-Zopfy and Nobel, 1982).

plier. The temperature at which thermal breakdown of chloroplast membranes occurs is indicated by a sudden increase in fluorescence. The average upper temperature limit for cacti under field conditions determined this way was 54.7±2.8°C, and Fig. 12.4 shows a trace obtained with *Opuntia basilaris*.

Didden-Zopfy and Nobel (1982) examined a number of techniques which indicate breakdown in the integrity of cell, rather than chloroplast, membranes. They compared electrolyte leakage, plasmolysis, vital staining, acid accumulation and survival of tissue (nonnecrosis) after heat treatments, and chose vital staining as the best. In *Opuntia bigelovii* the staining test correlates well with temperature limits determined by plasmolysis of tissues, and with the mean of temperature limits measured by electrolyte leakage and subsequent acid accumulation at 20°C (Fig. 12.4). The method depends on accumulation of Neutral Red (0.3 mM 3-amino-7-dimethylamino-2-methylphenazine HCl in 7 mM KH_2PO_4/Na_3PO_4 at pH 7.4) in vacuoles of cells in tissues stained for 24 h at 5°C, following 1 h treatment of stem sections at prescribed temperatures in a water bath. In each test at least 500 cells in the upper chlorenchyma are scored and the critical temperature determined as that which caused a 50% reduction in the number of cells taking up stain. When this method was applied to field populations of cacti, an average critical temperature of 55.5°C was obtained (Smith *et al.*, 1984), in close agreement with chlorophyll fluorescence data (Fig. 12.4).

Nobel and Smith (1983) found that the upper and lower thermal limits for a 50% inhibition of acidification were about 6° narrower than those indicated by stain uptake (cf. Fig. 12.4). When the upper and lower thermal tolerance limits in *Agave* were examined using vital staining, as much as 10°C acclimation was observed in either direction when plants were transferred between 10°C day/0°C night and 50°C day/40°C night at 10 day intervals. Much larger acclimation of the high temperature tolerances of cacti has been measured by this technique (up to 14.5°C), and if acidification is taken as the criterion more than 20°C acclimation is observed (Smith *et al.*, 1984).

It is increasingly clear that environmental stress factors do not act in isolation. Thus, in CAM plants under desert field conditions, high temperature stress is inevitably associated with high light, and in temperate conditions chilling injury at night is often followed by high light the following day. In tropical rain forests gap formation is likely to expose epiphytic CAM plants to a coincidence of high light, high temperature and water stress. The interaction of light with other environmental stresses is still poorly understood, especially in CAM plants, even though the phenomenon of light-dependent damage to photosynthetic systems (photoinhibition) is well established (Powles, 1984). It should be noted that evidence of chronic photoinhibition has been obtained in CAM plants exposed to daily total photon flux densities in excess of their natural environment. Nobel and Hartsock (1983) found that above 30 mol of photons m^{-2} day^{-1} *Opuntia ficus-indica* showed reduced nocturnal acidification and chlorophyll bleaching. Given that CAM plants are found in habitats of markedly contrasting light environments, it seems likely that photoinhibition may be a major factor in their ecophysiology.

There is as yet no universal agreement as to the best way to assess photoinhibitory damage. Quantum yield is one indicator of light-dependent damage during stress. The leaf disc oxygen electrode is especially useful in such studies, because it can be coupled with simple detectors of ambient-temperature chlorophyll *a* fluorescence to provide additional insights into photoinhibition in the course of quantum yield measurement (Adams *et al.*, 1987; Walker and Osmond, 1986). In addition, Björkman has developed a system for measuring low temperature fluorescence from intact tissues which provides a 'simple, rapid, sensitive and reproducible method for assessing photoinhibitory injury' to photosynthetic tissues (Powles and Björkman, 1982). The principles of these measurements, their relative merits and preliminary applications to the stress physiology of CAM plants are discussed below.

Room temperature chlorophyll *a* fluorescence is thought to indicate primarily fluorescence from photosystem II (Schreiber, 1983), and the quenching of this by complex interactions during photosynthetic carbon reduction (Walker, 1981; Krause and Weiss, 1984). Osmond (1982) used a Brancker S-10 Plant productivity meter (Richard Brancker Associates, Ottawa, Canada) and an oscillographic recorder to detect diurnal changes in these parameters on separate faces of *Opuntia stricta* cladodes. The slope of the rise in variable fluorescence, and the magnitude of the variable fluorescence are good indicators of photoinhibition (Critchley and Smillie, 1981) and other stress responses (Smillie and Hetherington, 1983). However, the fluorescence detector available as an accessory with the Hansatech leaf disc O_2 electrode is a more convenient and less expensive method (Winter *et al.*, 1986a). For the most part, high-speed recording to resolve the early structure of the fluorescence transient is not required, and the dark adaptation period of 10–20 min is conveniently incorporated into the respiratory measurement interval prior to the measurement of photosynthetic O_2 exchange and quantum yield.

Both of the above instruments use red

light-emitting diodes with peaks at approximately 650 nm to provide the exciting light. They are derived from systems described by Schreiber *et al.* (1975), the principles of which are reviewed in Schreiber (1983). Relatively high photon fluxes are used (up to 300 μmol of photons m^{-2} s^{-1} in the case of the Hansatech system). Interference filters are used to isolate long-wave fluorescence emission (730–760 nm) which is measured by photodiodes mounted directly behind the source and close to the leaf surface in the Brancker system, or remotely located at a 45° angle to the leaf surface in the Hansatech system. The original Brancker S-10, and subsequent versions suffer from sensor detection of some reflected excitation light, and are primarily designed for analysis of the fast initial kinetics of room temperature fluorescence using oscillographic or transient recorders. They are less flexible in their output characteristics than the Hansatech system, but the latter instrument is very sensitive to

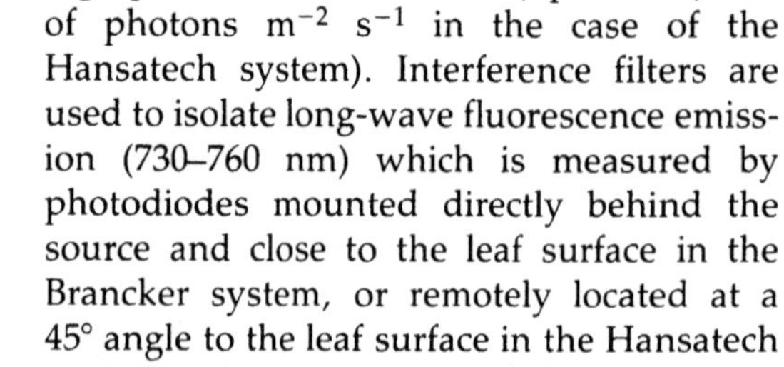

Fig. 12.5 Room temperature fluorescence in leaves of sun and shade-grown fronds of the epiphytic CAM fern *Pyrrosia longifolia* in a North Australian rain forest, measured using the fluorescence detector attached to a Hansatech leaf disc O_2 electrode system set up in an adjacent motel room. The initial rise to and decline from the peak, *P*, is unusually fast in this species. The fluorescence yield is higher in shade-grown plants and after 5 h treatment in bright light (800 μmol m^{-2} s^{-1}) photoinhibition is indicated by loss of the variable component of fluorescence ($P-T$) (redrawn from Winter *et al.*, 1986a).

orientation of tissues in the O_2 electrode chamber.

Fig. 12.5 shows traces of fluorescence emission from leaves of the epiphytic CAM fern *Pyrrosia longifolia* from an open habitat, a deeply shaded habitat and after exposure of the shaded plant to bright light. Measurements were made in the Hansatech O_2 electrode chamber, and the most reproducible parameters extracted from these curves are the peak fluorescence, *P*, and the steady state level, *T*, which are used to calculate the variable component F_{VRT} ($= P - T$). Following photoinhibition, the variable components of room temperature fluorescence emission, the quantum yield of photosynthetic O_2 exchange, and the light-saturated rate of O_2 exchange are depressed, and the light compensation point of photosynthesis is increased (Adams *et al.*, 1987b; Winter *et al.*, 1986a).

Chlorophyll fluorescence emission at liquid N_2 temperature (77 K) reflects the well being of primary photochemistry of photosystem II (fluorescence emission at 692 nm) or photosystem I (fluorescence emission at 734 nm), and is not complicated by quenching associated with metabolism (Krause and Weiss, 1984). Kitajima and Butler (1975) modeled fluorescence quenching in photosystem II, using observations of isolated chloroplast fluorescence, and concluded that the variable component of 77 K fluorescence (F_v) as a fraction of the maximum yield (F_m) is proportional to the quantum yield of primary photochemistry. The most useful fluorescence parameter measured *in vivo* is also the ratio of variable over maximum fluorescence yield (F_v/F_m). It is independent of the intensity of the illuminating beam or the units used in measuring the absolute levels of fluorescence. This parameter is also preferable because it will not be affected by the architecture of a particular leaf or photosynthetic organ, which might influence the absolute yields of fluorescence through reabsorption. Thus F_v/F_m can also be used to compare 77 K fluorescence between plants that have grown under different environmental conditions, between intact tissues and isolated chloroplasts, and between species.

The device for 77 K fluorescence analysis

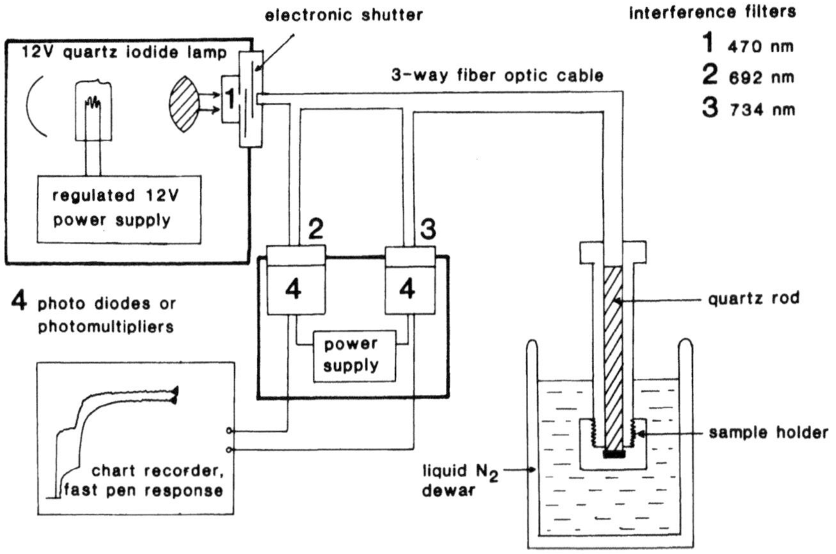

Fig. 12.6 77 K fluorescence detection system; after Powles and Björkman (1982) and Adams *et al.* (1987a).

in vivo, described by Powles and Björkman (1982), uses low-intensity (~1 μmol photon m^{-2} s^{-1}) narrow-band blue light (470 or 480 nm) obtained from a power-stabilized quartz–iodide source and interference filter. This is conducted to the surface of a leaf disc by optical fibers and a 10 mm diameter quartz rod. The fluorescence emission is conducted back to the detectors (photomultipliers or photodiodes) via the same cables and selected by interference filters. A high-speed shutter (1–2 ms) is used to admit light to the leaf disc which is previously dark-adapted at room temperature at the end of the quartz rod and then frozen to 77 K in liquid N_2. A relatively high-speed (10 mm s^{-1}) paper chart recorder with rapid pen response is needed to detect F_o. With CAM plants we have found it convenient to use detachable darkened quartz rods, fitted with adjustable sample chambers, which permit pre-adaptation of samples of varying thickness. The detachable units can be set up and discs dark-adapted under room temperature and light conditions, then fitted to the fiber cables for cooling in liquid N_2 prior to fluorescence measurement (see Fig. 12.6).

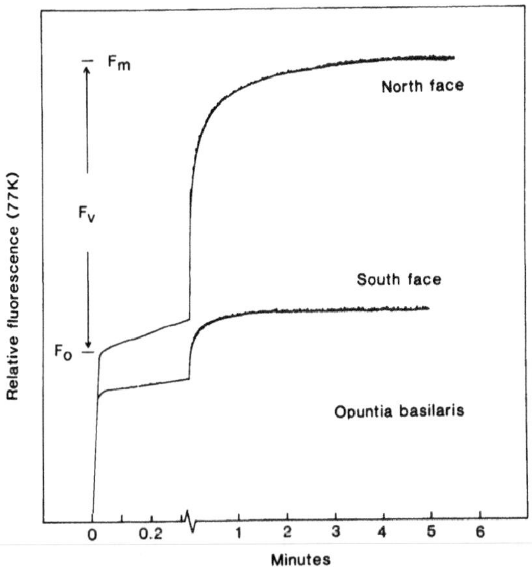

Fig. 12.7 Time course of low-temperature fluorescence from north and south faces of discs from *Opuntia basilaris* cladodes in Death Valley, California at 7 am, 5 October 1985. The reductions in F_o and the variable component, F_v, of fluorescence from the sun-exposed south face indicate photoinhibition on this face, relative to the sun-protected north face. Data were obtained with a device similar to that shown in Fig. 12.6 set up in a trailer at the field site (Adams *et al.*, 1987a).

Fig. 12.7 shows primary data obtained with cladodes of *Opuntia basilaris* in Death Valley in October, using a system based on the above design (Adams *et al.*, 1987a). It indicates the resolution of the instantaneous fluorescence (F_o), the maximum yield (F_m) and the variable component ($F_v = F_m - F_o$). Although the fluorescence yield at 77 K varies greatly with plant growth conditions, especially light intensity and probably with leaf anatomy, the ratio $F_v/F_m \approx 0.83$ in healthy leaves (Björkman and Demmig, 1987). Similar values have been found for a wide range of healthy CAM plant tissues. Following photoinhibition, the value F_v/F_m is reduced. In Fig. 12.7, the sun exposed south face of *Opuntia basilaris* shows much smaller values for F_v than the north face, suggesting greater photoinhibition on the south face. Demmig and Björkman (1987) found quantitatively similar reductions in F_v/F_m and quantum yield of O_2 exchange in photosynthesis when leaves of C_3 plants were subjected to various degrees of photoinhibition. Our preliminary studies with CAM plants exposed to water stress under desert conditions suggests a similar correlation may be obtained (Table 12.2). An interpretation of 77 K fluorescence changes *in vivo*, in terms of current models of photoinhibition, has been developed by Björkman and Demmig (1987).

Both room temperature and low-temperature fluorescence techniques, when used in conjunction with quantum yield measurements, are likely to be useful indicators of photoinhibition during environmental stress. The only systematic comparison of these methods so far is that of Adams *et al.* (1987b) in which photoinhibition of shade-grown *Kalanchoë daigremontiana* was examined. Fig. 12.8 compares the changes observed in leaves of plants grown in a range of shade treatments, then subjected to a standard 4 h photoinhibitory treatment in bright light. The effects of this treatment on quantum yield, 77 K fluorescence (F_v/F_m) and room temperature fluorescence ($P - T$) are shown as a percentage of the untreated controls. The trends in response of quantum yield and 77 K fluorescence are similar as one would expect from highly correlated parameters (Adams *et al.*, 1987b; Demmig and Björkman, 1987).

Application of a newly developed pulse amplitude modulation (PAM) chlorophyll fluorimeter (Schreiber *et al.*, 1986) to the analysis of photoinhibition in CAM as well as other plants seems promising. The instrument, which is available commercially from H. Walz, Mess- und Regeltechnik, D-8521 Effeltrich, BRD, uses a pulsed measuring beam and a selective pulse amplification system to detect the fluorescence. This modulated technique allows detection of fluorescence signals in the presence of actinic light of any quantity or quality, including full sunlight. Moreover, measurements can be made through the window of a gas-exchange cuvette allowing

Table 12.2 Indications of photoinhibition in cladodes of *Opuntia basilaris* at the end of the summer in Death Valley, California (October 1985). Quantum yield was measured using the leaf disc O_2 electrode (Adams *et al.*, 1987a) and 77 K fluorescence at 690 nm (pre-dawn) by a device similar to that depicted in Fig. 12.6. Field data are compared to the highest quantum yield and value for F_v/F_m measured in glasshouse stocks from this population (unpublished data, W.W. Adams)

Tissue and orientation	*Quantum yield (mol of O_2 mol^{-1} quanta)*	*77 K fluorescence (F_v/F_m)*
Glasshouse, well watered	0.101	0.81
Death Valley, north face	0.057	0.70
Death Valley, south face	0.033	0.48

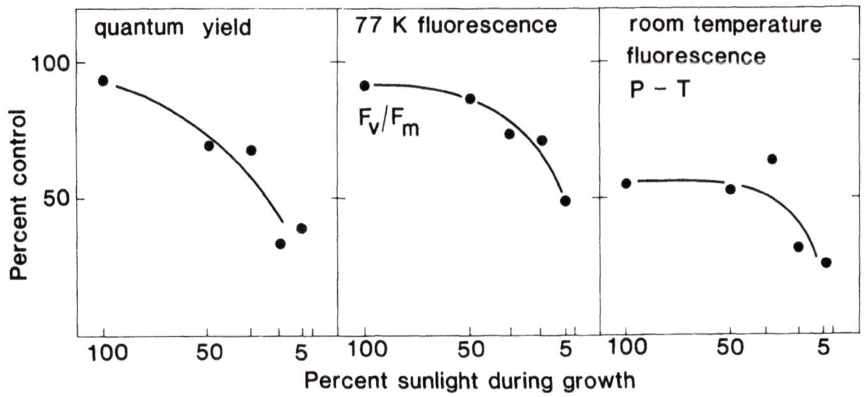

Fig. 12.8 Comparison of change in quantum yield of O_2 exchange, 77 K fluorescence at 690 nm (F_v/F_m), and room temperature chlorophyll *a* fluorescence (*P–T*) in *Kalanchoë daigremontiana* grown under a range of shade treatments following a 4 h photoinhibitory treatment in bright light (Adams *et al.*, 1987b).

simultaneous measurements of fluorescence and assimilation. This instrument has been used by Winter and Demmig (1987) to examine chlorophyll fluorescence throughout the 24 h cycle of CAM metabolism in *K. daigremontiana*. They detected minimal levels of energy-dependent quenching of fluorescence during deacidification, consistent with the high internal CO_2 concentrations and maximal rates of electron transport at this time. When high light intensities were applied in the afternoon following deacidification, the extent of this quenching increased in the afternoon to its maximum value, and the onset of conditions conducive to photoinhibition were detected. Application of this instrument to plants under field conditions, such as those described by Adams *et al.* (1987b) may well help distinguish between harmless nonradiative dissipation of excitation energy and the extent of damage to primary photochemistry due to photoinhibition (Demmig and Björkman, 1987; Adams and Osmond, 1988).

ACKNOWLEDGEMENTS

The application of techniques for the study of CAM outlined here was supported by NSF grant PCM-83 14980. We are grateful to Tom Sharkey for help with construction of the 77 K fluorescence system, and to Prudence Kell for technical assistance to obtain data for this review. Ulrich Lüttge and Park Nobel made helpful comments on the manuscript.

REFERENCES

Acevedo, E., Badilla, I. and Nobel, P.S. (1983) Water relations, diurnal acidity changes and productivity of a cultivated cactus, *Opuntia ficus-indica*. *Plant Physiol.*, **72**, 775–80.,

Adams III, W.W., Nishida, K. and Osmond, C.B. (1986) Quantum yields of CAM plants measured by photosynthetic O_2 exchange. *Plant Physiol.*, **81**, 297–300.

Adams III, W.W. and Osmond, C.B. (1988) Internal CO_2 supply during photosynthesis of sun and shade grown CAM plants in relation to photoinhibition. *Plant Physiol.*, **86**, 117–23.

Adams III, W.W., Smith, S.D. and Osmond, C.B. (1987a) Photoinhibition of the CAM succulent *Opuntia basilaris* growing in Death Valley: evidence from 77K fluorescence and quantum yield. *Oecologia*, **71**, 221–8.

Adams III, W.W., Osmond, C.B. and Sharkey, T.D. (1987b) Responses of two CAM species to different irradiances during growth and susceptibility to photoinhibition by high light. *Plant Physiol.*, **83**, 213–18.

Austin, R.B. and Longdon, P.C. (1967) A rapid method for the measurement of rates of photosynthesis using $^{14}CO_2$. *Ann. Bot.*, **31**, 245–53.

Barcikowski, W. and Nobel, P.S. (1984) Water relations of cacti during desiccation: distribution of water in tissues. *Bot. Gaz.*, **145**, 110–15.

Bartholomew, D.P. and Kadzimin, S.B. (1976) Porometer cup to measure leaf resistance of pineapple. *Crop Sci.*, **16**, 565–8.

Björkman, O. and Demmig, B. (1987) A survey of photon yield of O_2 evolution and chlorophyll 77 K fluorescence characteristics among vascular plants of diverse origins. *Planta*, **170**, 489–504.

Boston, H. and Adams, M.S. (1985) Seasonal diurnal acid rhythms in two aquatic crassulacean acid metabolism plants. *Oecologia*, **65**, 573–9.

Boston, H. and Adams, M.S. (1986) The contribution of crassulacean acid metabolism to the annual productivity of two aquatic vascular plants. *Oecologia*, **68**, 615–22.

Cockburn, W. (1985) Variation in photosynthetic acid metabolism in vascular plants: CAM and related phenomena. *New Phytol.*, **101**, 3–24.

Cockburn, W. and McAuley, A. (1975) The pathway of carbon dioxide fixation in Crassulacean plants. *Plant Physiol.*, **55**, 87–9.

Cockburn, W., Ting, I.P. and Sternberg, L.O. (1979) Relationship between stomatal behavior and internal carbon dioxide concentration in CAM plants. *Plant Physiol.*, **63**, 1029–32.

Critchley, C. and Smillie, R.M. (1981) Leaf chlorophyll fluorescence as an indicator of high light stress (photoinhibition) in *Cucumis sativus* L. *Austr. J. Plant Physiol.*, **8**, 133–41.

Crombie, W.M.L. (1952) Oxalic acid metabolism in *Begonia semperflorens*. *J. Exp. Bot.*, **5**, 173–83.

Delieu, T. and Walker, D.A. (1981) Polarographic measurement of photosynthetic O_2 evolution by leaf discs. *New Phytol.*, **89**, 165–75.

Delieu, T. and Walker, D.A. (1983) Simultaneous measurement of oxygen evolution and chlorophyll fluorescence by leaf discs. *Plant Physiol.*, **73**, 534–41.

Demmig, B. and Björkman, O. (1987) Comparison of the effect of excessive light on chlorophyll fluorescence (77K) and photon yield of O_2 evolution in leaves of higher plants. *Planta*, **171**, 171–84.

Didden-Zopfy, B. and Nobel, P.S. (1982) High temperature tolerance and heat acclimation of *Opuntia bigelovii*. *Oecologia*, **52**, 176–80.

Dinger, B.E. and Patten, D.T. (1974) Carbon dioxide exchange and transpiration in species of *Echinocereus* (Cactaceae), as related to their distribution within the Pinaleno Mountains, Arizona. *Oecologia*, **14**, 389–411.

Downton, W.J.S., Berry, J.A. and Seemann, J.R. (1984) Tolerance of photosynthesis to high temperature in desert plants. *Plant Physiol.*, **74**, 786–90.

Earnshaw, M.J., Carver, K.A. and Lee, J.A. (1985) Changes in leaf water potential and CAM in *Sempervivum montanum* and *Sedum album* in response to water availability in the field. *Oecologia*, **67**, 486–92.

Ekern, P.C. (1965) Evapotranspirations of pineapple in Hawaii. *Plant Physiol.*, **40**, 736–9.

Field, C. and Mooney, H.A. (1984) Measuring gas exchange of plants in the wet tropics. In *Physiological Ecology of Plants of the Wet Tropics* (eds E. Medina, H.A. Mooney and C. Vazquez-Yanes), Junk, The Hague, pp. 129–38.

Gerwick, B.C. and Williams III, G.J. (1978) Temperature and water regulation of gas exchange of *Opuntia polycantha*. *Oecologia*, **35**, 149–60.

Gibson, A.C. (1982) The anatomy of succulence. In *Crassulacean Acid Metabolism* (eds I.P. Ting and M. Gibbs), American Society of Plant Physiologists, Rockville, pp. 1–17.

Gibson, A.C. and Nobel, P.S. (1986) *The Cactus Primer*, Harvard University Press, Cambridge, MA, 286 pp.

Green, J.M. and Williams III, G.J. (1982) The subdominant status of *Echinocereus viridiflorus* and *Mammillaria vivipara* in the short grass prairie: the role of temperature and water effects on gas exchange. *Oecologia*, **52**, 43–8.

Griffiths, H., Lüttge, U., Stimmel, K.-H., Crook, C.E., Griffiths, N.M. and Smith, J.A.C. (1986) Comparative ecophysiology of CAM and C_3 bromeliads. III. Environmental influences on CO_2 assimilation and transpiration. *Plant, Cell Environ.*, **9**, 385–94.

Heyne, B. (1815) On the deoxidation of the leaves of *Cotyledon calycina*. *Trans. Linn. Soc. London*, **11**, 213–15.

Hohorst, H.J. (1965) L(−)-malate, determination with malate dehydrogenase and DPN. In *Methods of Enzymatic Analysis* (ed. H.U. Bergmeyer), Academic Press, London, pp. 328–34.

Jones, M.B. (1975) The effect of leaf age on leaf resistance on CO_2 exchange of the CAM plant *Bryophyllum fedtschenkoi*. *Planta*, **123**, 91–6.

Johnson, H.B., Rowlands, P.G. and Ting, I.P. (1979) Tritium and carbon-14 double isotope porometer for simultaneous measurements of transpiration and photosynthesis. *Photosynthetica*, **13**, 409–18.

Kausch, W. (1965) Beziehungen zwischen Wurzelwachstum, Transpiration und CO_2-Gaswechsel bei einigen Kakteen. *Planta*, **66**, 228–38.

Keeley, J.E. and Morton, B.A. (1982) Distribution of diurnal acid metabolism in submerged aquatics outside the genus *Isoetes*. *Photosynthetica*, **16**, 546–53.

Keeley, J.E., Osmond, C.B. and Raven, J.A. (1984) *Stylites*, a vascular land plant without stomata absorbs CO_2 via its roots. *Nature, London*, **310**, 694–95 (and correspondence **314**, 200).

Kitajima, M. and Butler, W.L. (1975) Quenching of chlorophyll fluorescence and primary photochemistry in chloroplasts by dibromothymoquinone. *Biochim. Biophys. Acta*, **376**, 105–15.

Kluge, M. and Ting, I.P. (1978) *Crassulacean Acid Metabolism: An Analysis of an Ecological Adaptation, Ecological Studies*, Vol. 30, Springer-Verlag, Heidelberg, 209 pp.

Krause, G.H. and Weis, E. (1984) Chlorophyll fluorescence as a tool in plant physiology. II Interpretation of fluorescence signals. *Photosynth. Res.*, **5**, 139–57.

Lange, O.L. and Medina, E. (1979) Stomata of the CAM plant *Tillandsia recurvata* respond directly to humidity. *Oecologia*, **40**, 357–63.

Lange, O.L., Schulze, E.-D., Kappen, L., Evenari, M. and Buschbom, U. (1975) CO_2 exchange pattern under natural conditions of *Caralluma negevensis*, a CAM plant of the Negev desert. *Photosynthetica*, **9**, 318–26.

Lange, O.L. and Zuber, M. (1977) *Frerea indica*, a stem succulent CAM plant with deciduous C_3 leaves. *Oecologia*, **31**, 67–72.

Lüttge, U. and Ball, E. (1974) Proton and malate fluxes in cells of *Bryophyllum daigremontianum* leaf slices in relation to potential osmotic pressure of the medium. *Zeitschr. Pflanzenphysiol.*, **73**, 326–38.

Lüttge, U. and Ball, E. (1977) Water relations parameters of the CAM plant *Kalanchoë daigremontiana* in relation to diurnal malate oscillations. *Oecologia*, **31**, 85–94.

Lüttge, U., Ball, E., Kluge, M. and Ong, B.L. (1986a) Photosynthetic light requirements of various tropical vascular epiphytes. *Physiol. Veg.*, **24**, 285–90.

Lüttge, U., Stimmel, K.-H., Smith, J.A.C. and Griffiths, H. (1986b) Comparative ecophysiology of CAM and C_3 bromeliads. II. Field measurements of gas exchange of CAM bromeliads in the humid tropics. *Plant, Cell Environ.*, **9**, 377–84.

Lüttge, U. (1987) Carbon dioxide and water demand: Crassulacean acid metabolism (CAM), a versatile ecological adaptation exemplifying the need for integration in ecophysiological work. *New Phytol.*, **106**, 593–629.

MacDougal, D.T. and Spalding, E.S. (1910) *The Water Balance of Succulent Plants*, Carnegie Institution of Washington Publication no. 141, 77 pp.

Martin, C.E., Christensen, N.L. and Strain, B.R. (1981) Seasonal patterns of growth, tissue acid fluctuations, and $^{14}CO_2$ uptake in the crassulacean acid metabolism epiphyte *Tillandsia usneoides* L. (Spanish moss). *Oecologia*, **49**, 322–8.

McWilliams, E.L. (1970) Comparative rates of dark CO_2 uptake and acidification in Bromeliaceae, Orchidaceae and Euphorbiaciae. *Bot. Gaz.*, **131**, 285–90.

Medina, E., Delgado, M., Troughton, J.H. and Medina, J.D. (1977) Physiological ecology of CO_2 fixation in Bromeliaceae. *Flora*, **166**, 137–52.

Medina. E. and Osmond, C.B. (1981) Temperature dependence of dark CO_2 fixation and acid accumulation in *Kalanchoë daigremontiana*. *Austr. J. Plant Physiol.*, **8**, 641–9 (and corrigendum **12**, 212.)

Medina, E. and Troughton, J.H. (1974) Dark CO_2 fixation and the carbon isotope ratio in Bromeliaceae. *Plant Sci. Lett.*, **2**, 357–62.

Meinzer, F.C. and Rundel, P.W. (1973) Crassulacean acid metabolism and water use efficiency in *Echeveria pumila*. *Photosynthetica*, **7**, 358–64.

Milburn, T.R., Pearson, D.J. and Ndegwe, N.A. (1968) Crassulacean acid metabolism under natural tropical conditions. *New Phytol.*, **67**, 883–97.

Morrow, P.A. and Slatyer, R.O. (1971) Leaf temperature effects of measurements of diffusive resistance to water vapor transfer. *Plant Physiol.*, **47**, 559–61.

Neales, T.F. (1975) The gas exchange patterns of CAM plants. In *Environmental and Biological Control of Photosynthesis* (ed. R. Marcelle), Junk, The Hague, pp. 299–310.

Neales, T.F., Hartney, V.J. and Patterson, A.A. (1968) Physiological adaptation to drought in the carbon assimilation and water loss of xerophytes. *Nature, London*, **219**, 469–72.

Nierhaus, D. and Kinzel, H. (1971) Vergleichende Untersuchungen über die organischen Säuren in Blättern höherer. *Pflanz. Zeitschr. Pflanzenphysiol.*, **64**, 107–23.

Nishida, K. (1963) Studies on stomatal movement of crassulacean plants in relation to acid metabolism. *Physiol. Plant.*, **16**, 281–98.

Nobel, P.S. (1976) Water relations and photosynthesis of a desert CAM plant *Agave deserti*. *Plant Physiol.*, **61**, 510–14.

Nobel, P.S. (1977) Water relations and photosynthesis of a barrel cactus, *Ferocactus acanthodes*, in the Colorado desert. *Oecologia*, **27**, 117–33.

Nobel, P.S. (1983) Spine influences on PAR interception, stem temperature, and nocturnal acid accumulation by cacti. *Plant, Cell Environ.*, **6**, 153–9.

Nobel, P.S. (1984) Productivity of *Agave deserti*: measurement by dry weight and monthly prediction using physiological responses to environmental parameters. *Oecologia*, **64**, 1–7.

Nobel, P.S. (1985) PAR, water, and temperature limitations on the productivity of cultivated *Agave fourcroydes* (Henequen). *J. Appl. Ecol.*, **22**, 157–73.

Nobel, P.S. (1988) *Environmental Biology of Agaves and Cacti*, Cambridge University Press, Cambridge, 270 pp.

Nobel, P.S. and Hartsock, T.L. (1978) Resistance analysis of nocturnal carbon dioxide uptake by a Crassulacean acid metabolism succulent, *Agave deserti*. *Plant Physiol.*, **61**, 510–14.

Nobel, P.S. and Hartsock, T.L. (1983) Relationships between photosynthetically active radiation, nocturnal acid accumulation and CO_2 uptake for a crassulacean acid metabolism plant *Opuntia ficus-indica*. *Plant Physiol.*, **71**, 71–5.

Nobel, P.S. and Hartsock, T.L. (1986) Temperature, water, and PAR influences on predicted and measured productivity of *Agave deserti* at various elevations. *Oecologia*, **68**, 181–5.

Nobel, P.S. and Jordan, P.W. (1983) Transpiration stream of desert species: resistances and capacitances for a C_3, a C_4, and a CAM plant. *J. Exp. Bot.*, **34**, 1379–91.

Nobel, P.S. and Meyer, S.E. (1985) Field productivity of a CAM plant, *Agave salmiana*, estimated using daily acidity change under various environmental conditions. *Physiol. Plant.*, **65**, 397–404.

Nobel, P.S. and Quero, E. (1986) Environmental productivity indices for a CAM plant in the Chihuahuan Desert, *Agave lecheguilla*. *Ecology*, **67**, 1–11.

Nobel, P.S. and Sanderson, J. (1984) Rectifier-like activities of two desert succulents. *J. Exp. Bot.*, **35**, 727–37.

Nobel, P.S. and Smith, S.D. (1983) High and low temperature tolerances and their relationships to distribution of agaves. *Plant, Cell Environ.*, **6**, 711–19.

Nobel, P.S., Zaragoza, L.J. and Smith, W.K. (1975) Relation between mesophyll surface area, photosynthetic rate, and illumination level during development for leaves of *Plectanthus parviflorus* Henckel. *Plant Physiol.*, **55**, 1067–70.

Osmond, C.B. (1976a) Ion absorption and carbon metabolism in cells of higher plants. In *Transport in Plants. Encyclopedia of Plant Physiology*, New Series (eds U. Lüttge and M.G. Pitman), Springer-Verlag, Berlin, Vol. IIA, pp. 347–72.

Osmond, C.B. (1976b) CO_2 assimilation and dissimilation in the light and dark in CAM plants. In *CO_2 Metabolism and Plant Productivity* (eds R.H. Burris and C.C. Black), University Park Press, Baltimore, pp. 217–33.

Osmond, C.B. (1978) Crassulacean acid metabolism: a curiosity in context. *Ann. Rev. Plant Physiol.*, **29**, 374–414.

Osmond, C.B. (1982) Carbon cycling and the stability of the photosynthetic apparatus in CAM. In *Crassulacean Acid Metabolism* (eds I.P. Ting and M. Gibbs), American Society of Plant Physiologists, Rockville, pp. 112–27.

Osmond, C.B., Austin, M.P., Berry, J.A., Billings, W.D., Boyer, J.S., Dacey, J.W.H., Nobel, P.S., Smith, S.D. and Winner, W.E. (1987) Stress physiology in the context of physiological ecology. *Bioscience*, **37**, 38–48.

Osmond, C.B., Bender, M.M. and Burris, R.H. (1976) Pathways of CO_2 fixation in the CAM plant *Kalanchoë daigremontiana*. III. Correlation with $\delta^{13}C$ value during growth and water stress. *Austr. J. Plant Physiol.*, **3**, 787–89 (and corrigendum **4**, 689).

Osmond, C.B. and Björkman, O. (1975) Pathways of CO_2 fixation in the CAM plant *Kalanchoë daigremontiana* II. Effects of O_2 and CO_2 concentration on light and dark CO_2 fixation. *Austr. J. Plant Physiol.*, **2**, 155–62.

Osmond, C.B. and Holtum, J.A.M. (1981) Crassulacean acid metabolism. In *The Biochemistry of Plants* (eds M.D. Hatch and N.K. Boardman), Academic Press, New York, Vol. 8, pp. 283–328.

Osmond, C.B., Ludlow, M.M., Davis, R., Cowan, I.R., Powles, S.B. and Winter, K. (1979b) Stomatal responses to humidity in *Opuntia inermis* in relation to control of CO_2 and H_2O exchange patterns. *Oecologia*, **41**, 65–76.

Osmond, C.B., Nott, D.L. and Firth, P.M. (1979a) Carbon assimilation patterns and growth of the introduced CAM plant *Opuntia inermis* in Eastern Australia. *Oecologia*, **40**, 331–50.

Osmond, C.B., Winter, K. and Ziegler, H. (1982) Functional significance of different pathways of CO_2 fixation in photosynthesis. In *Encyclopedia of Plant Physiology*, New Series. (eds O.L. Lange, P.S. Nobel, C.B. Osmond and H.

Ziegler), Springer-Verlag, Berlin, Vol. 12B, pp. 497–547.

Powles, S.B. (1984) Photoinhibition of photosynthesis by visible light. *Annu. Rev. Plant Physiol.*, **35**, 15–44.

Powles, S.B. and Björkman, O. (1982) Photoinhibition of photosynthesis: effect on chlorophyll fluorescence at 77K in intact leaves and in chloroplant membranes of *Nerium oleander*. *Planta*, **156**, 97–107.

Pucher, G.W., Vickery, H.B., Abrahams, M.D. and Leavenworth, C.S. (1949) Studies on the metabolism of crassulacean plants: diurnal variation in the organic acids and starch in excised leaves of *Bryophyllum calycinum*. *Plant Physiol.*, **24**, 610–20.

Rayder, L. and Ting, I.P. (1983) Shifts in the carbon metabolism of *Xerosicyos danguyi* H. Humb. (Cucurbitaceae) brought about by water stress. *Plant Physiol.*, **72**, 606–10.

Ruess, B.R. and Eller, B.M. (1985) The correlation between crassulacean acid metabolism and water uptake in *Senecio smedley-woodii*. *Planta*, **166**, 57–66.

Sale, P.J.M. and Neales, T.F. (1980) Carbon dioxide assimilation by pineapple plants, *Ananas comosus* (L.) Merr. I. Effects of daily irradiance. *Austr. J. Plant Physiol.*, **7**, 363–73.

Schreiber, U. (1983) Chlorophyll fluorescence yield changes as a tool in plant physiology I. The measuring system. *Photosynth. Res.*, **4**, 361–73.

Schreiber, U. and Berry, J.A. (1977) Heat-induced changes in chlorophyll fluorescence in intact leaves correlated with damage of the photosynthetic apparatus. *Planta*, **136**, 233–8.

Schreiber, U., Grobermann, L. and Vidaver, W. (1975) Portable solid-state fluorometer for the measurement of chlorophyll fluorescence induction in plants. *Rev. Sci. Instrum.*, **46**, 538–42.

Schreiber, U., Schliwa, U. and Bilger, W. (1986) Continuous recording of photochemical and nonphotochemical chlorophyll fluorescence quenching with a new type of modulation fluorimeter. *Photosyn. Res.*, **10**, 51–62.

Sharp, R.E., Matthews, M.A. and Boyer, J.S. (1984) Kok effect and the quantum yield of photosynthesis. *Plant Physiol.*, **75**, 95–101.

Sinclair, R. (1983a) Water relations of tropical epiphytes. I. Relationships between stomatal resistance, relative water content and the components of water potential. *J. Exp. Bot.*, **149**, 1652–63.

Sinclair, R. (1983b) Water relations of tropical epiphytes. II. Performance during droughting. *J. Exp. Bot.*, **149**, 1664–75.

Smillie, R.M. and Hetherington, S.E. (1983) Stress tolerance and stress-induced injury in crop plants measured by chlorophyll fluorescence *in vivo*. Chilling, freezing, ice-cover, heat and light. *Plant Physiol.*, **72**, 1043–50.

Smith, J.A.C., Griffiths, H., Bassett, M. and Griffiths, N.M. (1985) Day–night changes in the leaf water relations of epiphytic bromeliads in the rainforests of Trinidad. *Oecologia*, **67**, 475–85.

Smith, J.A.C., Griffiths, H. and Lüttge, U. (1986a) Comparative ecophysiology of CAM and C_3 bromeliads. I. The ecology of the Bromeliaceae in Trinidad. *Plant Cell Environ.*, **9**, 359–76.

Smith, J.A.C., Griffiths, H., Lüttge, U., Crook, C.E., Griffiths, N.M. and Stimmel, K.-H. (1986b) Comparative ecophysiology of CAM and C_3 bromeliads. IV. Plant water relations. *Plant, Cell Environ.*, **9**, 395–410.

Smith, J.A.C. and Lüttge, U. (1985) Day–night changes in leaf water relations associated with the rhythm of crassulacean acid metabolism in *Kalanchoë daigremontiana*. *Planta*, **163**, 272–82.

Smith, S.D., Didden-Zopfy, B. and Nobel, P.S. (1984) High temperature responses of North American cacti. *Ecology*, **65**, 643–51.

Steudle, E., Smith, J.A.C. and Lüttge, U. (1980) Water-relation parameters of individual mesophyll cells of the crassulacean acid metabolism plant *Kalanchoë daigremontiana*. *Plant Physiol.*, **66**, 1155–63.

Szarek, S.R., Johnson, H.B. and Ting, I.P. (1973) Drought adaptation in *Opuntia basilaris*. Significance of recycling carbon through crassulacean acid metabolism. *Plant Physiol.*, **52**, 539–41.

Szarek, S.R. and Ting, I.P. (1974) Seasonal patterns of acid metabolism and gas exchange in *Opuntia basilaris*. *Plant Physiol.*, **54**, 76–81.

Teeri, J.A., Tonsor, S.J. and Turner, M. (1981) Leaf thickness and carbon isotope composition in the Crassulaceae. *Oecologia*, **50**, 367–9.

Ting, I.P. (1985) Crassulacean acid metabolism. *Annu. Rev. Plant Physiol.*, **36**, 595–622.

Ting, I.P. and Hanscom III, Z. (1977) Induction of acid metabolism in *Portulacaria afra*. *Plant Physiol.*, **59**, 511–14.

Vernon, L.P. (1960) Spectrophotometric determination of chlorophylls and pheophytins in plant extracts. *Anal. Chem.*, **32**, 1144–50.

von Willert, D.J., Brinckmann, E., Scheitler, B. and Eller, B.M. (1985) Availability of water controls crassulacean acid metabolism in succulents of the Richtensveld (Namib Desert, South Africa). *Planta*, **164**, 44–55.

von Willert, D.J., Thomas, D.A., Lobin, W. and Curdts, E. (1977) Ecophysiological investigations in the family Mesembryanthemaceae: occurrence of CAM and ion contents. *Oecologia*, **29**, 67–76.

Walker, D.A. (1981) Secondary fluorescence kinetics of spinach leaves in relation to the onset of photosynthetic carbon assimilation. *Planta*, **153**, 273–8.

Walker, D.A. (1988) *The use of the oxygen electrode and fluorescence probes in simple measurements of photosynthesis*, Oxygraphics Ltd, Sheffield.

Walker, D.A. and Osmond, C.B. (1986) Measurement of photosynthesis *in vivo* with a leaf disc electrode: correlations between light dependence of steady state photosynthetic O_2 evolution and chlorophyll *a* fluorescence transients. *Proc. R. Soc. London Ser. B*, **227**, 267–80.

Wiebe, H.H. and Al-Saadi, H.A. (1976) Matric bound water of water tissue from succulents. *Physiol. Plant.*, **36**, 47–51.

Winter, K. (1980) Carbon dioxide and water vapor exchange in the crassulacean acid metabolism plant *Kalanchoë pinnata* during a prolonged light period. *Plant Physiol.*, **66**, 917–21.

Winter, K. (1985) Crassulacean acid metabolism. In *Photosynthetic Mechanisms and the Environment* (eds J. Barber and N.R. Baker), Elsevier, Amsterdam, pp. 329–87.

Winter, K. and Demmig, B. (1987) Reduction state of Q and non-radiative energy dissipation during photosynthesis in leaves of a crassulacean acid metabolism plant, *Kalanchoë daigremontiana* Hamet et Penn. *Plant Physiol.*, **85**, 1000–7.

Winter, K., Lüttge, U., Troughton, J.H. and Winter, E. (1978) Seasonal shift from C_3 photosynthesis to crassulacean acid metabolism in *Mesembryanthemum crystallinum* growing in its natural environment. *Oecologia*, **25**, 225–37.

Winter, K., Osmond, C.B. and Hubick, K.T. (1986a) Crassulacean acid metabolism in the shade. Studies on an epiphytic fern, *Pyrrosia longifolia*, and other rainforest species from Australia. *Oecologia*, **68**, 224–30.

Winter, K., Schröppel-Meier, G. and Caldwell, M.M. (1986b) Respiratory CO_2 as a carbon source for nocturnal acid synthesis at high temperatures in three species exhibiting crassulacean acid metabolism. *Plant Physiol.*, **81**, 390–4.

Winter, K., Wallace, B.J., Stocker, G. and Roksandic, Z. (1983) Crassulacean acid metabolism in Australian vascular epiphytes and some related species. *Oecologia*, **57**, 129–41.

13

Stable isotopes

James R. Ehleringer and C. Barry Osmond

13.1 INTRODUCTION

The use of stable isotopes at natural abundance levels is rapidly emerging as a powerful approach for understanding a number of physiological processes and food web and environmental interactions in ecology, especially in physiological ecology. The analysis of stable isotopes developed as an outgrowth from geochemical investigations and represents a relatively new approach within ecological studies. In the 1970s, the principal application of stable isotopes in physiological ecology was for the measurement of $^{13}C/^{12}C$ ratios to identify the photosynthetic pathway of a species. Today stable isotopes are being applied to a broader range of questions, including nitrogen fixation, water-use efficiency and water-source studies. Our ecological understanding of the applications of stable isotopes is still in its infancy, but recent advances suggest that the rapid expansion of stable isotope studies into new areas is likely to continue in the coming decade. The purpose of this chapter is to introduce stable isotopes and the techniques for their measurement, as well as sampling and preparation procedures. Although several ecological applications of stable isotopes are presented, the discussion is by no means complete and may represent only a fraction of the applications developed in the next several years.

13.2 NATURAL ABUNDANCES OF STABLE ISOTOPES OF ECOLOGICAL INTEREST

Most elements of biological interest have two or more stable isotopes, although one isotope is usually present in far greater abundance than other forms (Hoefs, 1980). Table 13.1 lists the average abundance of the elements used in ecological studies. Isotopic abundances of these elements are by no means uniform in nature, but most of the variation between biotic and abiotic components is within 1% of the values in Table 13.1. In addition to the five light elements, strontium isotopes are assuming greater importance in understanding ecological transport processes (Rundel *et al.*, 1988) and have therefore been included, although they will not be discussed further in this chapter.

Table 13.1 Average terrestrial abundance of the isotopes of major elements used in environmental studies (from Fritz and Fontes, 1980)

Element	*Isotope*	*Average terrestrial abundance (%)*
Hydrogen	^1H	99.985
	^2H	0.015
Carbon	^{12}C	98.89
	^{13}C	1.11
Oxygen	^{16}O	99.759
	^{17}O	0.037
	^{18}O	0.204
Nitrogen	^{14}N	99.63
	^{15}N	0.37
Sulfur	^{32}S	95.0
	^{33}S	0.76
	^{34}S	4.22
	^{36}S	0.014
Strontium	^{84}Sr	0.56
	^{86}Sr	9.86
	^{87}Sr	7.02
	^{88}Sr	82.56

13.3 STABLE ISOTOPE MASS SPECTROMETRY

13.3.1 Mass spectrometers

Mass spectrometers are instruments which measure the mass-to-charge ratio of a substance. In a mass spectrometer, the compound is first ionized under high-vacuum conditions and then deflection of its ions is measured while subject to a magnetic field. Most isotope ratio mass spectrometers are capable of measuring only low-molecular-weight compounds (usually <64). The compounds are introduced into the instrument as gases, most often as H_2, CO_2, N_2 and SO_2, permitting measurement of the isotope ratios of H, C, N, O and S in organic and inorganic materials. Heavier elements, such as strontium, are measured with a thermal emission isotope ratio mass spectrometer.

In an isotope ratio mass spectrometer, the pure gas (H_2, CO_2, N_2, SO_2) is introduced at an inlet at one end of a flight tube. At this point, the gas is then ionized in an ion source which knocks an electron from the outer shell of the compound. The beam of ionized gas is accelerated and deflected along the flight tube by a powerful magnet (Fig. 13.1). Because the ions have different mass-to-charge ratios, light and heavy ions containing different isotopes will be deflected differently and sorted by the magnetic field. At the opposite end of the flight tube are a series of collectors (Faraday cups) which are positioned to capture the charged ions of different mass (Fig. 13.2). DC amplifiers attached to the Faraday cups convert the ionic impacts into a voltage, which is then converted to a frequency. The absolute intensity of the signal on the Faraday cup is not critical because this depends on the amount of gas introduced into the mass spectrometer and other factors. Rather the critical parameter is the ratio of the signals measured by the different Faraday cups.

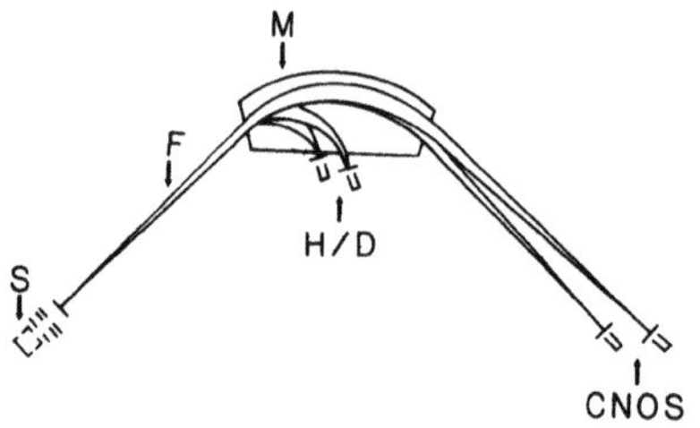

Fig. 13.1 Principal features of an isotope ratio mass spectrometer, including flight tube (F), ionizing source (S), magnet (M), Faraday cups for detecting hydrogen isotopes (H/D), and Faraday cups for detecting C, N, O and S isotopes (CNOS).

Only two Faraday cups are needed for hydrogen isotope measurements, since the only ionized gases are $^1H^1H$, $^2H^1H$. However, for CO_2 and N_2, three cups are used in the

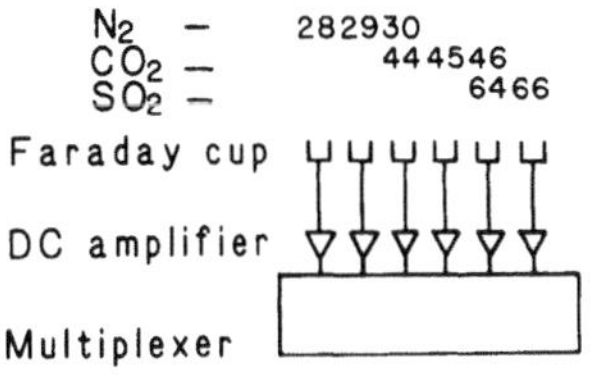

Fig. 13.2 Schematic diagram of Faraday cups and amplifier systems.

measurements. Three isotope forms of nitrogen gas ($^{14}N^{14}N$, $^{14}N^{15}N$ and $^{15}N^{15}N$) require one Faraday cup to detect each form. Carbon dioxide is more complicated since two elements and five isotopes can lead to light isotopic forms of the gas. Potential complications arise because several forms have the same mass and as many as six different masses (44 through 49) should conceivably be measured to detect all the forms. Fortunately ^{17}O is present in only trace amounts, assumed to be a constant percentage of the ^{18}O abundance and forms with more than two isotopes are present at low frequency. It is usually assumed that the different combinations are present in equal proportion to their isotopic abundances, so only three masses (44, 45 and 46) need be measured to calculate the $^{13}C/^{12}C$ and $^{18}O/^{16}O$ ratios.

One of the first requirements of an isotope ratio mass spectrometer is a good vacuum system which maintains extremely low levels of contaminating molecules. The flight tube operates at a vacuum of approximately 10^{-8} Torr ($\approx 10^{-5}$ Pa). Since the mean free path of a gas molecule is inversely proportional to the pressure, a vacuum that low will insure a mean free path length of over 500 m. This means that ions traveling down the flight tube (usually 0.5–1 m in length) will not collide with other gas molecules and be scattered.

The gas inlet into the mass spectrometer is symmetrically arranged for the introduction of either sample or standard gases. Gases are temporarily stored in a metal bellows and then are passed through a set of matched capillaries (one for each side) to ensure viscous flow of the gases. Precise matching of the capillaries ensures that fractionation of the gases prior to introduction into the mass spectrometer is small. A changeover valve is used to switch between the standard and sample gases. The difference in the signals between sample and standard gases is used to calculate the isotope ratio for the sample. Measurements of the absolute ratios (i.e., 45/44 and 46/44 for CO_2) on a given gas cannot be made with the same precision as a comparison between two samples, so the difference between the sample and standard ratios is of interest in the isotope ratio calculation.

13.3.2 Delta units and standards

The differences in the equilibrium and kinetic characteristics of isotopic species are usually small (on the order of a few percent), and thus absolute variations in isotopic abundances based on physical factors may be small. Enzymatic discrimination for or against an isotopic species will affect the absolute abundances, but again these variations are on the order of one or two percent. Therefore, in any isotopic analysis very precise analytical techniques are required. Most often, it has been found that measuring the absolute isotopic composition is not as reliable and/or convenient as measuring isotopic differences between a sample and a given standard. This is because, while obtaining high precision in absolute isotopic composition of a sample is not difficult over the short term, machine drift has a time base of minutes needed for replicate measurements (Hayes, 1983). In contrast, analyses based on the measurement of the differences between a defined standard and sample provide high precision and repeatability over both short-term and long-term periods. The differential analysis approach allows very small differences in the isotopic composition of two samples to be accurately and reliably determined.

Isotopic composition of a sample is therefore usually expressed with the differential notation (Friedman and O'Neil, 1978). That is:

$$\delta X_{std} = (R_{sam}/R_{std} - 1)1000 \quad [0/00]$$

where δX_{std} is the isotope ratio in delta units relative to a standard, and R_{sam} and R_{std} are the isotope abundance ratios of the sample and standard respectively. Multiplying by 1000 allows the values to be expressed in parts per thousand, or as more commonly expressed on a 'per mil' (‰) basis. Since the isotopic composition of two samples will not differ extensively in their absolute values, the differential notation allows one to focus on the differences between samples.

There are presently four accepted isotopic standards for the five principal light elements of biological interest. These are Standard Mean Ocean Water (SMOW) for hydrogen and oxygen, PeeDee Belemnite (limestone) (PDB) for carbon, atmospheric air for nitrogen and the Canyon Diablo meteorite (CD) for sulfur. Estimated absolute ratios of these standards are listed in Table 13.2. While there is some variance in the estimates of the absolute ratios in these standards, the use of the differential or deviation from standard measurement approach overcomes these and provides far greater precision and long-term reliability. The original supplies of both SMOW and PDB have been exhausted and replaced by other materials which had been carefully compared to the original standards. These standards are available to investigators for calibration of working standards in each mass spectrometer laboratory. The International Atomic Energy Agency in Vienna has mixed various waters together to produce V-SMOW (Vienna SMOW), which has an isotopic composition nearly identical to that of the original SMOW. The National Bureau of Standards provides a graphite, NBS–21, with a carbon isotope of −28.10‰ on the PDB scale. The latter standard is not easily combusted so working standards of a chemical composition similar to those of the unknowns should be selected. Greatest accuracy will be obtained with standards having an R value similar to that of the unknowns. Thus in carbon isotope analyses, inclusions of sucrose standards (such as the Australian National University Radiocarbon Dating sucrose standard −10.5‰; Chinese Radiocarbon Dating Charred Sucrose standard −24.4‰ or other beet sucrose source) among every ten or so unknowns insures contained appropriate calibration.

Table 13.2 Isotopic compositions of primary standards (from Hayes, 1983)

Primary standard	*Isotope ratio*	*Accepted value (×10⁶) (with 95% confidence interval)*
Standard Mean Ocean Water (SMOW)		
	$^{2}H/^{1}H$	155.76±0.10
	$^{18}O/^{16}O$	2005.20±0.43
	$^{17}O/^{16}O$	373±15
PeeDee Belemnite (PDB)		
	$^{13}C/^{12}C$	11237.2±9.0
	$^{18}O/^{16}O$	2067.1±2.1
	$^{17}O/^{16}O$	379±15
Air		
	$^{15}N/^{14}N$	3676±8.1

13.3.3 Resolution and precision

Resolution on modern mass spectrometers is approximately 0.005–0.01‰. However, this value often exceeds the precision that mass spectrometers can provide. The precision in isotopic measurements is dependent on the particular element of interest and depends on three factors. First is the precision of the isotope ratio mass spectrometer itself, which is calculated from repeated measurements of the same sample gas, second is the amount of gas injected, and third is the precision of the sample preparation (conversion of element from sample form to a gas which can be injected into the mass spectrometer), which will be discussed in a later section. The internal precision is usually defined as two times the standard error of 10 analyses of a single gas sample. Standard inlets for modern gas isotope mass spectrometers have an internal precision of approximately 0.01‰ for 100 μl of CO_2, 0.02‰ for 200 μl of N_2 and 0.2‰ for 200 μl of H_2.

13.3.4 Automation and other recent advances

A number of recent advances allow for analysis of small sample sizes and for large sample throughput. However, precise isotope measurements require adequate sample gas pressure in the mass spectrometer. This is achieved by reducing the volume within the inlet system, usually by means of a variable metal bellow in conjunction with a 'cold-finger' which freezes the gas into a small volume. With a cold finger attachment, the above precision can be achieved with only one-tenth the amount of gas.

In terms of sample throughput, there have been two recent advances. The first is the use of automatic gas-handling equipment at the front end of the mass spectrometer. Such systems are computer controlled and allow for 36 or more samples to be analyzed in succession with high precision and without operator intervention. The sample cycle time in such a situation will be approximately 20–30 min, meaning that as many as 50 or so samples could be analyzed per day. A very recent development is the use of CHN elemental analyzers coupled to the mass spectrometer. In this approach, an organic sample is first combusted and its elemental composition determined, then the gases are sent directly to the mass spectrometer. Such systems are still in their early stages of testing; it is thought that 100–300 samples per day could be analyzed using this approach without significant loss of precision.

13.4 SAMPLE PREPARATION

Few special precautions are necessary for preparing and storing plant samples for later determination of their isotopic composition. Leaves have been commonly used for most measurements in the past, although there is now increasing interest in the isotopic composition of other tissues and plant parts. Changes in chemical composition which accompany long storage and slow drying of living material should be avoided. It is best to freeze tissue at time of collection and then to freeze-dry it, or dry as quickly as possible at moderate temperature, to avoid loss of organic materials. The isotope ratios of organic material are determined on dried tissues that have been ground to pass a 40 mesh screen. Only a small amount of tissue is required for the analysis, and so the grinding ensures that the sample is homogeneous, and minimizes variation in isotope composition that might exist within the tissue or in any bulked sample. Finely ground material also burns more uniformly. Isotopic composition of water samples requires that the water be immediately sealed in filled glass vials to ensure that isotopic fractionation due to distillation does not occur between the time of collection and later analysis.

The necessity for sample homogeneity cannot be emphasized enough. In most cases there will be a greater variance in the repeated analysis of the same 'bulk sample' than in repeated analysis of an individual sample through the mass spectrometer. This is partly because the amount of tissue required for an analysis is usually quite small. In most cases, less than 3 mg of dried organic material or water is used for D/H, $^{13}C/^{12}C$ and $^{18}O/^{16}O$ measurements (slightly more is required for $^{15}N/^{14}N$ analyses). With a coldfinger option on the mass spectrometer, less than 0.1 mg will be used in the analysis, further compounding this source of variation.

13.4.1 $^{13}C/^{12}C$ in organic matter

The $^{13}C/^{12}C$ of organic materials is analyzed as CO_2 in the mass spectrometer. Originally, the quantitative production of CO_2 from organic matter required repeated cycling of an O_2 atmosphere and the combustion products through a furnace and traps to absorb contaminant products such as nitrogen and sulfur oxides (Craig, 1953). This was a slow process in which only one sample could be prepared at a time. The O_2 pressure bomb method described by Osmond *et al.* (1975) was similarly time-consuming and susceptible to incomplete combustion. These methods have been replaced by semiautomatic combustion trains such as the ISOPREP-13 (VG Instruments, Oxford, UK) and those based on elemental analyses in which high-efficiency combustion and small volume accelerate the process. Batch preparations of CO_2 for $^{13}C/^{12}C$ analysis can be done by an in-vial combustion technique similar to that described originally by Buchanan and Corcoran (1959). Dried organic material, cupric oxide and silver foil are sealed under vacuum in a Vycor glass tube. The sealed tubes are then heated in a furnace at 850°C for 4 h and then allowed to cool slowly for another 12 h. After combustion the sealed tube contains CO_2, H_2O and N_2. The tube is then cracked under vacuum and the gases are separated by passing them first through an ethanol–dry ice trap, to freeze out the water, and then through a liquid-nitrogen trap to freeze out the CO_2. The remaining gas (primarily diatomic nitrogen) is pumped away. The clean CO_2 is then cryogenically moved into a vial to be then transferred to the mass spectrometer. Organic samples should be analyzed soon after they have been combusted, because H_2O and CO_2 will slowly interact with the copper present in the tube to form copper carbonate and this may affect the isotopic composition of the remaining CO_2. The variance in sample preparation with this technique is less than 0.05‰.

13.4.2 D/H in water samples

Hydrogen isotope ratios are measured after the hydrogen in water is reduced to diatomic hydrogen through one of two possible procedures. The first involves the reduction of water to hydrogen by passing the water vapor through a uranium furnace at 750°C with a Toepler pump and then trapping the hydrogen on activated charcoal at liquid-nitrogen temperatures (Bigeleisen *et al.*, 1952). This is a relatively time-consuming process, in which all samples must be individually processed. A second speedier method uses zinc as the catalyst at temperatures of 420°C (Coleman *et al.*, 1982). A clear advantage of using zinc as the catalyst is that the process can be batched so that large numbers of samples can be prepared simultaneously. In the batch procedure developed by Hayes and Studley (personal communication), water samples (usually in a capillary) are inserted into Vycor tubes (previously backfilled with nitrogen), zinc catalyst is added, and the tubes are then frozen, evacuated and sealed. The tubes are then heated at 500°C for 30 min in a heating block. The only gas then remaining in the tubes is diatomic hydrogen, which can be directly taken to the mass spectrometer for analysis. The precision of this technique is approximately 0.8‰.

Water in plant tissues can be recovered by

lyophilization or azeotrophic distillation from dry toluene. In the final method, leaves or other tissues collected in the field are enclosed in small sealed containers (e.g. plastic bags) and frozen in liquid N_2 or on solid CO_2 (Sternberg *et al.*, 1986). Care should be taken to minimize transpiratory losses between removal from the plant and freezing. Tissue water is subsequently recovered by freeze-drying with a double liquid-N_2 trap in the vacuum line between the sample and the pump (Farris and Strain, 1978). A less convenient sampling procedure involves rapid transfer of the tissue to a flask of sodium-dried toluene, followed by azeotrophic distillation of the water in the laboratory (Leaney *et al.*, 1985). This method is useful, however, for the recovery of soil water for isotopic analysis.

13.4.3 D/H in organic matter

If the purpose of the investigation is to relate long-term average δD values of organic matter to average soil/meteoric water δD values, the most acceptable method is to purify cellulose and to nitrate it. This method ensures that the exchangeable hydroxyl groups of cellulose do not experience further isotopic fractionation during processing (Mann, 1971). The cellulose is extracted with sodium chlorite/acetic acid, then washed with sodium hydroxide and acetic acid (Wise, 1944). Finally, the cellulose is nitrated with nitric acid and acetic anhydride (Bennett and Timell, 1955; Epstein *et al.*, 1976; DeNiro, 1981; Yapp and Epstein, 1982). Nitration can be applied to other fractions such as sucrose (Dunbar and Schmidt, 1984). The nitrated material is then combusted using the same combustion technique described above for carbon. Following this, the combustion tube is cracked under vacuum and the water separated by passing the gases through an ethanol–dry ice trap to freeze out the water. The remaining gases are pumped away. The water from combustion is then reduced to diatomic hydrogen through one of two possible procedures as described previously for D/H measurements of water.

If the purpose of the investigation is to understand the dynamics of hydrogen isotope fractionation processes during photosynthesis in different plants and environments, then there is little point in concentrating on cellulose and its nitration. The δD value of lipids, for example, vary markedly from that of cellulose (Smith and Epstein, 1971a; Estep and Hoering, 1980; Sternberg *et al.*, 1984a). Estep and Hoering (1980) showed that saponification and separation procedures for lipids did not significantly alter the isotopic composition of standards. The δD value of sucrose from C_3 plants is different from that of C_4 plants (Smith, 1975) and it is unlikely that this is an artifact of commercial purification procedure. The δD value of organic hydrogen obtained from water of combustion of dried plant material varies markedly with metabolic pathway and environment (Ziegler *et al.*, 1976; Estep and Hoering, 1981). The δD value of cellulose nitrate from CAM plants (but not that of lipid), differs from that in C_3 and C_4 plants growing in the same location (Sternberg *et al.*, 1984a). These experiments show that water collected by combustion of dried organic fractions from plants, if checked with controls passed through the same purification procedures, is likely to yield valuable information on the relationship between the physiology and biochemistry of water in photosynthetic reduction.

13.4.4 $^{15}N/^{14}N$ in organic tissues

The $^{15}N/^{14}N$ of organic materials is determined from the isotope composition of N_2 produced from ammonium sulfate prepared by the Kjeldahl method (Kohl *et al.*, 1971; Hauck, 1982). In this reaction, the ammonia is mixed with sodium hypobromide to produce diatomic nitrogen via the Rittenberg reaction. This is a slow process, involving many steps and in which only one sample can be prepared at a time.

Organic material is now routinely prepared for $^{15}N/^{14}N$ analysis using a batch mode combustion technique similar to that described above for carbon (Minagawa *et al.*, 1984). Dried organic material, cupric oxide and silver foil are sealed under vacuum in a Vycor glass tube. The sealed tubes are then heated in a furnace at 850°C for 4 h and then allowed to cool slowly for another 12 h. After the combustion process has been completed, the sealed tube contains CO_2, H_2O and N_2 which are separated by passing them first through an ethanol–dry ice trap to freeze out the water and then through a liquid-nitrogen trap to freeze out the CO_2. After the CO_2 has been frozen out, the nitrogen gas is then trapped on activated charcoal in a vial at liquid-nitrogen temperatures. The clean N_2 is then transferred to the mass spectrometer for analysis.

13.4.5 $^{18}O/^{16}O$ in water samples

Largely because of adsorption and condensation problems, the isotopic composition of water is not directly measured in mass spectrometers. Instead, the $^{18}O/^{16}O$ composition of waters is usually determined by equilibration with CO_2 (Compston and Epstein, 1958); ^{18}O and ^{16}O composition at equilibrium is known. A known volume of wäter, typically 3–5 ml, is placed in a small vessel of approximately twice that volume, air is removed from the vessel and replaced by CO_2. After allowing the vessel to equilibrate for 8–36 h in a constant-temperature water bath, a portion of the CO_2 is withdrawn and analyzed in the mass spectrometer. Since the molar fraction of oxygen in the water is so much greater than that in the CO_2, the $^{18}O/^{16}O$ ratio of the CO_2 takes on the value of the water (after correcting for a known liquid–gas phase equilibrium fractionation). Although this procedure can be batched in that numerous vessels can be equilibrating simultaneously, it is nonetheless a relatively slow process.

A promising approach to measuring the $^{18}O/^{16}O$ ratios in small volumes of water is to react guanidine hydrochloride with water to produce CO_2. This technique has been used successfully by Dugan *et al.* (1985) and Wong *et al.* (1987) to measure $^{18}O/^{16}O$ ratios on 10 μl water samples. The guanidine hydrochloride and water are heated in an evacuated sealed tube at 260°C for 16 h. The two gases formed in this reaction are NH_3 and CO_2. Upon cooling, the CO_2 combines with NH_3 to form an ammonium carbamate. The CO_2 is released from the ammonium carbamate by reacting it with phosphoric acid, trapped by freezing the CO_2 at liquid-N_2 temperatures and injected into the mass spectrometer for analysis. The clear advantage of the guanidine hydrochloride method is that very small sample sizes can be used. The precision of this technique is similar to that of the H_2O–CO_2 equilibration method (Wong *et al.*, 1987). Both methods have a standard deviation for sample preparation of approximately 0.2‰.

13.4.6 $^{18}O/^{16}O$ in organic tissues

The $^{18}O/^{16}O$ ratios in plant organic matter are usually determined on purified cellulose (Burk, 1979). The best method, like those for $^{13}C/^{12}C$ and H/D, is an in-vial combustion technique, in which cellulose and mercuric chloride are combusted in sealed tubes at 850°C in a muffle furnace. The gases produced are CO_2, CO and HCl. The CO is converted to CO_2 by electric discharge and then the gaseous HCl is removed by trapping it in isoquinoline. The CO_2 is frozen out into a vial at liquid-nitrogen temperatures and then transferred to the mass spectrometer for analysis. The combustion process can be batched; however, the later steps involving conversion of CO to CO_2 and trapping out of the HCl are slow and cannot be batched. On-line methods, based on elemental analyses are being developed.

13.5 SAMPLE VARIABILITY

13.5.1 Variation among tissue types and within cellular components

Individual plant tissue components may vary considerably in their carbon and nitrogen isotope ratios. Pectins, hemicellulose, starches and sugars are typically heavier (have more ^{13}C) than cellulose and lignins; lipids tend to be lighter than other cell components (Deines, 1980; O'Leary, 1981). Amino acids and other nitrogenous compounds may differ in their nitrogen isotope ratios (Shearer and Kohl, 1988). Consequently, when carbon or nitrogen isotope ratios of different organs are compared, there can be systematic variations dependent on tissue composition (O'Leary, 1981; Farquhar *et al.*, 1988; Shearer and Kohl, 1988). However, since the correlations among different organ types (leaves, roots, seeds, wood, etc.) remain high, interplant comparisons will remain valid so long as the ecological comparison is made using similar tissues. For the most part, ecological comparisons have been made using whole leaf tissues.

13.5.2 Environmentally induced variation

It is now commonly accepted that the carbon isotope ratio of C_3 and CAM-inducible photosynthetic plants is very much influenced by environmental factors (discussed in greater detail below). This necessitates a certain caution in that if the interest is beyond simply determining the photosynthetic pathway of a plant, care must be taken in sampling to insure that the tissue samples are from similar environmental regimes. It is incorrect also to assume that carbon isotope ratios of leaves, for example, are constant within a single plant (such as a tree) if leaves are exposed to different microclimates. For example, Ehleringer *et al.* (1986) have shown that there is a strong correlation between leaf carbon isotope ratio and light environment, so that leaves developing within the canopy or under the shade of other canopies can differ widely in the isotopic values (Fig. 13.3).

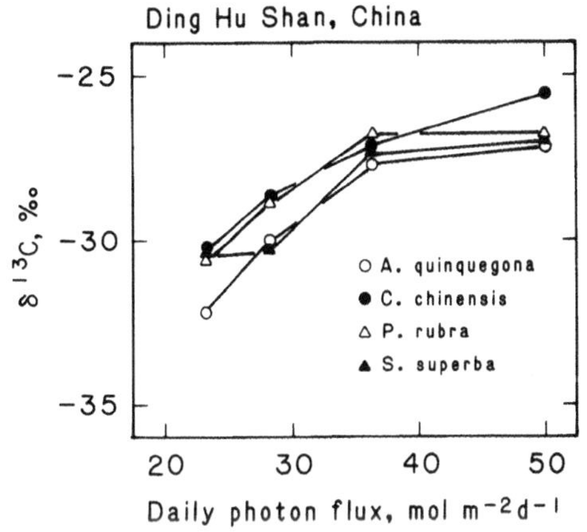

Fig. 13.3 The correlation between leaf carbon isotope ratio and the light environment in which that leaf developed for several tree and shrub species in a monsoonal tropical forest in China (redrawn from Ehleringer *et al.*, 1986).

13.5.3 Sample requirements

Sample sizes will of course depend on the specific research question posed, and will be very much affected by the magnitude of the differences to be resolved, and on the extent of both environmental and genetic heterogeneity. The precision of the mass spectrometer and preparation methods have been presented earlier and these values set the minimum error associated with a sample. Few studies have examined the isotopic variability in ecological situations, but of those available the indication is that three to five individual sample replicates may be needed to characterize an ecological situation. Since there are trade-offs between sample size and the cost of the research, one alternative is to bulk samples, thereby reducing interplant variability. However, it is likely that genetic variation in isotopic composition exists within natural populations, and that by lumping samples together to form a single sample observation, valuable information on the structure of the population will be lost.

13.6 APPLICATION OF STABLE ISOTOPES IN ECOLOGICAL STUDIES

13.6.1 Photosynthetic pathway determination

Perhaps the first ecophysiologically related uses of stable isotopes were those of Bender (1968, 1971) and Smith and Epstein (1971b), who showed that carbon isotopic composition could be used to distinguish between C_3 and C_4 photosynthetic pathway plants. This area developed rapidly and over the next several years the phylogenetic and ecological distributions of the C_3 and C_4 photosynthetic pathways were established (Smith and Brown, 1973; Card *et al.*, 1974; Troughton *et al.*, 1974; Osmond *et al.*, 1975, 1982; Smith and Turner, 1975; Webster *et al.*, 1975; Eickmeier and Bender, 1976; Winter *et al.*, 1976; Mooney *et al.*, 1977; Rundel *et al.*, 1979; Winter, 1979; Hattersley, 1982, 1983). During this time, it also became apparent that CAM plants were often intermediate between C_3 and C_4 $\delta^{13}C$ values (Bender *et al.*, 1973; Osmond *et al.*, 1973; Lerman *et al.*, 1974) and that a number of CAM succulents exhibited large environmentally related variations in their $\delta^{13}C$ values (Troughton *et al.*, 1977).

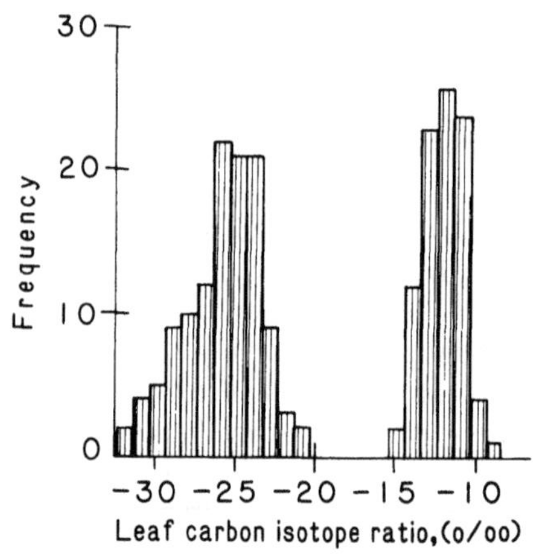

Fig. 13.4 Frequency histogram of carbon isotope ratios for different species of C_3 and C_4 plants (redrawn from Deines, 1980).

We now know that the $\delta^{13}C$ of plants can vary from −7 to −35‰ with C_4 plants having values of −7 to −15‰, CAM plants −10 to −22‰, and C_3 plants −20 to −35‰ (Fig. 13.4). The sources of these variations in carbon isotope composition in land plants are principally associated with the photosynthetic carboxylation enzymes, with second-order differences being due to diffusional fractionations, and differences in the $\delta^{13}C$ value of the atmospheric CO_2 fixed in photosynthesis (O'Leary, 1981). In aquatic plants often the latter factor may be the major source of variation (Osmond *et al.*, 1981; O'Leary, 1984). The primary carboxylase of C_3 photosynthesis, ribulose-1,5-bisphosphate carboxylase oxygenase (Rubisco), discriminates strongly against ^{13}C (approx. −29‰ with respect to the source CO_2; Whelan *et al.*, 1973; Roeske and O'Leary, 1984). The primary carboxylase of C_4 photosynthesis, phosphoenolpyruvate carboxylase (PEPCase), discriminates much less strongly against ^{13}C (approx. 2‰ with respect to source CO_2; Reibach and Benedict, 1977; O'Leary *et al.*, 1981). Diffusional contributions to carbon isotope fractionation during CO_2 exchange are observed during CAM (O'Leary and Osmond, 1980) and during CO_2/HCO_3^- uptake in aquatic plants (Raven *et al.*, 1982). Small variations in source isotope composition due to changing atmospheric CO_2 concentration (progressively enriched with CO_2 from fossil sources at about −30‰) and larger variations due to respiratory CO_2 sources in dense rainforest canopies (Medina and Minchin, 1980) or in aquatic plants (Osmond *et al.*, 1981) have been detected. These sources of variation have been integrated into functional models for CO_2 fixation in C_3 photosynthesis (Farquhar *et al.*, 1982b) and in C_4 photosynthesis (Farquhar, 1983).

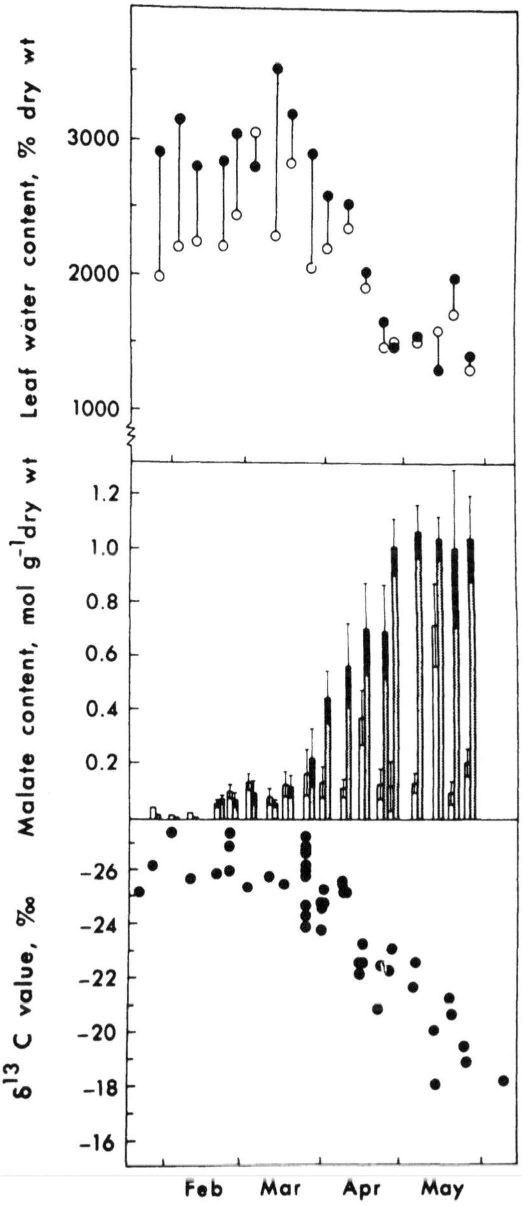

Fig. 13.5 Seasonal courses of leaf water content, malate content and leaf carbon isotope ratio of *Mesembryanthemum cristallinum* (redrawn from Winter *et al.*, 1978).

Within a species these sources of variation can indicate ecophysiologically significant changes in function. In CAM plants for example, the proportion of carbon fixed in the dark (by PEPCase) or in the light (by Rubisco) is directly indicated by changes in $\delta^{13}C$ values (Bender *et al.*, 1973; Osmond *et al.*, 1973). In field studies, the water-stress-dependent induction of dark CO_2 fixation by CAM (Fig. 13.5) has been elegantly shown with correlations between $\delta^{13}C$ value and nocturnal acidification (Winter *et al.*, 1978). In C_3 plants, changes in $\delta^{13}C$ value along environmental gradients such as salinity can be correlated with increased diffusional limitations associated with stomatal closure (Guy *et al.*, 1980; Farquhar *et al.*, 1982a). These relationships are discussed below. In some C_4 plants growth under low-nitrogen nutrition leads to more negative $\delta^{13}C$ values, indicating impaired function of the CO_2-concentrating mechanism (Wong and Osmond, 1988). Other correlations with nutrients (Bender and Berge, 1979), temperature and light (Smith *et al.*, 1976) have yet to be evaluated in terms of function. There are even examples of differences in photosynthetic pathways between organs of a species (e.g. C_3 leaves on CAM stems, Lange and Zuber, 1977), and of the relative contribution of different carboxylation pathways to the composition of different tissues such as guard cells and mesophyll (Nishida *et al.*, 1981) and root cells and nodules (Yoneyama and Ohtani, 1983).

Measurements of $\delta^{13}C$ value on individual different species of succulent plants, for example, have been used to indicate changes in ecophysiological functions along environmental gradients (Osmond *et al.*, 1975; Eickmeier and Bender, 1976). The $\delta^{13}C$ value of biomass has been used to monitor the contributions of C_3 and C_4 plants in communities along elevational gradients (Tieszen *et al.*, 1979b) and correlates closely with percent of species with the C_4 pathway (Fig. 13.6). One especially important application, where most other techniques fail, is estimation of belowground biomass due to C_3 and C_4 plants, such as in competition studies with C_3 and C_4 plants (Wong and Osmond, 1988).

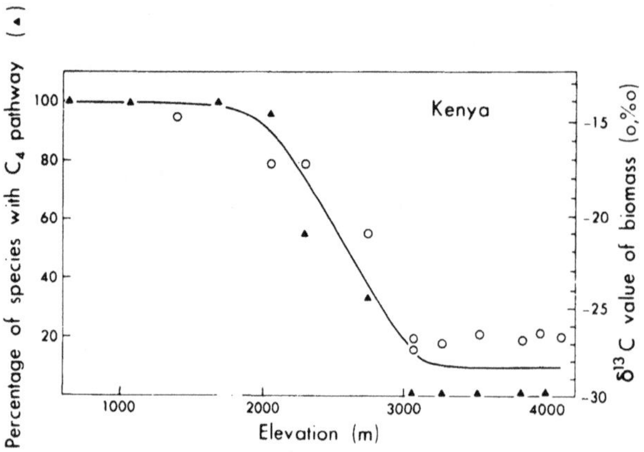

Fig. 13.6 Changes in the percentage of C_4 photosynthetic pathway plants and in carbon isotope ratio of plant biomass along an elevational transect in Kenya (redrawn from Tieszen *et al.*, 1979b).

13.6.2 Water-use efficiency in C_3 plants

The previously mentioned studies were largely survey-type investigations seeking to delineate biochemically based phenomena. In 1980 three independent approaches to the integration of physical (stomatal diffusion) and biochemical (discrimination of carboxylations) processes of carbon isotope discrimination were published (Farquhar, 1980; O'Leary and Osmond, 1980; Vogel 1980). These led to better models of carbon isotope discrimination (O'Leary, 1981), and to the recognition of a relationship between $\delta^{13}C$ value and intercellular CO_2 concentration (Farquhar *et al.*, 1982b). These theoretical interpretations can be tested by direct 'on-line' analysis of CO_2 fractionation in leaves during conventional gas exchange (Evans *et al.*, 1986). Supporting evidence has been obtained from several sources (Farquhar *et al.*, 1982b; Fig. 13.7; Bradford *et al.* (1983) for tomatoes; Farquhar and Richards (1984) for different wheat cultivars; Ehleringer *et al.* (1985) for desert shrubs and their parasitic mistletoes). Variation in the intercellular CO_2 concentration may account for much of the intraspecific isotopic variation observed, as well as the known variation that seems to be associated with water stress and growth humidity levels (Shomer-Ilan *et al.*, 1979; Winter *et al.*, 1982).

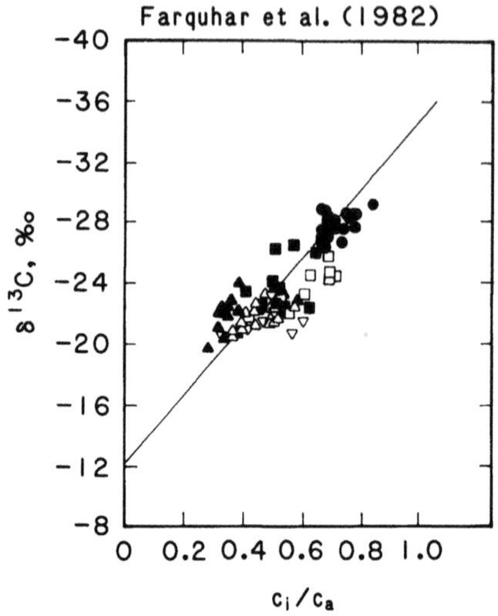

Fig. 13.7 Relationship between intercellular CO_2 concentration (c_i) and leaf carbon isotope ratio (redrawn from Farquhar *et al.*, 1982b). c_a is ambient CO_2 level. (Symbols represent different species.)

What is extremely useful about the relationship between intercellular CO_2 concentration (c_i) and carbon isotope ratio is that c_i is also related to water-use efficiency (molar ratio of photosynthesis to transpiration) as can be seen from the equations below:

$$A = [(c_a - c_i)g]/1.6$$
$$E = \Delta w g$$
$$A/E = (c_a - c_i)/(1.6\Delta w)$$

where A is photosynthetic rates, E is transpiration rate, g is leaf conductance to water vapor, c_a is ambient CO_2 level, c_i is intercellular CO_2 level, and Δw is leaf to air water vapor concentration gradient.

As c_a is essentially constant, then the carbon isotope ratio should depend only on c_i and Δw. This gives us a powerful tool for estimating integrated long-term water-use efficiency by a plant. Farquhar and Richards

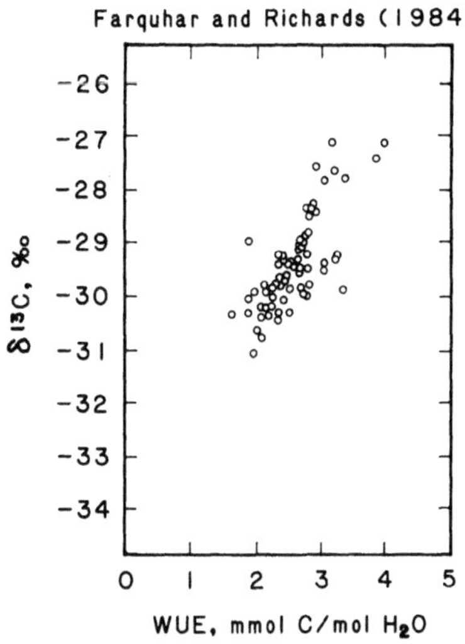

Fig. 13.8 Correlations between leaf carbon isotope ratio and the measured whole-plant water-use efficiency (WUE) (plant mass to soil water extracted) (redrawn from Farquhar and Richards, 1984).

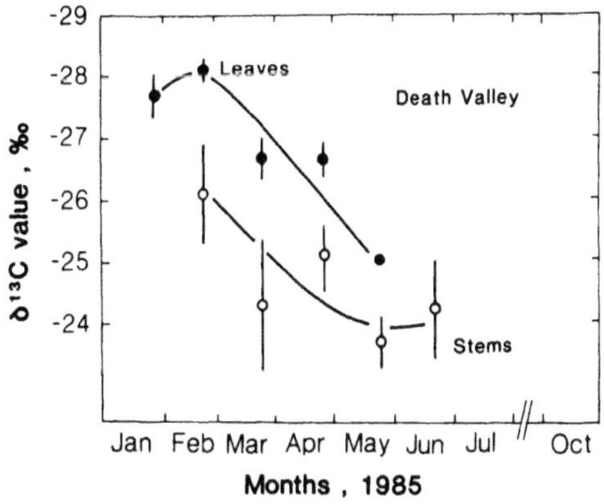

Fig. 13.9 Seasonal changes in carbon isotope ratios of leaves and photosynthetic stems of *Eriogonum inflatum* (redrawn from Smith and Osmond, 1987).

(1984) confirmed this with wheat cultivars grown under different watering regimes (Fig. 13.8). Other data suggest that carbon isotopic composition can be used to investigate growth irradiance conditions, short-term versus long-term leaf responses and variations in isotopic composition of photosynthetic structures within a single plant. For example, $\delta^{13}C$ values of leaves and stems of *Eriogonum inflatum* measured throughout the growing season (Fig. 13.9) are consistent with and confirm the lower stomatal conductance and higher Δw conditions which characterize stem photosynthesis compared with leaf photosynthesis (Smith and Osmond, 1987).

In aquatic plants much the same principles of diffusional and biochemical fractionation apply. Smith and Walker (1980) defined the problems of CO_2 and HCO_3^- diffusion in solution, and subsequent authors have demonstrated the usefulness of these approaches in relation to anatomy and water movement and biochemical pathway (Osmond *et al.*, 1981; Raven *et al.*, 1982; Keeley *et al.*, 1984). Sharkey and Berry (1985) developed an 'on-line' system to assess the significance of CO_2 concentrating mechanisms in algae using carbon isotope discrimination.

13.6.3 Water sources used by plants

There has been less research into the eco-physiological applications of hydrogen and oxygen isotope fractionation. Ehhalt *et al.* (1963) and Schiegl and Vogel (1970) identified differences in the deuterium content of organic matter that were highlighted by large variations in the isotopic composition of rain water. Perhaps for these reasons isotopic studies of plant water relations have not progressed very far. However, recent evidence suggests that isotopic analyses of xylem sap for either element may provide a signature of the source of soil moisture which a plant is using (White *et al.*, 1985). This approach has been used in studies of water balance in pines, where the objective was to separate the uptake of ground water from recent precipitation during the growing season (Fig. 13.10).

The relationships between the isotopic content of leaf water, determined by factors such as ground water isotope content, atmospheric humidity and stomatal conductance,

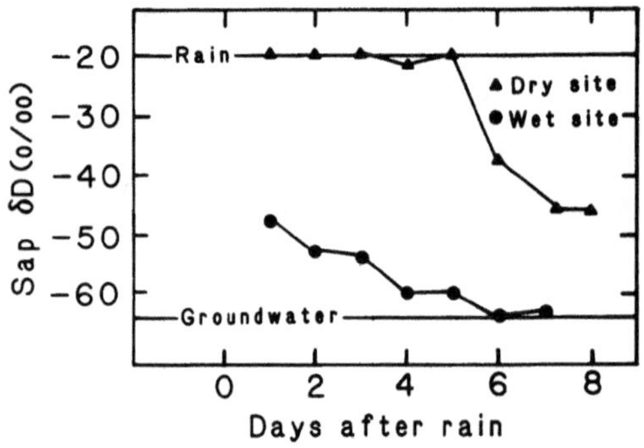

Fig. 13.10 Daily changes in the hydrogen isotope ratio of xylem water in pines following summer rains (redrawn from White *et al.* (1985).

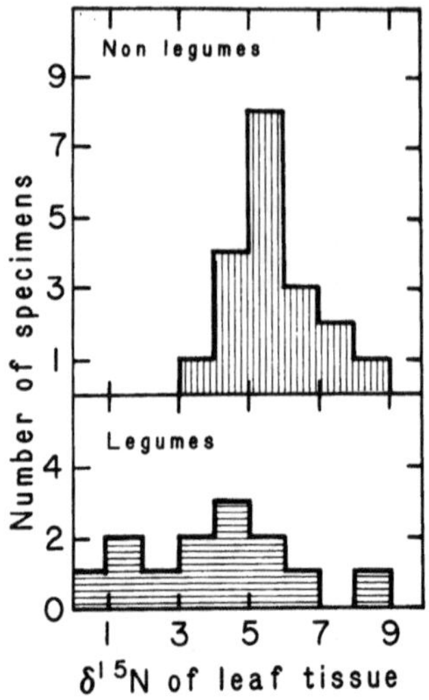

Fig. 13.11 Frequency histogram of nitrogen isotope ratios for different species of nitrogen-fixing and nonnitrogen-fixing plants (redrawn from Shearer *et al.*, 1983).

and fractionations during photosynthesis in this water are being assessed. Leaf water δD and $\delta^{18}O$ values are very dynamic (Dongmann *et al.*, 1974; Farris and Strain, 1978; Förstel, 1978; Zundel *et al.*, 1978; Leaney *et al.*, 1985), yet it seems C_3 photosynthesis fractionates to a rather constant extent in spite of the exchanges with H_2O which are potentially possible (Estep and Hoering, 1980, 1981). Comparative studies suggest (Sternberg and DeNiro, 1983) and direct measurements show (Leaney *et al.*, 1985) that different photosynthetic pathways discriminate differently against deuterium, in spite of vastly different diurnal changes in leaf water δD. However, the more negative δD values of organic hydrogen in *Sedum* spp. at higher elevation suggest that photosynthesis fractionations can reflect the changing isotope composition of ground water (Ziegler *et al.*, 1976). The δD value of organic hydrogen in CAM plants becomes less negative with water stress (Ziegler *et al.*, 1976). Recent studies show that this is a characteristic of the biochemistry of CAM not a reflection of the transpiration strategy of these plants (Sternberg, *et al.*, 1984b, 1986).

13.6.4 Nitrogen-fixation studies

Symbiotic nitrogen fixation in natural ecosystems is difficult to estimate by conventional means and it is even more difficult to obtain long-term estimates of the contribution of fixed nitrogen to the total nitrogen content of a plant. However, stable nitrogen isotopes can provide integrated estimates of nitrogen sources for plants (Shearer *et al.*, 1978, 1983; Sweeney *et al.*, 1978). This is because there are small differences between the natural abundance of ^{15}N between atmospheric N_2 and soil sources of nitrogen. Soil nitrogen tends to be enriched in ^{15}N (mean surface value of $\delta^{15}N$ = 9.2‰) whereas bacterial fixation of N_2 does not discriminate against ^{15}N. Thus, legumes which fix N_2 have less negative $\delta^{15}N$ values than species that do not (Fig. 13.11). The fraction of the nitrogen in a legume derived from nitrogen fixation activity can be estimated as the ratio of the difference between the $\delta^{15}N$ of the leaf minus the $\delta^{15}N$ values expected if nitrogen were derived solely from the atmosphere ($\delta^{15}N$ = 0) divided by the difference in $\delta^{15}N$ of nonnitrogen-fixing plants minus the atmospheric value. Such approaches indicate that legumes can differ widely in the proportion of nitrogen derived from soil versus nitrogen fixation sources.

13.6.5 Food web studies

DeNiro and Epstein (1976) in documenting the influence of diet on carbon isotope ratios pointed out that you are what you eat (plus a few ‰). A variety of single and multiple isotope signatures have been used to study patterns of plant–herbivore interactions and energy transfer along food chains. The majority of these have involved the use of carbon isotope ratios to investigate patterns of food selection in the diet of animals (Ludlow *et al.*, 1976; DeNiro and Epstein, 1976, 1978). Since the carbon isotope ratio in animal tissues closely parallels the ratio of the food eaten, diet selectivity between foods of different isotopic composition can be assessed. Recent studies analyzing vertebrate herbivore food preferences have further illustrated the utility of isotopic analyses for quantitatively determining the feeding preferences of different species over time. In particular, a study of large herbivores in Kenyan grasslands by Tieszen *et al.* (1979a) demonstrated that reliable estimates of both long- and short-term feeding preferences can be obtained for large numbers of animals with limited sampling efforts.

Carbon isotope ratios have also been used to explore the nature of ancient human diets. Bone collagen provides a permanent record of the diet at the time it was laid down and can be used to estimate the relative amounts of marine and terrestrial foods in prehistoric diets. Such studies have been used to trace the introduction of corn among different tribes of North American Indians (Chisholm, *et al.*, 1982; DeNiro and Hastorf, 1985).

To date there has been only limited interest in the use of nitrogen isotope ratios in food chain studies. $\delta^{15}N$ values in animals reflect the composition of their diets, but are characteristically 2–4‰ more positive at each trophic level (Schoeninger and DeNiro, 1984). This characteristic elevation in nitrogen isotope ratios along food chains is thought to be due to isotopic fractionation associated with catabolic metabolism.

One of the most exciting future developments in food chain studies using stable

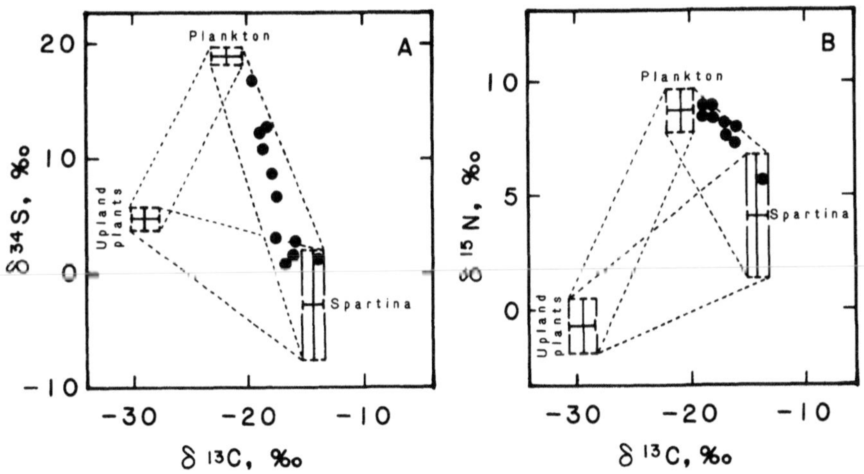

Fig. 13.12 Plots of stable isotope ratios (means and standard deviations) of C, N, and S for plankton, upland plants and marsh grass from a salt marsh and the stable isotope ratio signatures of a ribbed mussel (solid dots) feeding on detritus within this ecosystem (redrawn from Peterson *et al.*, 1985).

isotope ratios will almost certainly come from multiple element studies. Organisms that are similar in isotope ratios for one element may well differ in another. Using stable isotopes of C, N and S, Peterson *et al.* (1985) have been able to trace the flow of organic matter within a salt marsh ecosystem and to indicate clearly the detritial substances utilized by mussels downstream (Fig. 13.12). This approach may require sophisticated treatment of mathematical data in addition to the analytical needs, but it shows great promise for a wide variety of applications.

REFERENCES

Bender, M.M. (1968) Mass spectrometric studies of carbon-13 variations in corn and other grasses. *Radiocarbon*, **10**, 468–72.

Bender, M.M. (1971) Variations in the $^{13}C/^{12}C$ ratios of plants in relation to the pathway of carbon dioxide fixation. *Phytochemistry*, **10**, 1239–44.

Bender, M.M. and Berge, A.J. (1979) Influence of N and K fertilization and growth temperature on $^{13}C/^{12}C$ ratios of Timothy grass (*Phleum pratense* L.). *Oecologia*, **44**, 117–18.

Bender, M.M., Rouhani, I., Viner, H.M. and Black, C.C. (1973) $^{13}C/^{12}C$ ratio changes in Crassulacean acid metabolism plants. *Plant Physiol.*, **42**, 427–30.

Bennett, C.F. and Timell, T.E. (1955) Preparation of cellulose trinitrate. *Sven. Papperstidn.*, **58**, 281–6.

Bigeleisen, J., Perlman, M.L. and Prosser, H.C. (1952) Conversion of hydrogenic materials to hydrogen for isotopic analysis. *Anal. Chem.*, **24**, 1356–7.

Bradford, K.J., Sharkey, T.D. and Farquhar, G.D. (1983) Gas exchange, stomatal behavior and $\delta^{13}C$ values of the flacca tomato mutant in relation to abscisic acid. *Plant Physiol.*, **72**, 245–50.

Buchanan, D.L. and Corcoran, B.J. (1959) Sealed tube combustions for the determination of carbon-14 and total carbon. *Anal. Chem.*, **31**, 1635–8.

Burk, R.L. (1979) Factors affecting $^{18}O/^{16}O$ ratios in cellulose. Ph.D. Dissertation, University of Washington, Seattle.

Card, K.A., Mahall, B. and Troughton, J.H. (1974) Salinity and carbon isotope ratios in C_3 and C_4 plants. *Carnegie Inst. Wash. Yrbk.*, **73**, 784–5.

Chisholm, B.S., Nelson, D.E. and Schwarz, H.P., (1982) Stable-carbon isotope ratios as a measure of marine versus terrestrial protein in ancient diets. *Science*, **216**, 1131–2.

Coleman, M.L., Shepherd, T.J., Durham, J.J., Rouse, J.E. and Moore, G.R. (1982) Reduction of water with zinc for hydrogen isotope analysis. *Anal. Chem.*, **54**, 993–5.

Compston, W. and Epstein, S. (1958) A method for the preparation of carbon dioxide from water vapor for oxygen isotope analysis. *Trans. Am. Geophys. Union*, **39**, 511–12.

Craig, H. (1953) The geochemistry of the stable carbon isotopes. *Geochim. Cosmochim. Acta*, **3**, 53–92.

Deines, P. (1980) The isotopic composition of reduced organic carbon. In *Handbook of Environmental Isotope Geochemistry* (eds P. Fritz and J.Ch. Fontes), Elsevier, Amsterdam, pp. 329–406.

DeNiro, M.J. (1981) The effects of different methods of preparing cellulose nitrate on the determination of the D/H ratios of non-exchangeable hydrogen of cellulose. *Earth Planet. Lett.*, **54**, 177–85.

DeNiro, M.J. and Epstein, S. (1976) You are what you eat (plus a few ‰): the carbon isotope cycle in food chains. *Geol. Soc. Am. Abs. Prog.*, **8**, 834–5.

DeNiro, M.J. and Epstein, S. (1978) Influence of diet on the distribution of carbon isotopes in animals. *Geochim. Cosmochim. Acta*, **42**, 495–506.

DeNiro, M.J. and Hastorf, C.A. (1985) Alteration of $^{15}N/^{14}N$ and $^{13}C/^{12}C$ ratios of plant matter during the initial stages of diagenesis: studies utilizing archaeological specimens from Peru. *Geochim. Cosmochim. Acta*, **49**, 97–115.

Dongmann, G., Nurnberg, H.W., Förstel, H. and Wagner, K. (1974) On the enrichment of $H_2{}^{18}O$ in the leaves of transpiring plants. *Radiat. Environ. Biophys.*, **11**, 41–52.

Dugan, J.P., Borthwick, J., Harmon, R.S., Gagnier, M.A., Glahn, J.E., Kinsel, E.P., MacLeod, S., Viglino, J.A. and Hess, J.W. (1985) Guanidine hydrochloride method for determination of water oxygen isotope ratios and the oxygen-18 fractionation between carbon dioxide and water at 25° C. *Anal. Chem.*, **57**, 1734–6.

Dunbar, J. and Schmidt, H.-L. (1984) Measurement of the $^2H/^1H$ ratios of the carbon bound hydrogen atoms in sugars. *Fresenius Z. Anal. Chem.*, **317**, 853–7.

Ehhalt, D., Knott, K., Nagel, J.F. and Vogel, J.C.

(1963) Deuterium and oxygen-18 in rainwater. *J. Geophys. Res.*, **68**, 3775.

Ehleringer, J.R., Field, C.B., Lin, Z.F. and Kuo, C.Y. (1986) Leaf carbon isotope and mineral composition in subtropical plants along an irradiance cline. *Oecologia*, **70**, 520–6.

Ehleringer, J.R., Schulze, E.D., Ziegler, H., Lange, O.L., Farquhar, G.D. and Cowan, I.R. (1985) Xylem-tapping mistletoes: water or nutrient parasites? *Science*, **227**, 1479–81.

Eickmeier, W.G. and Bender, M.M. (1976) Carbon isotope ratios of Crassulacean acid metabolism species in relation to climate and phytosociology. *Oecologia*, **25**, 341–7.

Epstein, S., Yapp, C.J. and Hall, J. (1976) The determination of the D/H ratio of non-exchangeable hydrogen in cellulose extracted from aquatic and land plants. *Earth Planet. Lett.*, **30**, 241–51.

Estep, M.F. and Hoering, T.C. (1980) Biogeochemistry of the stable hydrogen isotopes. *Geochim. Cosmochim. Acta*, **44**, 1197–206.

Estep, M.F. and Hoering, T.C. (1981) Stable hydrogen isotope fractionation during autotrophic and mixotrophic growth of microalgae. *Plant Physiol.*, **67**, 474–7.

Evans, J.R., Sharkey, T.D., Berry, J.A. and Farquhar, G.D. (1986) Carbon isotope discrimination measured concurrently with gas exchange to investigate CO_2 diffusion in leaves of higher plants. *Austr. J. Plant Physiol.*, **13**, 281–92.

Farquhar, G.D. (1980) Carbon isotope discrimination by plants and the ratio of intercellular and atmospheric CO_2 concentrations In *Carbon Dioxide and Climate: Australian Research* (ed. G.I. Pearman), Australian Academy of Science, Canberra, pp. 105–10.

Farquhar, G.D. (1983) On the nature of carbon isotope discrimination in C_4 species. *Austr. J. Plant Physiol.*, **10**, 205–26.

Farquhar, G.D., Ball, M.C., von Caemmerer, S. and Roksandic, Z. (1982a) Effect of salinity and humidity on $\delta^{13}C$ value of halophytes – evidence for diffusional isotope fractionation determined by the ratio of intercellular/atmospheric partial pressure of CO_2 under different environmental conditions. *Oecologia*, **52**, 121–4.

Farquhar, G.D., Hubick, K.T., Condon, A.G. and Richards, R.A. (1988) Carbon isotope fractionation and plant water-use efficiency. In *Stable Isotopes in Ecological Research* (eds P.W. Rundel, J.R. Ehleringer and K.A. Nagy), Springer-Verlag, New York, pp. 21–40.

Farquhar, G.D., O'Leary, M.H. and Berry, J.A. (1982b) On the relationship between carbon isotope discrimination and the intercellular carbon dioxide concentration in leaves. *Austr. J. Plant Physiol.*, **9**, 121–37.

Farquhar, G.D. and Richards, R.A. (1984) Isotopic composition of plant carbon correlates with water-use efficiency of wheat genotypes. *Austr. J. Plant Physiol.*, **11**, 539–52.

Farris, F. and Strain, B.R. (1978) The effects of water-stress on $H_2{}^{18}O$ enrichment. *Radiat. Environ. Biophys.*, **15**, 167–202.

Förstel, H. (1978) The enrichment of ^{18}O in leaf water under natural conditions. *Radiat. Environ. Biophys.*, **15**, 323–44.

Friedman, I. and O'Neil, J.R. (1978) Isotopes in Nature. In *Handbook of Geochemistry* (ed. K.H. Wedepohl), Springer-Verlag, Berlin, pp. 1B1–1B8.

Fritz, P. and Fontes, J.C. (eds) (1980) *Handbook of Environmental Isotope Geochemistry*, Vol. I, *The terrestrial environment*, Elsevier, Amsterdam.

Guy, R.D., Reid, D.M. and Krouse, H.R. (1980) Shifts in carbon isotope ratio of two C_3 halophytes under natural and artificial conditions. *Oecologia*, **44**, 241–7.

Hattersley, P.W. (1982) $\delta^{13}C$ values of C_4 types in grasses. *Austr. J. Plant Physiol.*, **9**, 139–54.

Hattersley, P.W. (1983) The distribution of C_3 and C_4 grasses in Australia in relation to climate. *Oecologia*, **57**, 113–28.

Hauck, R.D. (1982) Nitrogen–isotope ratio analysis, *Methods of Soil Analysis, Part 2*, 2nd edn, American Society of Agronomy, Madison, pp. 735–79.

Hayes, J.M. (1983) Practice and principles of isotopic measurements in organic geochemistry. In *Organic Geochemistry of Contemporaneous and Ancient Sediments* (ed. W.G. Meinschein), SEPM, Bloomington, Indiana, pp. 5–31.

Hoefs, J. (1980) *Stable Isotope Geochemistry*, Springer-Verlag, Berlin, 208 pp.

Keeley, J.E., Osmond, C.B. and Raven, J.A. (1984) *Stylites*, a vascular land plant without stomata absorbs CO_2 via its roots. *Nature, London*, **310**, 694–5.

Kohl, D.H., Shearer, G.B. and Commoner, B. (1971) Fertilizer nitrogen: contribution to nitrate in surface water in a cornbelt watershed. *Science*, **174**, 1331–4.

Lange, O.L. and Zuber, M. (1977) *Frerea indica*, a stem succulent CAM plant with deciduous C_3 leaves. *Oecologia*, **31**, 67–72.

Leaney, F.W., Osmond, C.B., Allison, G.B. and Ziegler, H. (1985) Hydrogen-isotope composition of leaf water in C_3 and C_4 plants: its relationship to the hydrogen-isotope composition of dry matter. *Planta*, **164**, 215–20.

Lerman, J.C., DeLeens, E., Nato, A. and Moyse, A. (1974) Variations in the carbon isotope composition of a plant with Crassulacean acid metabolism. *Plant Physiol.*, **53**, 581–4.

Ludlow, M.M., Froughton, J.H. and Jones, R.J. (1976) A technique for determining the proportion of C_3 and C_4 species in plant samples using natural isotopes of carbon. *J. Agric. Sci. Canb.*, **87**, 625–32.

Mann, J. (1971) Deuteration and tritiation. In *Cellulose and Cellulose Derivatives V*, Part IV (eds N. Bikales and L. Segal), Interscience Publ., New York, 89 pp.

Medina, E. and Minchin, P. (1980) Stratification of $\delta^{13}C$ values of leaves in Amazonian rainforest. *Oecologia*, **45**, 377–8.

Minagawa, M., Winter, D.A. and Kaplan, I.R. (1984) Comparison of Kjeldahl and combustion methods for measurement of nitrogen isotope ratios in organic matter. *Anal. Chem.*, **56**, 1859–61.

Mooney, H.A., Troughton, J.H. and Berry, J.A. (1977) Carbon isotope ratio measurements of succulent plants in South Africa. *Oecologia*, **30**, 295–305.

Nishida, K., Roksandic, Z. and Osmond, B. (1981) Carbon isotope ratios of epidermal and mesophyll tissues from leaves of C_3 and CAM plants. *Plant Cell Physiol.*, **22**, 923–6.

O'Leary, M.H. (1981) Carbon isotope fractionation in plants. *Phytochemistry*, **20**, 553–67.

O'Leary, M.H. (1984) Measurement of isotopic fractionation associated with diffusion of carbon in aqueous solution. *J. Phys. Chem.*, **88**, 823–5.

O'Leary, M.H. and Osmond, C.B. (1980) Diffusional contribution to carbon isotope fractionation during dark CO_2 fixation in CAM plants. *Plant Physiol.*, **66**, 931–4.

O'Leary, M.H., Rife, J.E. and Slater, J.D. (1981) Kinetic and isotope effect studies of maize phosphoenolpyruvate carboxylase. *Biochemistry*, **20**, 7308–14.

Osmond, C.B., Allaway, W.G., Sutton, B.G., Troughton, J.H., Queiroz, O., Lüttge, U. and Winter, K. (1973) Carbon isotope discrimination in photosynthesis of CAM plants. *Nature, London*, **246**, 41–2.

Osmond, C.B., Valaane, N., Maslam, S.M., Uotila, P. and Roksandic, Z. (1981) Comparisons of $\delta^{13}C$ values in leaves of aquatic macrophytes from different habitats in Britain and Finland; some implications for photosynthetic processes in aquatic plants. *Oecologia*, **50**, 117–24.

Osmond, C.B., Winter, K. and Ziegler, H. (1982) Functional significance of different pathways of CO_2 fixation in photosynthesis. In *Encyclopedia of Plant Physiology New Series, Physiological Plant Ecology II* (eds O.L. Lange, P.S. Nobel, C.B. Osmond and H. Ziegler), Springer-Verlag, New York, Vol. 12B.

Osmond, C.B., Ziegler, H. Stichler, W. and Trimborn, P. (1975) Carbon isotope discrimination in alpine succulent plants supposed to be capable of Crassulacean acid metabolism. *Oecologia*, **18**, 209–17.

Peterson, B.J., Howarth, R.W. and Garritt, R.H. (1985) Multiple stable isotopes used to trace the flow of organic material in estuarine food webs. *Science*, **227**, 1361–3.

Raven, J., Beardall, J. and Griffiths, H. (1982) Inorganic C-sources for *Lemanea cladophora* and *Ranunculus* in a fast-flowing stream: Measurements of gas exchange and of carbon isotope ratio and their ecological implications. *Oecologia*, **53**, 68–78.

Reibach, P.H. and Benedict, C.R. (1977) Fractionation of stable carbon isotopes by phosphoenolpyruvate carboxylase from C_4 plants. *Plant Physiol.*, **59**, 564–8.

Roeske, C. and O'Leary, M.H. (1984) Carbon isotope effects on the enzyme catalyst carboxylation of ribulose bisphosphate. *Biochemistry*, **23**, 6275–84.

Rundel, P.W., Ehleringer, J.R. and Nagy, K.A. (eds) (1988) *Stable Isotopes in Ecological Research. Ecological Studies*, Vol. 68, Springer-Verlag, New York.

Rundel, P.W., Stichler, W., Zandler, R.H. and Ziegler, H. (1979) Carbon and hydrogen isotope ratios of bryophytes from arid and humid regions. *Oecologia*, **44**, 91–4.

Schiegl, W.E. and Vogel, J.C. (1970) Deuterium content of organic matter. *Earth Planet. Sci. Lett.*, **7**, 307–13.

Schoeninger, M.J. and DeNiro, M.J. (1984) Nitrogen and carbon isotopic composition of bone collagen from marine and terrestrial animals. *Geochim. Cosmochim. Acta*, **48**, 625–39.

Sharkey, T.D. and Berry, J.A. (1985) Carbon isotope fractionation of algae as influenced by an inducible CO_2 concentrating mechanism. In *Inorganic Carbon Uptake by Aquatic Photosynthetic Organisms* (eds W.J. Lucas and J.A. Berry), American Society of Plant Physiology, Rockville, pp. 389–401.

Shearer, G. and Kohl, D.H. (1988) Estimates of N_2 fixation in ecosystems: the need for and basis of the ^{15}N natural abundance method. In *Stable Isotopes in Ecological Research* (eds P.W. Rundel, J.R. Ehleringer and K.A. Nagy), Springer-Verlag, New York, pp. 342–74.

Shearer, G., Kohl, D.H. and Chien, S.H. (1978)

The nitrogen-15 abundance in a wide variety of soils. *Soil Sci. Soc. Am. J.*, **42**, 899–902.

Shearer, G., Kohl, D.H., Virginia, R.A., Bryan, B.A., Skeeters, J.L., Nilsen, E.T., Sharifi, M.R. and Rundel, P.W. (1983) Estimates of N_2-fixation from variation in the natural abundance of ^{15}N in Sonoran Desert ecosystems. *Oecologia*, **56**, 365–73.

Shomer-Ilan, A., Nissenbaum, A., Galun, M. and Waisel, Y. (1979) Effect of water regime on carbon isotope composition of lichens. *Plant Physiol.*, **63**, 201–5.

Smith, B.N. (1975) Carbon and hydrogen isotopes of sucrose from various sources. *Naturwissenschaften*, **62**, 390.

Smith, B.N. and Brown, W.V. (1973) The Kranz syndrome in the Graminae as indicated by carbon isotopic ratios. *Am. J. Bot.*, **60**, 505–13.

Smith, B.N. and Epstein, S. (1971a) Biogeochemistry of the stable isotopes of hydrogen and carbon in salt marsh biota. *Plant Physiol.*, **46**, 738–42.

Smith, B.N. and Epstein, S. (1971b) Two categories of $^{13}C/^{12}C$ ratios for higher plants. *Plant Physiol.*, **47**, 380–4.

Smith, B.N., Oliver, J. and McMillan, C. (1976) Influence of carbon source, oxygen concentration, light intensity, and temperature on $^{13}C/^{12}C$ ratios in plant tissues. *Bot. Gaz.*, **137**, 99–104.

Smith, B.N. and Turner, B.L. (1975) Distribution of Kranz syndrome among Asteraceae. *Am. J. Bot.*, **62** 541–5.

Smith, F.A. and Walker, N.A. (1980) Photosynthesis by aquatic plants: effects of unstirred layers in relation to assimilation of CO_2 and HCO_3^- and to carbon isotopic discrimination. *New Phytol.*, **86**, 245–59.

Smith, S.D. and Osmond, C.B. (1987) Stem photosynthesis in a desert ephemeral, *Eriogonum inflatum*. Morphology, stomatal conductance and water-use efficiency in field populations. *Oecologia*, **72**, 533–41.

Sternberg, L., DeNiro, M.J. and Johnson, H.B. (1986) Oxygen and hydrogen isotopes ratios of water from photosynthetic tissues of CAM and C_3 plants. *Plant Physiol.*, **82**, 428–31.

Sternberg, L. and DeNiro, M.J. (1983) Isotopic composition of cellulose from C_3, C_4 and CAM plants growing near one another. *Science*, **220**, 947–9.

Sternberg, L., DeNiro, M.J. and Ajie, H. (1984a) Stable hydrogen isotope ratios of saponifiable lipids and cellulose nitrate from CAM, C_3 and C_4 plants. *Phytochemistry*, **23**, 2475–7.

Sternberg, L., DeNiro, M.J. and Keeley, J.E. (1984b) Hydrogen, oxygen, and carbon isotope ratios of cellulose from submerged aquatic Crassulacean acid metabolism and non-Crassulacean acid metabolism plants. *Plant Physiol.*, **76**, 68–70.

Sweeney, R.E., Liu, K.K. and Kaplan, I.R. (1978) Oceanic nitrogen isotopes and their uses in determining the source of sedimentary nitrogen. In *Stable Isotopes in the Earth Science* (ed. B.W. Robinson), Division of Scientific and Industrial Research Bull. 220.

Tieszen, L.L., Hein, D., Qvortrup, S.A., Troughton, J.H. and Imbamba, S.K. (1979a) Use of $\delta^{13}C$ values to determine vegetation selectivity in East African herbivores. *Oecologia*, **37**, 351–9.

Tieszen, L.L., Senyimba, M.M., Imbamba, S.K. and Troughton, J.H. (1979b) The distribution of C_3 and C_4 grasses and carbon isotope discrimination along an altitudinal and moisture gradient in Kenya. *Oecologia*, **37**, 337–50.

Troughton, J.H., Card, K.A. and Hendy, C.H. (1974) Photosynthetic pathways and carbon isotope discrimination by plants. *Carnegie Inst. Wash. Yrbk.*, **73**, 768–80.

Troughton, J.H., Mooney, H.A., Berry, J.A. and Verity, D. (1977) Variable carbon isotope ratios of *Dudleya* species growing in a natural environment. *Oecologia*, **30**, 307–11.

Vogel, J.C. (1980) Fractionation of the carbon isotopes during photosynthesis. In *Sitzungsberichte der Heidelbergwer Akademie der Wissenschaften Mathematisch-Naturwissenschaftliche Klasse Jahrgang 1980, 3. Abhandlung*, Springer-Verlag, Berlin, pp. 111–34.

Webster, G.L., Brown, W.V. and Smith, B.N. (1975) Systematics of photosynthetic carbon fixation pathways in *Euphorbia*. *Taxon*, **24**, 27–33.

Whelan, T., Sackett, W.M. and Benedict. C.R. (1973) Enzymatic fractionation of carbon isotopes by phosphoenol pyruvate carboxylase from C_4 plants. *Plant Physiol.*, **51**, 1051–4.

White, J.W.C., Cook, E.R., Lawrence, J.R. and Broecker, W.S. (1985) The D/H ratios of sap in trees: implications for water sources and tree ring D/H ratios. *Geochim. Cosmochim. Acta*, **49**, 237–46.

Winter, K. (1979) $\delta^{13}C$ values of some succulent plants from Madagascar. *Oecologia*, **40**, 103–12.

Winter, K., Holton, J.A.M., Edwards, G.E. and O'Leary, M.H. (1982) Effect of low relative humidity on $\delta^{13}C$ value in two C_3 grasses and in *Panicum miliodes*, a C_3-C_4 intermediate species. *J. Exp. Bot.*, **132**, 88–91.

Winter, K., Lüttge, U., Winter, E. and Troughton, J.H. (1978) Seasonal shift from C_3 photosynthesis to Crassulacean acid metabolism in *Mesembryanthemum crystallinum* growing in its natural environment. *Oecologia*, **34**, 225–37.

Winter, K., Troughton, J.H. and Card, K.A. (1976) $\delta^{13}C$ values of grass species collected in the Northern Sahara Desert. *Oecologia*, **25**, 115–23.

Wise, L.E. (1944) *Wood Chemistry*, Reinhold, New York, 900 pp.

Wong, S.C. and Osmond, C.B. (1988) Elevated atmospheric partial pressure of CO_2 and plant growth. III. Measurement of root biomass in mixed-culture using $\delta^{13}C$ values, and its importance during interactions between wheat and Japanese millet in response to N-nutrition and irradiance. *Austr. J. Plant Physiol.*, in press.

Wong, W.W., Lee, L.S. and Klein, P.D. (1987) Oxygen isotope ratio measurements on carbon dioxide generated by reaction of microliter quantities of biological fluids with guanidine hydrochloride. *Anal. Chem.*, **59**, 690–3.

Yapp, C.J. and Epstein, S. (1982) A reexamination of cellulose carbon-bound hydrogen δD measurements and some factors affecting plant-water D/H relationships. *Geochim. Cosmochim. Acta*, **46**, 955–65.

Yoneyama, T. and Ohtani, T. (1983) Variations in the natural ^{13}C abundances in leguminous plants. *Plant Cell Physiol.*, **24**, 971–7.

Ziegler, H., Osmond, C.B., Stichler, W. and Trimborn, P. (1976) Hydrogen isotope discrimination in higher plants: correlations with photosynthetic pathway and environment. *Oecologia*, **128**, 85–92.

Zundel, G., Miekeley, W., Grisi, B.M. and Förstel, H. (1978) The $H_2\,^{18}O$ enrichment of leaf water of tropic trees: comparison of species from the tropical rain forest and the semi-arid region in Brazil. *Radiat. Environ.*, **15**, 203–12.

14

Canopy structure

John M. Norman and Gaylon S. Campbell

14.1 INTRODUCTION

Descriptions of canopy structure are essential to achieving an understanding of plant processes because of the profound influence that structure has on plant–environment interactions. The vegetation architecture not only affects exchanges of mass and energy between the plant and its environment, but it also may reveal a strategy of the plant for dealing with long-lasting evolutionary processes, such as adaptation to physical, chemical or biotic factors, by reflecting the organism's vital activity or peculiarities in growth and development. Plant morphological studies, which are mostly qualitative, have long recognized this fact. Unfortunately quantitative descriptions of geometric features of canopies, plants or individual organs are difficult because canopies are spatially and temporally variable. The level of complexity is ever increasing as we proceed from individual organs to plants to pure stands to heterogeneous plant communities, since each higher level contains elements of the lower levels. For example, Sitka spruce needles are organized along a twig with a determined orientation distribution that varies with depth in the canopy; these shoots are organized into branches in a way that reflects developmental strategy, environmental stimulation and growth restrictions (Norman and Jarvis, 1974); these branches are organized along a stem to reveal a tree and of course trees are distributed throughout a forest. The elegance of this structure challenges the imagination to its limits. Although this elegance may teach us humility, it quickly overwhelms our quantitative aspirations so we must resort to the expediency of statistics for quantitative relations.

The influence of canopy structure on wind and radiation environments within the canopy is perhaps the most obvious. The effect of canopy structure on wind is usually described using measured normalized mean wind profiles within the canopy (Fritschen, 1985). Effects of canopy characteristics such as leaf area distribution or foliage clumping on wind usually are not quantified because of the obvious complexities associated with measurements and modeling. However, Pereira and Shaw (1980) have considered the effect of the vertical distribution of leaf area on wind profiles. Other interesting interactions between canopy structure and wind are described in Hutchison and Hicks (1985).

The relation between radiation environment within a canopy and canopy structure is much better quantified than the interaction between structure and wind (Ross, 1981). In fact the coupling between radiation exchange and canopy structure is so strong that measurements of radiation may be used to infer canopy features. This relationship forms the basis for indirect measurement techniques.

Canopy structure affects other environmental factors such as air temperature, leaf temperature, atmospheric moisture, soil evaporation below the canopy, soil heat storage and soil temperature, precipitation interception, leaf wetness duration and others. However, these effects may be subtle and may require complex models to quantify (Goudriaan, 1977; Norman and Campbell, 1983). Canopy structure, through its impact on canopy environment, affects not only plants, but also other organisms that may live within or below the canopy. Thus plant pests are affected by canopy structure (Toole *et al.*, 1984).

What is canopy structure? This is a most difficult question to answer, and a comprehensive answer would take a sizeable volume to complete. Simply stated, canopy structure may be thought of as the amount and organization of aboveground plant material. Perhaps the root system could be thought of as a canopy suspended in the soil matrix, but in this chapter the word canopy is used to refer to the aboveground portion of the plant. The question of whether the root system is considered as 'canopy' is a useful one because many of the same formulations that are used to describe the aerial architecture can be applied to root architecture.

A more detailed definition of canopy structure might include the size, shape, orientation and positional distributions of various plant organs such as leaves, stems, branches, flowers and fruits. As previously stated the acquisition of such information for each organ in a canopy is not, at present, feasible, so quantitative descriptions are statistical in character. Often we may refer to a representative plant whose characteristics are derived from observations averaged over many actual plants. The simplest mathematical descriptions assume organs to be randomly distributed in space. However, descriptions are available for regular, semiregular and clumped distributions as well. The amount of leaf material may be represented by the leaf area index (or stem or branch area index also) and orientation by the leaf angle distribution (or branch, stem, flower, or fruit angle distributions also). Leaves and branches have the greatest impact on the canopy environment so in this chapter we will emphasize these organs.

The most obvious method for obtaining canopy structure information is by direct measurement of plant organs. These measurements might include areas, shapes, angles or even positions. Although all direct methods are not destructive, they all disturb the canopy, and, at least in the case of leaf angle distributions, this disturbance usually compromises the quality of the data.

Indirect methods, in contrast to direct methods, usually involve the measurement of radiation from within or above the canopy. The data from these radiation methods must be combined with an appropriate radiative transfer theory so that canopy structural estimates can be obtained from an inversion procedure. This inversion is usually accomplished with a computer program.

Direct methods usually require much labor in the field and very simple data reduction. By contrast indirect methods usually require simple and rapid field measurements but complex algorithms for the reduction of data. However, once the appropriate computer programs are available, data reduction can be simple. The distinction between direct and indirect measurement can be 'fuzzy'. The term 'indirect measurement' usually refers to an observation of radiation that is used to infer some physical characteristic: therefore using one's eye to read the light reflected from a ruler could be interpreted as an indirect measurement. However, this is not what we have in mind. Although the preponderance of evidence suggests that the eye is a rather peculiar sensor, for the purposes of this chapter we shall assume it to be a direct sensor as most scientists prefer.

The advantages of indirect methods are so overwhelming that in the future direct methods will only be used when there is no indirect alternative. The object of this chapter is to review a few direct measurement procedures and discuss in detail some indirect canopy structural measurement techniques.

14.2 DIRECT METHODS

The measurement of canopy architecture by direct methods involves an attempt to obtain a representative description of the whole canopy by observations on individual plant organs. For the purposes of this chapter, direct measurements refer to measurements made directly on plant organs by manual methods. The material in this section has been organized according to a section in Ross's (1981) book entitled *A Rational Method for Determining Phytometric Characteristics of Stand Architecture and Productivity*. The general strategy is to make a series of very simple measurements on a very large number of plants (hundreds); then choose a modest number (tens) of plants, which are deemed representative of the population based on their simple characteristics, and make detailed measurements on these plants. In general the survey of very simple measurements is well worth the effort regardless of whether one chooses to use direct or indirect techniques on the detailed measurements.

14.2.1 General characteristics

The description of any vegetation must reasonably include a survey of general characteristics. This list of characteristics should include species, cultivar, plant density (trees ha^{-1} or plants m^{-2}), percentage of canopy cover and an estimate of the active phenological stages (vegetative, flowering, seed filling or bud development). Some measure of the time of initiation of vegetative growth always is appropriate; for example planting date, grafting date, emergence date, or date of bud break. Sometimes more than one of these dates is appropriate such as date of bud break and tree age. Heterogeneous stands require a list of general characteristics for each important species of the community.

14.2.2 Primary statistical characteristics

Primary statistical characteristics are obtained from relatively simple measurements that can be made on 100–300 plants. The following characteristics should be included: plant height, uppermost and lowermost foliage levels, stem height, stem diameter at the base and top node, total number of nodes, total number of leaves if possible, and dimensions of an imaginary canopy envelope that would just include nearly all of the organs of the plant. Appropriate canopy envelope dimensions may depend on the shape of the individual plant, but truncated ellipsoids (Whitfield, 1980; Norman and Welles, 1983) can be used to describe a wide variety of plant shapes. Thus six measurements may be needed: the length of the three axes of the ellipsoid that just contains most of the foliage, the height of the center of the ellipsoid, and the top and bottom truncation heights if the ellipsoid has been truncated. In heterogeneous canopies such canopy envelope dimensions are essential in studies of radiative and convective transfer. Depending on the canopy, envelope dimensions may be considered to be a more detailed measurement to be made only on representative plants.

Statistical summaries of these primary characteristics include means, standard deviations, ranges and in some cases histograms.

Usually a histogram of some primary characteristic is used to select 10–30 plants to represent all the plants in the distribution. This primary characteristic might be plant height (agricultural crops), envelope volume (prairie bunch grasses) or stem diameter at breast height (forests).

14.2.3 Detailed measurements on representative plants

Detailed measurements typically are made on 10–30 plants that are chosen to be representative of the population of plants based on the primary statistical characteristics. Since these 10–30 plants are to represent the spectrum of plants in the distribution, the number of samples with a given characteristic should be proportional to the fraction of the original population with that characteristic. The detailed measurements include all of the measurements of the two previous sections along with a detailed phenological description and the approximate height of the most dense region of the canopy. In addition, the height of each node, the height of some identifiable part of each organ, (or appropriate groups of organs) and the dimensions, area and orientation of each organ should be measured. For a complete description, the horizontal position of each organ also would be measured. Although this rarely is done because of the obvious difficulty, Lang (1973) has designed an instrument for measuring position and orientation distributions in smaller canopies.

The detailed canopy measurements usually reduce to estimates of area and orientation for stems, branches and leaves as a function of height. Studies of canopy architecture are often combined with productivity studies so dry matter sampling frequently is combined with foliage area determinations.

(a) Foliage area

Two basic sampling strategies are used to measure foliage area of a canopy: the stratified-clip method and the dispersed individual plant (DIP) method.

The stratified clip method usually is used with plants that have very small leaves such as turf grasses, prairie grasses, alfalfa, etc. A rectangular or circular area is identified and all the foliage within this area is cut. The vertical distribution of leaf area can be obtained by clipping in appropriate horizontal layers and measuring the foliage area in each layer separately. Large scissors or pruning shears are used to cut the leaves, stems and branches. Labeled polyethylene bags should be prepared beforehand to reduce water loss, because leaf area can decrease as tissue dries. Usually a wire frame is used to identify the sampling volume. With small crops 0.5 m by 0.5 m areas are used with two to ten replicates (usually four in relatively uniform crops). With row crops of larger herbs the row spacing is usually one of the dimensions and the area sampled approximately 1 m^2. The size of the clip area requires some judgment, and in general smaller clip areas and more replicates provide better sampling for a given amount of cutting. However, the clip area should be large enough so that minimal leaf area changes occur when the area is moved a distance roughly equal to its shortest dimension.

The DIP method involves detailed measurements as a function of height on individual plants selected randomly from the canopy. This method has several advantages over the stratified-clip method and is preferable when vegetation characteristics permit. Besides providing a more representative sample for a given amount of cut foliage and a less disruptive destructive sample, the DIP method emphasizes the characteristics of individual plants instead of groups of plants; this is highly desirable when obtaining canopy structural information for models of plant–environment interaction. For example node height or leaf area per node are rarely measured with the stratified-clip method but are obvious measurements with the DIP method.

The measurement of the area of the foliage elements themselves has been discussed extensively in the literature (Kvet and Marshall, 1971; Ross, 1981). Essentially three methods are used widely: (1) automatic planimeters, (2) area from leaf dimensions and (3) leaf area to weight ratios.

Automatic planimeters may be of the laboratory type, which require leaves to be destructively sampled and brought to the machine (for example LI-COR model LI-3100 – LI-COR, P.O. 4425, Lincoln, NE 68504, USA, or the Delta-T area meter type AMS, Delta-T Devices, 128 Low Road, Burwell, Cambridge CB5 0EJ, UK), or field portable and thus non-destructive (such as LI-COR model LI-3000). These machines can be accurate to better than 1% if properly maintained and calibrated. Furthermore they measure true leaf area not including holes from insects or other sources of leaf damage. However, these machines 'flatten' leaves during measurement and thus underestimate the area of leaves with 'ripples'. For example with corn the leaf area measurement on a single intact leaf may be 4–8% less than when the leaf is cut into pieces and the area of the pieces measured.

The estimation of leaf area from two simple leaf dimensions can provide non-destructive area estimates within 5%. However, in most cases calibrations should be checked frequently. The simple formula

$$\text{Area} = k\ (\text{length} \times \text{maximum width})$$

usually is adequate although more elaborate formulae have been used (Ross, 1981). The coefficient, k, is of course 0.50 for a triangle, 1 for a rectangle, 0.75 for grasses such as maize, sorghum and wheat, and near 2/3 for many dicots. McKee (1964) found a value of 0.73 for maize and Bonhomme *et al.* (1974) found a value of 0.74 for sugar cane and 0.64 for cow pea.

The leaf-area-to-dry-weight ratio (specific leaf area) usually is determined by selecting a subsample of leaves or punching disks from many leaves. When weight is easier to measure than area, such as with conifers, then a representative subsample of leaves can be selected from a much larger group of leaves (such as 2-year-old needles from node 12 on tree 9) for measurement of leaf-area-to-fresh-weight ratio so that the area of the entire group can be estimated from the fresh weight of the group. When such subsampling is done the whole sample should be thoroughly mixed before drawing the subsample. This can be done with dry weight also. If leaf area is easier to measure, then subsampling specific leaf area may provide estimates of weight. Great care must be used with these subsampling methods because of spatial and temporal variations. For example the leaf dry weight per unit area of potatoes can change by 40% on a diurnal basis (Ross, 1981). Although the leaf dry weight per unit area of corn may be reasonably stable on a diurnal basis, it can vary by a factor of three over the length of the corn mid-rib and 40% over the length of the leaf blade (Ross, 1981). In Sitka spruce the specific leaf area varies dramatically with shoot age and height in the canopy (Norman and Jarvis, 1974). Ross (1981) has suggested punching a sampling disk at a location where the local leaf dry weight per unit area is equal to the average for the whole leaf. In maize this was 0.63 of the way along the leaf and in cotton 0.92. A subsampling procedure for a conifer forest is outlined in Waring *et al.* (1982). Another subsampling method requires measuring the ratio of the area of the largest leaf to the total leaf area on the corn plant for a limited number of plants; then only the largest leaves on a large number of plants are measured to obtain an estimate of leaf area index (LAI) (Daughtry and Hollinger, 1984).

Daughtry and Hollinger (1984) evaluated the cost of measuring leaf area index (leaf area per unit ground area) by the four methods briefly described above. The most efficient method depends on the desired precision of the leaf area estimate, and whether dry weight is desired or not. However, the difference in effort required by the various methods is less than one might expect.

Leaf area measurement of trees is a formidable task. The direct sampling methods described in this section can be applied to trees, but the effort required may be tremendous. An alternative method that relates sapwood conducting area at breast height (1.37 m) to foliage weight or foliage area has been studied (Grier and Waring, 1974). The results from many conifer species (Waring *et al.*, 1982) vary from 0.15 to 0.75 m^2 leaf area per cm^2 sapwood area. Therefore this method requires calibration for a particular site and species. A refinement of this method is described by Whitehead *et al.* (1984). By plotting foliage area versus the product of sapwood basal area times water permeability, differences between *Picea sitchensis* and *Pinus contorta* disappear so that a single relationship can be used to relate sapwood conducting area to foliage area for both species. In fact a single relation can be used to derive the vertical distribution of foliage area for both trees with species. Anderson (1981) summarizes forest LAI data from around the world.

The vertical distribution of leaf area can be measured directly by the stratified-clip method if leaves from different layers are measured separately. This vertical distribution can be calculated from the DIP method if measurements of leaf height are included. With canopies that have large leaves relative to their height such as corn or wheat, the leaf height is difficult to determine and the stratified-clip method is more accurate for vertical foliage distributions. However, the vertical distribution of leaf area does not normally need to be defined with great precision. For most applications an assumed triangular distribution of leaf area density (m^2 leaf area per m^3 canopy volume) can be fitted to three points: the uppermost leaf level, the lowermost leaf level and the height of the most dense region of vegetation (Norman, 1979). More elaborate equations for describing the vertical distribution of leaf area can be found in Ross (1981).

(b) Foliage orientation

The distribution of inclination (angle between perpendicular to the leaf surface and the vertical) and azimuth (angle between a horizontal projection of the leaf perpendicular and North) angles for leaves is termed the leaf angle distribution. In general the fraction of leaf area in each angle class is the desired quantity. If the fraction of leaf area in leaf inclination angle class j and azimuth angle class k is given by g_{jk}, then

$$\sum_{j=1}^{N_j} \sum_{k=1}^{N_k} g_{jk} = 1.0 \qquad (14.1)$$

Thus if we divide the canopy into five inclination classes ($N_j = 5$) and eight azimuth classes ($N_k = 8$), $g_{jk} = 0.025$ if each angle class contains an equal fraction of the total leaf area. Classes with fractional areas greater than 0.025 have more leaf area than the average class, and classes with a value of less than 0.025 have a sparcity of leaf area. The symbol g_{jk} can be thought of as $g(\alpha_j, \beta_k)$ since j refers to inclination angle class and k to azimuth angle class.

Direct measurement of the leaf angle distribution is exceedingly time-consuming and fraught with potential errors. An excellent review of leaf angle distribution measurements is given in Ross (1981).

The most common instrument for measuring the leaf angle distribution directly is a compass–protractor. This is a simple device for measuring leaf inclination with a protractor and azimuth with a magnetic compass. Surely there are as many variations in the details of design of this device as there are individuals using it. Most compass–protractor instruments bear some resemblance to the device shown in Fig. 14.1. This is an exceedingly easy device to build in a few hours by raiding your children's toy box. It is doubtful

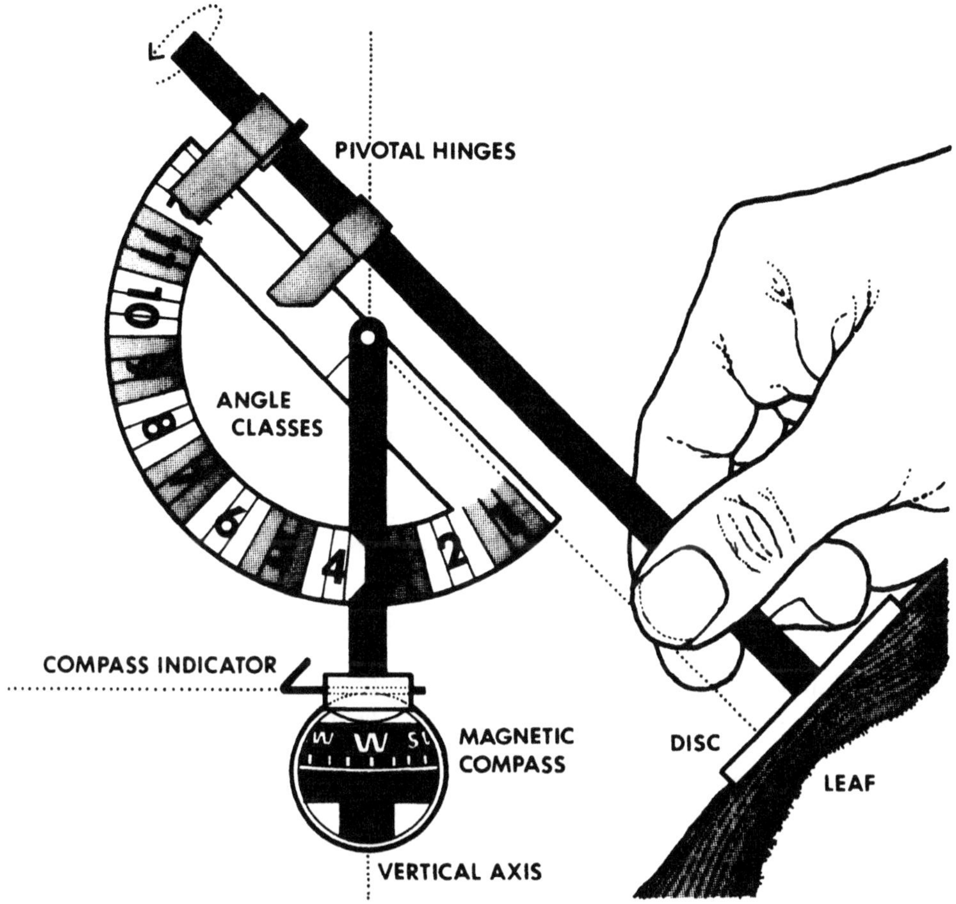

Fig. 14.1 Sketch of simple compass–protractor devise for measuring leaf inclination and leaf azimuth angles. See text for description and use.

whether a more elaborate device is worth the effort considering the problems with direct measurements of leaf angle distributions. Some care must be taken in obtaining a reasonably good compass. The instrument in Fig. 14.1 is used by placing the disc, which is at the end of the main rod and perpendicular to it, parallel to a bit of leaf surface. The readings from the compass and protractor can be recorded audibly with a voice-actuated tape recorder. Usually five inclination classes and eight azimuth classes are adequate. Some authors have suggested aligning a protractor edge along the slope of steepest descent of a bit of leaf; this is very difficult to do quickly in the field and aligning a flat surface parallel to the leaf surface is easier, faster and more reliable.

The objective of the leaf angle distribution measurement is to obtain the area distribution of leaf angles; therefore measurements must be done on many bits of leaf area. In fact, one may make many measurements on a single leaf. Each measurement can be assumed appropriate for an amount of leaf area equal to the disc area on the device in Fig. 14.1.

Typically about 1000 individual randomly selected measurements are considered minimal. Since random selection of bits of leaves can be difficult, often the complete area of 200–500 leaves is sampled. Obviously measuring the leaf angle distribution is a formidable task, especially when one realizes that measurements must be done with minimal disturbance of the canopy; a near impossible task. With small leaves such as turf grasses, wheat or alfalfa, these measurements are exceedingly difficult. With grasses one cannot simply measure the angles of the leaf midribs, as has been done in the past, because the leaf blades are not flat. The presence of such a bias would void the measurement of the leaf angle distribution, and one might just as well guess at the leaf angle distribution based on appearance and the literature as measure it, thus saving a lot of time.

The angular distribution of stems and branches is often ignored; this is not desirable because stems and branches do intercept light and in some plants they carry on photosynthesis. The inclination and azimuth of the longitudinal axis of the stem or branch is all that need be measured with cylindrical elements.

The leaves of most canopies approximate azimuthal symmetry (Lemur, 1973; Ross, 1981); therefore only leaf inclination angle may often be measured. If azimuthal symmetry is assumed, then the sum over k, or azimuth angle, can be represented by a dot so

$$g_{j\cdot} = \sum_{k=i}^{N_k} g_{jk} \tag{14.2}$$

and

$$\sum_{j=1}^{N_i} g_{j\cdot} = 1.0 \tag{14.3}$$

When azimuthal symmetry is not appropriate, the measured data are averaged over all inclinations and thus the azimuthal distribution is given by $g_{\cdot k}$. In general, different inclination classes could have different azimuthal preferences; however, to our knowledge, data are not available for testing such a hypothesis.

Direct measurement of leaf angle distributions requires much determination and extraordinary care to be worth the effort; *let the uninitiated beware.*

14.3 SEMIDIRECT METHODS

This subheading is intended to separate the direct brute-force methods of the last section from somewhat more elegant methods that require physical contact with leaves and stems. Furthermore these semidirect approaches require considerably more elaborate methods for reduction of the direct-measurement data. These methods therefore fall between the direct and indirect methods. The two methods included in this section are as follows: (1) inclined point quadrats as formulated by Warren Wilson (1960), including the drop-line method used in forests, and (2) the coordinate digitizing method of Lang (1973). Some might argue that the method of Lang (1973) is strictly direct, and though we might agree with them, Lang's method is so much more elegant that it deserves to be separated from the methods of the last section.

14.3.1 Inclined point quadrats

The inclined-point-quadrat method originated in New Zealand in the 1930s (Levy and Madden, 1933) and was refined by Warren Wilson for canopies of full cover (Warren Wilson, 1960) as well as row crops and individual plants (Warren Wilson, 1965). The method consists of inserting a probe with a sharp point into the canopy at a known inclination and azimuth angle and counting the number of times the point contacts leaves and stems. Determining when the probe point contacts a leaf or stem, without disturbing the canopy, is not a trivial part of this method. Usually it is necessary to be close enough to the probe to see contacts and this may disturb the canopy. A very clever contact detection system has been devised by Caldwell *et al.* (1983a). Using a commercially available fiber-optics-based light source and photocell, when the probe nears a leaf, light from the source is scattered from the leaf to the detector, and the contact with a leaf recorded without operator intervention. Caldwell *et al.* (1983b) used a motor-driven probe shaft with encoder, so the distance traversed by the probe into the canopy was also measured automatically. This system was used very successfully by Caldwell and co-workers to measure nondestructively the canopy structure of desert bunch grasses, which have very thin leaves.

Automatic recording is most useful with the point-quadrat method because a large number of contacts must be measured. For example, to obtain a 1% estimate of the number of contacts per insertion, 1000 contacts are required (Ross, 1981). The relationship between the number of contacts per insertion and the canopy structure depends on the angle of insertion of the probe. For a single insertion angle, the LAI can be best estimated using a zenith angle of 57° and

$$\mathrm{LAI} = 1.1\, N(57) \qquad (14.4)$$

where $N(57)$ is the number of contacts per insertion in the canopy with a probe zenith angle of 57 degrees. Optimistically Equation 14.4 could provide estimates of LAI within about 10%, with enough data, the uncertainty being the result of uncertainty in the 1.1 factor in Equation 14.4. If the basic assumption of randomness is not met, then inclined point quadrats may not provide 10% estimates. From Equation 14.4, the number of contacts per insertion is approximately equal to the LAI; thus for a canopy of LAI = 2, approximately 500 insertions of the probe would be required for a 1% estimate of $N(57)$.

The assumption of a random position distribution of foliage in the horizontal can be assessed by calculating the ratio of the variance of the number of contacts to the mean number of contacts. For small leaves in a canopy of large horizontal extent, the probability distribution for contacts of the probe with randomly positioned foliage elements is a Poisson distribution; for such a distribution the variance and mean are equal. The ratio of variance to mean should therefore be unity if the canopy elements are randomly dispersed. One should be aware, however, that even larger samples are required to estimate the contact frequency variance than are needed for means.

Using two or more insertion angles provides better estimates of LAI and also permits estimates of the mean leaf inclination angle (see Anderson, 1971, for a summary of equations).

Measurements from inclined point quadrats can be analyzed using the methods discussed in the section on indirect methods later in this chapter. The average number of contacts per insertion from a zenith angle, θ_i, is given by

$$N(\theta_i) = N_i = uK_{i.}Z \qquad (14.5)$$

where the subscript i refers to the probe zenith angle, u is the leaf area density (leaf area per unit volume of canopy), Z is the vertical distance from the top of the canopy, and $K_{i.}$ is the extinction coefficient given by

$$K_{i\cdot} = \sum_{j=i}^{N_j} g_{j\cdot} \cos\delta_{ij\cdot} \qquad (14.6)$$

for a canopy approximating azimuthal symmetry. For such a canopy

$$\cos\delta_{ij\cdot} = \cos\alpha_j \qquad \text{for } \theta \le 90-\alpha$$

$$\cos\delta_{ij\cdot} = \cos\alpha_j \frac{1 + 2(\tan x - x)}{\pi} \qquad \text{for } \theta > 90-\alpha$$

$$(14.7)$$

where

$$x = \cos^{-1}(\cot\alpha_j \cot\theta_i)$$

and $\cos\delta_{ij.}$ is the fraction of leaf area inclined at angle α_j to the horizontal that is projected in the direction θ_i assuming azimuthal symmetry. The product, uZ, is the leaf area index (LAI) over the height interval Z so that

$$N_i = \text{LAI } K_{i.} \qquad (14.8)$$

where $K_{i.}$ can be thought of as the fraction of foliage area projected on to the horizontal plane. If values for N_i are measured, inclined point quadrats can be used to estimate LAI and leaf angle distribution for a canopy. The details for this calculation will be given later when indirect methods are discussed. One probably would prefer to use indirect methods over point quadrats for estimating LAI and leaf angle distribution because they are much faster.

Obviously, the information obtained from inclined point quadrats is very similar to that obtained from analyses of direct-beam transmittance of canopies. The relationship can be shown as follows. If the fraction of the horizontal area under the canopy that is in sunflecks is given by $T(\theta_i) = T_i$, where the dependence of the direct-beam transmittance on the solar zenith angle θ_i is analogous to the dependence on probe incidence angle in point quadrats, then the fraction of horizontal area in flecks illuminated by the direct beam is given by

$$T_i = T(\theta_i) = \exp(-\text{ LAI } K_{i.}) = \exp[-N(\theta)] = \exp(-N_i) \qquad (14.9)$$

Therefore, indirect sunfleck methods are linked with point quadrats because, $N(\theta_i) = -\ln[T(\theta_i)]$. The objective then of inclined point quadrats or indirect methods is to use measurements of $N(\theta_i)$ or $T(\theta_i)$ versus θ_i to estimate u (or LAI) and $g_{j.}$. In the next section we will discuss this in more detail.

A method similar to inclined point quadrats, and termed a drop-line method, is reported by Miller and Lin (1985) for a red maple forest; however, the text of this paper is not clear about whether the authors properly weighted the results of the drop-line method to obtain the area distribution of angles. Miller and Lin describe the appropriate weighting for estimating LAI from the drop-line method but say nothing about the angle distribution weighting. The distributions in the upper canopy might be expected to have more vertical leaves, and omission of this weighting could explain this. Although the point-quadrat method usually is assumed to be practical only for short canopies, Ford and Newbould (1971) used it on an 8 m deciduous forest canopy to estimate LAI, leaf angle distribution and aggregation patterns.

Inclined point quadrats can be a useful nondestructive method for characterizing canopies, especially those canopies with mixed species; however, because of the considerable effort involved in its use, its main value is becoming historical as a predecessor of the rapid indirect methods to be discussed later.

14.3.2 The method of Lang

Lang (1973) devised a very clever instrument for measuring canopy architecture. He used ultra-high-precision potentiometers that record the angles of three arms to permit the measurement of three Cartesian coordinates that define the position of any chosen point on a foliage element. By selecting an appropriate array of points on any given leaf, the position, inclination, azimuth and area of any triangle, which is enclosed by three of these points, can be measured directly. This probably is the best method for measuring canopy structure directly, and it is fast enough, with an appropriate computer to interrogate the potentiometers, to follow heliotropic movements (Shell and Lang, 1975, 1976; Lang and Shell, 1976). Appropriate hand-held computers are available from several companies. Lang's method probably is the best method for testing the indirect methods described in the next section. The reader is referred to the papers of Lang for details on the construction and use of his instrument, since it is not for the faint hearted.

14.4 INDIRECT METHODS

The term 'indirect' will be used in this section to refer to the use of radiation measurements to obtain estimates of canopy structure. Several methods will be discussed which have a common feature; the need to accomplish an 'inversion'. The term, inversion, refers to a mathematical process whereby the original independent variable is transformed to a dependent variable. One could argue that inversion is a necessary part of any kind of measurement because interpretation always requires the conversion of the quantity actually measured into the desired scientific information. However, for most measurements the linkage between the observations and the desired result is so straightforward that the inversion process is simply a division, and we might not be fully aware of its nature.

This mathematical process of inversion might best be understood with a simple example. Consider the measurement of solar radiation using a sensor that is assumed to produce an output voltage which increases linearly with radiant flux density. If the radiant flux density is represented by E, then the voltage output, V, of this sensor can be represented as follows:

$$V = CE \tag{14.10}$$

where C is the calibration factor. In this case the radiant flux density, E, is the independent variable and the voltage, V, the dependent variable. In defining this problem the voltage is chosen as the dependent variable because, in the context of this instrument, voltage is simply dependent on incident flux density and radiation may take on any value, thus making radiation the true and original independent variable. Equation 14.10 is a simple statement of a forward problem: given a model (Equation 14.10) and model parameters (C), find the output voltage (V) for some specified input radiation (E). But suppose we want to use the instrument, in which case we choose to consider the voltage as the independent variable and the radiation as the variable dependent on the voltage. Thus we want to invert Equation 14.10 for the purpose of estimating solar radiation from some measured voltage. If $\hat{E}$ is an estimate of solar radiation, then

$$\hat{E} = C^{-1}V \tag{14.11}$$

Some would consider $C^{-1} = 1/C$ to be the calibration factor and thus not be aware of the inversion that is implicit in using the radiometer. In practice both V and C have errors that produce uncertainty in the estimated radiation, E. For example the radiometer may respond to temperature as well as radiation, or there may be electrical noise from a nearby radio station inducing spurious voltages on the sensor lead wires. If a single voltage corresponds to only one radiation level, then we cannot assess errors and each result is unique. However, if numerous voltages can be associated with a single radiation level, then errors can be evaluated. This is another characteristic of inversion problems: we need to find a method that will accommodate measurement errors and still yield the best estimate of the desired quantities.

Perhaps an analogy might be useful here. Suppose one has an egg and a prescribed set of motions for a fork, such as 100 precise circles. After executing these 100 precise circular motions with the tines of the fork in the egg, we have a scrambled egg. This is the 'forward' problem. Now, given the state of this egg and a kernel that describes mathematically how raw egg flows around the tines of a rotating fork, we should, in principle, be able to mathematically reassemble the original egg. This reassembly is the 'inversion' problem. Clearly we may not know the kernel for describing the effect of a rotating fork on a raw egg well enough to accomplish this inversion because unquantifiable 'noise', or in this case turbulence, introduces chaos into the process so the original characteristics are not recoverable. Thus the practicality of accomplishing an inversion depends on whether the inherent noise that is present is small enough not to overwhelm the 'information content' of the input data and appropriateness of the kernel. In other words, errors determine whether an inversion will be practical to accomplish or not.

The estimation of canopy structure from radiation measurements may require many measurements to obtain estimates of several canopy characteristics such as leaf area index and mean leaf inclination angle. Under these conditions V, C and E in Equations 14.10 and 14.11 are matrices and Equation 14.11 thus requires a matrix inverse. In this chapter we denote a matrix variable with an underline so Equation 14.11 in matrix form becomes

$$\underline{V} = \underline{C}^{-1}\,\underline{E} \tag{14.12}$$

If the number of data values measured is equal to the number of output values desired (or the number of model values), then the solution requires solving simultaneous equations and the inverse problem is unique. If the number of observations is greater than the number of output factors (or model values) desired, then the method of least squares usually is used. The key to usefulness of any measurement is the quality of the inversion and this is particularly true for indirect estimates of canopy architecture.

Three general groups of indirect methods are discussed in this section, spectral methods, gap fraction methods and bidirectional reflectance distribution function (BRDF) methods. Their inversions vary from trivial to difficult to nearly impossible. Other indirect methods have been used but will not be discussed in this chapter (Smith *et al.*, 1977; Smith and Berry, 1979).

14.4.1 Spectral methods

Spectral methods rely on a difference between the spectral reflectance of foliage and other materials such as soil (Table 14.1). Live leaves absorb about 85% of visible wavelengths of radiation (photosynthetically active radiation, PAR) and scatter about 85% of near-infrared (NIR) wavelengths with a sharp transition at 700 nm. However, soil absorbs roughly similar amounts of visible and near-infrared radiation. Therefore as a foliage cover forms over the soil the ratio of near-infrared to visible radiation increases. A useful form of this ratio is termed the normalized difference vegetation index (NDVI), and is given by

$$\text{NDVI} = \frac{\text{NIR rad} - \text{PAR rad}}{\text{NIR rad} + \text{PAR rad}} \tag{14.13}$$

The NDVI varies between zero and unity (Choudhury, 1987). The near-infrared and visible measurements usually are made in relatively narrow wavebands near 600 nm and 800 nm with a restricted-view instrument having a field-of-view of 2–15°. The instrument is pointed down toward the canopy from directly overhead. (This zenith angle of view of zero degrees is termed the nadir view.) Normal-incident narrow-view sensors work better for these measurements than hemispherical-view sensors.

Table 14.1 Reflectance of leaves (Gausman, 1974) and soils (Stoner *et al.*, 1980) in several wavelength bands of the solar spectrum

Wavelength band (μm)	*Leaf reflectance*	*Soil reflectance*	
		Sandy loam	*Clay*
0.5 – 0.6	0.12	0.05	0.03
0.75 – 0.85	0.42	0.19	0.05
1.2 – 1.3	0.39	0.31	0.10
1.7 – 1.8	0.28	0.28	0.12
2.1 – 2.2	0.17	0.18	0.10

A simple inexpensive instrument for measuring NDVI is described in the Experimenter's Corner section of *Popular Electronics* (1981). This instrument uses visible and near-infrared light emitting diodes for sensors. For normalized difference measurements the sensors need only be calibrated relative to each other by pointing them at a uniformly lit surface with similar or known reflectances in the PAR and NIR. Barium sulfate paint is used as a reference surface and the instrument is pointed at this surface, which is illuminated by the direct sun. The best measurements are made by normalizing canopy measurements to reference-surface measurements frequently in the field.

This spectral method has been used extensively in recent years because of applications to remote sensing from satellites. Thus commercial instruments are available with the same wavelength characteristics as several satellite radiometers. For example the Exotec model 100 (Exotech Inc., 1200 Quince Orchard Blvd., Gathersburg, MD 20878, USA) simulates the Multispectral Scanner on early Landsat satellites and the Barnes model MMR (Barnes Engineering Co., 30 Commerce Rd., Stamford, CN 06904, USA) simulates the Thematic Mapper on later Landsat satellites. The normalized difference has been used successfully to estimate LAI in wheat up to values of three (Asrar *et al.*, 1984). The useful range depends on solar zenith angle and canopy leaf angle distribution. Horizontal leaf canopies have the narrowest range of LAI sensitivity. Results for wheat from Asrar *et al.* (1984) and for soybean from Holben *et al.* (1980) are shown in Fig. 14.2. Typical uncertainties in LAI estimates from the NDVI are about a factor of two.

This spectral method may be usable on small isolated plants provided that the view of the instrument is filled with leaves. This would mean that if the plant is 20 cm across and the instrument view is 10°, then observations should be made from a height less than 1 m (height = 20 cm/tan (10) = 113 cm). Some calibration is likely to be necessary with this method and it has not been tested.

Some attempts have been made to use the PAR/NIR ratio below a canopy to estimate LAI (Jordan, 1969). For a given site the PAR/NIR transmittance ratio is linear with LAI; however, the slope of this relationship may vary with the canopy type and thus require calibration (Norman and Jarvis, 1974). Although this transmittance-ratio method has not been exhaustively explored, other indirect methods are probably more suitable from below the canopy.

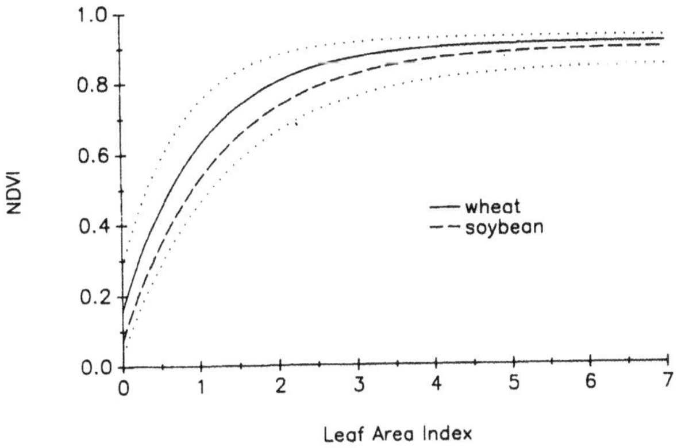

Fig. 14.2 Mean curves fit to measurements of normalized difference vegetation index (NDVI) versus leaf area index for wheat (Asrar *et al.*, 1984) and soybean (Holben *et al.*, 1980). The dotted curves, which represent typical scatter associated with this kind of data, were drawn by eye to include most of the data points in the original figures.

14.4.2 Gap-fraction methods

Gap-fraction methods offer a powerful tool for estimating LAI and leaf inclination angles for canopies of full cover, isolated single canopies and even heterogeneous canopies. In this section full-cover canopies are discussed in detail because the gap-fraction method is best established for this case.

If the canopy elements are assumed to be randomly distributed in azimuth, the transmitted fraction (T_i) of a beam of radiation incident on the canopy at zenith angle θ_i is given by

$$-\ln T_i = \sum_{j=1}^{N_i} f_j K_{ij} \qquad (14.14)$$

where

$$L = \sum_{j=1}^{N_j} f_j$$

and N_j is the number of leaf angle classes, f_j is the LAI in inclination angle class, j, (assuming azimuthal symmetry) and K_{ij} is the extinction coefficient for a beam zenith angle, θ_i and a leaf inclination angle, α_j, calculated as in Equation 14.6 where $g_{j.} = f_j/L$. Using measurements of T_i at several zenith angles, in principle values of f_j that are consistent with measured values of T_i can be derived by 'inverting' Equation 14.14. The term K_{ij} is referred to as the kernel, and its form may depend on what is known about the canopy in question. For isolated canopies, such as bunch grass or widely spaced trees, the kernel may be more complicated than the one in Equation 14.14.

Equation 14.14 represents a system of linear equations with unknowns f_j, since $-\ln T_i$ is measured and K_{ij} are calculated from known values of θ_i and α_j. If $N_i = N_j$ (number of observation zenith angles equals the number of leaf angle classes), then the unknown f_j values are uniquely determined, and the LAI and leaf angle distribution can be found.

If $N_j < N_i$, then there are more equations than unknowns and the system of equations is over determined. A set of values for f_j can be obtained by standard linear least squares. However, as a result of errors in the measurements of T_i, the possibility exists for some of the f_j, which result from the least squares calculation, to be negative. Since negative values of LAI are impossible, the solution must be constrained so f_j must be positive. In other words, additional information must be added to the solution. A technique for doing this is illustrated in the next section.

(a) Inversion procedure

One way to understand the gap-fraction inversion procedure is with an example that has a known answer. We will consider a canopy of randomly positioned leaves with a spherical leaf angle distribution ($K_{ij} = 0.5/\cos\theta_i$) and a leaf area index (L) of 2.0. Fig. 14.3 contains the gap fraction as a function of zenith angle for the spherical leaf angle distribution as well as several other distributions. The basic equation that we must invert is Equation 14.14 and for $N_j = 2$, or two leaf angle classes

$$-\ln T_i = f_1 K_{i1} + f_2 K_{i2} \qquad (14.15)$$

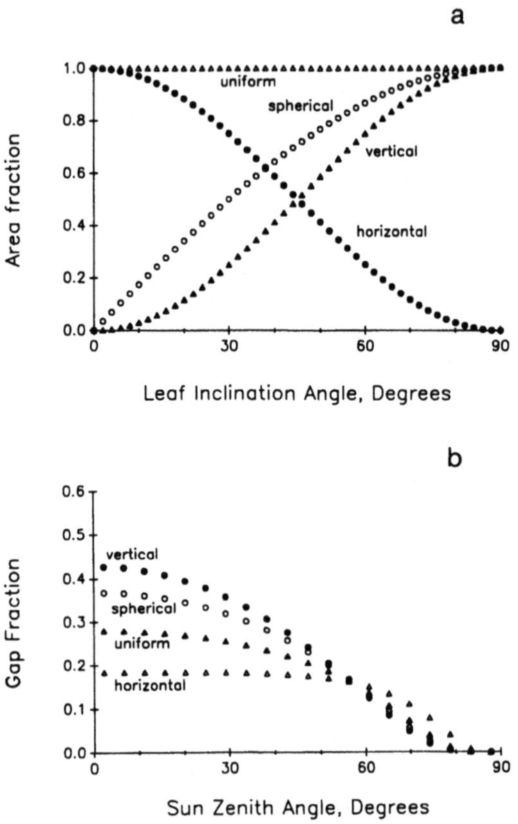

Fig. 14.3 (a) Leaf angle distributions for canopies representing mainly vertically, horizontally, spherically and uniformly distributed leaves assuming azimuthal symmetry (deWit, 1965). (b) Gap fraction versus solar zenith angle for the four leaf angle distributions above for LAI = 2.0.

For this example consider gap fractions at three solar zenith angles, which we shall calculate from the spherical leaf angle distribution with LAI = 2.0:

i	θ_i	$T(\theta_i)$	$-\ln [T(\theta_i)]$
1	20	0.345011	1.06418
2	45	0.243117	1.41421
3	65	0.093836	2.36620

Ordinarily the $T(\theta_i)$ would be measured for a canopy and would not be known to six significant figures, perhaps only two. Because of such errors, more than three observation angles usually would be used; here we include six digits so the reader can appreciate the nature of inversions and reproduce this example for himself. The angles that we will use for the leaf distribution are $\alpha_1 = 22.5°$, $\alpha_2 = 67.5°$, because $j = 2$ and we choose the midpoint angle of each class.

Equation 14.15 can be written in matrix form as

$$\underline{Y} = \underline{A}\underline{G} \qquad (14.16)$$

where the contents of the three matrices are given as below for input of gap fractions at three solar zenith angles ($i = 3$) and two leaf angle classes ($j = 2$):

$$\underline{Y} = \begin{vmatrix} -\ln [T(\theta_1)] \\ -\ln [T(\theta_2)] \\ -\ln [T(\theta_3)] \end{vmatrix} = \begin{vmatrix} 1.06418 \\ 1.41421 \\ 2.36620 \end{vmatrix}$$

$$\underline{A} = \begin{vmatrix} K_{11} & K_{12} \\ K_{21} & K_{22} \\ K_{31} & K_{32} \end{vmatrix} = \begin{vmatrix} 0.923880 & 0.382683 \\ 0.923880 & 0.639378 \\ 0.923880 & 1.284915 \end{vmatrix}$$

$$\underline{G} = \begin{vmatrix} f_1 \\ f_2 \end{vmatrix}$$

In principle the solution to this matrix equation for $\underline{G}$ by least squares is

$$\underline{G} = [\underline{A}^T \underline{A}]^{-1} [\underline{A}^T\underline{Y}] \qquad (14.17)$$

Where $\underline{A}^T$ is the transpose of matrix $\underline{A}$. Substituting numbers for $\underline{A}^T\underline{A}$ and $\underline{A}^T\underline{Y}$, we obtain

$$\underline{G} = \begin{vmatrix} 2.56066 & 2.13137 \\ 2.13137 & 2.20626 \end{vmatrix}^{-1} \begin{vmatrix} 4.47582 \\ 4.35182 \end{vmatrix}$$

or

$$\underline{G} = \begin{vmatrix} 1.99346 & -1.92580 \\ -1.92580 & 2.31368 \end{vmatrix} \begin{vmatrix} 4.47582 \\ 4.35182 \end{vmatrix}$$

Finally, carrying out the indicated multiplication results in

$$\underline{G} = \begin{vmatrix} 0.541625 \\ 1.44919 \end{vmatrix}$$

and therefore the LAI at leaf angles between zero (horizontal) and 45° is 0.54 and between 45° and 90° (vertical) is 1.45 for a total LAI = 1.99, very near to the 2.00 we began with. For a spherical leaf angle distribution the leaf area in each angle class is 0.59 and 1.41 which is in reasonably good agreement with values obtained by inversion. The difference arises because the spherical leaf angle distribution is being approximated with a conical distribution having only two leaf angle classes.

Consider an example where the gap-fraction measurements are in error. Assume the transmittance is in error so that $T(\theta_1) = T(\theta_2) = 0.2$ and $T(\theta_3) = 0.1$, then the resulting LAI = 2.05 and $f_1 = 0.86$, $f_2 = 1.19$ and the inverted leaf angle distribution approaches uniform instead of spherical.

The errors inherent in least-squares inversion can be reduced by constraining the inversion. From experience the most suitable constraint appears to be a constraint toward a uniform leaf angle distribution. Thus if the input data are not consistent with the one-dimensional kernel or contain too much noise, then a large constraint may be necessary to avoid negative f_j values and the output leaf angle distribution will approach the uniform distribution that was input as a constraint. Under such conditions only an estimate of the LAI is available from the inversion. In general the LAI is less sensitive to error than the leaf angle distribution.

The constraint matrix, which is represented

by $\underline{H}$, is a $N_j \times N_j$ tridiagonal matrix with off-diagonal values of -1, diagonal values of 2 except for upper left and lower right corners which are unity. If $N_j = 4$, then

$$\underline{H} = \begin{vmatrix} 1 & -1 & 0 & 0 \\ -1 & 2 & -1 & 0 \\ 0 & -1 & 2 & -1 \\ 0 & 0 & -1 & 1 \end{vmatrix}$$

The matrix equation for the leaf angle distribution, $\underline{G}$ is

$$\underline{G} = (\underline{A}^T\underline{A} + \gamma\underline{H})^{-1}\, \underline{A}^T\underline{Y} \qquad (14.18)$$

where γ is a scalar that dictates the importance of the constraint. Normally γ is chosen as small as is consistent with keeping all f_i values positive. For our example

$$\underline{G} = \left|\begin{vmatrix} 2.56066 & 2.13137 \\ 2.13137 & 2.20626 \end{vmatrix} + \begin{vmatrix} \gamma & -\gamma \\ -\gamma & \gamma \end{vmatrix}\right|^{-1} \begin{vmatrix} 4.47582 \\ 4.35182 \end{vmatrix}$$

and if $\gamma = 1$, a relatively large value, then $f_1 = 0.92$, $f_2 = 1.04$ and LAI $= 1.96$; this would approximate a uniform leaf angle distribution.

Program 1 at the end of this chapter solves for the LAI and leaf angle distribution using this method of constrained least squares.

An alternative method for accomplishing this integral inversion of gap-fraction data uses an ellipsoidal leaf angle distribution proposed by Campbell (1986). The basis of this method is that the leaf angle distributions of a canopy can be represented by the distribution of the area on the surface of an ellipsoid of revolution. A single parameter, x, determines the shape of the distribution. Horizontal canopies have large values of x, and vertical canopies have small values. If $x = 1$, the canopies have a spherical distribution. Extremeophile and plagiophile distributions (deWit, 1965) are not represented very well, but this does not cause serious limitations because extremeophile have never been observed and plagiophile approximations by the elliptical distribution normally would not cause serious errors. A better representation of leaf angle distributions can be achieved with the beta distribution (Goel and Strebel, 1984). The procedure used here is to search for values of x, which represent leaf angle distributions, that will produce a reasonable fit to the input data of gap fraction ($T(\theta_i)$) versus incident sun angle. For an elliptical leaf angle distribution, the extinction coefficient as defined by Equation 14.6 is given by Campbell (1986):

$$K_{i.} = K(\theta_i,x) = [(x^2 + \tan^2\theta_i)^{1/2}]/D$$

where

$$D = x + 1.774\,(x + 1.182)^{-0.733} \qquad (14.19)$$

The leaf angle distribution itself is given by Campbell and Norman (1988). We would like to find values for x and L (leaf area index) which minimize

$$F = \Sigma(\ln T_i + K(\theta_i,x)\,L)^2 \qquad (14.20)$$

subject to the constraint, $x>0$. This is a nonlinear constrained least-squares problem. The minimum can be found by solving $\partial F/\partial L = 0$ and $\partial F/\partial x = 0$, simultaneously. From the first equation we obtain

$$L = -\Sigma\,[K(\theta_i,x)\ln T_i]/\Sigma\,K(\theta_i,x)^2 \qquad (14.21)$$

and from the second,

$$L = \frac{-\Sigma\,[\ln T_i\,\partial K(\theta_i,x)/\partial x]}{\Sigma\,[K(\theta_i,x)\,\partial K(\theta_i,x)/\partial x]} \qquad (14.22)$$

To solve the equations, L can be eliminated between them, and x found by the bisection method. Once x is known, L is found from Equation 14.21. Program 2 is a BASIC computer program which illustrates the bisection method and finds L and x from measured transmission coefficient data.

(b) Gap-fraction applications

The input data that must be measured to

apply the gap-fraction inversion in the last section can be obtained by several methods:

1. The measurement of sunflecks with a sensor that is scanned through the canopy for spatial sampling (Perry, 1985; Lang *et al.*, 1985);
2. A meterstick to measure the size of sunflecks on the ground;
3. A lighted rod that is placed below a canopy and photographed with a camera from above the canopy at night (Kopec *et al.*, 1987);
4. Spatially averaged measurements of light penetration with quantum light-bar sensors. This instrument measures direct and diffuse radiation so special care must be taken to eliminate light scattered from leaves;
5. Hemispherical lens (fisheye) measurements in the visible waveband from below the canopy looking upward (Anderson, 1971);
6. Hemispherical lens measurements in the near-infrared wave band from above the canopy looking downward.

All of these methods represent an attempt to measure the canopy gap fraction as a function of zenith angle for any given canopy. Each of these measurement techniques has potential errors and advantages as well as disadvantages.

Scanning sensors can be used in canopies large enough to permit their passage such as corn (Perry, 1985; Perry *et al.*, 1988) or sorghum (Lang *et al.*, 1985). In this case the sun is used as the light source and measurements are made at various times of the day to determine canopy beam transmittance at various solar zenith angles. These sensors are designed to measure either the fraction of the transect with light levels above some threshold or the average flux density along a transect. In tall canopies, or when leaves are small, overlapping penumbra can cause errors in measuring gap fraction with threshold sensors or shadows on a meterstick (Norman *et al.*, 1971). As penumbral effects become important, the sunflecks appear more dimly lit and larger. This also is a limitation of the yardstick. A threshold sensor, which is 0.8 m long, is commercially available from Decagon Devices, Inc., Pullman, WA, USA. The sensor described in Lang *et al.* (1985), which averages beam radiation along a transect, can be used in canopies of any height because penumbral illumination levels are appropriately accommodated in the measurement. Scan path lengths of 9 m have been used with a threshold-detecting light sensor in corn and soybean row crops and have produced excellent spatial samples (Fig. 14.4). The sensor described by Lang *et al.* (1985) was scanned approximately 50–100 m below a *Eucalyptus* canopy of about 50 m height by walking with the sensor mounted on a gimbal. The indirect measurement of LAI was 1.8 and the direct measurement 1.7. Fig. 14.5 shows the variability from scan to scan. Lang and Xiang Yueqin (1986) suggest that canopy transmittances should not be averaged over entire scan paths but only over distances about 10 times the leaf size; then the log of these subscan transmittances averaged over the entire path. An instrument that performs this task and based on the Lang *et al.* (1985) sensor design is built by Assembled Electrics, 66 Smith St, Lagoona, N.S.W. 2199, Australia.

Lighted rods of three different sizes have been used with good results. In this set-up the rod is the light source, a camera the detector and the measurements are done in the dark either at night or in an enclosure. A 3 m rod (2.5 cm in diameter) has been used in corn and soybean, a 1 m rod in alfalfa and a 0.3 m rod (6 mm diameter) in fescue grass. The 3 m and 0.3 m rods were made from clear lucite with a light bulb glued to one or both ends to provide light for the rod. The 1 m lighted rod was a fluorescent tube (Lee, 1985). Each rod must be coated along the bottom half to avoid lighting up the ground beneath the rod. The lighted rod is most useful in short grass canopies such as fescue

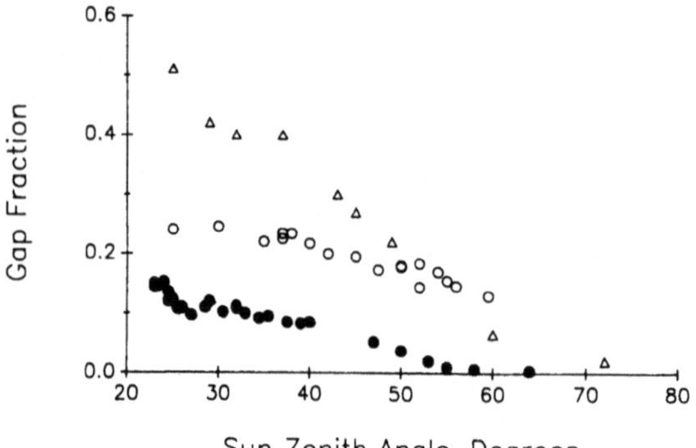

Fig. 14.4 Measured sunfleck or gap fraction as a function of solar zenith angle for two different planting densities of corn in Pennsylvania (Perry, 1985): Δ, Measured LAI = 2.3, $\bar{\alpha}$ = 42°, inverted LAI = 2.1, $\bar{\alpha}$ = 44°; O, Measured LAI = 4.7, $\bar{\alpha}$ = 58°, inverted LAI = 4.8, $\bar{\alpha}$ = 70°. Measured sunfleck or gap fraction in soybean in Nebraska (●): Measured LAI = 2.0, no direct angle measurement, inverted LAI = 2.1, $\bar{\alpha}$ = 82°. $\bar{\alpha}$ is the mean leaf inclination angle. In corn the scan path length was 12 m and in soybean 9 m and both had row spacings of 0.76 m.

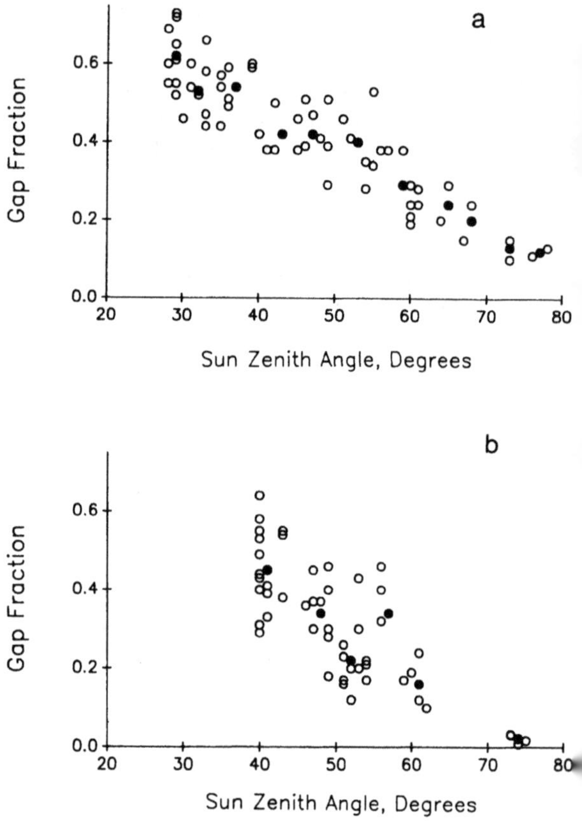

Fig. 14.5 (a) Measured gap fraction as a function of solar zenith angle in a *Eucalyptus* (Blue Gum) forest with trees about 10 m tall, and (b) in *Eucalyptus delicatensis* about 40 m tall with direct measured LAI=1.7. The filled dots represent averages of groups of raw data points that were used in inversion program 1. All data points were used in program 2.

where no other method is available. Measuring LAI in fescue directly is a very laborious task, and indirect nondestructive methods are an attractive alternative. The agreement within 20% of indirect and direct measurements is good.

Quantum light-bar sensors, such as the LI-191 from LI-COR, Lincoln, NE, USA, and the SF-80 from Decagon Devices, Pullman, WA, USA, are suitable for measuring the gap fraction of canopies of low LAI. Since the gap-fraction methods are based on measurements of transmitted beam radiation, scattered light must be eliminated from the measurements. At high LAI, most of the light on the quantum sensor may be scattered from leaves and a method must be devised for eliminating this if gap fraction is to be found. This can be done with four measurements using a rectangular shade (approximately 5 cm by 120 cm in size) for shading the light bar: (1) With the light bar above the canopy measure total incident radiation; (2) then shade the light bar to obtain diffuse light only so direct beam can be estimated by subtraction; (3) place the light bar below the canopy and measure total transmitted radiation; (4) then shade the light bar below the canopy and obtain diffuse transmitted radiation; the difference is the direct beam transmitted through the canopy. The ratio of direct beam transmitted to incident approximates the gap fraction. This procedure is repeated for several sun angles. To correct for the diffuse radiation blocked by the shade, the ratio of unshaded light-bar output to shaded light-bar output can be measured on a cloudy day with the shade at the desired distance (approx. 20 cm). Each shaded light-bar measurement is then multiplied by this correction factor. Transmittance of PAR has been used to estimate LAI in wheat on clear days with reasonable results (Fuchs *et al.*, 1984).

Hemispherical lenses have also been used to characterize canopy structure (Anderson, 1971, 1981; Bonhomme and Chartier, 1972). Diffuse sky conditions are most suitable for hemispherical lens image analysis whether the camera is 'looking' upward from below the canopy or downward from above. If high-contrast film is used, negatives or positives can be digitized by machine and gap fractions obtained as a function of angle on a given image. This multistep procedure of taking the picture, developing the film, digitizing the image by machine or by hand and inverting the result is cumbersome. In recent years solid-state cameras (CCD cameras) have become available at modest prices with 500 × 500 pixel resolution or more. These cameras are very small and can be hand carried to the field with a portable video cassette recorder (VCR) to collect large volumes of data. The video cassette can then be taken back to a microcomputer equipped with a video digitizing board to digitize the image and invert the result. Normally visible wavelengths are used for 'looking' skyward from below the canopy on an overcast day. An alternative, used by Bonhomme *et al.* (1974) for crops with low LAI, is to look downward on an overcast day. For this application the near-infrared waveband gives the best contrast between leaves and soil, and photographic-infrared film or an infrared filter on a CCD camera is used. Fig. 14.6 is a CCD camera image of a sorghum row crop, in the near-infrared waveband, taken from above the canopy. The LAI of this sorghum crop is about 1.5. Clearly the leaves in Fig. 14.6 are not randomly distributed in space and the kernel used in Equation 14.4 is not appropriate. Inversion of gap-fraction data from Fig. 14.6 still is possible, but a more complex model, which includes row structure, is needed.

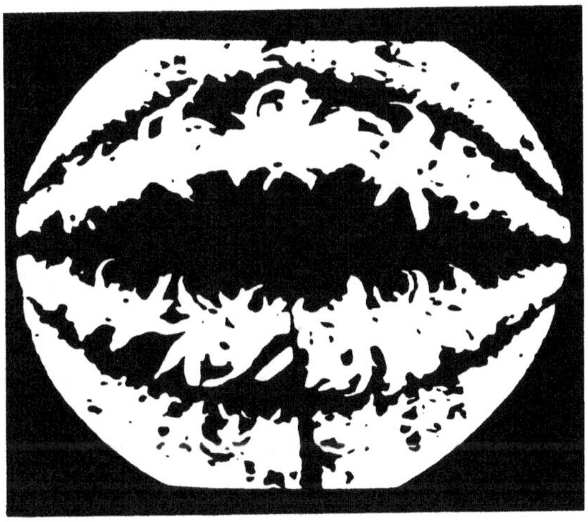

Fig. 14.6 Near-infrared image of a sorghum field taken with a 512 × 512 pixel resolution CCD camera with direct measured LAI = 1.4 from about 1.5 m above the ground.

(c) Evaluation of gap-fraction methods

The suitability of gap-fraction methods depends on two considerations: (1) adequacy of spatial sampling in defining the precision of gap-fraction measurements and (2) the adequacy of the canopy model that assumes random leaf positioning in the horizontal. These limitations can be evaluated by comparing direct and indirect measurements, provided that direct measurements are adequate.

The adequacy of spatial sampling can be assessed from scatter on a graph of gap fraction measured as a function of sun zenith angle (Figs 14.4 and 14.5). The scatter in Fig. 14.4 for measurements in corn and soybeans, which have scan paths of about 16 and 12 row spacings respectively, is modest and quite acceptable for the inversion procedures discussed in this chapter. The results from *Eucalyptus* (Fig. 14.5) show more scatter because of the greater heterogeneity of natural

PROGRAM 1. INVERSION OF CANOPY TRANSMISSION DATA TO FIND LEAF AREA INDEX AND LEAF ANGLE DISTRIBUTION BY CONSTRAINED LEAST SQUARES.

```
10 PI=3.14159
20 NA=3 ' NUMBER OF LEAF ANGLE CLASSES
30 INPUT "NUMBER OF ZENITH ANGLES";NZ
40 DIM Z(NZ),T(NZ),X(NA),G(NZ,NA),A(NA,NA),W(NA,NA),B(NA),C(NA)
50 FOR I=1 TO NZ
60   PRINT "ZENITH ANGLE";I;" - DEGREES";:INPUT Z(I)
70   PRINT "TRANSMISSION AT ";Z(I);"DEG";:INPUT T(I)
75   Z(I)=Z(I)*PI/180:T(I)=-LOG(T(I))
80 NEXT
85 '********** SET UP KERNEL MATRIX
90  FOR I=1 TO NA:LI=(I-.5)*PI/(2*NA)
100    FOR J=1 TO NZ:ZA=Z(J)
110   GOSUB 2000:G(J,I)=KB
120 NEXT:NEXT
125 '********** GT*G AND GT*D
130 FOR I=1 TO NA:FOR J=1 TO NA:A(I,J)=0:NEXT:C(I)=0:NEXT
140 FOR I=1 TO NA:FOR J=I TO NA:FOR K=1 TO NZ:A(I,J)=A(I,J)+G(K,I)*G(K,J)
150 NEXT:A(J,I)=A(I,J):NEXT:NEXT
160 FOR I=1 TO NA:FOR K=1 TO NZ:C(I)=C(I)+G(K,I)*T(K):NEXT:NEXT
165 '********** ADD THE CONSTRAINT MATRIX
170 INPUT "CONSTRAINT";GA
180 FOR I=1 TO NA:FOR J=1 TO NA:W(I,J)=A(I,J):NEXT:B(I)=C(I):NEXT
190 FOR I=2 TO NA-1
200   W(I,I)=W(I,I)+2*GA:W(I,I-1)=W(I,I-1)-GA:W(I,I+1)=W(I,I+1)-GA
210 NEXT
220 W(1,1)=W(1,1)+GA:W(1,2)=W(1,2)-GA
230 W(NA,NA-1)=W(NA,NA-1)-GA:W(NA,NA)=W(NA,NA)+GA
240 GOSUB 1000 '********** SOLVE FOR LAI'S
250 SUM=0:FOR I=1 TO NA:SUM=SUM+X(I):PRINT "LAI AT";(I-.5)*90/NA;"DEG=";X(I)
260 NEXT:PRINT "TOTAL LAI=";SUM
270 GOTO 170
280 END
999 '********** GAUSS ELIMINATION SUBROUTINE
1000 FOR J=1 TO NA-1
1010   PIVOT=W(J,J)
1020   FOR I=J+1 TO NA
1030     MULT=W(I,J)/PIVOT
1040     FOR K=J+1 TO NA:W(I,K)=W(I,K)-MULT*W(J,K):NEXT
1050     B(I)=B(I)-MULT*B(J)
1060 NEXT:NEXT
1070 X(NA)=B(NA)/W(NA,NA)
1080 FOR I=NA-1 TO 1 STEP -1
1090   TOP=B(I)
1100   FOR K=I+1 TO NA:TOP=TOP-W(I,K)*X(K):NEXT
1110   X(I)=TOP/W(I,I)
1120 NEXT
1130 RETURN
1999 '********** EXTINCTION COEFFICIENT SUBROUTINE
2000 IF LI<(PI/2-ZA+.01) THEN KB=COS(LI):GOTO 2020
2010 TB=SQR((TAN(LI)*TAN(ZA))^2-1):B=ATN(TB):KB=COS(LI)*(1+2*(TB-B)/PI)
```

EXAMPLE RUN FOR PROGRAM 1.

```
RUN
NUMBER OF ZENITH ANGLES? 3
ZENITH ANGLE 1  - DEGREES? 40
TRANSMISSION AT  40 DEG? .4
ZENITH ANGLE 2  - DEGREES? 50
TRANSMISSION AT  50 DEG? .35
ZENITH ANGLE 3  - DEGREES? 60
TRANSMISSION AT  60 DEG? .3
CONSTRAINT? 0
LAI AT 15 DEG= .8764444
LAI AT 45 DEG=-.4456763
LAI AT 75 DEG= .7093708
TOTAL LAI= 1.140139
CONSTRAINT? .001
LAI AT 15 DEG= .567206
LAI AT 45 DEG= .1868077
LAI AT 75 DEG= .4574701
TOTAL LAI= 1.211484
CONSTRAINT? .0001
LAI AT 15 DEG= .7848009
LAI AT 45 DEG=-.258048
LAI AT 75 DEG= .6345338
TOTAL LAI= 1.161287
CONSTRAINT? .0005
LAI AT 15 DEG= .6318379
LAI AT 45 DEG= 5.489773E-02
LAI AT 75 DEG= .5098436
TOTAL LAI= 1.196579
```

PROGRAM 2. INVERSION OF CANOPY TRANSMISSION DATA TO OBTAIN LEAF AREA INDEX AND LEAF ANGLE DISTRIBUTION USING THE ELLIPSOIDAL LEAF ANGLE DISTRIBUTION

```
10 REM ******* ELLIPSOIDAL EXTINCTION COEFFICIENT
20 DEF FNK(Z,X)=SQR(X*X+Z*Z)/(X+1.774*(X+1.182)^-.733)
25 REM *******  Z IS TAN(ZENITH ANGLE)
30 REM *******
40 PI=3.14159:DX=.01
50 INPUT "NUMBER OF ZENITH ANGLES";NZ
60 DIM Z(NZ),T(NZ)
70 FOR I=1 TO NZ
80   PRINT "ZENITH ANGLE";I;" - DEGREES";:INPUT Z(I)
90   PRINT "TRANSMISSION AT ";Z(I);"DEG";:INPUT T(I)
100   Z(I)=TAN(Z(I)*PI/180):T(I)=LOG(T(I))
110 NEXT
120 REM ******* FIND X USING BISECTION METHOD
130 XMAX=10:XMIN=.1:X=1
140 S1=0:S2=0:S3=0:S4=0
150 FOR J=1 TO NZ:TZ=Z(J)
160    KB=FNK(TZ,X):DK=(FNK(TZ,X+DX)-KB)
170    S1=S1+KB*T(J):S2=S2+KB*KB:S3=S3+KB*DK:S4=S4+DK*T(J)
180 NEXT
190 F=S2*S4-S1*S3  :PRINT X,F
200 IF F<0 THEN XMIN=X ELSE XMAX=X
210 X=(XMAX+XMIN)/2
220 IF (XMAX-XMIN)>.01 THEN GOTO 140
230 REM ******* FIND LAI AND PRINT RESULTS
240 L=-S1/S2:PRINT "LEAF AREA INDEX=";L
250 PRINT "RATIO OF VERTICAL TO HORIZONTAL PROJECTIONS=";X
260 PRINT "MEAN LEAF ANGLE - DEGREES=";90*(.1+.9*EXP(-.5*X))
270 PRINT
280 PRINT "ZENITH ANG.","MEASURED T","PREDICTED T"
290 FOR J=1 TO NZ
300    PRINT ATN(Z(J))*180/PI,EXP(T(J)),EXP(-FNK(Z(J),X)*L)
310 NEXT
320 END
```

EXAMPLE RUN FROM PROGRAM 2.

```
RUN
NUMBER OF ZENITH ANGLES? 3
ZENITH ANGLE 1  - DEGREES? 40
TRANSMISSION AT  40 DEG? .4
ZENITH ANGLE 2  - DEGREES? 50
TRANSMISSION AT  50 DEG? .35
ZENITH ANGLE 3  - DEGREES? 60
TRANSMISSION AT  60 DEG? .3
 1             -4.529432E-04
 5.5            3.976567E-05
 3.25           1.04144E-04
 2.125          1.099415E-04
 1.5625        -1.912634E-05
 1.84375        6.964733E-05
 1.703125       3.301492E-05
 1.632813       9.127427E-06
 1.597656      -4.420989E-06
 1.615234       2.495479E-06
 1.606445      -9.271316E-07
LEAF AREA INDEX= 1.254089
RATIO OF VERTICAL TO HORIZONTAL PROJECTIONS= 1.61084
MEAN LEAF ANGLE - DEGREES= 45.19892

ZENITH ANG.   MEASURED T    PREDICTED T
 40            .4            .3941366
 50.00001      .35           .3580211
 60.00001      .3            .2974503
Ok
```

stands. The scan paths of about 100 m were not adequate so the gap-fraction measurements of 5–15 scan paths were averaged together and these means used as input for the inversion methods discussed earlier. A simple rule for assuring adequate spatial sampling in the wide range of canopies for which the gap-fraction inversion method is suitable (grasses a few centimeters tall to forests tens-of-meters tall) is somewhat evasive. Approximately 10–20 times the spacing of plants is a starting point. Clearly frequent large gaps, which may be considerably larger than a mean plant spacing, will be extremely important in defining spatial sampling requirements.

The suitability of the one-dimensional radiation model, which assumes random leaf positioning, can be evaluated by comparing direct and indirect methods. Fig. 14.7 contains a comparison of direct and indirect methods for a wide range of canopy types. Considering the uncertainties in both methods, agreement is reasonably good.

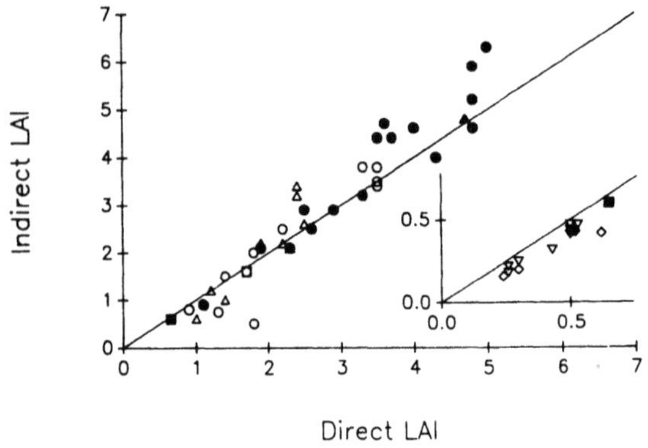

Fig. 14.7 Comparison of indirect versus direct estimates of LAI in corn in Nebraska (●) and Pennsylvania (▲) (Perry, 1985), Fescue grass (○), (Kopec *et al.*, 1987), Soybean (△), *Eucalyptus* (□), a leafless mixed hardwood (■) (Hutchison *et al.*, 1986), sorghum (◇) and sunflower (▽) (Lang *et al.*, 1985).

The effect of row structure and heliotropism on indirect LAI measurement was evaluated by Lang *et al.* (1985). They found that sizeable errors in the LAI estimates can occur when these effects are not accounted for in the model used for the inversion. We used the three-dimensional model of Norman and Welles (1983) to compute LAI of the canopy shown in Fig. 14.6, and compared this with the directly measured LAI and LAI predicted using program 2. Program 2 yielded an inverted LAI of 0.9, the Norman–Welles inversion yielded an LAI of 1.9 and the direct measurement gave an LAI of 1.5 based on six plants. Clearly when row structure is as prominent as that shown in Fig. 14.6, more complex inversions should be used, or the indirect method 'calibrated' to direct measurements.

Another limitation of gap-fraction inversion methods is that the leaf area associated with individual species present in a mixed canopy cannot be separated easily unless they are stratified vertically. With vertical separation of species, scans at various heights would permit estimation of LAI profiles.

14.4.3 BRDF methods

The bidirectional reflectance distribution function (BRDF) of plant canopies is related to their architecture (Norman *et al.*, 1985). The BRDF represents the brightness of a canopy when that canopy is viewed from various view angles for a given angle of the sun. Thus the BRDF is the distribution of canopy brightness with view angle and this distribution depends on sun incidence angle. Because the BRDF is more difficult to model than the gap fraction, the computational requirements of the BRDF method are greater. Further, the inversion procedure is much more difficult. This method has potential for applications to satellite remote sensing as well as ground-based indirect measurements, but applications require considerable computational effort. The BRDF method has been applied to several agricultural crops (Goel and Thompson, 1984) with mixed results. This brief description of the BRDF method is included here mainly for reference because it is not yet ready for widespread application.

14.5 SUMMARY

The measurement of canopy structure can be divided into two methodologies: direct and indirect. Direct methods usually require sampling, cutting and measuring various organs, and they are straightforward, arduous and any level of detail can be obtained with enough patience and labor. Indirect methods involve the measurement of radiation and inference about the canopy structure that may have been associated with the measured radiation characteristics. Indirect methods are nondestructive, rapid and sample large areas of canopy. However, because of the need for a model that relates radiation characteristics to canopy structure, some care must be exercised in the use of indirect methods and limited information is obtained. The indirect methods discussed in this chapter can be used to obtain total LAI and mean leaf inclination angle in homogeneous canopies.

Indirect methods are not likely to replace direct methods but should reduce the need for many labor-intensive canopy structure measurements.

ACKNOWLEDGEMENTS

The data on *Eucalyptus* was gathered in Australia in co-operation with A.R.G. Lang in 1983. The stimulus and financial support for that research was provided by Dr Joe Landsberg of the CSIRO Division of Forestry, Canberra, Australia.

REFERENCES

Anderson, M.C. (1971) Radiation and crop structure. In *Plant Photosynthetic Production, Manual of Methods* (eds Z. Sestak, J. Catsky and P.G. Jarvis), Junk, The Hague, pp. 412–66.

Anderson, M.C. (1981) The geometry of leaf distribution in some South-eastern Australian forests. *Agric. Meteorol.*, **25**, 195–205.

Asrar, G., Fuchs, M., Kanemasu, E.T. and Hatfield, J.L. (1984) Estimating absorbed photosynthetic radiation and leaf area index from spectral reflectance in wheat. *Agron. J.*, **76**, 300–6.

Bonhomme, R. and Chartier, P. (1972) The interpretation and automatic measurement of hemispherical photographs to obtain sunlit foliage area and gap frequency. *Isr. J. Agric. Res.*, **22**, 53–61.

Bonhomme, R., Varlet Grancher, C. and Chartier, P. (1974) The use of hemispherical photographs for determining the leaf area index of young crops. *Photosynthetica*, **8**, 299–301.

Caldwell, M.M., Dean, T.J., Nowak, R.S., Dzurec, R.S. and Richards, J.H. (1983a) Bunchgrass architecture, light interception and water-use efficiency: Assessment by fiber optic point quadrats and gas exchange. *Oecologia*, **59**, 178–84.

Caldwell, M.M., Harris, G.W. and Dzurec, R.S. (1983b) A fiber optic point quadrat system for improved accuracy in vegetation sampling. *Oecologia*, **59**, 417–18.

Campbell, G.S. (1986) Extinction coefficients for radiation in plant canopies calculated using an ellipsoidal inclination angle distribution. *Agric. For. Meteorol.*, **36**, 317–21.

Campbell, G.S. and Norman, J.M. (1988) The description and measurement of plant canopy structure. In *Plant Canopies: Their Growth, Form and Function* (ed. G. Russell), Society for Experimental Biology, Seminar Series 29, Cambridge University Press, New York.

Choudhury, B.J. (1987) Relationships between vegetation indices, radiation absorption and net photosynthesis evaluated by a sensitivity analysis. *Remote Sens. Environ.*, **22**, 209–33.

Daughtry, C.S.T. and Hollinger, S.E. (1984) Costs of measuring leaf area index of corn. *Agron. J.*, **76**, 836–41.

deWit, C.T. (1965) Photosynthesis of leaf canopies. *Agric. Res. Rep.*, No. 663, Center for Agriculture Publication and Documentation, Wageningen, The Netherlands, 57 pp.

Ford, E.D. and Newbould, P.J. (1971) The leaf canopy of a coppiced deciduous woodland. *J. Ecol.*, **59**, 843–62.

Fritschen, L.J. (1985) Characterization of boundary conditions affecting forest environmental phenomena. In *The Forest–Atmosphere Interaction* (eds

B.A. Hutchison and B.B. Hicks), Reidel, Boston, pp. 3–23.

Fuchs, M., Asrar, G. and Kanemasu, E.T. (1984) Leaf area estimates from measurements of photosynthetically active radiation in wheat canopies. *Agric. For. Meteorol.*, **32**, 13–22.

Gausman, H.W. (1974) Leaf reflectance of near-infrared. *Photogrammetric Engr.*, 183–91.

Goel, N.S. and Strebel, D.E. (1984) Simple beta distribution representation of leaf orientation in vegetation canopies. *Agron. J.*, **76**, 800–2.

Goel, N.S. and Thompson, R.L. (1984) Inversion of vegetation canopy reflectance models for estimating agronomic variables. V. Estimation of LAI and average leaf angle using measured canopy reflectances. *Remote Sensing Environ.*, **16**, 69–85.

Goudriaan, J. (1977) *Crop Micrometeorology: A Simulation Study*, Center for Agriculture Publication Documentation, Wageningen, The Netherlands.

Grier, C.C. and Waring, R.H. (1974) Conifer foliage mass related to sapwood area. *For. Sci.*, **20**, 205–6.

Holben, B.N., Tucker, C.J. and Fan, C.J. (1980) Spectral assessment of soybean leaf area and leaf biomass. *Photogramm. Eng. Remote Sensing*, **46**, 651–6.

Hutchison, B.A. and Hicks, B.B. (eds) (1985) *The Forest–Atmosphere Interaction*, Reidel, Boston, 684 pp.

Hutchison, B.A., Matt, D.R., McMillen, R.T., Gross, L.J., Tajchman, S.J. and Norman, J.M. (1986) The architecture of a deciduous forest canopy in Eastern Tennessee. *J. Ecol.*, **74**, 635–46.

Jordan, C.F. (1969) Derivation of leaf area index from quality of light on the forest floor. *Ecology*, **50**, 663–6.

Kopec, D.M., Norman, J.M., Shearman, R.C. and Peterson, M.P. (1987) An indirect method for estimating turfgrass leaf area index. *Crop Sci.*, **27**, 1298–1301.

Kvet, J. and Marshall, J.K. (1971) Assessment of leaf area and other assimilating plant surfaces. In *Plant Photosynthetic Production: Manual of Methods* (eds Z. Sestak, J. Catsky and P.G. Jarvis), Junk, The Hague, pp. 517–55.

Lang, A.R.G. (1973) Leaf orientation of a cotton plant. *Agric. Meteorol.*, **11**, 37–51.

Lang, A.R.G. and Shell, G.S.G. (1976) Sunlit areas and angular distributions of sunflower leaves for plants in single and multiple rows. *Agric. Meteorol.*, **16**, 5–15.

Lang, A.R.G. and Xiang Yueqin (1986) Estimation of leaf area index from transmission of direct sunlight in discontinuous canopies. *Agric. For. Meteorol.*, **37**, 229–43.

Lang, A.R.G., Xiang Yueqin and Norman, J.M. (1985) Crop structure and the penetration of direct sunlight. *Agric. For. Meteorol.*, **35**, 83–101.

Lang, A.R.G. and Xiang Yueqin (1986) Estimation of leaf area index from transmission of direct sunlight in discontinuous canopies. *Agric. For. Meteorol.*, **37**, 229–43.

Lang, A.R.G., Xiang Yueqin and Norman, J.M. (1985) Crop structure and the penetration of direct sunlight. *Agric. For. Meteorol.*, **35**, 83–101.

Lee, K.H. (1985) Leaf Area Index Measurements of Alfalfa Canopies, M.Sc. Thesis, Department of Soil Science, University of Wisconsin, Madison.

Lemur, R. (1973) A method for simulating the direct solar radiation regime in sunflower, Jerusalem artichoke, corn and soybean canopies using actual stand structure data. *Agric. Meteorol.*, **12**, 229–47.

Levy, E.B. and Madden, E.A. (1933) The point method of pasture analysis. *New Zeal. J. Agric.*, **46**, 267–79.

McKee, G.W. (1964) A coefficient for computing leaf area in hybrid corn. *Agron. J.*, **56**, 240–1.

Miller, D.H. and Lin, J.D. (1985) Canopy architecture and a red maple edge stand measured by a point-drop method. In *The Forest–Atmosphere Interaction* (eds B.A. Hutchison and B.B. Hicks), Reidel, Boston.

Norman, J.M. (1979) Modeling the complete crop canopy. In *Modification of the Aerial Environment of Crops* (eds B.J. Barfield and J. Gerber), American Society of Agricultural Engineers, St. Joseph, MI, pp. 249–77.

Norman, J.M. and Campbell, G.S. (1983) Application of a plant-environment model to problems in irrigation. In *Advances in Irrigation* (ed. D. Hillel), Academic Press, New York, Vol. 2, pp. 155–88.

Norman, J.M. and Jarvis, P.G. (1974) Photosynthesis in Sitka Spruce (*Picea sitchensis* (Bong.) Carr.). III. Measurements of canopy structure and interception of radiation. *J. Appl. Ecol.*, **11**, 375–98.

Norman, J.M., Miller, E.E. and Tanner, C.B. (1971) Light intensity and sunflect-size distributions in plant canopies. *Agron. J.*, **63**, 743–8.

Norman, J.M. and Welles, J.M. (1983) Radiative transfer in an array of canopies. *Agron. J.*, **75**, 481–8.

Norman, J.M., Welles, J.M. and Walter, E.A. (1985) Contrasts among bidirectional reflectance of leaves, canopies and soils. *IEEE Trans. Geoscience and Remote Sensing*, GE23, 695–704.

Pereira, A.R. and Shaw, R.H. (1980) A numerical experiment on the mean wind structure inside canopies of vegetation. *Agric. Meteorol.*, **22**, 303–18.

Perry, S.G. (1985) Remote Sensing of Canopy Structure With Simple Radiation Measurements, Ph.D. Thesis, Penn. State Univ., University Park, Pa. Univ. Microfilms, Ann Arbor, Mich.

Perry, S.G., Fraser, A.B., Thomson, D.W. and Norman, J.M. (1988) Indirect sensing of plant canopy structure with simple radiation measurements. *Agric. Forest Meteorol.*, **42**, 255–78.

Popular Electronics, Experimenter's Corner (1981) pp. 75–8.

Ross, J. (1981) *The Radiation Regime and Architecture of Plant Stands*, Junk, The Hague, 391 pp.

Shell, G.S.G. and Lang, A.R.G. (1975) Description of leaf orientation and heliotropic response of sunflower using directional statistics. *Agric. Meteorol.*, **15**, 33–48.

Shell, G.S.G. and Lang, A.R.G. (1976) Movements of sunflower leaves over a 24-h period. *Agric. Meteorol.*, **16**, 161–70.

Smith, J.A., Oliver, R.E. and Berry, J.K. (1977) A comparison of two photographic techniques for estimating foliage angle distribution. *Austr. J. Bot.*, **25**, 545–53.

Smith, J.A. and Berry, J.K. (1979) Optical diffraction analysis for estimating foliage angle distribution in grassland canopies. *Austr. J. Bot.*, **27**, 123–33.

Stoner, E.R., Baumgardner, M.F., Biehl, L.L. and Robinson, B.F. (1980) Atlas of soil reflectance properties. *Res. Bull.* 962, Purdue University Agricultural Experimental Station, West Lafayette, Indiana.

Toole, J.L., Norman, J.M., Holtzer, T. and Perring, T. (1984) Simulating Banks grass mite population dynamics as a subsystem of a crop canopy-micro-environment model. *Environ. Entomol.*, **13**, 329–37.

Waring, R.H., Schroeder, P.E. and Oren, R. (1982) Application of the pipe model theory to predict canopy leaf area. *Can. J. For. Res.*, **12**, 556–60.

Warren Wilson, J. (1960) Inclined point quadrats. *New Phytol.*, **59**, 1–8.

Warren Wilson, J. (1965) Point quadrat analysis of foliage distributions for plants growing singly or in rows. *Austr. J. Bot.*, **13**, 405–9.

Whitehead, D., Edwards, W.R.N. and Jarvis, P.G. (1984) Conducting sapwood area, foliage area, and permeability in mature trees of *Picea sitchensis* and *Pinus contorta*. *Can. J. For. Res.*, **14**, 940–7.

Whitfield, D.M. (1980) Interaction of single tobacco plants with direct-beam light. *Austr. J. Plant Physiol.*, **7**, 435–47.

15

Growth, carbon allocation and cost of plant tissues

Nona R. Chiariello, Harold A. Mooney and Kimberlyn Williams

15.1 INTRODUCTION

The capacity to change in size, mass, form and/or number is an essential feature of life, and the term 'growth' can refer to any or all of these types of change. In this chapter, we focus on methods to analyze one type of growth – the increase in dry mass of plants or plant parts through time. We consider components of growth that occur over time periods ranging from minutes to years, and at structural levels ranging from tissues to the whole plant. Our central theme is that a variety of processes at different temporal and structural scales contribute to plant growth and success. In some studies, the control of photosynthate partitioning may be of critical interest in understanding growth, while in others, it may be the relative costs of twigs versus leaves.

The importance of understanding whole plant growth rate has been emphasized recently in several ecological contexts: (1) Plants of resource-poor environments tend to have lower relative growth rates than plants of resource-rich environments, even under conditions of high resource availability (Grime, 1966; Chapin, 1980; but see Tilman, 1988). The basis for this inherent limitation is not understood, and will require studies of the cost and control of growth under natural conditions. (2) Because growth rate determines plant biomass (size) at any point in time, and because many ecological phenomena are size dependent [e.g. competition (Hutchings and Budd, 1981), reproductive maturation (Werner, 1975; Gross, 1981) and reproductive output], plant growth studies can provide an environmental basis for aspects of plant demography. (3) At the scale of the microenvironment, the formation of size (biomass) hierarchies in populations has important evolutionary implications but the controls generating these hierarchies are poorly understood (Turner and Rabinowitz, 1983). (4) Ecological interpretations of photosynthetic characteristics often assume a link between photosynthetic capacity and growth, and studies are needed to determine when such an assumption is valid.

There are several levels at which growth of plants and organs may be analyzed (Fig. 15.1). The first level we consider is 'growth analysis', a set of techniques for studying weight changes in whole plants and plant parts with time. Growth analysis elucidates ontogenetic trends in structural and functional aspects of growth and allows comparisons across a broad range of life forms.

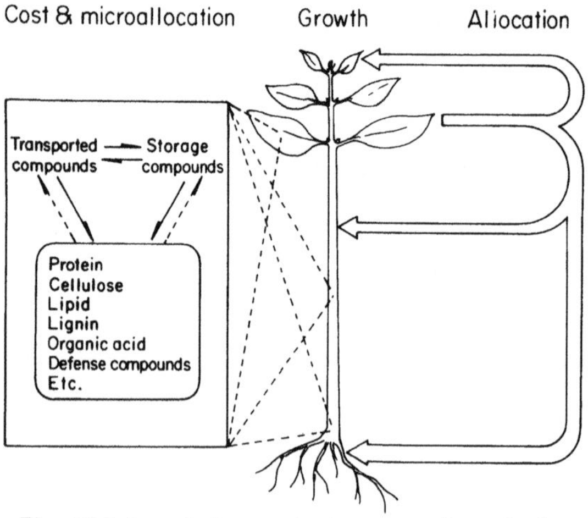

Fig. 15.1 Levels for analyzing growth and allocation in plants. Growth analysis (center) examines biomass changes at the whole-plant level; carbon allocation (right) considers the movement of carbon from sources to sinks; cost and microallocation (left) refer to the carbon and energy used in building and maintaining functionally different tissues.

Growth analysis, however, does not address many aspects of resource investment. Overall growth reflects the net movement of many resources in and out of the plant and its various organs. Each resource may be invested differently and provide different insights into the plant's adaptive mechanisms and physiological balance (Abrahamson and Caswell, 1982). As a second level of analysis, we discuss methods for studying the investment of one resource, carbon, which provides direct links between the physiological processes of photosynthetic carbon fixation and observed growth. One can study carbon investment in terms of the instantaneous fate of photosynthetically fixed carbon or the long-term distribution of carbon, termed 'allocation' and 'partitioning' respectively. However, ecologists commonly use allocation to mean the final carbon distribution, making the two terminologies at odds. To avoid confusion, we use 'allocation' as a general term that includes both the instantaneous fate of carbon and the final distribution. Although resources other than carbon (e.g. nitrogen, water, phosphorus) may limit growth, their acquisition and allocation can be translated into carbon units (Bloom *et al.*, 1985). Chapter 10 discusses methods of studying nutrient allocation.

As a third level of analysis, we examine techniques for calculating the amount of carbon and energy required to construct and maintain a tissue of measured composition. Cost calculations include respiratory losses as well as the carbon required for incorporation into biomass. This level of analysis forges the link between carbon fixation and biomass growth. It also provides an analysis of resource investment in terms of metabolic functions.

15.2 GROWTH ANALYSIS

Ecological studies examine growth in two different, but complementary, ways. The first emphasizes productivity and views growth as the change in mass of live biomass through time. The second emphasizes demographic processes and views growth as the difference between the production of new biomass units, or 'modules' (such as leaves, stems, twigs, roots), and the death or loss of old modules. Growth during a time interval can be calculated by simple subtraction – biomass or module number at the end of the interval minus that at the beginning.

Going beyond the simple calculation of growth, a variety of approaches, commonly called 'growth analysis', can be used to account for growth in terms that have functional or structural significance. Historically, 'growth analysis' has referred to the analysis of biomass growth, a field that developed in the early 1900s. This type of growth analysis requires measurements of plant biomass and assimilatory area (usually leaf area), and methods of computing certain parameters that describe growth. We use the term 'traditional' growth analysis to include this set of parameters and three different approaches for calculating them: classical (or interval), integral and functional growth analysis. Growth analysis can also describe the application of demographic theory to module dynamics. Demographic growth analysis can be used to calculate leaf lifespan, leaf age structure and plant growth. Other types of growth analysis include physiological growth analysis and yield component analysis, which we will discuss only briefly since they have not been widely used in ecological studies. Yield component analysis evaluates productivity as the algebraic product of morphological or developmental components; it uses many of the same parameters as traditional growth analysis. Physiological approaches emphasize physiological processes and use parameters with explicit mechanistic significance.

Our discussion of growth analysis focuses on traditional and demographic growth analysis, and on the measurement of biomass and modules. We examine the fundamental parameters of growth analysis, approaches for computing parameters and the measurements on which the analysis is based. Other reviews have covered some of these topics in greater depth. Experimental methods are covered in detail in G.C. Evans' (1972) *The Quantitative Analysis of Plant Growth*, the definitive text on many aspects of classical plant growth analysis. Evans' volume was prompted in part by the International Biological Programme and is addressed primarily to ecologists. It is unique in its scope and treatment of practical details such as 'where does the shoot stop and the root begin'. A very concise treatment with less emphasis on experimental details is Hunt's (1978b) *Plant Growth Analysis*. A review paper by Květ *et al.* (1971) emphasizes the interpretation and application of growth parameters, including statistical aspects of growth analysis. In the same volume, Ondok and Květ (1971) survey indirect biomass measurements based on reference plants and regression techniques.

We do not discuss the analysis of growth by dense stands (often termed crop growth analysis), which expresses growth per unit ground area (Watson 1958; Warren Wilson, 1981). Techniques used in this area were reviewed recently by Roberts *et al.* (1985) and Biscoe and Jaggard (1985). Agronomists often apply individual and crop growth analysis sequentially – first, individual growth analysis while plants are widely spaced, and later, crop growth analysis, as plants become progressively crowded and the canopy closes. It is also possible to merge the two approaches (Hunt *et al.*, 1984).

15.2.1 Traditional (classical, integral, functional) growth analysis

Traditional growth analysis began with agronomic approaches developed in the early 1900s by Blackman (1919) and Briggs *et al.* (1920a,b). Some of the primary concepts of traditional growth analysis are still taught using seminal experiments as examples (e.g. Kreusler *et al.*, 1879, in Hunt, 1978b). Classical, integral and functional growth analysis share a set of parameters for describing growth, but differ in the method of estimating the parameters.

(a) Parameters

The fundamental parameter of traditional growth analysis is the relative growth rate (RGR, R_W), also termed specific growth rate (SGR), which is the instantaneous rate of increase relative to the productive mass of the plant. Introduced as the 'efficiency index' by Blackman (1919), RGR provides one of the most ecologically significant and useful indices of plant growth. RGR is defined by the differential equation:

$$\text{RGR} = \frac{dW}{dt}\frac{1}{W} = \frac{d(\ln W)}{dt} \quad [d^{-1}] \quad (15.1)$$

where W is the total individual plant dry weight (g) and t is time (d^{-1}). Theoretically, W represents the total dry mass of living tissue, but for most investigations it is not practical to separate living from dead tissue, and total mass is generally used. The error introduced by this simplification increases with the amount of woody (dead) tissue. In growth analysis of mature trees, W can be measured as the biomass of productive tissues (Ledig, 1974) or the annual increment of production (Brand *et al.*, 1987).

Under SI guidelines, the second should be the unit of time in growth analysis, but longer time units are used and are more meaningful biologically and experimentally. Because physiological processes contributing to growth have diurnal rhythms, the typical (and minimum) time unit for growth parameters is the day. The week is often closer to the measurement interval and is also used. Variables other than time follow SI guidelines (see Table 15.1 for the units of growth parameters).

The net assimilation rate (NAR), also called unit leaf rate (ULR, E), is defined by the equation

$$\text{NAR} = \frac{dW}{dt}\frac{1}{A} \quad [g\ m^{-2}\ d^{-1} \text{ or } g\ m^{-2}\ w^{-1}] \quad (15.2)$$

where A is the surface area of assimilatory organs. NAR is a measure of the efficiency of assimilatory organs in producing new growth, and reflects both resource availability (especially light) and leaf display. Therefore, A is usually measured as the area of one surface of the leaf lamina, together with any photosynthesizing stem surface, and does not account for hypostomy versus amphistomy. Studies of needle-like leaves often substitute assimilatory mass for assimilatory area. For comparing very different leaf morphologies, one can also use chlorophyll content, leaf nitrogen, leaf protein or carboxylating enzyme instead of leaf area or leaf mass.

For some investigations, NAR is more meaningful than RGR because it partially factors out large differences in biomass relationships (e.g. leaf/stem ratio) from the calculation of RGR (Fig. 15.2). For example, changes in net assimilation rate through time may reflect the physiological implications of plant architecture, independent of those associated with allometric changes in the ratio of support to photosynthetic tissue. However, because respiration by nonphotosynthetic tissues affects plant growth rate, net assimilation rate does not entirely control for the nonassimilatory fraction of total biomass.

Leaf area ratio (LAR) is the ratio of leaf area to plant dry mass, and reflects the size of the photosynthetic surface relative to the respiratory mass. Generally, it is factored into two terms, leaf weight ratio (LWR) and specific leaf area (SLA):

$$\text{LAR} = \frac{A}{W} = \frac{W_L}{W}\frac{A}{W_L} = \text{LWR} \times \text{SLA} \quad [m^2\ g^{-1}] \quad (15.3)$$

As with unit leaf rate, A generally represents the area of photosynthetic structures, but other measures of assimilation capacity can be used. Leaf weight ratio is the fraction of total biomass in leaves.

Specific leaf area is the ratio of leaf area to leaf mass. It varies considerably between species and is plastic within individuals. Being largely a function of leaf thickness and the amount of mechanical tissue in leaves, SLA is an important index of leaf structure. The inverse of SLA, generally called specific

Table 15.1 Summary of traditional growth analysis parameters, units, instantaneous values and methods of calculation

Parameter	*Units*	*Instantaneous*	*Classical (interval)**	*Functional*	*Integral†*
Relative growth rate, Specific growth rate (RGR, *R*, SGR)	d^{-1} w^{-1}	$\frac{1}{W}\frac{dW}{dt}$	$\frac{\ln W_2 - \ln W_1}{t_2 - t_1}$ (15.1b)	$\frac{1}{f(t)}\frac{df(t)}{dt}$ (15.1c) where $W = f(t)$	$\frac{W_2 - W_1}{\text{BMD}}$ (15.1d)
Net assimilation rate, Unit leaf rate (NAR, ULR, *E*)	$g\ m^{-2}\ d^{-1}$ $g\ m^{-2}\ w^{-1}$	$\frac{1}{A}\frac{dW}{dt}$	$\frac{1}{t_2 - t_1}\int_{t_1}^{t_2}\frac{1}{A}\frac{dW}{dt}\,dt$ (15.2b)	$\frac{1}{g(t)}\frac{df(t)}{dt}$ (15.2c) where $A = g(t)$	$\frac{W_2 - W_1}{\text{LAD}}$ (15.2d)
Leaf area ratio (LAR, *F*)	$m^2\ g^{-1}$	$\frac{A}{W}$	$\frac{1}{t_2 - t_1}\int_{t_1}^{t_2}\frac{A}{W}\,dt$ (15.3b)	$\frac{g(t)}{f(t)}$ (15.3c)	
Specific leaf area (SLA)	$m^2\ g^{-1}$	$\frac{A}{W_L}$	$\frac{1}{t_2 - t_1}\int_{t_1}^{t_2}\frac{A}{W_L}\,dt$	$\frac{g(t)}{h(t)}$ where $W_L = h(t)$	
Leaf weight ratio (LWR)		$\frac{W_L}{W}$	$\frac{1}{t_2 - t_1}\int_{t_1}^{t_2}\frac{W_L}{W}\,dt$	$\frac{h(t)}{f(t)}$	

W = dry biomass (g); t = time (d or w); A = assimilatory area (m^2); W_L = dry leaf biomass (g); BMD = biomass duration (g d); LAD = leaf area duration (m^2d).
* Explicit general formulas; see Table 15.2 for specific formulas for calculating net assimilation rate.
† Approximate formulas; formulas for BMD and LAD are given in Section 15.2.1.c.

Table 15.2 Equations for calculating net assimilation rate (NAR) in classical growth analysis given various relationships between W and A (from Radford, 1967)

W versus A relationship	*Equation for NAR*
$W = c + bA$	$\dfrac{2(W_2 - W_1)}{(A_2 + A_1)(t_2 - t_1)}$
$W = c + bA^2$	$\left(\dfrac{W_2 - W_1}{A_2 - A_1}\right)\left(\dfrac{\ln A_2 - \ln A_1}{t_2 - t_1}\right)$
$W = c + bA^{\alpha}$	$\left(\dfrac{\alpha}{\alpha - 1}\right)\left(\dfrac{W_2 - W_1}{t_2 - t_1}\right)\left(\dfrac{A_2^{\alpha-1} - A_1^{\alpha-1}}{A_2^{\alpha} - A_1^{\alpha}}\right)$
$W = c + bA + dA^2$	$\dfrac{b(\ln A_2 - \ln A_1)}{(t_2 - t_1)} + \dfrac{2d(A_2 - A_1)}{(t_2 - t_1)}$

leaf weight, frequently is used because it is positively related to leaf thickness. Jarvis (1985) points out, however, that specific leaf weight is a misnomer because 'specific' indicates 'per unit mass', and 'specific leaf weight' must always equal 1. He suggests W_Z is the appropriate symbol for the inverse of SLA.

At any instant, LAR is the ratio of RGR to NAR:

$$\text{LAR} = \frac{A}{W} = \frac{A}{(\mathrm{d}W/\mathrm{d}t)}\ \frac{(\mathrm{d}W/\mathrm{d}t)}{W} = \frac{1}{\text{NAR}}\ \text{RGR} = \frac{\text{RGR}}{\text{NAR}} \qquad (15.4)$$

The growth parameters defined above are instantaneous terms. Typically the parameters are dynamic and may vary on a number of time scales, e.g. RGR varies hourly, diurnally, seasonally and yearly. Growth parameters can be calculated with any of these time scales in mind.

(b) Classical (interval) approach

The classical or interval approach estimates mean values for the parameters during time intervals between successive pairs of harvests (Table 15.1). The second harvest for one set of calculations becomes the first harvest for the next calculations. Equations for the mean values of parameters are derived by integrating the instantaneous expressions for the parameters and then dividing by the time between harvests. The history, development and application of the approach are thoroughly covered by Evans (1972).

Mean relative growth rate is one of the most useful parameters in this approach because the instantaneous expression (Equation 15.1) can be explicitly integrated, and Equation 15.1b (Table 15.1) gives the true mean RGR (Radford, 1967). If growth is exponential [as it often is during early vegetative growth (Hunt, 1978b)], then RGR is constant throughout the interval between harvests and is equal to the mean calculated from Equation 15.1b.

Problems arise in the calculation of mean net assimilation rate because NAR is a function of two time-dependent variables. The instantaneous expression for NAR (Equation 15.2) cannot be integrated without knowing the relationship between biomass (W) and assimilatory area (A) or the relationships be-

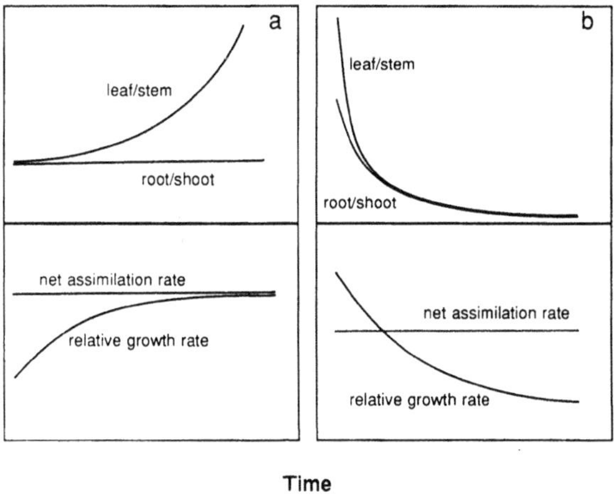

Fig. 15.2 The effect of allometric changes in leaf/stem ratio on relative growth rate in simulations that result in constant net assimilation rate (NAR). Because NAR expresses growth relative to leaf area, a constant NAR can be associated with either an increasing (a) or decreasing (b) relative growth rate, depending on whether the leaf/stem ratio increases or decreases.

tween W and t and A and t. In the development of the classical approach, W and A were assumed to be linearly related (Briggs *et al.*, 1920a,b; Williams, 1946), but other functions (e.g. quadratic, exponential) may describe their relationship better. The choice of W versus A relationship can significantly affect the determination of mean NAR and other parameters.

Similar considerations apply to the calculation of mean leaf area ratio, because the instantaneous expression for LAR (Equation 15.3) cannot be integrated without knowing A/W versus t, or A versus t and W versus t. Most calculations of mean LAR assume that between harvests, LAR increases linearly with time; this assumption may conflict with the W versus A relationship used to calculate mean NAR (Radford, 1967). Also, one generally cannot calculate mean LAR by substituting means into the instantaneous relationships in Equations 15.3 and 15.4. Means can be substituted under the special case where A and W increase exponentially with time and have the same exponent (Radford, 1967). In this case, LAR is a constant. In general, however, one should use empirically determined relationships to evaluate the integral in Equation 15.3b (Table 15.1).

In summary, unless one knows the empirical relationships between W and $\dot{A}$ through time, many unrecognized and potentially conflicting assumptions enter the calculation of mean values for NAR, LAR, SLA and LWR. Therefore, when using the classical approach, these terms should be presented simply as instantaneous values at each harvest, or the appropriate formulas should be derived from the measured variables, W, W_L and A.

(c) *Integral approach*

The integral approach to growth analysis adds two integrated parameters – leaf area duration (LAD, D) and biomass duration (BMD, Z) – to the instantaneous parameters. Defined as

$$\text{LAD} = \int_{t_1}^{t_2} A \, \mathrm{d}t \quad [\mathrm{m}^2\ \mathrm{d}\ \text{or}\ \mathrm{m}^2\ \mathrm{w}] \quad (15.5)$$

leaf area duration accounts for several aspects of assimilatory potential, such as duration of the growing period, and the dynamics of canopy area. Substituting for A using the equation for relative growth rate of leaf area (R_A):

$$R_A = \frac{\mathrm{d}A}{\mathrm{d}t}\frac{1}{A}$$

LAD can be calculated over the time interval t_1 to t_2 as:

$$\text{LAD} = \frac{1}{R_A}\int_{t_1}^{t_2} \frac{\mathrm{d}A}{\mathrm{d}t}\,\mathrm{d}t$$

$$= \frac{(A_2 - A_1)(t_2 - t_1)}{\ln A_2 - \ln A_1} \quad (15.6)$$

An analogous set of equations describes biomass duration, BMD:

$$\text{BMD} = \frac{1}{\text{RGR}} \int_{t_1}^{t_2} \frac{dW}{dt}\, dt = \frac{(W_2 - W_1)(t_2-t_1)}{\ln W_2 - \ln W_1} \quad \text{[g d or g w]} \quad (15.7)$$

The primary parameters of growth analysis can be approximated using LAD and BMD according to Equations 15.1d and 15.2d (Table 15.1), but the formulas are only approximate.

(d) Functional approach

The functional approach to growth analysis calculates growth parameters from functions fitted to the time trends of the primary values W and A, or their logarithms. Fitted equations for $W(t)$ and $A(t)$ essentially unveil the central W versus A relationship (a hidden and complicating assumption in the classical approach). As shown in Table 15.1, this allows one to make full use of the instantaneous relationships for deriving RGR, NAR and LAR from the fitted functions.

A variety of functions can describe the time trends of W and A, such as exponential or logistic equations or polynomial expressions of different orders. The functions can be fitted to raw data or logarithmic transformations. Hunt and Parsons (1974) list many of the functions that have been used.

The results of functional analysis are sensitive to the choice of curve fitting procedure (e.g. Nicholls and Calder, 1973; Hunt and Parsons, 1974). Often there are no *a priori* biological grounds for choosing a particular mathematical function for $f(t)$, $g(t)$ or $h(t)$ (see Table 15.1). Generally there is a trade-off between goodness of fit and minimizing standard errors on the derived growth analysis parameters (Nicholls and Calder, 1973). Stepwise regression procedures, such as those developed by Hunt and Parsons (1974) employ an explicit criterion – the polynomial includes only terms significant at $P < 0.05$. This standardizes the degree of fit, but may lead to polynomials of different order for different parameters.

Venus and Causton (1979) argue that the Richards growth function has more of a biological basis than polynomials, is very flexible and allows one to use a single function. The general form of the Richards function is:

$$W = a(1 \pm \exp(b - kt)^{-1/n}) \quad \text{[g]} \quad (15.8)$$

where a, b, k and n are constants.

(e) Analysis of variance (ANOVA) of relative growth rate

An analysis of variance can be used to test for differences between species or treatments in RGR. Although this technique is not a part of traditional growth analysis, we include it here because it provides a means of testing for differences in RGR, the primary variable of traditional growth analysis.

The ANOVA can be applied to the interval between two harvests, or can be extended to look at linear changes in RGR with time. The analysis of variance is performed using ln-transformed plant weight as the dependent variable; time and treatment (or species) can be considered factors in the ANOVA. The interaction between time and treatment can be expressed such that the sum of squares due to interaction ($SS_{interaction}$) in the ANOVA provides a test for differences in RGR (Poorter and Lewis, 1986). With more than two harvests, a time trend analysis of $SS_{interaction}$ can be used to detect a difference in RGR that persists through time, as well as a temporal effect on RGR.

15.2.2 Measurements used in traditional growth analysis

The raw data for growth analysis are often

difficult to acquire from naturally growing plants. Successive measurements of plant biomass are difficult to obtain because direct measures of size, mass, etc. are destructive – the individual plant is killed in the process of making the first measurement and there is no subsequent growth to measure. There are two ways to circumvent this problem. If the individuals of a population are sufficiently alike in size and growth rate, one can calculate growth statistics for the population from sequential harvests of randomly selected (or, in some cases, paired) individuals. Alternatively, one can develop a nondestructive method of measuring biomass and apply the method repeatedly through time to individual plants. These two approaches – destructive versus nondestructive (or indirect) – constitute the major dichotomy in methods of measuring biomass. Historically, classical growth analysis has involved sequential destructive harvests, but it can also use indirect methods of measuring mass and area.

The selection of measurement technique and analytical approach must be considered together because both affect the statistical aspects of growth analysis. Because the key elements in growth analysis are parameters that are not measured directly, decisions about the design and frequency of measurements affect the form of the variation in data. Statistical considerations for traditional growth analysis are discussed by Promnitz (1974) and Květ *et al.* (1971).

For example, often one may want to test for differences in RGR. In order to do this in the classical approach, variances first must be calculated, either at random (which may overestimate variances) or by pairing plants of successive harvests according to their size ranking (which may underestimate variances). Computer programs used for functional growth analysis generally calculate 95% confidence intervals on RGR, but because of the procedures used, the confidence intervals are unequal with time – they flare out for the earliest and latest harvests. These factors should be considered in choosing both an analytical approach and a measurement technique.

(a) Direct methods

Methods for assessing growth by means of sequential destructive harvests have developed through nearly a century of investigations, both agronomic and ecological. Experimental methods are covered in detail in G.C. Evans' (1972) *The Quantitative Analysis of Plant Growth*.

Sample population. In order to characterize individual growth in a population from harvest data, it is necessary to sample enough individuals so that significant trends can be determined. Therefore approaches based on sequential harvests work best when growth conditions are controlled and uniform over a treatment. In general, growth chambers most closely meet these criteria, followed by greenhouses, common gardens, controlled field conditions and, finally, natural communities. Sources of variation in individual plant size include differences in germination time, genetic differences among individuals, differences in individual microsites and resource availability, and differences in interactions with other organisms (competitors, herbivores, parasites, etc.). In natural communities, conspecifics of the same age are often tremendously different in size. In nearly synchronous populations of annual plants, for example, individual biomass at flowering can span two orders of magnitude. In such cases, growth analysis based on harvested material at best provides a rough idea of interspecific differences, seasonal trends, etc.

Because of these factors, individual growth analysis of harvested plants has rarely been applied to natural field situations. Principles useful for growth analysis of pot-grown plants can help extend the techniques to the field. For example, designating the experimental population from the start is important in growth analysis of naturally growing plants, and can be done by assigning numbers to individuals for random harvesting throughout the study. Otherwise, any nonrandom mortality in the population will change the experimental population through time, and early harvests will include individuals that would have had a higher probability of mortality than those of later harvests. This results in an analysis that is progressively biased towards the most successful individuals and probably an overestimate of growth during periods of peak mortality (if smaller individuals have higher mortality). Designating the experimental population from the start does not factor out nonrandom mortality, but it at least allows one to track the level and timing of mortality. Obviously, the initial experimental population must take into account the number of plants to be harvested and the anticipated mortality between germination and the end of the experiment.

In some cases, it may be possible to constitute the experimental population from individuals with a low risk of early mortality, e.g. the earliest or largest germinants. The appropriateness of this depends on the hypothesis being tested and the predictability of plant survival. In communities where plants compete primarily for light, aboveground size at the seedling stage may be a good predictor of fate. If belowground processes or community interactions control growth, success in reaching the reproductive state may be poorly related to aboveground size, at least at the seedling stage (Fig. 15.3).

Because of the problems associated with nonrandom mortality and large variation in individual size, harvest-based growth analysis has clear limitations in natural communities. Indirect measures of biomass are often more successful. For some questions, naturally growing individuals may not be necessary, however, and it may be appropriate to sow plants in pots placed in the field, i.e. phytometers. This technique controls for variation in germination time and belowground environment, yet allows one to test the effects of aboveground environment on species native to the site or from an entirely different environment (e.g. Warren Wilson, 1966; Smith and Walton, 1975). Because belowground harvesting is feasible, this approach takes advantage of most of the strengths of harvest-based traditional growth analysis, and avoids many of its pitfalls.

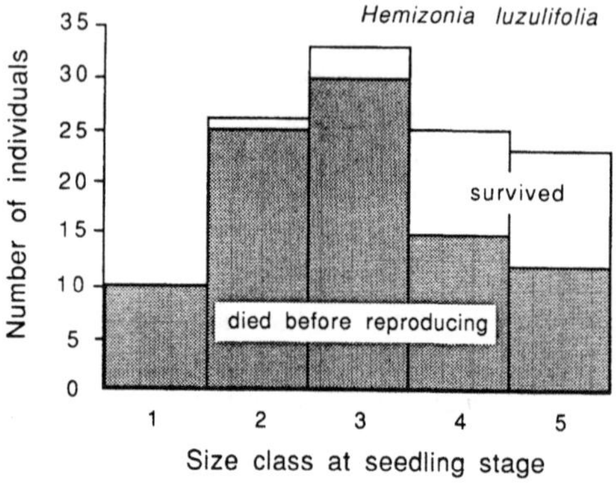

Fig. 15.3 Distribution of plant biomass (estimated from indirect measurements) at seedling stage in relation to pre-reproductive mortality. Plants were growing naturally under field conditions. None of the smallest seedlings (class 1) survived to reproduction, which began in survivors 7 months after the seedling census (from Chiariello, unpublished).

Harvesting. The ideal data base for growth analysis would be one with both large sample sizes and frequent harvests. Usually, however, these are conflicting goals because of time, space, labor, etc., and one strikes a compromise between the two. The form of the compromise depends on the type of analysis. For analyses based on interval methods, the general rule of thumb is that the harvest interval should be less than the doubling time for leaf area and the harvests should be evenly spaced. For functional approaches, the harvests should be more frequent and need not be evenly spaced.

Because there may be pronounced diurnal cycles in nonstructural carbohydrate contents of leaves, harvests should occur at roughly the same time of day. After harvesting, the plants should be kept cold (e.g. in plastic bags on ice) to reduce wilting and respiratory losses.

Drying. 'Dry biomass' is tissue that has been dried to a constant mass, typically by heat-drying in a forced-draft oven. Drying procedures affect both the absolute amount and the biochemical composition of the tissue, so the choice of drying method should take into account the type of tissue under study and whether the tissue will be used for analysis of carbon stores (Section 15.3.5) or for studies of biochemical cost (Section 15.4). During heat-drying, respiratory losses of nonstructural carbohydrates (which are rich in oxygen) occur until the temperature of the tissue is high enough to denature enzymes. However, temperatures high enough to denature enzymes rapidly may also volatilize high-energy compounds (with low oxygen content). Freeze-drying minimizes both types of losses, but is often unavailable.

Studies of growth and biomass allocation typically use a two-stage heat-drying process with 60–90 min at 100°C followed by drying to constant mass at 70°C. This procedure approximates the respiratory losses during freeze-drying (Raguse and Smith, 1965; Smith, 1969). Continuous drying at 70°C prolongs the period of respiratory losses and can reduce total nonstructural carbohydrates (TNC) by about 25%, while drying at air temperature (27°C) can reduce TNC by 40% (Raguse and Smith, 1965). The significance of respiratory losses during drying depends on the concentration of TNC in the tissues. Leaf TNC concentrations vary diurnally with a maximum of 10% or less. Roots, stems and twigs may have TNC concentrations as high as 30 to 50% of dry biomass (Mooney and Billings, 1960; Mooney and Bartholomew, 1974). If a plant has a root/shoot ratio of 1 : 1, with 40% TNC in the root and 10% TNC in the leaves, drying tissues continuously at 70°C may reduce the dry biomass by about 6% (assuming 25% loss of TNC). Errors introduced in this way would lead to an underestimate of the plant growth rate during periods of carbohydrate storage and to an exaggeration of respiratory losses during plant dormancy.

Studies of biochemical cost (Section 15.4) based on elemental analysis or heat of combustion may require different methods of drying plant tissues. This is an important consideration if the same samples are used to analyze both growth and cost. Cost studies generally must strike a balance between maximizing tissue dryness (minimizing residual water) and minimizing compositional changes due to excessive heat or prolonged drying. Residual water can have two effects on cost analyses. First, it increases the apparent mass of the sample. Second, if hydrogen and oxygen determinations are made on separate samples, differences between samples in residual water will cause an error in the measured hydrogen/oxygen ratio. (This error does not occur, of course, if both determinations are made on a single sample.) The methods of choice are freeze-drying, drying over strong desiccant and drying in a vacuum at low temperature (<60°C). If these procedures are unavailable, drying at moderate temperature (60–80°C) with forced ventilation is a suitable alternative (Allen, 1974; Lieth, 1975; Paine, 1971).

(b) Indirect methods

Indirect (or nondestructive) methods of measuring the primary variables in growth analysis (e.g. W, A) permit the investigator to follow the same individual plants through time and avoid many of the problems associated with harvest-based studies. However, in order to relate indirect measures to the primary variables of traditional growth analysis, some destructive harvesting is needed. The major methods for indirectly estimating biomass of individual plants use (1) linear dimensions related to biomass through regressions derived from calibration harvests, (2) parallel harvests of matching or paired plants, or (3) module counts converted to biomass through calibration harvests.

Linear dimensions and plant biomass. For a number of species, single nondestructive measures can provide a good index of biomass of the whole plant or some organs. For four herbaceous rosette species, each varying in size over three orders of magnitude, Gross (1981) found that 83–95% of the variation in whole-plant biomass could be predicted from rosette diameter (broad rosettes) or root crown diameter (erect rosettes). In three conifers, Grier and Waring (1974) found that the area of sapwood in a transverse section (estimated from an increment core) provided a good estimate of the total mass of leaves supplied by the stem.

As these studies suggest, the growth form of a plant may suggest what traits are correlated with biomass, but often the relationship is not simple. For example, one would expect that for round relatively flat rosettes, aboveground biomass would be related to rosette area squared; adding belowground structures should increase the exponent. This would explain exponents greater than 2.5 in Gross's (1981) study, and also interspecific differences in the exponent. Age and developmental changes (e.g. bolting, flowering, leaf senescence) are likely to affect the relationship, so one should calculate a new relationship at each developmental stage.

Simple measurements such as these may be sufficient indicators of size for qualitative predictions of survival or reproduction, but biomass estimates used for growth studies usually require more detailed measurements. For example, King and Roughgarden (1983) followed growth in the rosette annual *Plantago erecta* using weekly measurements of the length of each leaf. The measurements were converted to biomass using regressions derived from plants harvested nearby, measured similarly, then dried and weighed. For the erect and branching *Clarkia rubicunda*, they measured both leaves and stems. Pearcy (1983) combined linear dimensions of both stems and leaves, together with node recording, to keep track of aboveground growth by understory trees. Although his studies did not provide estimates of belowground growth, he was relating growth to light availability, and it is unlikely that belowground growth patterns would alter the conclusions of the study.

15.2.3 Demographic growth analysis

Now we turn to the major alternative to traditional growth analysis – demographic analysis. Demographic analysis of growth treats the individual plant as an assemblage of units or modules that fall into classes uniform enough that each class can be modeled as a population. Typically all the plant's leaves are considered one population (e.g. Bazzaz and Harper, 1977), but leaves of different branch order can be considered separate populations (e.g. Maillette, 1982). Twigs, stems, reproductive structures and roots can all be considered populations. Because the module (e.g. leaf) size is fairly determinate, module counts can be converted to biomass by multiplying module number and unit module biomass.

Demographic growth analysis is an offshoot of plant population biology and uses classical ideas that were reintroduced to plant ecologists by Harper (1968) and White (1979). Because it is a relatively new approach (first applied by Bazzaz and Harper, 1977), demographic growth analysis is not as standardized as traditional biomass-based growth analysis. The potential applications are well established from demographic theory, such as calculating the intrinsic rate of population growth of module number, survivorship curves for modules and expected module lifespan. Applying demographic techniques to all the types of modules on a plant can provide a complete accounting of aboveground growth. Often, however, demographic analysis is used to obtain an index of growth, not a complete accounting (e.g. Hobbs and Mooney, 1987).

Leaf production rate, lifespan and turnover rate are the demographic parameters most often studied in relation to physiological aspects of growth. We can illustrate these parameters by considering leaf production by an idealized plant (Fig. 15.4). As a plant

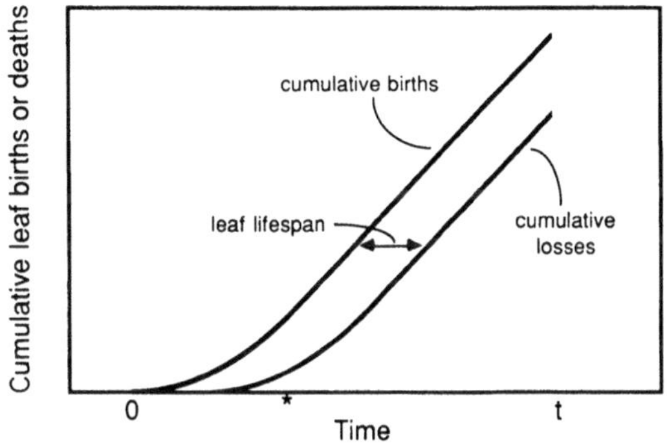

Fig. 15.4 Cumulative leaf births and leaf losses by a plant from time 0 to time *t*. The asterisk indicates a transition from exponential to linear leaf accumulation. Cumulative leaf death follows the same trajectory but is offset to the right by an amount equal to leaf lifespan, as indicated by the arrow.

grows, leaf production follows a trajectory that may include exponential phases, linear phases and/or other types of dynamics. If leaf lifespan is constant, cumulative leaf losses follow an identical trajectory, offset from production by a time lag equal to leaf lifespan. Lifespan can be measured by (1) identifying a cohort at birth and recording the time until their death, or (2) constructing plots of cumulative leaf births and deaths, and measuring the lag between curves. The latter may be necessary where leaf lifespan exceeds the length of a study (Fig. 15.5).

The life history of modules such as leaves or shoots can also be used to construct models that predict the growth of module populations or plant biomass. Module demography has borrowed from demographic theory two types of matrix models: the

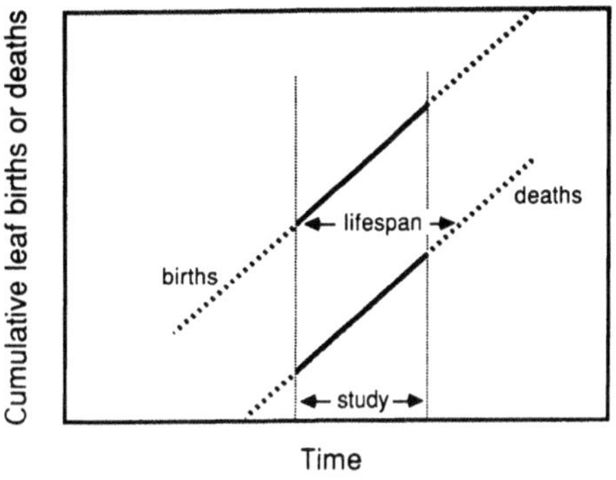

Fig. 15.5 For the case where leaf lifespan exceeds the duration of a demographic study, leaf lifespan can be calculated by plotting measured cumulative leaf births and deaths through time (solid lines) and extrapolating (dashed lines) until a horizontal line intersects both trajectories (e.g. Williams *et al.*, 1989).

Lefkovitch (1965) matrix, which is based on size, and the Leslie matrix, which is based on age classes. Because growth in plants depends more on plant size (biomass) than plant age, size-based matrices are more useful for determining individual growth, while age-based matrices are useful for standard demographic parameters such as expected module lifespan and survivorship. Both size-based and age-based matrices are also used to analyze the growth of populations of plants.

If the modules comprising an individual are divided into equal size classes, the growth of the module population can be projected from an equation consisting of a matrix of transition probabilities for each size class to any other, and a vector representing the distribution of modules among size classes. If $M_j(t)$ is the number of modules of size j at time t, and a_{ij} represents the probability of a module of size j at time t becoming size i at time $t+1$, the equation for predicting population size is (see Lefkovitch, 1965; McGraw and Antonovics, 1983).

Approximate total biomass is obtained by

$$\begin{pmatrix} a_{11} & a_{12} & \cdot & \cdot & a_{1j} \\ a_{21} & a_{22} & \cdot & \cdot & \cdot \\ \cdot & \cdot & \cdot & \cdot & \cdot \\ a_{i1} & a_{i2} & \cdot & \cdot & a_{ij} \end{pmatrix} \times \begin{pmatrix} M_1(t) \\ M_2(t) \\ \cdot \\ M_j(t) \end{pmatrix} = \begin{pmatrix} M_1(t+1) \\ M_2(t+1) \\ \cdot \\ M_j(t+1) \end{pmatrix} \quad (15.9)$$

summing over all size classes the product of module population and the mid-point biomass. For projecting growth over one time interval, Equation 15.9 is calculated once. For projections over longer time periods, Equation 15.9 can be iterated as many times as necessary.

The same type of matrix model can be applied to modules classed by age. This approach follows the Leslie model standardly used in demographic theory of populations (see Roughgarden, 1979). The span of each age class is set equal to the census interval for the study. From repeated censusing, transition probabilities for dying and for producing new modules can be determined. Because chronological aging is constant for all modules, a Leslie matrix has a characteristic form with survival probabilities on the subdiagonal of the matrix and 'fertility' terms across the top row. All other elements in the matrix are zero because modules cannot regress or leap-frog in age. This is a major difference between the age-based and size-based models – modules classed by size theoretically can move to any other size class during the census interval.

Similar techniques are used to analyze both types of matrices. Both can be iterated, and after some number of iterations, the distribution of modules among size classes or age classes becomes stable, provided the transition probabilities stay constant. This occurs even though total module population size may continue to change. At this point, growth can be projected with the size-classed model by multiplying the size distribution (M) by a single number, γ, which is the growth rate of the module population. λ is the corresponding parameter for the age-based model. γ and λ are calculated as the largest positive eigenvalue of the transition matrix.

The first application of demographic growth analysis calculated standard demographic parameters such as birth and death rates, age distribution and intrinsic rate of increase – all from a life table of leaves, rather than individuals (Bazzaz and Harper, 1977). This study showed that plant density and light availability affect rates of leaf birth and death. More recent applications have evaluated meristem limitation as a control on growth and form. For example, transition probabilities for meristem switches from vegetative to reproductive functions can be used to examine costs of reproduction (Watson, 1984; Porter, 1983), and relationships between branch order and meristem number can act as a control on plant form (Porter, 1983).

(a) Demographic models

McGraw and Antonovics (1983) have developed a general model of module population growth based on field measurements of size (leaf area) and the probability of growing, branching, flowering and dying by shoots of *Dryas octopetala*. Total leaf area was determined by nondestructively measuring the length of each leaf and converting to area using regressions based on harvested material. Using Lefkovitch matrices with measured transition probabilities, they projected the population growth of shoot modules classed by leaf area. They also conducted a sensitivity analysis of the model to determine the sensitivity of growth to small changes in particular transition probabilities. With this approach McGraw and Antonovics (1983) found that growth by modules was size dependent and that whole plant growth was particularly sensitive to the growth rate of the smaller modules.

(b) Census techniques

Following individual plants through a study

provides the greatest accuracy in determining demographic parameters for individual growth. This may not be feasible if field time is very limited or if the material to be censused is very inaccessible, such as leaves of trees (Maillette, 1982). In such cases, material harvested periodically can be used and should provide an accurate picture of certain aspects of growth such as when leaves flush on different branch orders (Jow *et al.*, 1980) or the age distribution of leaf biomass (Maillette, 1982).

Species differ in the opportunities they provide for keeping track of modules. One widely used technique is to slip bands (color-coded for each census date) around leaves as they emerge to establish reference points. If censusing is not very frequent, it may be necessary to develop a numbering system for branch order and leaf placement or to 'map' the plant's modules. Whatever approach one uses, the data one gathers with demographic analyses are likely to be voluminous, but can often be recorded and analyzed with computer spreadsheets and programs that allow one to write short subroutines.

15.2.4 Physiological growth analysis

Physiological approaches for analyzing growth are also less standardized than traditional growth analysis and share a philosophy rather than a common set of parameters. They emphasize a more mechanistic view of energy capture, carbon fixation and resource allocation, together with environmental control of plant growth. Physiological approaches are often coupled with models of plant growth. The models range from relatively simple models with less than ten parameters (e.g. Charles-Edwards and Fisher, 1980) to very detailed models with dozens of parameters (Wann *et al.*, 1978; Wann and Raper, 1979). The parameters represent many of the same concepts recognized by the traditional parameters of growth analysis, but they are more specific and more amenable to physiological interpretation. For example, while traditional approaches emphasize assimilatory potential as NAR or LAR, physiological approaches consider the light extinction coefficient of the canopy and photosynthetic capacity. Other important parameters are the partitioning coefficients for allocation of new biomass, developmental switches, abscission rates and retranslocation. As a result of this difference, it is possible with physiological analysis to account for environmental effects on growth, such as changes in daily irradiance.

The physiological approach that is most comparable to the methods of traditional growth analysis is one in which curve-fitting procedures are used with sequential harvest data to estimate the parameters of a growth model (Charles-Edwards and Fisher, 1980; Fisher *et al.*, 1980). Although similar to the functional approach in its use of regression procedures, this approach has parameters that are mechanistic and can be tested with independent measurements. This makes it possible to evaluate certain parameters independently, constrain their values and then use curve-fitting to obtain other parameters.

15.2.5 Yield component analysis

Yield component analysis factors growth into multiplicative, developmental or morphological components and is usually applied to stands rather than individual plants. For example, Joliffe *et al.* (1982) expressed yield of pod beans as:

$$\text{Yield} = \frac{\text{number of plants}}{\text{land area}} \times \text{leaf area per plant}$$

$$\times \frac{\text{individual biomass}}{\text{leaf area per plant}} \times \frac{\text{pod biomass}}{\text{individual biomass}}$$

A log transformation of this equation produces an arithmetic series which can be used with regression analysis to identify controls on yield. Because of its concern with yield (rather than growth), this approach emphasizes characteristics that are typically omitted from growth analysis, such as harvest index (pod biomass/individual biomass).

15.2.6 Discussion of growth analysis approaches

Each analytical technique has inherent strengths and liabilities, which are weighted differently according to the type of investigation. Long considered the paradigm, classical growth analysis is seriously limited by the often conflicting assumptions made about the W versus A relationship necessary for calculating mean NAR, LAR, etc. Various precautions can reduce the impact of one's choice for the W versus A relationship. For example, if the harvest interval is less than the doubling time for A, then one can use either a linear or quadratic relationship without significantly affecting the calculation of mean NAR (Coombe, 1960). Alternatively, one can empirically determine the relationship between W and A, either directly or from the relationships between A and t, and W and t. One can evaluate the W versus A relationship from all the harvests, and then apply this relationship to each harvest interval, or the relationship can be evaluated in preliminary experiments. In many cases, one can combine a short harvest interval with an experimental determination of W versus A from multiple harvests.

Although these precautions can minimize the errors of classical analysis, functional analysis is now the method of choice in traditional growth analysis. Most investigators using traditional growth analysis switched from classical to functional approaches as microcomputers and curve-fitting programs became more available. Problems with the functional approach have improved as analytic techniques improved. For example, computer analysis makes it possible to test the significance of the assumptions inherent in choosing a particular curve-fitting procedure.

A more significant decision is whether to use traditional growth parameters or an alternative such as demographic or physiological analysis. Analytical techniques based on the traditional parameters (classical, integral and functional approaches) express growth in terms that are commensurate across a wide range of life forms and a large body of existing literature. Developers of the functional approach stress that their empirical models provide a concise way of handling data and testing hypotheses, and that curve-fitting reduces experimental noise without asserting particular mechanistic relationships (Richards, 1959; Hunt, 1979). The relatively few simple parameters of traditional growth analysis (e.g. Table 15.1) facilitate comparisons between studies and also monitoring of ontogenetic trends (Hunt, 1979).

The same attributes can be viewed as liabilities. Traditional growth analysis factors growth into simple parameters (e.g. LAR, NAR) that are algebraically independent, but not biologically independent (orthogonal) (Hardwick, 1984). This weakens growth analysis as a tool for tracing the causes of growth differences because the interdependence of parameters is itself a factor in understanding growth (Hardwick, 1984). Also, the parameters are not explicitly mechanistic, leading to inconsistencies in interpretation. For example, 'leafiness' is measured as LAR by some authors (Ledig, 1974; Hunt and Bazzaz, 1980), but as SLA by others (Beadle, 1985). Finally, the calculated parameters tend to have larger standard errors than the primary variables (Hardwick, 1984), making it possible that significant differences in yield will be traced to insignificant differences in parameters. Thus, the choice of analytic approach is based on one's perspective on mechanistic versus empirical analyses, which may relate to the questions being posed and subsequent levels of study.

The ANOVA approach (Poorter and Lewis, 1986) offers several advantages if one is interested in differences in RGR. It avoids pairing of plants (typically used in classical analysis to calculate deviations for testing for differences in RGR), and the problem of uneven confidence intervals found in functional growth analysis. As a result, it may be better at detecting the effects of time, treatment or species on RGR.

The choice of direct versus indirect measurements may alter the relative advantages of different analytical approaches. The ability of indirect procedures to follow individual plants through time gives indirect techniques three advantages over sequential harvests. Because sampling errors associated with harvesting random individuals (e.g. occasional extreme outliers) are avoided, growth curves from indirect methods tend to be smoother, and better at detecting transient changes in growth parameters. Second, hindsight can be used in analyzing growth. For example, one can census plants through a growing season and then, at the end of the season, relate individual differences in early growth to subsequent performance. One can, for example, limit growth analysis to plants that successfully reproduced, or compare their performance with others that died during the vegetative period. This advantage has been discussed for demographic analyses (McGraw and Wulff, 1983), but it also applies to traditional growth analysis based on indirect biomass measurements. Third, although some harvesting is needed with indirect procedures, it is less destructive to a population than a fully harvest-based study. This is especially true for populations that are highly variable (and would require large sample sizes if destructive harvests were used) and where random destructive harvesting would disrupt the community or stand. The harvesting necessary for relating indirect measures to primary variables often can be targeted in a nonrandom way, preserving the study population.

A disadvantage of indirect procedures is that often they work only for aboveground tissues. This is a problem because growth analysis should be based on whole-plant biomass. One can extend the estimate of aboveground biomass to whole-plant biomass provided that root/shoot allometry is relatively fixed, but this must be verified for each study. A second disadvantage is that the study population must be large enough to absorb any mortality and still leave enough plants for final censuses. Finally, indirect procedures typically require long periods of time in the field.

Some plants or experimental questions are better suited to demographic or physiological approaches than traditional growth analysis. Biomass turnover is an important factor in growth but escapes accounting in sequential analyses of standing biomass. Classical and functional analyses based on harvest data generally do not account for turnover and, therefore, measure only net growth. Repeated modular demography can track leaf and branch abscission, permitting an assessment of turnover. Demographic analyses also reveal plant responses that may have direct links with leaf level physiology, e.g. changes in leaf production, leaf lifespan or the age structure of leaves.

In principle, demographic analyses can be translated into units of biomass or cost; however, the success depends on the uniformity of modules. If time or the experimental treatment affects the biomass or cost of a module (e.g. changes in leaf area, thickness, protein content, etc.), demographic analysis by itself has limited ability to describe biomass growth and should be coupled to some type of harvest-based analysis. Perhaps the most significant drawback of demographic approaches is that the time they require is roughly proportional to the number of modules censused. For large plants (e.g. conifer trees, Maillette, 1982), this limits the number of individual plants one can follow and the fraction of the plant that can be censused.

Like demographic analysis, physiological analysis can be useful for plants with high rates of leaf turnover and where links with leaf-level studies are desired. However, physiological regression approaches treat most parameters as constants. To look for ontogenetic changes in parameters, one may have to divide the experimental period into two or more separate series of harvests, e.g. at the switch in allocation at reproduction.

Through their common emphasis on relative growth rate, links can be made between approaches, such as traditional growth analysis, yield component analysis and demographic analysis (Jolliffe *et al.*, 1982; Jolliffe and Courtney, 1984). For example, the RGR of a plant is equal to the sum of the relative growth rates of the multiplicative components identified in yield component analysis; it is also equal to the sum of the relative growth rates of additive demographic components weighted by their fractional mass (Jolliffe and Courtney, 1984). A second approach can thus compensate for the weaknesses of one's primary analytical approach (Hunt, 1978a). Although one should plan an experiment with an analytical approach in mind, the same harvest data may be suitable for various analyses.

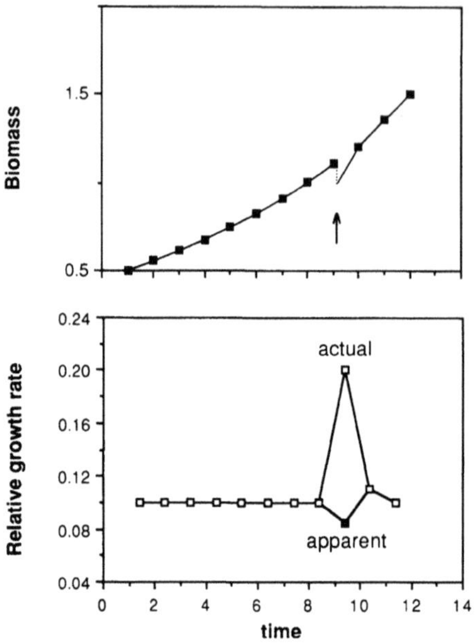

Fig. 15.6 Simulation of plant growth (top, solid line) in which an episode of herbivory (at the arrow) reduces plant biomass, but biomass largely recovers by the time of the next harvest (solid squares). Using classical growth analysis on the sequential harvest values, the measured relative growth rate (bottom) shows an apparent drop due to herbivory, because the herbivory occurred after the census at time 9, but the recovery occurred before the census at time 10. The actual relative growth rate calculated from the solid line in the top panel shows a compensatory jump in RGR, which is responsible for the recovery.

Ecological studies use growth measurements in conjunction with so many other types of approaches and levels of analysis that it is not practical to identify the best method(s) for every situation. We use some field examples, however, to illustrate how the choice of technique might be made.

If there is significant herbivory during a study, turnover dynamics in relation to growth may be very important. Because traditional growth analysis assumes that biomass and assimilatory area are continuous functions of time, growth following herbivory may appear as a decrease in RGR, when in fact there might be a compensatory increase in RGR (Fig. 15.6). To assess growth completely in this situation one must identify the new tissue (census modules), determine whether recovery tissue differs from earlier tissue in mass or SLA (calibration harvests), and determine where the resources for recovery tissues originated (levels of reserves, photosynthesis). Like many ecological questions about growth, this one requires multiple techniques.

As a second example, ecologists may be interested in the formation of size hierarchies in plant populations. Often in growth analysis studies, one assumes that relative ranking of individuals is preserved through time (e.g. Evans, 1972), so that sequential harvests should enable one to calculate the growth of individuals at different positions in the size hierarchy. However, this is a case where simple indirect measurements may be useful, and can show that relative rankings are not fixed (Fig. 15.7). This possibility should be examined in studies of hierarchy formation

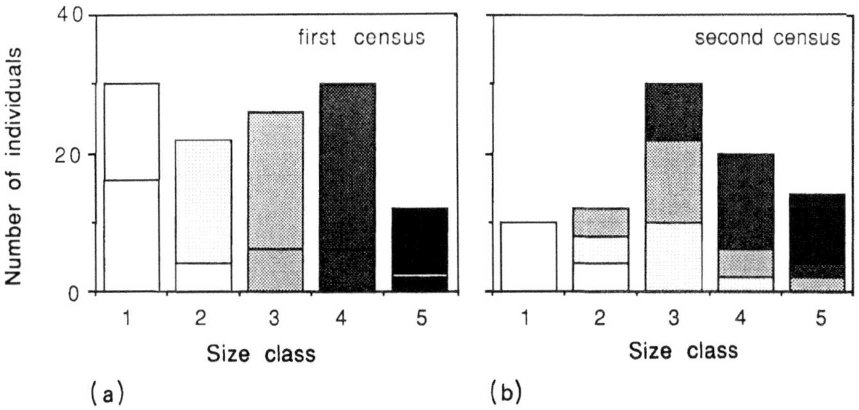

Fig. 15.7 Changes in the size hierarchy through time for mapped individuals in monocultures of *Plantago erecta* seeded into cleared quadrats in otherwise undisturbed grassland. Initial biomass distribution is indicated in (a), with the smallest size on the left, and the lower portion of each bar indicates the fraction of the size class that died between the first and second censuses. (b) shows the biomass distribution at the second census. Shading within each bar indicates the composition of the size class in terms of the earlier size class (e.g. class 5 at census 2 was mostly individuals from class 5 at census 1 but also included some individuals from classes 3 and 4). Biomass was estimated from indirect measurements (from Chiariello and Mooney, unpublished).

because it indicates that RGR is not simply a function of size or ranking.

15.3 FATE OF CARBON

One of the major determinants of plant growth rate is the relative growth of structures differing in function, such as photosynthetic leaves versus storage roots versus stems. Studies of the distribution of photosynthetically fixed carbon provide a major link between whole-plant photosynthesis and growth. Fixed carbon is a unique resource in the sense that it not only dominates the composition of plant tissues but also represents the form in which energy is stored and transported. Both of these factors make it the most logical currency for ecological studies of resource allocation.

The initial fate of carbon as determined with ^{14}C can be compared with its long-term or final distribution using growth analysis (Heilmeier and Whale, 1986). A partitioning coefficient, PC, is defined as $(dW_c/dt)/(dW/dt)$, where W_c is the dry weight of a given plant component and W is the dry weight of the whole plant. The PC of a given organ can be compared with the allocation percentages found with ^{14}C. The values will differ somewhat since the final distribution will account for respiratory losses, redistribution of carbon and tissue losses.

In this section we discuss methods by which carbon movement is traced through the plant using radioisotopes of carbon, principally ^{14}C. Depending on one's objectives, the carbon can be traced to specific organs, cell fractions or classes of chemical compounds. We also introduce procedures for separating samples in chemical constituents.

15.3.1 Introduction of ^{14}C into the plant

The commonest way of introducing ^{14}C into a plant is to enclose a leaf in a chamber that is connected to a reaction flask in a closed loop (Fig. 15.8). A chamber with a side-arm reaction chamber may also be used. Radioactive carbon in the form of a carbonate (e.g. $NaH^{14}CO_3$) is placed in the reaction chamber and, at the start of the experiment, acid is added to the carbonate. The $^{14}CO_2$ that is generated is then circulated over the leaf long enough to get sufficient amount of label into the leaf. Exposure time is in part determined by the specific radioactivity of the label. The air in the chamber is then circulated through a CO_2 absorbant (e.g. KOH) to remove all ^{14}C before opening the apparatus. The plant is then placed back into its growth environment for a period before it is harvested for determining carbon allocation patterns. Generally, within several days the primary allocation within the plant has taken place.

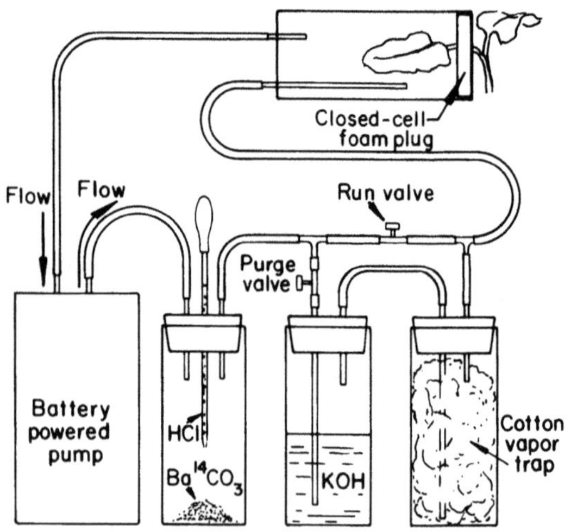

Fig. 15.8 Simple system for supplying ^{14}C to leaves in the field (from McCrea *et al.*, 1985).

Depending on objectives, there are innumerable variations on this general theme. The tissue may be exposed to ^{14}C for very brief periods and then immediately harvested to determine initial products of photosynthesis within the leaf or the time course of translocation (Fig. 15.9). The amount of carbon taken up by the leaf can be calculated by removing leaf discs immediately after labeling, measuring their specific radioactivity and multiplying by leaf area (Pearen and Hume, 1982). Alternatively, the chamber may be placed in a closed loop with an infrared gas analyzer and the rate of CO_2 depletion noted. Knowing the amount of ^{14}C supplied, the total amount of label assimilated can be calculated. The amount supplied depends on the specific activity of the carbonate utilized in the experiment (specific activity = amount of radioactivity/mass or more precisely the ratio of the number of radioactive atoms of an element to the total number of that element in a given compound). The unit describing the amount of radioactivity is the becquerel (Bq), defined as 1 disintegration s^{-1}. The amounts of radioactivity utilized in these kinds of experiments are generally in the kBq

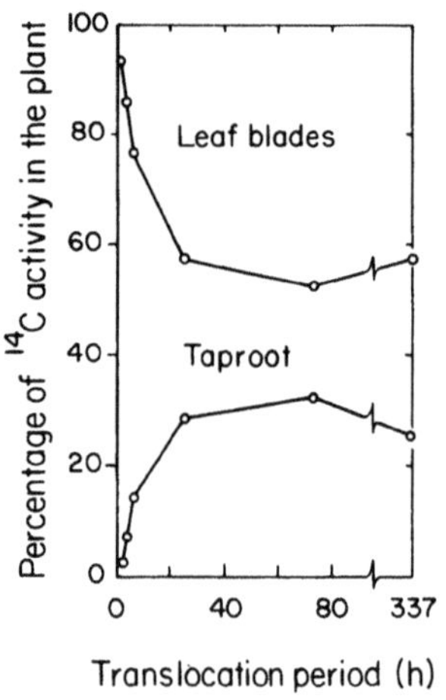

Fig. 15.9 Changes in relative ^{14}C activity with time of leaves and tap root of a carrot plant (from Benjamin and Wren, 1978).

range. The specific isotope amount required depends on a number of things including the dilution of the label in the experimental system and the efficiency of sample recovery and counting. Other considerations are the accuracy required and the time which can be allowed for counting a given sample. Guidelines are given in general isotope textbooks such as Chapman and Ayrey (1981) and Wang *et al.* (1975).

To assure normal uptake of CO_2 the chamber should receive adequate illumination

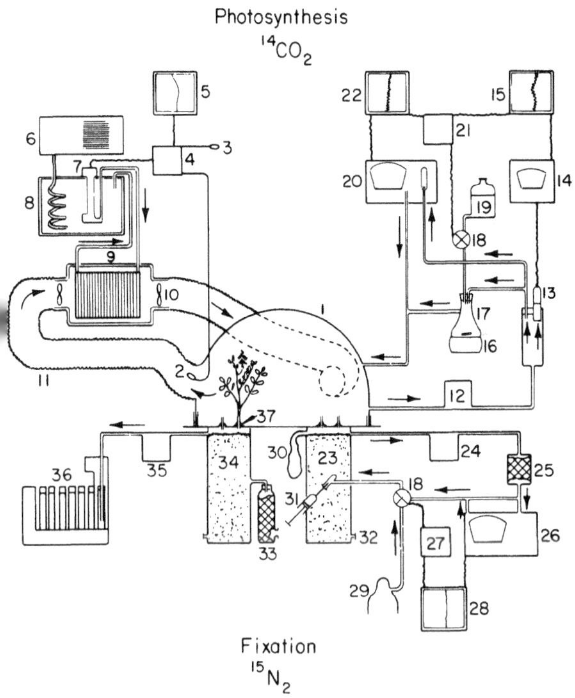

Fig. 15.10 System for simultaneous labeling of whole plants with ^{14}C and ^{15}N (from Warembourg *et al.*, 1982). 1. Biosynthesis canopy; 2,3. inside and outside temperature sensors; 4. temperature regulator; 5. temperature recorder; 6. refrigeration unit; 7. water pump; 8. cooling bath; 9. heat exchanger; 10. fans; 11. flexible tubing; 12. air pump; 13. Geiger–Müller probe; 14. G.M. counter; 15. radioactivity recorder; 16. magnetic stirrer; 17. H_2SO_4 solution; 18. solenoid valve; 19. $Na_2{}^{14}CO_3$ solution; 20. infrared CO_2 analyzer; 21 CO_2 regulator; 22. CO_2 recorder; 23. root container for $^{15}N_2$; 24. soil air pump; 25. CO_2 trap; 26. O_2 analyzer; 27. O_2 regulator; 28. O_2 recorder; 29. O_2 cylinder; 30. rubber expansion bag; 31. $^{15}N_2$ inlet; 32. drainage hole; 33. CO_2 trap (soda lime); 34. root container for respiration measurements; 35. pump; 36. automatic CO_2 collector; 37. air tight seal.

and temperatures should be maintained in the normal operating range of the leaf. Complex systems where both ^{14}C and ^{15}N can be supplied simultaneously to shoots and roots of plants growing outside under controlled conditions have been described (Warembourg *et al.*, 1982) (Fig. 15.10).

Because very small amounts of radioactivity can be detected using scintillation counters, as described below, it is possible to use quite small tissue sample sizes (1–100 mg) to determine relative proportions of the various constituents that have been synthesized in the sample by photosynthetically fixed ^{14}C. Dickson (1979) describes a separation and analysis protocol for quantifying the major plant biochemical components (Fig. 15.11).

15.3.2 Locating the radioactive label

There are a number of different ways to locate and quantify the carbon that has been translocated within the plant after labeling. The fate of carbon can be seen visually, using autoradiographs, or it can be quantitatively assessed by a variety of radioactivity-counting techniques. For the latter, the total carbon present may be counted, or it may be separated into various organic fractions for analysis (see Chapter 10).

(a) Autoradiography

Typically this technique involves drying the sample and then placing it in darkness in direct contact with the emulsion of X-ray film. The length of time before development of the film depends on the amount of radioactivity present, but also on temperature and film emulsion characteristics. Thus, the determination of exposure time may have to be determined empirically by exposing the sample for varying time periods. After developing, the sample is repositioned on the film to show what particular organ has received the labeled carbon (Fig. 15.12). Using autoradiographs of sectioned material, the fate of carbon can also be determined at the cellular level, such as studying the vascular bundles involved in transport from specific leaf regions (Vogelmann *et al.*, 1982). Details of autoradiography are given in standard texts such as Chapman and Ayrey (1981) or Rogers (1979).

Autoradiography has been widely utilized

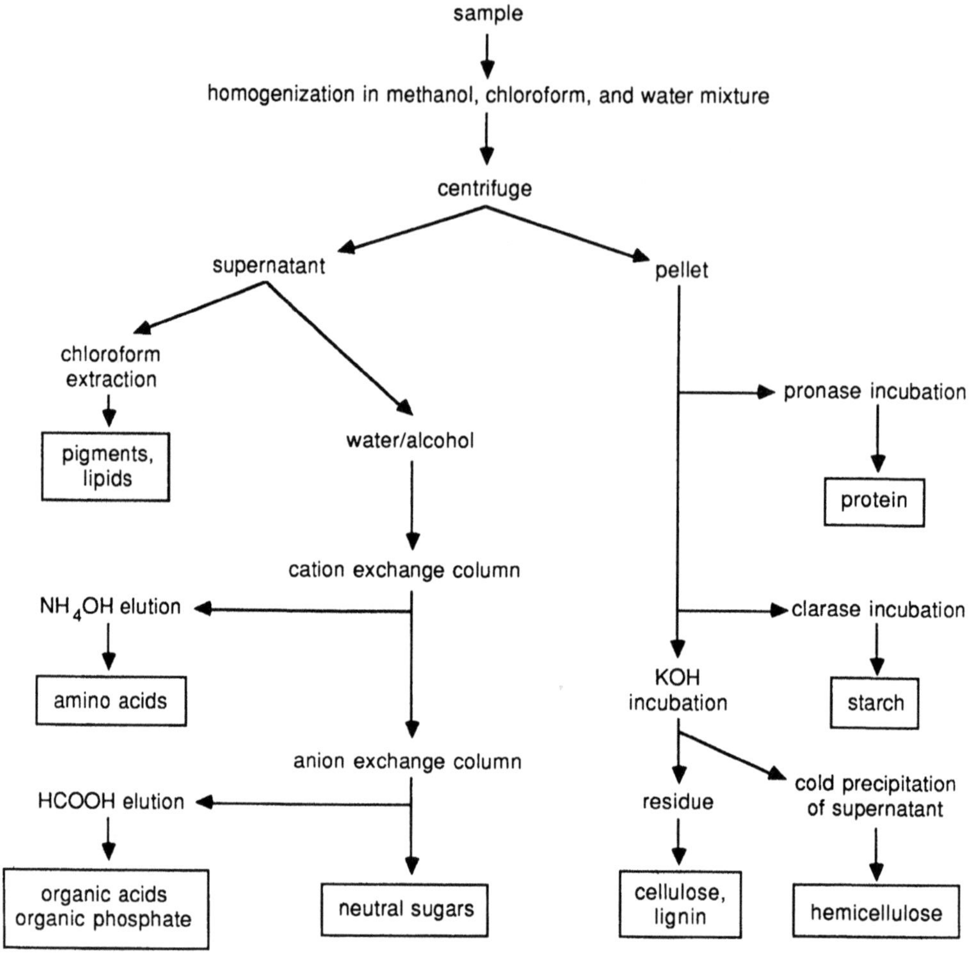

Fig. 15.11 Schematic of procedures for extracting and separating photosynthetically fixed ^{14}C in up to eight chemical fractions from small (1–100 mg) samples of leaf material (from Dickson, 1979).

in physiological ecology to determine such things as connections among ramets of rhizomatous plants (Pitelka and Ashmun, 1985), source–sink relationships, etc.

(b) Scintillation counting

For quantitative analysis of the amount of radioactivity in a plant part the sample is generally placed, after suitable preparation, in a scintillation counter. For ^{14}C, liquid-scintillation counting is utilized, since it gives the greatest accuracy for this low-energy beta emitter.

Sample digestion. Generally the sample is digested and then treated to eliminate organic compounds that may interfere with the counting procedure, namely photosynthetic pigments. Ground samples or, if the tissue is not too tough, the entire tissue may be placed in commercially available digestion medium (e.g. Lumasolve, Lumac B.V., the Netherlands; Protosol, New England Nuclear, Boston, MA, USA) overnight and then bleached with a peroxide (Recalcati *et al.*, 1982). The samples are then placed in a scintillation 'cocktail' that serves to uniformly disperse the sample (e.g.

a b

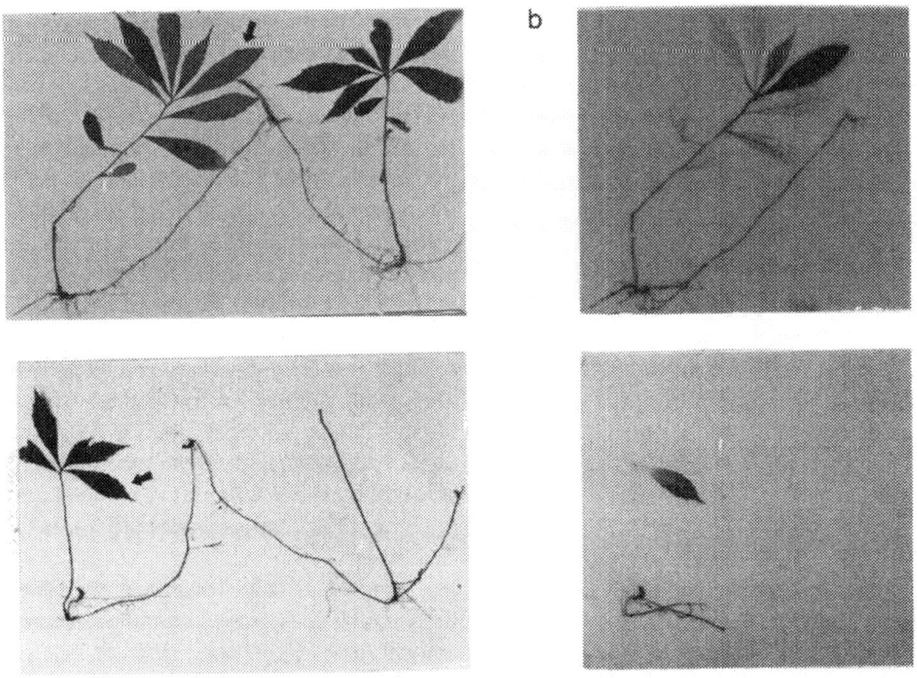

Fig. 15.12 Herbarium specimen (a) of a shoot which had received ^{14}C at leaf indicated by the arrow and an autoradiograph (b) showing movement of ^{14}C. These samples are of two ramets of *Aster acuminatus*. This experiment indicated no movement of carbon between ramets (from Pitelka and Ashmun, 1985. Reprinted by permission of Yale University Press).

Aquasol, New England Nuclear) and counted in a scintillation counter. Some investigators have utilized dry finely ground sample material directly in a scintillation mixture and reported high counting efficiencies (McCrea *et al.*, 1985). Glenn (1982) provides more details on sample preparation for liquid-scintillation counting.

Sample combustion. It is sometimes desirable to assure that there is a complete recovery of the label from a sample or sometimes the label amount is low and large sample sizes are required. In these cases the samples may be totally oxidized and the carbon is evolved as $^{14}CO_2$. The gas is trapped in a basic solution that is mixed in a scintillation cocktail and counted. The sample may be oxidized chemically with strong acids (wet combustion) (Coughtrey *et al.*, 1986) or combusted in oxygen (dry combustion). There are commercially available sample combusters (Packard Sample Oxidizer, Packard Instruments).

15.3.3 Special applications

In some cases it is not possible to retrieve parts of plants for which information on allocation amounts are desired, such as large underground parts. Allocation in these cases can be estimated knowing the original quantity of label given to the total shoot minus that lost in respiration. Label not accounted for by shoot concentration and respiration is assumed to have been transported to the root (Mordacq *et al.*, 1986).

15.3.4 Other carbon tracers

Another radioactive form of carbon, ^{11}C, has been utilized for allocation studies. This isotope has a higher-energy beta emitter than

^{14}C and thus can be detected directly through several cm of tissue. *In vivo* experiments of allocation can be performed using this isotope (Magnuson *et al.*, 1982). However, at present, these experiments can only be performed at specialized facilities where this short-lived isotope (20 min half-life) can be produced.

The stable carbon isotope ^{13}C also has been used for carbon allocation studies (Mordacq *et al.*, 1986). This isotope, of low natural abundance (1.1% of natural total atmospheric carbon), is measured in a mass spectrometer on combusted organic material (see Chapter 13). Pure compressed $^{13}CO_2$ can be obtained commercially and mixed with $^{12}CO_2$ in N to supply enriched $^{13}CO_2$ as a label to plants. Mordacq *et al.* used a feeding concentration of 23% ^{13}C for a sufficient period to supply 0.5 mg of ^{13}C g^{-1} plant dry matter (4 h for their sample size).

15.3.5 Carbon stores

A major variable in linking carbon allocation and plant growth is the extent to which fixed carbon is stored before being incorporated in biomass or metabolized. Especially for perennials, but also for some annuals, storage of carbon is a major mechanism by which naturally growing plants are adapted to their environments.

The principal storage energy reserves used by plants are carbohydrates, particularly sugars, starches and fructosans. In combination, these carbohydrates are generally considered the total carbohydrates that are available for plant functions and are usually referred to either as total nonstructural carbohydrates (TNC) or total available carbohydrates (TAC). Collection and preparation of plant samples for analysis of TNC is relatively simple since dried plant tissue is preferred for analysis (Koch and Hehl, 1975). Tissues should be freeze-dried or dried in a forced-draft oven (see Section 15.2.2).

TNC analysis is best performed on tissue that has been ground to a 40-mesh size in a mill. Quantitative chemical analysis of TNC can be performed at various levels of resolution. For most ecological studies, knowledge of the total TNC pool size is all that is required. To accomplish this, all the carbohydrate present is hydrolyzed to simple monosaccharides, which are readily detected colorimetrically. Procedures are available to accomplish this either manually (Smith, 1969) or with auto-analytical techniques (Weier *et al.*, 1977; Gaines, 1973).

The 'free' sugars of plants (monosaccharides, such as glucose and fructose, disaccharides, such as sucrose, and the lightly polymerized polysaccharides, fructosans) are all soluble in cold water. Starch polysaccharides are not. Hot alcohol/water (75%) extracts all carbohydrates.

A variety of techniques are used to reduce the extracted carbohydrates to monomers for colorimetric detection. Enzymes, such as takadiastase, will hydrolyze sucrose and starch to monomers, but fructosans are not affected and must be hydrolyzed with weak acid (Smith, 1969). Fructosans occur in a variety of vascular plants, including both monocots and dicots, and clearly serve as reserves (Lewis, 1984).

15.4 CARBON AND ENERGY COSTS OF GROWTH AND MAINTENANCE

A somewhat different approach to studying allocation and growth involves dividing carbon and energy requirements of plant tissue into the functional categories of growth (or construction) costs and maintenance costs. Growth costs include the carbon or energy that produces a net gain in dry weight. These costs include carbon actually incorporated into new biomass as well as the carbohydrate metabolized to produce ATP and reductant for biosynthetic processes, transport processes and nutrient uptake and reduction. Maintenance costs include all processes requiring energy but not resulting in a net increase in dry matter, such as maintenance of ion gradients across membranes and the turnover of organic compounds, especially protein. This division of costs has been questioned by some authors. Veen (1980) and DeVisser and Lambers (1983) recognize a third component of cost covering net ion uptake. Thornley (1977) relates growth and respiration to the storage, degradable structural and nondegradable structural components of biomass.

The concepts of carbon cost and energy cost in plant tissues are related because fixed carbon is the form in which energy is stored and transported. Costs are usually reported as grams of CO_2 or grams of glucose. Although valid for nonphotosynthetic tissues, this relationship between carbon and energy costs breaks down for photosynthetic tissue. In such tissue, ATP and NADPH produced directly from light-driven electron transport may contribute to some processes involved in growth and maintenance.

Three methods for assessing construction costs of plant tissues are described here. Maintenance costs are less tractable, but two techniques for their estimation are also presented. Methods have been developed for determining the cost of ion uptake in laboratory-grown plants (Veen, 1980; DeVisser and Lambers, 1983). However, this cost is much more difficult to assess in field situations and is not addressed in the techniques described below.

15.4.1 Gas exchange and growth analysis

Formulas for partitioning respiration into components associated with growth and maintenance have been presented by Pirt (1975), McCree (1970), Thornley (1970) and Hesketh *et al.* (1971). The maintenance respiration rate is considered to be proportional to the dry weight of the plant or plant part (assumed to be the amount of biomass requiring 'maintenance') and the growth respiration rate is proportional to the growth rate or gross photosynthetic rate. Hesketh *et al.* (1971) gave total respiration (R) as:

$$R = R_g \left(\frac{dW}{dt}\right) + R_m W \qquad (15.10)$$

where W is the dry weight, R_g is a coefficient for growth respiration and R_m is a coefficient for maintenance respiration. Rearrangement of this equation gives:

$$\frac{R}{W} = R_g \frac{(dW/dt)}{W} + R_m \qquad (15.11)$$

Plotting specific respiration rate of a plant or plant part against its specific growth rate should yield a straight line with a slope of R_g and intercept of R_m (Fig. 15.13).

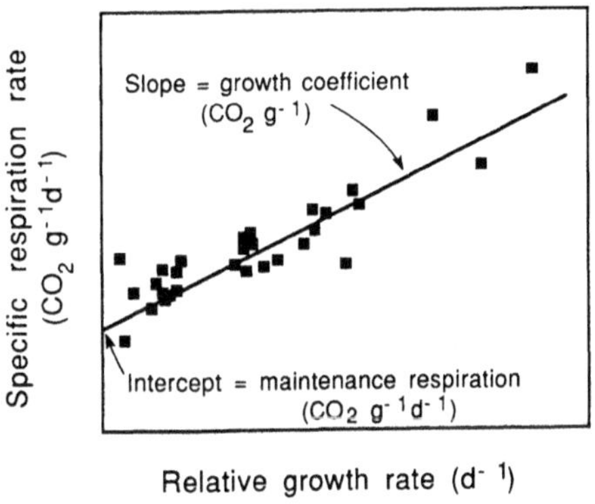

Fig. 15.13 Determination of maintenance respiration rate and growth coefficient of respiration by plotting specific respiration rate against relative growth rate. The data are for *Helianthus tuberosus* (from Kimura *et al.*, 1978).

In laboratory studies, this approach has been applied both to whole plants and to plant parts. In field studies it is difficult to measure respiration rates on a whole-plant basis and the analysis is limited to plant parts. Experimentally, simultaneous measurements of specific respiration rate and specific growth rate are required. If one is able to obtain a cohort of organs whose developmental stage, genetic background and micro-environment are sufficiently similar to insure a uniform trajectory of growth, one may obtain specific growth rates by harvesting subsamples of the cohort at time intervals of one to several days.

In field situations it is usually difficult to obtain large uniform cohorts of plants or plant organs. For this reason, nondestructive methods of estimating dry matter are frequently employed (e.g. Kimura *et al.*, 1978; Merino *et al.*, 1982). Relationships between dry mass and parameters which may be measured nondestructively (e.g. leaf area, twig volume, etc.) may be obtained by harvesting one set of organs. Dry mass may then be estimated repeatedly on the same organ for calculations of specific growth rate and specific respiration rate. This approach requires that there be little variability in the relationship between dry matter and the parameter measured (e.g. leaf area). In cases where substantial variability does exist in this relationship, paired organs may be used for determining the specific growth rate and the specific respiration rate. Kimura *et al.* (1978), working with a plant having opposite leaves, measured leaf area expansion over 3 or 4 days on one leaf and measured respiration rate, midway through this period, on the opposite leaf. The former leaf was harvested at the end of three or four days for determination of dry mass and, combined with data on leaf expansion, calculation of specific growth rate. The latter leaf was harvested at the time of respiration measurement for calculation of specific respiration rate. Such a procedure still relies on a relationship between leaf area and leaf mass. It has advantages under conditions in which mass may be poorly predicted from nondestructive measurements because variation due to differences in genetic background and, to some extent, micro-environment are reduced. It has certain disadvantages, such as possible effects of harvesting one organ on the growth rate of a nearby organ.

Methods for measuring gas exchange on intact leaves in the field are described by Field *et al.* (Chapter 11). Dark respiration has generally been measured at a single controlled temperature. For laboratory or greenhouse experiments in which plants are grown at a single controlled temperature, this approach is valid. However, for application of this method to field situations an attempt should be made to determine the average respiration rate over the time interval encompassed by the growth measurements, taking into account diurnal and day-to-day fluctuations in environmental conditions, as well as night-time declines in respiration due to declining carbohydrate reserves (Azcon-Bieto and Osmond, 1983).

Estimation of respiratory rates in photosynthetic tissue in the light is not straightforward. Some evidence (e.g. Peisker and Apel, 1980; Brooks and Farquhar, 1985) indicates that rates of biochemical processes involved in dark respiration decrease in the light. Photosynthetic electron transport may partially counterbalance the decline in dark respiration by providing ATP and NADPH for some growth and maintenance processes. In the absence of quantitative information on these various processes, most workers have assumed that respiration rates in the light and dark are similar and that they are affected primarily by changes in temperature.

Using the general approach described here, maintenance carbon costs are simply the maintenance respiration rate, as determined from the intercept of the graph of specific respiration rate vs. specific growth rate. Construction costs are a sum of the growth respiration rate, as determined from the slope of the same graph, and the carbon included in newly formed biomass per time. This quantity is calculated from the growth rate, as determined above, and a measurement of the carbon content of the biomass.

There are several assumptions underlying the use of the relationship between growth and respiration to determine the carbon costs of growth and maintenance. Some assumptions are inherent in the approach, others are peculiar to applying the method to plant parts, rather than whole plants. A primary assumption is that growth respiration is proportional to the mass of tissue formed. Individual biochemical compounds differ in their carbon cost and, therefore, the amount of growth respiration associated with their construction. The assumption that growth respiration is proportional to the mass of tissue formed implies, therefore, that the tissue cost does not differ appreciably among the samples used in plotting specific respiration rate against specific growth rate. In reality, variation in specific growth rate is frequently obtained by studying plants or organs of different ages (younger organs often exhibiting higher specific growth rates), and some investigators have found the concentration of relatively expensive compounds declines with tissue age (e.g. Merino *et al.*, 1984; LaFitte, 1985).

A second principal assumption is that maintenance respiration is proportional to the dry mass of a plant or organ. By its very nature, maintenance respiration reflects metabolic activity of the tissue (e.g. protein turnover) which in turn is subject to environmental influence. In extending analysis of growth and respiration to field studies, possible effects of microenvironmental variation, both spatially and seasonally, on maintenance respiration rates should be kept in mind. Variation in specific maintenance respiration among samples used to determine the growth/respiration relationship will either create scatter in the relationship or, if maintenance respiration rate varies with growth rate, yield a tight relationship which does not accurately reflect growth and maintenance costs. Some evidence (e.g. McCree, 1982) indicates that maintenance respiration does vary with growth rate.

Because this type of analysis is limited to individual plant organs, and because some organic compounds may be translocated into the growing organ as something more complex than simply carbohydrate, the 'growth respiration' measured on the organ in question may underestimate that actually associated with the construction of that organ. In these instances, the respiration associated with part of the organ's biosynthesis occurs elsewhere in the plant. As in the other methods addressed here, the cost of mineral uptake and most of the costs of mineral transport are not measured, as most of these costs are incurred in relatively inaccessible parts of field-grown plants.

A good discussion of the assumptions and drawbacks of this type of approach in general is presented by Lambers *et al.* (1983). It should be noted that, despite the many caveats and arguments presented above for violations of the assumptions inherent in this approach, several investigators have obtained linear relationships between specific respiration rate and specific growth rate (e.g. Hesketh *et al.*, 1971; Kimura *et al.*, 1978).

15.4.2 Biosynthetic pathway analysis

In the mid-1970s, F.W.T. Penning de Vries and his co-workers introduced a method for determining construction cost from the biochemical composition of plant tissue (Penning de Vries, 1972; Penning de Vries *et al.*, 1974). Taking glucose as the standard substrate for biosynthesis, one calculates the amount of glucose required to provide the carbon skeletons, reducing equivalents and ATP for forming the biochemical compounds in a tissue via probable biosynthetic pathways. Estimates of the cost of transporting organic substrates across membranes and the cost of accumulating minerals are commonly included in these cost calculations. (The mass of biochemical product produced by a unit mass

Table 15.3 Glucose requirements (1/PV) for the synthesis of a few classes of organic compounds. Selected values are from (a) Penning de Vries *et al.* (1974), (b) Williams *et al.* (1987; values adjusted to include an estimated membrane transport cost of 1 mol of ATP per of mol glucose), from (c) de Wit (1978) and (d) Chung and Barnes (1977)

Class of compound	*Glucose requirement (g of glucose g^{-1})*
Lipid	3.03 (a)
Nitrogenous compounds	
Ammonia = nitrogen source	1.62 (a)
Nitrate = nitrogen source	2.48 (a)
Carbohydrate	1.21 (a)
Sucrose	1.13 (b)
Cellulose or starch	1.21 (b)
Hemicellulose	1.24 (b)
Lignin	2.58 (b)
Organic acid	0.70 (c)
Phenolics	1.92 (d)

of glucose is termed the production value, PV.) In addition to calculating the grams of glucose required, it is possible to calculate the amount of oxygen, reductant and ATP required, as well as the carbon dioxide produced during biosynthesis of the tissue.

In practice it is unrealistic to attempt to determine the amounts of every biochemical compound present in a tissue. Proximate analysis is usually employed in which the plant is analyzed for broad categories of compounds such as carbohydrates, nitrogenous compounds, lipids, lignin, organic acids and mineral or ash content. Average production values for the classes of compounds are then employed for the calculation of the overall production value for the tissue. The glucose requirements calculated by various authors for several classes of compounds are listed in Table 15.3. These values represent the grams of glucose required to synthesize a gram of product and, as such, are the inverse of the corresponding production values.

In the event that one is examining a plant tissue with a significant fraction of compounds not listed above (e.g. tissues with large amounts of secondary compounds), the tissue must be analyzed for the compounds in question and production values calculated for these compounds. This requires some knowledge of the most probable biochemical pathway used to synthesize these compounds. In calculating biochemical pathway costs, 1 mol of glucose is generally assumed to yield 36 mol of ATP, 12 mol of NADH (24 mol of electrons) or somewhat less than 12 mol of NADPH. The cost of any excess ATP, NADH or NADPH produced in the biosynthetic pathway, as well as the cost of any additional organic compounds produced, should be deducted from the total amount of glucose required for the pathway, since these molecules are supposedly available for use in other biosynthetic pathways in the tissue. The amount of glucose required for the construction of some organic compounds is listed in Appendix Tables A15 and A16.

Various chemical methods are required for carrying out proximate analysis. Explanation of the details of these methods is beyond the scope of this book and the reader is referred to Allen (1974), Williams (1984) or other methodological texts. For the crudest categorization of organic compounds (i.e. into crude fiber, crude protein, crude fat and ash) all determinations may be carried out on air-dried tissue. However, when classifying organic compounds on a finer scale, determinations of different classes of compounds may require different sample preparation techniques. Since this precludes the use of exactly the same tissue for all analyses, care should be taken to make samples for the different analyses as alike as possible.

This method estimates the construction cost of all of the organic compounds present in a tissue at the time of sampling. In utilizing this method, it should be kept in mind that tissue composition generally changes with age. The costs of certain processes associated with growth (e.g. ion uptake and the transport of ions and monomers) are estimated.

Penning de Vries *et al.* (1974) used transport costs of 0.3 mol of ATP per mol of cation and 1 mol of ATP per mol of organic monomer (e.g. glucose).

The major assumptions underlying the approach are that glucose is the sole substrate for all ATP, reductant and carbon skeletons and that the biochemical pathways used to convert glucose to final product are sufficiently well known. When considering resource use, the production of ATP and reductant directly from photosynthetic electron transport may make certain compounds 'cheaper' to synthesize in photosynthetic tissue than in nonphotosynthetic tissue, especially in high-light environments. In certain nonphotosynthetic tissues, notably roots (Lambers *et al.*, 1983), anaerobic tissues and thermogenic tissues (e.g. aroid spadices), alternative pathways of respiration may be employed, reducing the ATP yield of glucose catabolism from the figures cited above. In such tissues, compounds are more expensive to produce than elsewhere. Finally, the retranslocation of compounds such as amino acids from senescing organs to other parts of the plant should reduce the actual construction cost of plant tissues because it means that all components in growing tissues do not have to be synthesized from simple carbohydrate.

(a) Estimates of maintenance cost from proximate analysis

Penning de Vries (1975) pointed out that the most important metabolic processes contributing to maintenance respiration were protein turnover and active transport processes which were required to maintain ion gradients. Reviewing the data available at the time, he estimated that, in leaves, protein turnover costs approximately 28–53 mg of glucose per g of protein per day and maintenance of ion gradients costs 6–10 mg of glucose per g dry weight per day. These figures are approximations since variation in the rate of protein turnover will cause variation in the maintenance cost of protein.

15.4.3 Elemental analysis

A method for estimating the construction cost from the elemental composition of tissue was proposed by McDermitt and Loomis in 1981. The amount of glucose required to provide the carbon skeletons and reductant for transforming glucose, nitrogen and sulfur from their substrate forms into biomass can be calculated from the amounts of carbon, hydrogen, oxygen, nitrogen and sulfur in a sample of biomass. The formula for this calculation is:

$$\mathrm{GE} = \frac{c}{6} + \frac{(h - 2x + kn + ms)}{24} \tag{15.12}$$

where GE represents the glucose requirement as 'glucose equivalents' in mol of glucose/100 g of tissue and c, h, x, n and s are the mol of carbon, hydrogen, oxygen, nitrogen and sulfur, respectively, in 100 g of tissue. The parameters k and m represent the oxidation states or valences of nitrogen and sulfur in their substrate forms. These values are +5 for nitrate, −3 for ammonium, +6 for sulfate and −2 for sulfide. The first term in Equation 15.12 represents the mol of glucose required to provide carbon for building tissue. The term in parentheses represents the glucose required to produce the net mol of electrons (as NADH or NADPH) required to reduce nitrogen and sulfur from their substrate forms to the forms present in the organic matter and to adjust the oxygen and hydrogen content of the biomass relative to glucose. One mol of glucose provides 6 mol of carbon or 24 mol of electrons.

The construction costs not covered by this formula (in addition to costs of monomer transport and mineral uptake and transport) include costs of providing ATP for biosynthesis and the reductant which is occasionally required to reduce molecular oxygen. To cover these costs an efficiency factor is required. McDermitt and Loomis (1981) calculated a fairly precise 'growth efficiency' for plant tissues of 0.884 (±0.010 S.D., $n = 13$), where 'growth efficiency' was the ratio of the glucose cost as determined from elemental analysis to that determined from biosynthetic pathway analysis. Omitting costs of substrate transport, Williams *et al.* (1987) estimated that this factor actually fell within a range of 0.84–0.95, depending on the composition of the plant tissue. The uncertainty of this factor limits the precision with which construction costs may be determined using this method. The total construction cost is obtained by dividing glucose equivalents by an estimate of growth efficiency or biosynthetic efficiency.

(a) Elemental analysis – methods

The most common methods of carbon determination involve combusting the sample to convert carbon to CO_2 and determining the amount of CO_2 released. This may be accomplished by passing a gas stream containing the combustion products through a removable trap which absorbs carbon dioxide. In general, the gas is passed first through a tube containing magnesium perchlorate (anhydrone or dehydrite) to absorb water vapor, and then through a tube containing ascarite (or soda asbestos) to absorb the CO_2. The amount of CO_2 released from the sample may then be determined gravimetrically as the mass gain of the CO_2 absorption tube (e.g. Allen, 1974, p. 140; Williams, 1984, p. 985).

CHN analyzers (e.g. Hewlett-Packard Model 185 CHN Analyzer; Perkin–Elmer 240 Elemental Analyzer, Carlo Erba Elemental Analyzer) also measure the CO_2 release upon combustion of plant material. In these analyzers, material is combusted rapidly at high temperature and the combustion products are carried in a helium stream through reducing chemicals in order to reduce oxides of nitrogen to N_2 and remove O_2 and other interfering gases. The CO_2, H_2O and N_2 in the remaining gas stream serve as measurements of the carbon, hydrogen and nitrogen in the combusted sample. These instruments are relatively expensive to operate, requiring costly reagents, but require a very small sample size (less than 5 mg). Some instruments can be configured to determine oxygen or sulfur. The oxygen determination procedure is a modification of a method developed by Schütze (1939), Unterzaucher (1940) and subsequent workers in which the sample is pyrolyzed in a stream of inert gas and the resulting gases passed over various solid chemicals to convert all oxygen-containing gases to CO or CO_2. In this configuration, the amount of CO or CO_2 obtained is a measure of the oxygen released by the sample rather than carbon. Sulfur may be determined by several methods discussed by Allen (1974), Chapman and Pratt (1961) and Williams (1984).

15.4.4 Heat of combustion

Heat of combustion of a tissue can be used to predict GE. The heat released during the oxidative reactions in combustion depends on the reduction state of the fuel. Kharasch (1929) and Kharasch and Sher (1925) showed that the heat of combustion of a compound appeared to be related to the number of electrons displaced from their position in a methane-type (C–H) bond to that held in a carbon dioxide-type (C : O) bond upon combustion. Energy release per electron undergoing this transition was 26.05 kg cal mol^{-1} of electron. Subsequent development of their ideas added various corrections for specific types of bonds in the fuel. For compounds containing only carbon, hydrogen and oxygen, the number of valence electrons experiencing this transition (N) are

$$N = 4c + h - 2x \qquad (15.13)$$

where c, h and x are as in Equation 15.12. Equation 15.13, divided by 24, is identical to McDermitt and Loomis' formulation for glucose equivalents (Equation 15.12) when k and m equal zero.

For compounds containing sulfur and nitrogen, Kharasch and co-workers applied various corrections for individual types of nitrogen- and sulfur-containing bonds (Kharasch, 1929). McDermitt and Loomis (1981) showed that, for compounds containing nitrogen, as well as carbon, hydrogen and oxygen, heat of combustion was well predicted from elemental composition. They obtained two regression lines for GE versus heat of combustion: one for carbohydrates and glycerol and another for other organic compounds. For ease in using this method without determining the amount of carbohydrate in a sample, Williams *et al.* (1987) obtained a single relationship for organic compounds:

$$\Delta H_c = 14.352\text{GE}_0' + 0.929 \qquad (15.14)$$

where ΔH_c is heat of combustion in kJ g^{-1} and GE_0' is glucose equivalents, expressed in g of glucose per g where $k = 0$, $m = 6$ and sulfuric acid is taken as the standard end product of sulfur in combustion calorimetry. Compounds in which nitrogen is bonded directly to oxygen deviate from this relationship. Fortunately, organic nitro-compounds are rare in plant tissues (Robinson, 1980).

Using Equations 15.12 and 15.14 the glucose equivalent of plant tissue is estimated from heat of combustion as:

$$\text{GE}' = (0.06968\Delta H_c - 0.065)(1-A) + \left[\frac{kN}{14.0067} + \frac{(m-6)S}{32.06}\right]\frac{180}{24} \qquad (15.15)$$

where GE′ is glucose equivalents expressed in g of glucose per g of dry mass of tissue, ΔH_c is the heat of combustion of the organic fraction of the tissue (ash-free heat of combustion) in kJ g^{-1}, A is the inorganic fraction of the dry weight (roughly, the ash fraction), k and m are the oxidation numbers of the nitrogen and sulfur substrates and N and S are the nitrogen and sulfur content in g g^{-1} dry mass. The first half of Equation 15.15 represents the estimated GE_0' (Equation 15.14). The second half represents the difference between the general formulation for GE′ (from Equation 15.12) and GE_o' (GE′ calculated with $k = 0$ and $m = 6$). Since sulfate ($m = 6$) is the most common form of sulfur taken up by plants, Equation 15.15 usually reduces to:

$$\text{GE}' = (0.06968\Delta H_c - 0.065)(1-A) + \left(\frac{kN}{14.0067}\right)\frac{180}{24} \qquad (15.16)$$

and construction cost (C) is then estimated as:

$$C = \frac{(0.06968\Delta H_c - 0.065)(1-A) + \left(\frac{kN}{14.0067}\right)\frac{180}{24}}{\text{efficiency}} \qquad (15.17)$$

For Equation 15.17, determinations of organic nitrogen, ash-free heat of combustion and ash content are required. For most purposes, a measurement of Kjeldahl nitrogen is sufficient for measurement of organic nitrogen. In tissues which accumulate inorganic nitrate, however, other procedures may be desirable. Since the Kjeldahl procedure frequently detects some of the nitrogen in nitrate, steps may be taken to either minimize the nitrate nitrogen detected in the Kjeldahl procedure (e.g. Pace *et al.*, 1982) or to detect all of the nitrogen in the sample (e.g. Cataldo *et al.*, 1974; Williams, 1984) and to subtract the amount of nitrate nitrogen determined separately.

Heat of combustion is determined as the heat released upon combustion of a measured mass of sample. Ash-free heat of combustion is calculated as that heat released per gram of ash-free sample mass (i.e. the initial sample mass minus the ash mass). It is generally recommended that, rather than using the residue in the bomb calorimeter as an estimate of the ash content, the percentage ash in a sample be determined separately by burning approximately 1 g of sample in a muffle furnace at approximately 500°C for 3–4 h (Paine, 1971). Differences between the ash obtained in this manner and the residue obtained in bomb calorimetry may make a significant difference in the calculations of ash-free heat of combustion. Ash-free heat of combustion is the recommended form of expressing heat of combustion in the literature for the purposes of comparing the energy content between species, organs, etc. These differences in ash estimates, however, make little difference to the calculation of cost when expressed on a total weight basis.

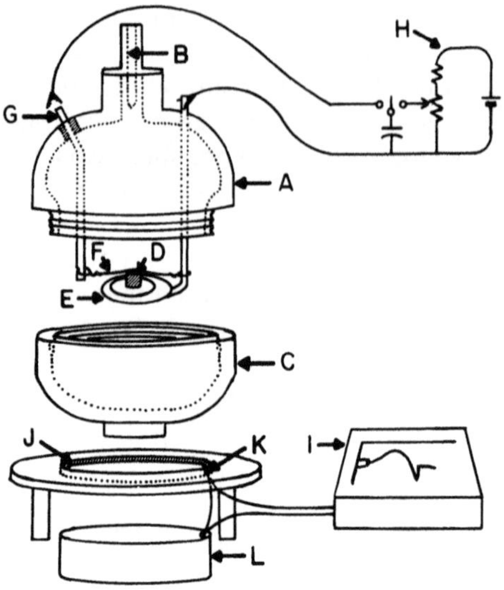

Fig. 15.14 Schematic of a Phillipson-type microbomb calorimeter. Top of chamber (A) has valve (B) to allow pressurization of the chamber with oxygen. Bottom of chamber (C) contains 1 ml of water. Pelleted sample (D), on a platinum dish, is placed on the sample holder (E). It is held in place with a platinum fuse wire (F). The positive firing terminal (G) is insulated from the body of the microbomb chamber. Ignition of the sample is achieved by passing a burst of current from the firing circuit (H) across the platinum fuse wire. (The firing circuit illustrated here is after Phillipson, 1964; firing can also be achieved through circuits employing the control features of modern data loggers.) Heat release is calculated from a temperature trace (I), produced by monitoring temperature changes of a brass ring (J) on which the microbomb chamber rests. Temperature changes are monitored with several thermocouples (e.g. K). An aluminum slug (L) provides the reference thermocouple temperature. The chamber, stand and aluminum slug must be protected from air currents and changes in external temperature during the determination.

(a) Heat of combustion – methods

The instruments used in calorimetry are many and varied (Hemminger and Hohn, 1984). A diagram of a Phillipson-type microbomb calorimeter is shown in Fig. 15.14. The use of oxygen bomb calorimetry in ecological studies has also received considerable attention (e.g. Paine, 1971; Lieth, 1975; Golley, 1961). Here we provide a brief introduction to oxygen bomb calorimetry and its use in ecological work.

In general, a sample is placed in a chamber which is then pressurized with oxygen gas. The sample is ignited by means of an electrical charge through a fuse wire and the heat released is measured. The chamber may or may not be purged with oxygen before pressurization in order to remove the original air. The means of measuring the heat released during combustion is highly variable and depends on the instrument in use.

The amount of heat released by oxidation of a sample in a calorimeter depends not only on the amount and composition of the sample but also on the identity of the combustion end products. In calorimetry, CO_2(gas), H_2O(liquid), N_2(gas) and SO_2(gas) are generally accepted as the standard end products for the carbon, hydrogen, nitrogen and sulfur in the fuel respectively. In reality, further oxidation of nitrogen and sulfur occurs under the conditions of bomb calorimetry and nitrogen- and sulfur-containing acids are formed.

In order to correct the heat release for the formation of these acids and to calculate the heat which would have been released had the standard end products been formed instead, 'acid corrections' are used. This is usually accomplished by titration of water placed in the bottom of the calorimeter chamber. A correction of the heat of combustion is applied based on the amount of acids formed (e.g. Barker *et al.*, 1955; Lieth, 1975). In some work (e.g. Cox and Pilcher, 1970), dilute sulfuric acid is taken as the standard combustion product of sulfur. For the purposes of using Equations 15.16–15.17, this convention should be employed.

The requirements for precision in calorimetry for ecological work are often less strenuous than the requirements for precision in other fields. The development of micro- and semimicro-calorimeters has encouraged the use of calorimetry in ecological studies because of the small amounts of sample required (1–200 mg as opposed to approximately 1 g for more traditional calorimeters) and the relative rapidity of determination (approximately two samples per hour). These advantages increase the number of samples which can be analyzed and preclude the necessity of combining samples, thereby masking natural variation in energy content. These micro- and semimicro-calorimeters are, however, less precise than most traditional calorimeters. Schroeder (1977) reports that replicate measurements using the Parr semimicrocalorimeter fall within ± 1% of the mean while those for a Phillipson microbomb calorimeter fall within ± 2.8% of the mean. Paine (1971) considers the calorimeters to be of similar accuracy and reproducibility and points out that a major source of nonreproducibility is the natural heterogeneity in biological material which may exist even in finely powdered samples. This variation would be more obvious the smaller the sample size. In view of such variation, Golley (1961) suggests that a 3% range be acceptable variation for replicate measurements of heat of combustion of biological material, rather than the more strenuous requirements often applied in bomb calorimetry.

This lack of precision has implications for the necessity of performing acid corrections when using microbomb calorimeters. The heat released by acid formation has been shown to be a small fraction of the total amount of heat released (less than 1.6%, Paine, 1971; 1.06%, Schroeder, 1977; 0.75%, Williams, 1986; 0.1%, Lieth, 1975). It is generally concluded that, while acid corrections are desirable, especially in work requiring precision, they have a relatively minor effect on the calculated heat of combustion.

15.4.5 Discussion

The goals of a particular study will largely determine the method of choice in determining construction cost. As discussed in the preceding sections, gas exchange coupled with growth analysis quantifies carbon costs, while other methods quantify a combination of carbon, ATP and reductant costs. While these costs may be equivalent in nonphotosynthetic tissues, they may not be equivalent in photosynthetic tissues. Most studies addressing cost in an ecological context make inferences based on the assumption that plant performance is affected by the amount of resources gained by the plant and the manner in which the plant allocates these resources. Because environmental conditions may determine whether carbon alone or the combination of carbon and energy is limiting, different approaches to quantifying the carbon-related resource may be most appropriate in different environments.

In arid high-light environments, carbon availability may be limited by stomatal closure, whereas light-driven electron transport may be able to produce ATP and reductant in excess of a photosynthetic tissue's needs. For photosynthetic tissues in this situation, it may be more appropriate to consider only carbon costs, using gas exchange and growth analysis. This approach entails assumptions about respiration rates in the light.

In mesic habitats with low availability of light, carbon fixation may compete with other growth and maintenance processes for ATP and reductant produced from light-driven electron transport. Under such circumstances, both carbon and energy costs are relevant. Any of the remaining techniques – pathway analysis, elemental analysis or heat of combustion – may then be used.

Different methods for determining carbon/energy costs are appropriate for different studies, depending on the aim of the study. For questions based on construction cost, the method based on heat of combustion is least expensive and possibly more accurate than pathway analysis. Although the cost estimate obtained from heat of combustion may deviate by up to 6% from that calculated by detailed pathway analysis, the difficulty of obtaining full and accurate biochemical analysis of tissues frequently renders pathway analysis less accurate. If, as in many published ecological studies employing pathway analysis, the 'unrecovered' fraction of plant dry mass exceeds 20% (e.g. Merino *et al.*, 1984; Miller and Stoner, 1979), then costs based on pathway analysis will be less precise than costs estimated from heat of combustion.

Questions involving more than just the construction cost of tissue require methods other than heat of combustion. The breakdown of cost among different classes of compounds requires biochemical analysis. For example, in studies of chemical defense against herbivores, studies frequently ask (1) what is the investment of resources in defense compounds and (2) what is the protein content of the tissue. For these questions, biochemical analysis is required. However, the proportion of construction cost residing in herbivore defense compounds can be most easily quantified by determining the cost of the defense compounds via pathway analysis and determining the total construction cost of the tissue via heat of combustion.

Maintenance costs are more difficult to determine than construction costs. This is partly due to conflicting definitions of 'maintenance respiration'. Following Penning de Vries (1975), some authors define respiration strictly as the respiration from processes that maintain cellular structures, metabolite pools, ion gradients and processes that allow cells to adapt to new conditions and maintain their viability. This differs from the operational definition used in gas exchange and growth analysis, i.e. the respiration rate of non-growing organs. CO_2 evolution associated with specialized functions of tissues (ion uptake in roots, heat production in thermogenic tissues and photorespiration in leaves) occurs in nongrowing organs. It would be included in the gas exchange/growth analysis approach, but does not conform to the Penning de Vries (1975) perspective. Some studies (e.g. Veen, 1980; DeVisser and Lambers, 1983) have separated the respiration associated with ion uptake from other respiration in roots. Comparable situations occur in other tissues.

Both methods discussed for determining maintenance respiration have serious drawbacks. Estimates from biochemical composition rely on knowledge of turnover rates of such compounds as protein, nucleic acid, lipids and some secondary compounds. Data are available for protein turnover rates in some tissues and indicate that rates vary considerably with the type of protein and tissue. Therefore estimates of maintenance respiration based on biochemical composition may be inaccurate if the biochemical turnover rates of the tissue are not known.

Gas exchange and growth analysis may yield better estimates of maintenance respiration because respiration rates are actually measured. However, each plot of specific respiration rate versus relative growth rate yields a single value for maintenance respiration rate. Since tissue composition (especially protein content) changes with age in almost all tissues, maintenance respiration should vary through time.

REFERENCES

Abrahamson, W.G. and Caswell, H. (1982) On the comparative allocation of biomass, energy, and nutrients in plants. *Ecology*, **63**, 982–91.

Allen, S.E. (ed.) (1974) *Chemical Analysis of Ecological Materials*, John Wiley and Sons, New York.

Azcon-Bieto, J. and Osmond, C.B. (1983) Relationship between photosynthesis and respiration. The effect of the carbohydrate status on the rate of CO_2 production by respiration in darkened and illuminated wheat leaves. *Plant Physiol.*, **71**, 574–81.

Barker, J.E., Mott, R.A. and Thomas, W.C. (1955) Studies in bomb calorimetry. IV. Corrections. *Fuel*, **34**, 303–16.

Bazzaz, F.A. and Harper, J.L. (1977) Demographic analysis of the growth of *Linum usitatissimum*. *New Phytol.*, **78**, 193–207.

Beadle, C.L. (1985) Plant growth analysis. In *Techniques in Bioproductivity and Photosynthesis*, 2nd edn (eds J. Coombs, D.O. Hall, S.P. Long and J.M.O. Scurlock), Pergamon Press, Oxford, pp. 20–5.

Benjamin, L.R. and Wren, M.J. (1978) Root development and source-sink relationships in carrot, *Daucus carota* L. *J. Exp. Bot.*, **29**, 425–33.

Biscoe, P.V. and Jaggard, K.W. (1985) Measuring plant growth and structure. In *Instrumentation for Environmental Physiology* (eds B. Marshall and F.I. Woodward), Cambridge University Press, Cambridge, pp. 215–28.

Blackman, V.H. (1919) The compound interest law and plant growth. *Ann. Bot.*, **33**, 353–60.

Bloom, A.J., Chapin, F.S. III and Mooney, H.A. (1985) Resource limitation in plants – an economic analogy. *Ann. Rev. Ecol. Syst.*, **16**, 363–92.

Brand, D.G., Weetman, G.F. and Rehsler, P. (1987) Growth analysis of perennial plants: the relative production rate and its yield components. *Ann. Bot.*, **59**, 45–53.

Briggs, G.E., Kidd, R. and West, C. (1920a) Quantitative analysis of plant growth. *Ann. Appl. Biol.*, **7**, 103–23.

Briggs, G.E., Kidd, R. and West, C. (1920b) Quantitative analysis of plant growth. *Ann. Appl. Biol.*, **7**, 202–23.

Brooks, A. and Farquhar, G.D. (1985) Effect of temperature on the CO_2/O_2 specificity of ribulose-1,5-bisphosphate carboxylase/oxygenase and the rate of respiration in the light. Estimates from gas-exchange measurements on spinach. *Planta*, **165**, 397–408.

Cataldo, D.A., Schrader, L.E. and Youngs, V.L. (1974) Analysis by digestion and colorimetric assay of total nitrogen in plant tissues high in nitrate. *Crop Sci.*, **14**, 854–6.

Chapin, III, F.S. (1980) The mineral nutrition of wild plants. *Annu. Rev. Ecol. System.*, **11**, 233–60.

Chapman, H.D. and Pratt, P.F. (1961) *Methods of Analysis for Soils, Plants and Waters*, Division of Agricultural Sciences, University of California, Berkeley.

Chapman, J.M. and Ayrey, G. (1981) *The Use of Radioactive Isotopes in the Life Sciences*, George Allen and Unwin, London, 148 pp.

Charles-Edwards, D.A. and Fisher, M.J. (1980) A physiological approach to the analysis of crop growth data. I. Theoretical considerations. *Ann. Bot.*, **46**, 413–23.

Coombe, E.E. (1960) An analysis of the growth of *Trema guineensis*. *J. Ecol.*, **48**, 219–31.

Chung, H.H. and Barnes, R.L. (1977) Photosynthate allocation in *Pinus taeda*. I. Substrate requirements of synthesis of shoot biomass. *Can. J. For. Res.*, **7**, 106–11.

Coughtrey, P.J., Nancarrow, D.J. and Jackson, D. (1986) Extraction of carbon-14 from biological samples by wet oxidation. *Commun. Soil Sci. Plant Anal.*, **17**, 393–9.

Cox, J.D. and Pilcher, G. (1970) *Thermochemistry of Organic and Organometallic Compounds*, Academic Press, London.

DeVisser, R. and Lambers, H. (1983) Growth and the efficiency of root respiration of *Pisum sativum* L. as dependent on the source of nitrogen. *Physiol. Plant.*, **58**, 533–43.

de Wit, C.T. (1978) *Simulation of Assimilation, Respiration and Transpiration of Crops*, John Wiley and Sons, New York.

Dickson, R.R. (1979) Analytical procedures for the sequential extraction of ^{14}C-labelled constituents from leaves, bark and wood of cottonwood plants. *Physiol. Plant.*, **45**, 480–8.

Evans, G.C. (1972) *The Quantitative Analysis of Plant Growth*, Blackwell Scientific Publications, Oxford.

Fisher, M.J., Charles-Edwards, C.A. and Campbell, N.A. (1980) A physiological approach to the analysis of crop growth data. II Growth of *Stylosanthes humilis*. *Ann. Bot.*, **46**, 425–34.

Gaines, T.P. (1973) Automated determination of reducing sugars, total sugars, and starch in plant tissue from one weighed sample. *J. Association of Official Analytical Chemists*, **56**, 1419–24.

Glenn, H.J. (1982) Preparation of samples for liquid scintillation and gamma counting. In *Biologic Applications of Radiotracers* (ed. H.J. Glenn), CRC Press, Ohio, pp. 151–69.

Golley, F.B. (1961) Energy values of ecological materials. *Ecology*, **42**, 581–4.

Grier, C.C. and Waring, R.H. (1974) Conifer foliage mass related to sapwood area. *For. Sci.*, **20**, 205–6.

Grime, J.P. (1966) Shade avoidance and shade tolerance in flowering plants. In *Light as an Ecological Factor* (eds G.C. Evans, R. Bainbridge and O. Rackham), Blackwell Press, London, pp. 187–207.

Gross, K.L. (1981) Predictions of fate from rosette size in four "biennial" plant species: *Verbascum thapsus, Oenothera biennis, Daucus carota*, and *Tragopogon dubius. Oecologia*, **48**, 209–13.

Hardwick, R.C. (1984) Some recent developments in growth analysis – a review. *Ann. Bot.*, **54**, 807–12.

Harper, J.L. (1968) The regulation of numbers and mass in plant populations. In *Population Biology and Evolution* (ed. R.C. Lewontin), Syracuse University Press, Syracuse, New York, pp. 139–58.

Heilmeier, H. and Whale, D.M. (1986) Partitioning of ^{14}C labelled assimilates in *Arcticum tomentosum. Ann. Bot.*, **57**, 655–66.

Hemminger, W. and Hohne, G. (1984) *Calorimetry: Fundamentals and Practice*, Verlag-Chemie, Weinheim.

Hesketh, J.D., Baker, D.N. and Duncan, W.G. (1971) Simulation of growth and yield in cotton: respiration and the carbon balance. *Crop Sci.*, **11**, 394–8.

Hobbs, R.J. and Mooney, H.A. (1987) Leaf and shoot demography in *Baccharis* shrubs of different ages. *Am. J. Bot.*, **74**, 1111–15.

Hunt, R. (1978a) Demography versus plant growth analysis. *New Phytol.*, **80**, 269–72.

Hunt, R. (1978b) *Plant Growth Analysis, Studies in Biology*, no. 96, Edward Arnold, London.

Hunt, R. (1979) Plant growth analysis: the rationale behind the use of the fitted mathematical function. *Ann. Bot.*, **43**, 245–9.

Hunt, R. and Bazzaz, F.A. (1980) The biology of *Ambrosia trifida* L. V. Response to fertilizer, with growth analysis at the organismal and sub-organismal levels. *New Phytol.*, **84**, 113–21.

Hunt, R. and Parsons, I.T. (1974) A computer program for deriving growth functions in plant growth-analysis. *J. Appl. Ecol.*, **11**, 297–307.

Hunt, R., Warren Wilson, J., Hand, D.W. and Sweeney, D.G. (1984) Integrated analysis of growth and light interception in winter lettuce. I. Analytical methods and environmental influences. *Ann. Bot.*, **54**, 743–57.

Hutchings, M.J. and Budd, C.S.J. (1981) Plant competition and its course through time. *BioScience*, **31**, 640–5.

Jarvis, P.G. (1985) Specific leaf weight equals 1.0 – always! *HortScience*, **20**, 812.

Jolliffe, P.A. and Courtney, W.H. (1984) Plant growth analysis: additive and multiplicative components of growth. *Ann. Bot.*, **54**, 243–54.

Jolliffe, P.A., Eaton, G.W. and Lovett Doust, J. (1982) Sequential analysis of plant growth. *New Phytol.*, **92**, 287–96.

Jow, W.M., Bullock, S.H. and Kummerow, J. (1980) Leaf turnover rates of *Adenostoma fasciculatum* (Rosaceae). *Am. J. Bot.*, **67**, 256–61.

Kharasch, M.S. (1929) Heats of combustion of organic compounds. *J. Res. Bur. Stand.*, **2**, 359–430.

Kharasch, M.S. and Sher, B. (1925) The electronic conception of valence and heats of combustion of organic compounds. *J. Phys. Chem.*, **29**, 625–58.

Kimura, M., Yokoi, Y. and Hogetsu, K. (1978) Quantitative relationships between growth and respiration. II. Evolution of constructive and maintenance respiration in growing *Helianthus tuberosus* leaves. *Bot. Mag.*, **91**, 43–56.

King, D. and Roughgarden, J. (1983) Energy allocation patterns of the California grassland annuals *Plantago erecta* and *Clarkia rubicunda. Ecology*, **64**, 16–24.

Koch, K. and Hehl, G. (1975) Influence of different preparation and extraction methods on changes in the content of carbohydrates, amino acids and nitrate of plant fresh and dry matter. *Z. Anal. Chem.*, **273**, 203–8.

Květ, J., Ondok, J.P., Nečas, J. and Jarvis, P.G. (1971) Methods of growth analysis. In *Plant Photosynthetic Production* (eds Z. Šesták, J. Čatský and P.G. Jarvis), Junk, The Hague, pp. 343–91.

LaFitte, H.R. (1985) Physiological investigations of nitrogen use efficiency in grain sorghum (*Sorghum bicolor* (L.) Moench), Ph.D. thesis. University of California, Davis.

Lambers, H., Szaniawski, R.K. and de Visser, R. (1983) Respiration for growth, maintenance and ion uptake. An evaluation of concepts, methods, values and their significance. *Physiol. Plant.*, **58**, 556–63.

Ledig, F.T. (1974) Concepts of growth analysis. In *Proceeding of the Third North American Forest Biology Workshop* (eds C.P.P. Reid and G.H. Fechner), Colorado State University, Fort Collins, Colorado, pp. 166–82.

Lefkovitch, L.P. (1965) The study of population

growth in organisms grouped by stages. *Biometrics*, **21**, 1–18.

Lewis, D.H. (ed.) (1984) *Storage Carbohydrates in Vascular Plants. Distribution, Physiology, and Metabolism*, Cambridge University Press, Cambridge.

Lieth, H. (1975) Measurement of caloric values. In *Primary Productivity of the Biosphere* (eds H. Lieth and R. Whittaker), Springer-Verlag, New York.

Magnuson, C.E., Fares, Y., Goeschl, J.D., Nelson, C.E., Strain, B.R., Jaeger, C.H. and Bilpuch, E.G. (1982) An integrated tracer kinetics system for studying carbon uptake and allocation in plants using continuously produced $^{11}CO_2$. *Radiat. Environ. Biophys.*, **21**, 51–65.

Maillette, L. (1982) Needle demography and growth pattern of Corsican pine. *Can. J. Bot.*, **60**, 105–16.

McCrea, K.D., Abrahamson, W.G. and Weis, A.E. (1985) Goldenrod ball gall effects on *Solidago altissima*: ^{14}C translocation and growth. *Ecology*, **66**, 1902–7.

McCree, K.J. (1970) An equation for the rate of respiration of white clover plants grown under controlled conditions. In *Prediction and Measurements of Photosynthetic Productivity* (ed. I. Setlik), Proceeding of IBP/PP Technical Meeting, Trebon 1969, PUDOC, Wageningen.

McCree, K.J. (1982) Maintenance requirements of white clover at high and low growth rates. *Crop Sci.*, **22**, 345–51.

McDermitt, D.K. and Loomis, R.S. (1981) Elemental composition of biomass and its relation to energy content, growth efficiency and growth yield. *Ann. Bot.*, **48**, 275–90.

McGraw, J.B. and Antonovics, J. (1983) Experimental ecology of *Dryas octopetala* ecotypes. II. A demographic model of growth, branching and fecundity. *J. Ecol.*, **71**, 899–912.

McGraw, J.B. and Wulff, R.D. (1983) The study of plant growth: a link between the physiological ecology and population biology of plants. *J. Theoret. Biol.*, **103**, 21–8.

Merino, J., Field, C. and Mooney, H.A. (1982) Construction and maintenance costs of Mediterranean climate evergreen and deciduous leaves. I. Growth and CO_2 exchange analysis. *Oecologia*, **53**, 208–13.

Merino, J., Field, C. and Mooney, H.A. (1984) Construction and maintenance costs of Mediterranean-climate evergreen and deciduous leaves. II. Biochemical pathway analysis. *Acta Oecol. Plant.*, **5**, 211–23.

Miller, P.C. and Stoner, W.A. (1979) Canopy structure and environmental interactions. In *Topics in Plant Population Biology* (eds O.T. Sobrig, S. Jain, G.B. Johnson and P.H. Raven), Columbia University Press, New York, pp. 428–58.

Mooney, H.A. and Bartholomew, B. (1974) Comparative carbon balance and reproductive modes of two Californian *Aesculus* species. *Bot. Gaz.*, **135**, 306–13.

Mooney, H.A. and Billings, W.D. (1960) The annual carbohydrate cycle of alpine plants as related to growth. *Am. J. Bot.*, **47**, 594–8.

Mordacq, L., Mousseau, M. and Deleeno, E. (1986) A ^{13}C method of estimation of carbon allocation to roots in a young chestnut coppice. *Plant, Cell Environ.*, **9**, 735–9.

Nicholls, A.O. and Calder, D.M. (1973) Comments on the use of regression analysis for the study of plant growth. *New Phytol.*, **72**, 571–81.

Ondok, J.P. and Květ, J. (1971) Indirect estimation of primary values used in growth analysis. In *Plant Photosynthetic Production* (eds Z. Šesták, J. Čatský and P.G. Jarvis), Junk, The Hague, pp. 392–411.

Pace, G.M., MacKown, C.T. and Volk, R.J. (1982) Minimizing nitrate reduction during Kjeldahl digestion of plant tissue extracts and stem exudates. *Plant Physiol.*, **69**, 32–6.

Paine, R.T. (1971) The measurement and application of the calorie to ecological problems. *Ann. Rev. Ecol. System.*, **2**, 145–64.

Pearcy, R.W. (1983) The light environment and growth of C_3 and C_4 tree species of the understory of a Hawaiian forest. *Oecologia*, **58**, 19–25.

Pearen, J.R. and Hume, D.J. (1982) Non-destructive estimation of ^{14}C in soybeans immediately after labelling. *Crop Sci.*, **22**, 669–71.

Peisker, M. and Apel, P. (1980) Dark respiration and the effect of oxygen on CO_2 compensation concentration in wheat leaves. *Zeitschr. Pflanzenphysiol.*, **100**, 389–95.

Penning de Vries, F.W.T. (1972) Respiration and growth. In *Crop Processes in Controlled Environments* (eds A.R. Rees, K.E. Cockshull, D.W. Hand and R.G. Hurd), Academic Press, London.

Penning de Vries, F.W.T. (1975) The cost of maintenance processes in plant cells. *Ann. Bot.*, **39**, 77–92.

Penning de Vries, F.W.T., Brunsting, A.H.M. and van Laar, H.H. (1974) Products, requirements and efficiency of biosynthesis: a quantitative approach. *J. Theoret. Biol.*, **45**, 339–77.

Phillipson, J. (1964) A miniature bomb calorimeter for small biological samples. *Oikos*, **15**, 130–39.

Pirt, S.J. (1975) *Principles of Microbe and Cell Cultivation*, Blackwell Scientific Publications, Oxford.

Pitelka, L.F. and Ashmun, J.W. (1985) The physiology and ecology of connections between ramets in clonal plants. In *The Population Biology and Evolution of Clonal Organisms* (eds J. Jackson, L. Buss and R. Cook), Yale University Press, New Haven.

Poorter, H. and Lewis, C. (1986). Testing differences in relative growth rate: A method avoiding curve fitting and pairing. *Physiol. Plant.* **67**, 223–6.

Porter, J.R. (1983) A modular approach to analysis of plant growth. II. Methods and results. *New Phytol.*, **94**, 191–200.

Promnitz, L.C. (1974) Sampling and statistical problems in growth analysis. In *Proceeding of the Third North American Forest Biology Workshop* (eds C.P.P. Reid and G.H. Fechner), Colorado State University, Fort Collins, Colorado, pp. 183–95.

Radford, P.J. (1967) Growth analysis formulae – their use and abuse. *Crop Sci.*, **7**, 171–5.

Raguse, C.A. and Smith D. (1965) Carbohydrate content in alfalfa herbage as influenced by methods of drying. *J. Agri. Food Chem.*, **13**, 306–9.

Recalcati, L.M., Basso, B., Albergoni, F.G. and Radice, M. (1982) On the determination of ^{14}C-labelled photosynthesis products by liquid scintillation counting. *Plant Sci. Lett.*, **27**, 21–7.

Richards, F.J. (1959) A flexible growth function for empirical use. *J. Exp. Bot.*, **10**, 290–300.

Roberts, M.J., Long, S.P., Tieszen, L.L. and Beadle, C.L. (1985) Measurement of plant biomass and net primary production. In *Techniques in Bioproductivity and Photosynthesis*, 2nd edn (eds J. Coombs, D.O. Hall, S.P. Long and J.M.O. Scurlock), Pergamon Press, Oxford, pp. 1–25.

Robinson, T. (1980) *The Organic Constituents of Higher Plants*, 4th edn, Cordus Press, North Amherst, MA.

Rogers, A.W. (1979) *Techniques of Autoradiography*, 3rd edn, Elsevier, Amsterdam, 429 pp.

Roughgarden, J. (1979) *Theory of Population Genetics and Evolutionary Ecology*, Macmillan, New York.

Schroeder, L.A. (1977) Caloric equivalents of some plant and animal material: The importance of acid corrections and comparison of precision between the Gentry-Weigert micro and the Parr semi-micro bomb calorimeters. *Oecologia*, **28**, 261–7.

Schütze, M. (1939) *Z. Anal. Chem.*, **118**, 245–58.

Smith, D. (1969) *Removing and Analyzing Total Nonstructural Carbohydrates from Plant Tissue*, Research Division, College of Agricultural and Life Sciences, Madison, WI, Research Report 41, 11 pp.

Smith, R.I.L. and Walton, D.W.H. (1975) A growth analysis technique for assessing habitat severity in tundra regions. *Ann. Bot.*, **39**, 831–43.

Thornley, J.H.M. (1970) Respiration, growth and maintenance in plants. *Nature, London*, **227**, 304–5.

Thornley, J.H.M. (1977) Growth, maintenance and respiration: a re-interpretation. *Ann. Bot.*, **41**, 1191–203.

Tilman, D. (1988) Plant strategies and the dynamics and structure of plant communities. Princeton University Press, Princeton, New Jersey.

Turner, M.D. and Rabinowitz, D. (1983) Factors affecting frequency distributions of plant mass: the absence of dominance and suppression in competing monocultures of *Festuca paradoxa*. *Ecology*, **64**, 469–75.

Unterzaucher, J. (1940) Microanalytical determination of oxygen. *Ber. Dtsch. Chem. Ges.*, **73B**, 391–404.

Veen, B.W. (1980) Energy cost of ion transport. In *Genetic Engineering of Osmoregulation; Impact on Plant Productivity for Food, Chemicals and Energy* (eds D.W. Rains, R.C. Valentine and A. Hollaender), Plenum Press, New York, pp. 187–95.

Venus, J.S. and Causton, D.R. (1979) Plant growth analysis: the use of the Richards function as an alternative to polynomial exponentials. *Ann. Bot.*, **43**, 623–32.

Vogelmann, T.C., Larson, P.R. and Dickson, R.E. (1982) Translocation pathways in the petioles and stem between source and sink leaves of *Populus deltoides* Batn. ex Marsh. *Planta*, **156**, 345–58.

Wang, C.H., Willis, D.L. and Loveland, W.D. (1975) *Radiotracer Methodology in the Biological, Environmental, and Physical Sciences*, Prentice-Hall, New Jersey.

Wann, M. and Raper, C.D. Jr (1979) A dynamic model for plant growth: adaptation for vegetative growth of soybeans. *Crop Sci.*, **19**, 461–7.

Wann, M., Raper, C.D. Jr, and Lucas, H.L. Jr (1978) A dynamic model for plant growth: simulation of dry matter accumulation for tobacco. *Photosynthetica*, **12**, 121–36.

Warembourg, F.R., Montange, D. and Bardin, R. (1982) The simultaneous use of $^{14}CO_2$ and $^{15}N_2$ labeling techniques to study the carbon and nitrogen economy of legumes grown under natural conditions. *Physiol. Plant.*, **56**, 46–55.

Warren Wilson, J. (1966) An analysis of plant growth and its control in arctic environments. *Ann. Bot.*, **30**, 393–402.

Warren Wilson, J. (1981) Analysis of growth, photosynthesis and light interception for single plants and stands. *Ann. Bot.*, **48**, 507–12.

Watson, D.J. (1958) The dependence of net assimilation rate on leaf area index. *Ann. Bot.*, **22**, 37–54.
Watson, M.A. (1984) Developmental constraints: effect on population growth and patterns of resource allocation in a clonal plant. *Am. Nat.*, **123**, 411–26.
Weier, K.L., Wilson, J.R. and White, R.J. (1977) A semi-automated procedure for estimating total non-structural carbohydrates in grasses, and comparison with two other procedures. CSIRO Aust. Div. Trop. Crops Past. Tech. Pap. No. 20, Melbourne, Australia, 10 pp.
Werner, P.A. (1975) Predictions of fate from rosette size in teasel. (*Dipsacus fullonum* L.). *Oecologia*, **20**, 197–201.
White, J. (1979) The plant as a metapopulation. *Annu. Rev. Ecol. System.*, **10**, 109–45.
Williams, K. (1986) *Estimating Carbon and Energy Costs of Plant Tissues*, Ph.D. thesis, Stanford University, Stanford.
Williams, K., Field, C.B. and Mooney, H.A. (1989) Relationships among leaf construction costs, leaf longevity and light environment in rain-forest plants of the genus *Piper*. *Am. Nat.* (in press).
Williams, K., Percival, F., Merino, J. and Mooney, H.A. (1987) Estimation of tissue construction cost from heat of combustion and organic nitrogen content. *Plant, Cell Environ.*, **10**, 725–34.
Williams, R.F. (1946) The physiology of plant growth with special reference to the concept of net assimilation rate. *Ann. Bot.*, **10**, 41–62.
Williams, S. (ed.) (1984) *Official Methods of Analysis of the Association of Official Analytical Chemists*, Association of Official Analytical Chemists, Arlington, VA.

16

Root systems

Martyn M. Caldwell and Ross A. Virginia

16.1 INTRODUCTION

Although root system studies are generally conceded to be important, the difficulties in such study are sufficiently daunting that root systems have received comparatively little attention in physiological ecology. The problems involve, in part, the sizable labor and time investment, the variability in root locations and activity, and the inadequacies of many root measures. The difficulties are often exacerbated for very fine roots and deep root systems. In spite of these difficulties, a judicious selection of experimental approaches, including indirect assessments, can yield meaningful results and can contribute significantly to an understanding of plant function in the field. Many new techniques have become available in the last few years. This chapter outlines a broad range of techniques for assessing root system structure and function in the field. Boehm (1979) has thoroughly described the classical, and still useful, root evaluation techniques. Emphasis here is, therefore, directed to newer techniques not covered to a great extent by Boehm (1979). This chapter treats methods as they relate primarily to the study of root systems in the field.

Individuals working with root systems have displayed a certain proclivity to coin new terms for the ecological lexicon. In this chapter one encounters terms such as a 'hydropneumatic elutriation system', a 'microrhizotron' and a 'perforon'. Of course, we feel a certain obligation to define such terms but at the same time to employ more conventional English descriptors.

16.2 ASSESSING ROOT SYSTEM STRUCTURE AND BIOMASS IN THE FIELD – DETERMINING WHAT IS THERE

16.2.1 Profile wall root mapping, monoliths, soil core extraction

The only way to examine the entire root system of the plant is by direct excavation and mapping or recovery of roots from the soil. This approach has been widely applied to describe the depth and lateral extent of root systems (Boehm, 1979). Excavation can provide an accurate description of the morphology of the root system and may be useful in studying root overlap between neighboring plants (Kummerow, 1981). This approach, however, is very labor intensive and provides an observation of the root system at only one point in time. Therefore it is of a limited value in studying root function and root growth. For deeply rooted plants, excavation may not be possible as roots may follow cracks in rocks or move through stony soil difficult to excavate without specialized equipment or root disturbance.

Investigators have taken advantage of eroded stream banks and gullies or excavations for roads to observe deep root systems since large volumes of soil have already been removed. Large-scale root excavations have also been conducted. Kummerow *et al.* (1977), Hellmers *et al.* (1955) and Hoffmann and Kummerow (1978) used water under pressure to expose roots of woody chaparral and matorral plants. Air pressure and vacuum have also been used to expose roots in forest studies (i.e. Boehm, 1979). The above techniques can remove large volumes of soil, but recovery of fine roots may be incomplete and can be difficult in clay soils.

The depth and size class distribution of roots can be determined by digging a trench adjacent to the plant and mapping the exposed roots. A trench approximately 1 m wide is dug at the point of interest. Care must be taken to extend the trench below the maximum depth of rooting for a complete description. Deep trenches may require shoring for safe use. If the root system is to be excavated, the soil is carefully removed using hand tools to expose the roots. The profile wall may be smoothed and the location and size of roots can be mapped or photographed against a grid system. It may be possible to use image analysis techniques to process photographic images of roots if the contrast in color between the roots and the substrate is sufficient. Trenching approaches require considerable labor and are time consuming. Particular care needs to be taken in preparation of the wall surface so as to expose the fine roots. However, no specialized or expensive equipment is required and this approach provides a direct and complete description of a plant's root morphology. It is especially useful to determine if roots are following cracks or discontinuities in the soil. Root descriptions for many systems have been made using this technique (Boehm, 1979).

When root recovery is required, monoliths can be extracted from the soil and the roots separated by washing. Monoliths can be cut from the soil using a grid to gain a three-dimensional view of the root system. If the soil lacks sufficient structure to extract intact monoliths, a metal frame can be driven into the soil to support the monolith during extraction. Other modifications include the box method where a large monolith encased in a box is removed from the face of a trench and washed free of roots to provide a view of the depth distribution of roots. A needle board can be driven into the monolith to keep roots in their natural position prior to washing the soil free. This maintains three-dimensional information from the monoliths. Further details are in Boehm (1979).

Coring methods have increased in popularity as better equipment has been developed. This is especially true for deep drilling (> 1 m, Boehm, 1979). The Viehmeyer tube (Viehmeyer, 1929) can be rammed into the soil using a drop weight and extracted with jacks or pulleys to give a relatively intact core up to 180 cm in length if the soil is not too stony. Larger mechanical equipment has been developed for deeper sampling (e.g. Kelley *et al.*, 1947; Boehm, 1977).

Undisturbed cores can be obtained with equipment that contains at least two sets of bits, usually an outer rotating bit which cuts the hole and a nonrotating inner coring bit (Fig. 16.1). Usually the coring bit slightly precedes the outer bit in order not to disturb the core. The outer bit cuts and/or removes the soil (if it contains auger flights) and allows the inner bit to enter the soil with relatively little pressure required. Single core

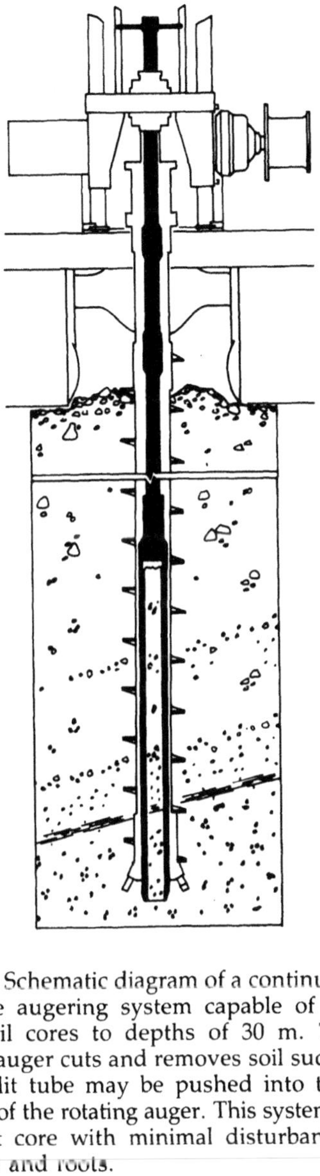

Fig. 16.1 Schematic diagram of a continuous flight split-tube augering system capable of obtaining intact soil cores to depths of 30 m. The outer rotating auger cuts and removes soil such that the inner split tube may be pushed into the soil in advance of the rotating auger. This system provides an intact core with minimal disturbance to soil structure and roots.

increments may be limited to 2 m lengths to minimize disturbance at the soil/core interface. Depths below 30 m may be reached with commercially available soil engineering equipment. However, for an intact 10 cm-diameter core, a hole of approximately 16 cm-diameter is drilled, and the required truck is large and heavy. Nonetheless, such equipment, coupled with improved root separation and measurement techniques, may greatly increase the rate at which data on deep samples may be collected.

Because coring devices are typically 2–10 cm in diameter, each core represents samples of a limited volume of the root zone. Where rooting densities are relatively uniform, such as in fibrous-rooted grasses, relatively few cores may be required to characterize the system. Where rooting patterns vary in a systematic fashion, such as in and between crop rows, appropriate sampling protocols may be established to account for this consistent variation. However, with irregularly spaced taprooted species, root distribution may be highly variable, especially for deep samples. A small-diameter core will have a small probability of encountering a significant portion of the deep root system.

16.2.2 Removing roots from soil

Soil cores or monoliths often yield the best quantitative information on root system biomass and rooting length per volume of soil. Quantitative removal of the roots from soil cores would seem to be straightforward. However, the degree of recovery of the very fine roots (in the range of 50–250 μm diameter) may be very dependent on the technique employed. The most appropriate method depends on root and soil characteristics and the quantity of organic debris. Usually a process involving flotation is most effective in the recovery of the very fine roots. For example, compared to dry sieving (pore size 0.2 mm^2) a flotation procedure using a saturated NaCl solution to increase buoyancy and disperse soil aggregates followed by wet sieving (pore size 0.03 mm^2) yielded a greater quantity of fine roots by a factor of 2.4 (Caldwell and Fernandez, 1975). These roots ranged in size between 60 and 350 μm diameter. Boehm (1979) recommended dry sieving only for roots greater than 2 mm in diameter, and then only in sandy soils.

Various washing and flotation procedures, as well as chemical dispersing agents, are reviewed by Boehm (1979). Smucker *et al.* (1982) tested numerous root-washing approaches before devising what they term a hydropneumatic elutriation device. The equipment employs a high-kinetic-energy first stage in which water jets erode the soil from the roots and a second low-kinetic-energy flotation stage which deposits the roots on a submerged sieve (Fig. 16.2). Whereas the other techniques were reported to be semiquantitative at best, they claim their device to be rapid (3–10 min per sample) and to yield a high recovery of roots without severing laterals. (A commercially available version of this device is available from Gillison's Variety Fabrications, Benzonia, MI, USA.) The most important advantage of the mechanical system is the greatly improved consistency in sample handling. Even when used by different operators, the degree of root recovery is comparable.

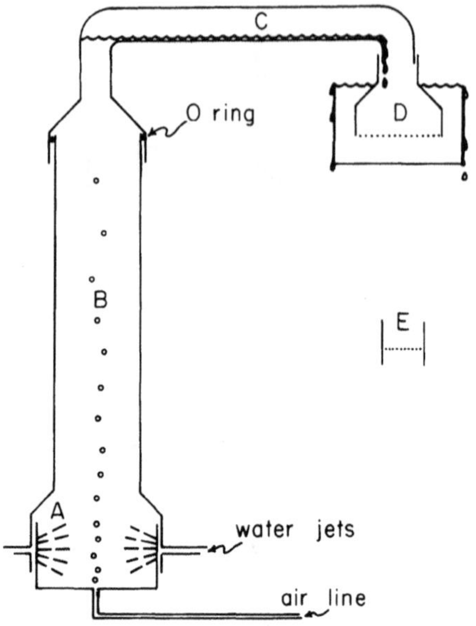

Fig. 16.2 Schematic representation of a hydropneumatic elutriation system for separating roots from soil. (A) A high kinetic-energy washing chamber, (B) elutriation chamber, (C) transfer tube, (D) low kinetic-energy primary sieve and (E) secondary sieve. The transfer tube (C) can be removed from the elutriation chamber (B) in order to remove soil and fragments from the previous sample and to insert a new soil–root sample. After the primary sieving (D), the roots can be placed on the secondary sieve (E) by inverting and washing on to the fine secondary sieve. (Adapted from Smucker *et al.*, 1982.)

16.2.3 Separating roots

If there is appreciable organic debris, some tedious hand separation of roots is usually necessary even with a mechanical elutriation device. Distinguishing between living and dead roots is usually a subjective decision based on criteria such as color, tendency to stain and elasticity. Techniques for separation of live and dead roots using autoradiography of roots previously labeled with ^{14}C or ^{32}P (Singh and Coleman, 1973; Svoboda and Bliss, 1974) can provide a more definitive determination of living roots under some circumstances, but this is usually impractical for large quantities of material. A tetrazolium dye technique has been reported as a suitable means of estimating the proportion of live material in batch samples of roots (Knievel, 1973; Joslin and Henderson, 1984). The differential buoyancy of live and dead roots in different organic solutions might also present a means of separating live and dead material for some species (Caldwell and Fernandez, 1975), but this would need to be developed for each species. Dye techniques coupled with photoelectric scoring and computer image analysis, described in a later section (16.3), also offer possibilities. Many workers, however, still rely on subjective separation for large numbers of root samples. Consistency in these separations is obviously important.

Even after washing, there always is a certain amount of remaining soil contamination. This clinging soil can constitute a large proportion of the mass of root samples. To avoid this problem, the ash content of root samples is often determined and subtracted from total dry mass. This allows root samples to be expressed as ash-free organic matter.

Separating roots of different species is also usually a matter of a manual, subjective approach based on criteria such as color, size and branching pattern. This is not only tedious, but can be particularly difficult with very fine young roots. Simple and rapid chemical tests may be employed for separating certain species combinations. For example, J. Manwaring has found distinct differences in fluorescence from concentrated basic extracts of dried roots of *Artemisia* and *Agropyron* species (Caldwell *et al.*, 1987). These differences are apparent for all size classes of roots and for roots extracted from the soil at different times of year. Whether such simple and unequivocal tests can be developed for other species combinations must be explored.

Stable isotope analysis may offer possibilities for determining proportions of roots from different species. For mixtures of plants with C_3 and C_4 photosynthetic pathways, a determination of the fraction of the root biomass contributed by C_3 and C_4 plants is possible because of the large difference in stable carbon isotope composition of plants with these photosynthetic pathways (Svejcar and Boutton, 1985). By using other stable isotope ratios, such as those of nitrogen and sulfur, in addition to carbon, it may be possible to establish stable isotope signatures for each species. For example, roots of N_2-fixing plants might be expected to have a different N isotope ratio than roots from nonsymbiotic plants (see Section 16.7.4). Although it would be impractical to determine the isotope composition of each root, one could determine the proportion of root mass represented by species in a mixture. Such approaches have, for example, been used in the analysis of filter feeder diets in estuarine food web studies (Peterson *et al.*, 1985). Of course, such isotope approaches would only provide the proportion by mass and not by root length since species vary considerably in root length/mass ratios.

Near-infrared reflectance spectroscopy has also been used to estimate the proportion of roots belonging to two species in mixed root samples (Rumbaugh *et al.*, 1988).

16.3 DETERMINATION OF ROOT LENGTH AND SURFACE AREA

Although root mass provides a useful measure of plant investment in root systems, root length and surface area are more appropriately related to absorptive capacity of root systems (Nye and Tinker, 1977). Total root length in samples is most frequently determined by a line-intercept approach. Newman (1966) provided a relationship to predict root length based on the number of intercepts of randomly placed lines (transects) with roots spread over a surface. This was modified by Marsh (1971) and Tennant (1975). The approach is to spread the roots fairly uniformly with minimal overlap in a shallow pan. A shallow layer of water facilitates positioning of the roots. A grid is then used for scoring the intercepts.

This sampling approach has been mechanized with photoelectric counters (Rowse and Phillips, 1974). A commercial unit patterned after Richards *et al.* (1979) slowly rotates the root material on a large glass disk and the scanner moves across the disk field much like a phonograph player (Comair Corp., Melbourne, Australia). Such a mechanical device greatly facilitates the sampling process. The availability of image analyzing computers which can more rapidly determine length and area is a welcome advance. (These are available from firms such as Skye Instruments, Buckingham, Pennsylvania or Decagon Devices, Pullman, Washington, USA.) Root length determinations can also be conducted on photographic or photocopy images of roots suitably dispersed on a flat plane. The photographic images can be enhanced by staining the roots and photographing them under special lighting. By use of Trypan Blue stain or a fluorescent dye which preferentially stains living roots (e.g. fluorescein diacetate), it has been reported that the length and area determinations of live roots can be conducted on photographs of root samples contaminated by organic debris without large errors (McGowan *et al.*, 1983; Ottman and Timm, 1984).

Apart from total rooting density, more detailed information on root system structure, such as the length of individual roots belonging to different branching orders, can also be obtained by a semiautomatic computer digitization system using photocopied images of roots (Costigan *et al.*, 1982). The operator must trace the individual roots with a stylus and note the branching order, but the digitized length computations are handled by a microcomputer.

Finally, at a fine resolution scale, scanning electron micrographs have been shown to be a useful means of determining root surface area and volume as well as root hair dimensions (Lamont, 1983). It is necessary to remove roots from soil samples in order to employ all of the foregoing methods of root length and area determinations. One technique that avoids root removal and still provides some estimate of rooting density is the core-break approach (Drew and Saker, 1980). Normal soil cores are taken, broken at intervals and the number of exposed root ends on the broken core surface counted.

16.4 MICROSCALE DISTRIBUTIONS OF ROOTS

Although profile wall mapping, monoliths and soil cores provide general information on root system distributions, the location of individual roots with respect to neighboring roots requires more refined techniques. Questions concerning the distance between adjacent roots or the degree of mixing of individual roots of different species have not been extensively addressed in the field. A few techniques have been described for such microscale root distribution determinations which may be suitable for field use.

Litav and Harper (1967) wished to determine the degree to which roots of neighboring plants intermixed. They labeled one neighbor with $^{14}CO_2$ and subsequently determined which of the individual roots in the soil interspace between the neighbors were radioactive. Unlabeled roots were assumed to belong to the other plant. Although this was a study of potted plants, this approach could be conceivably used under field conditions when only two individual plants are clearly involved. Although this approach is straightforward, it is tedious and the identification must be conducted on individual roots by autoradiography. Furthermore, it is often difficult to determine the exact position of the roots and to identify the nearest neighboring roots. A certain bias would also be involved in that apparently nonradioactive roots of the ^{14}C-labeled plant would be attributed to the unlabeled neighbor plant.

Fusseder (1983, 1985) developed a similar technique for use with potted plants or for plants growing under field conditions. Plants were labeled with radioisotopes and the position of roots determined by autoradiography. However, in order to preserve the location of the individual roots, the rooted soil parcel to be investigated was first frozen with liquid nitrogen and then cut in the frozen state into thin sections (5 mm) with a diamond saw. The autoradiography was then conducted at low temperature on the cut surfaces. The severed ends of the roots appeared as spots on the autoradiographs. To determine the location of roots from neighboring plants, the neighbors were labeled with isotopes that emit very different beta energy radiation, for example, a high-energy beta-emitting isotope such as ^{32}P and a low-energy beta-emitting isotope such as ^{33}P or ^{35}S. The roots of the two plants could be distinguished by filtering the autoradiograph during the exposure period (Fusseder, 1983). This technique, though obviously laborious and not without some radiological hazard especially during the sawing process, can provide considerable information on the exact spacing and distribution of roots on a microscale (Fusseder, 1985). A similar approach involving freezing by liquid nitrogen and sawing, but without isotope labeling, has been used (Fig. 16.3) (Caldwell *et al.*, 1987). In this case, roots were visually located on the cut surfaces and identified by an extract fluorescence technique mentioned in Section 16.2.3.

Belford *et al.* (1986) removed monoliths of wheat roots from a rock-free loessial silt loam soil. Using a water stream, they washed the soil away from the roots in alternate 20 mm bands so that the roots were held in place by the soil in the unwashed bands. The washed bands were filled by a clear casting resin and after this had set, the remaining soil bands were washed away and set in resin. If neighboring roots are to be distinguished, this is done by tracing the individual roots from the individual tiller bases. If one wished to determine interroot distances, the resin block could be sectioned. How well suited this technique might be in other soils is difficult to predict.

Direct vacuum impregnation of dried soil cores with a thermosetting resin has also been reported (Melhuish and Lang, 1969). They accomplished this without first washing out bands of soil. The resin penetrated the cylindrical voids (holes) in the soil left by the roots which had shrunk during the drying process. By preparing sections of this impregnated soil–resin block using a grinding machine, they could map the exact positions of the roots based on the location of the resin-filled root channels. Obviously, this technique would not be suited for the separation of neighboring plant roots. Other techniques described in the soils literature for soil pore analysis may also be appropriate for describing the channels made by roots (Pagliai *et al.*, 1983; Ringrose-Voase and Bullock, 1984).

16.5 ROOT SYSTEM TURNOVER AND PRODUCTION

Root systems are dynamic. The fine roots continually grow, senesce and die. Therefore, a determination of root system biomass at one point in time usually does not provide a measure of root system production, and certainly not a measure of fine root turnover. Estimates of root production are important in studies of total plant productivity and

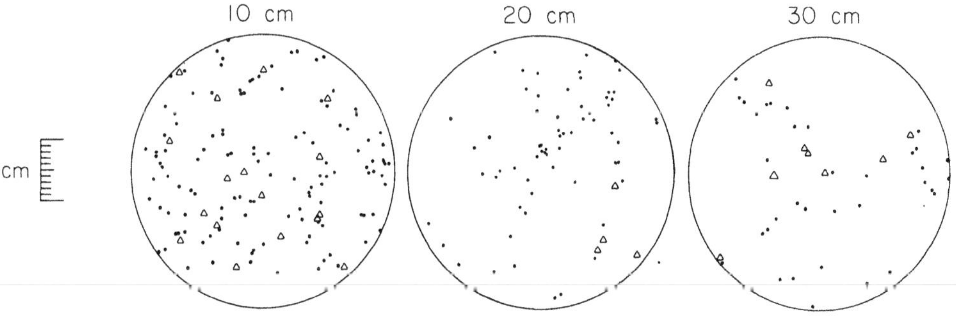

Fig. 16.3 The distribution of individual roots in cross-sections of soil between *Artemisia tridentata* (△) and *Agropyron desertorum* (●) at three depths. Parcels of soil previously frozen with liquid nitrogen were excavated and then sectioned at −70°C with a lapidary saw. The position where individual roots intersected the cut plane was mapped from the cut soil sections and from enlarged photographs of them. To determine whether the individual roots belong to *Artemisia* or the grass, a fluorescence technique has been developed and tested on roots of *Artemisia* and *Agropyron* of different ages and phenological status. On partial thawing of the soil, individual root segments (~1 cm) were removed from the soil and dried. A basic extract of the roots (2 M NaOH) was dried on chromatography paper and the spots observed under a UV-A lamp. The *Artemisia* and *Agropyron* roots are clearly distinguished by color and intensity of fluorescence (from Caldwell *et al.*, 1987).

nutrient cycling. In most stands of perennial vegetation, root system production is considerably greater than aboveground production (Caldwell, 1987). Several techniques have been used to estimate the magnitude of root production. The most common techniques will be discussed with respect to the problems and sources of error.

16.5.1 Sequential root biomass sampling

The most frequently used approach involves estimates of root system production based on a series of root biomass measurements through time. These are usually done by soil coring. The sequence of samples exhibits periods of increasing and decreasing root biomass. If the sampling has been properly planned to minimize spatial variation in root system biomass, these differences through time are considered to reflect the net result of new root growth and loss of roots by death and decay. However, during a period of increasing root biomass, there is also likely a loss of roots; therefore, the net increase should be an underestimate of the true root production. During the course of the growing season, the sample series usually exhibits a series of peaks and troughs, especially if the sampling interval is reasonably short.

Numerous approaches have been taken to calculate root production from sequential harvest data. The two primary approaches have been simply to take the difference in maximum and minimum root biomass over the course of a year, or to accumulate all the apparent positive biomass increments over the course of a year. The latter approach will usually result in a considerably larger estimation of root system production. Sometimes the calculations are restricted to an estimate of the live root fraction from these samples, provided that some method for determining live and dead components is available (see Section 16.2.3). Examples of more elaborate calculation approaches can be found in Persson (1980) and Hansson and Steen (1984). Correction factors for concomitant root death and decay based on litter-bag studies have also been employed to adjust production estimates (Hansson and Andren, 1986).

Another concern is to factor out sampling errors in the calculation process. These errors result from unavoidable spatial variation and other sampling errors. Since troughs and peaks are being expressly selected out of the time series of biomass samples, this sample variation contributes to an apparent increase in temporal variation of root biomass through time and, thus, to an overestimate of root production. The opportunities for error are greater if all positive biomass increments are summed to estimated production, particularly if the sampling interval is short (Singh *et al.*, 1984). Statistical approaches have been developed to estimate the error component of this apparent temporal variation (Kelley *et al.*, 1974; Sims and Singh, 1978; Hansson and Steen, 1984). Use of available information on root growth patterns for the particular system in question to plan the schedule of sequential biomass samples will improve the efficiency of sampling and help to distinguish between sampling variation and real changes in root biomass through time (Vogt *et al.*, 1986). These authors also point out that use of only the live component of the root biomass samples in calculating production greatly improves the signal-to-noise ratio in distinguishing change through time. Although there are numerous problems inherent in the sequential sampling approach, these can be mitigated to some extent. This approach continues to be the most frequently used method for estimating root system production.

16.5.2 Root ingrowth techniques

The second frequently used technique for estimating root production is the ingrowth technique where root-free soil in mesh bags is placed in holes and subsequently removed at prescribed intervals. The root mass invading the mesh bags is used as an index of root productivity (Persson, 1979; Fabiao *et al.*, 1985). The assumption plainly is that growth into the root-free soil in the mesh bag is the same as root production in the normal undisturbed soil. This is an assumption that is difficult to test and probably not correct. If the soil becomes compacted in the process of boring the hole, root growth could be influenced (Shierlaw and Alston, 1984). The disturbance of the soil in the process of removing the roots before the soil is placed in the mesh bags can cause changes in the mineral nutrient status of the soil (Runge, 1983). This can, in turn, alter root morphology and patterns of root growth (Hackett, 1968; Fitter, 1982; Robinson and Rorison, 1983). For example, one reaction of roots to increased available nutrients is proliferation of root growth (St. John *et al.*, 1983). Of course, other alterations of the soil environment such as loss of soil structure and reduced bulk density resulting from root extraction, could modify the pattern and quantity of root growth.

Although there are many factors that might contribute to either under- or over-estimation of root production by this technique, there are few reports of comparisons of this approach with other methods in the same ecosystem. Hansson and Andren (1986) reported reasonable correspondence of production estimates based on the ingrowth and sequential harvesting methods in a fescue pasture for a single year. Similarly, Persson (1979) found reasonable agreement of these approaches in *Pinus sylvestris* forests in Sweden, although the two methods were not used in the same growing seasons.

St. John (unpublished) developed a similar technique which he terms a plane-intercept technique. Fiberglass screens are inserted into the soil at a 45° angle using a spade. At subsequent intervals the screens are removed to determine the number of roots which have grown through the screens. This provides an estimate of root length growth through time. If soil conditions permit the installation and removal of the screens without much disturbance, this approach would have distinct advantages over the mesh bag technique. Root growth through the screen would be less artificial than growth into bags of completely disturbed soil.

16.5.3 ^{14}C-dilution technique

A third approach to root production estimates involves a change in $^{14}C/^{12}C$ in the structural carbon of the root system through time following a one-time $^{14}CO_2$ labeling of the plant shoot system, or what has been termed a ^{14}C-dilution technique (Caldwell and Camp, 1974). Plants are labeled with the radioisotope ^{14}C by photosynthetic fixation of $^{14}CO_2$. After most of the ^{14}C has been translocated throughout the plant, the structural carbon of the root tissues is sampled for the relative $^{14}C/^{12}C$ ratio. At a later point in time, usually at the end of the growing season, the root structural tissues are again sampled for the relative $^{14}C/^{12}C$ content. From the change in relative ^{14}C in the structural tissues of the root system, a turnover coefficient is calculated. This represents the incorporation of new ^{12}C into the root system. The turnover coefficient can be multiplied by the root biomass determined at about the time of ^{14}C labeling to calculate belowground production.

There are several advantages to this approach since it avoids the problems of peak-and-trough sampling and calculation. Root biomass must only be determined once and it is not necessary to separate live and dead roots. The major disadvantage is that at the first time of root sampling, most of the ^{14}C translocation within the plant must be complete. If there is flow of ^{14}C into the structural tissues subsequent to the first sampling, the turnover estimate will be invalid. In the original use of this technique with semiarid steppe shrub stands, the first sampling was done one week following labeling (Caldwell and Camp, 1974). The calculated values for carbon investment in root production coincided with other components of the total carbon balance assessed independently for these shrub stands (Caldwell *et al.*, 1977). However, Milchunas *et al.* (1985) found that wheat and blue grama grass (*Bouteloua gracilis*) had appreciable quantities of ^{14}C incorporated into the structural components of the root system over a 62-day period. They devised a set of correction factors based on a ^{14}C-balance study that could be used to correct for this added ^{14}C flow. If it is not feasible to derive these correction factors, they suggest delaying the first $^{14}C/^{12}C$ sampling until a full year following the initial ^{14}C labeling. Of course, it would be necessary that sufficient ^{14}C activity exists in the structural components of the root system by the time of the second sampling in order to make the necessary determinations.

With these refinements, Milchunas *et al.* (1985) suggest the $^{14}C/^{12}C$ ratio technique to have considerable promise in field studies. They compared the types of errors that can beset the sequential harvest and ^{14}C-dilution techniques for estimates of belowground production. Of the eight sources of error they identified to be involved in production estimates, seven were fully involved in the harvest technique while only four were involved in the ^{14}C-dilution approach and these were usually only partial errors, some of which would tend to cancel one another. There are, of course, practical limitations to the use of the approach in many systems. Large shrubs and trees could be labeled with $^{14}CO_2$ only with great difficulty. There also may be many areas where ^{14}C disposal problems would prevent use of this approach. Further discussion of this technique is found in Caldwell and Eissenstat (1987).

Although these three methods represent the most commonly used techniques for root production estimates, two indirect approaches have been reported recently. One involves a total nitrogen budget analysis (Aber *et al.*, 1985) and the other an estimate based on changes in root starch concentrations and soil temperatures (Marshall and Waring, 1985). Both of these approaches depend on several assumptions that must be established for a particular ecosystem. Both may have promise in certain situations, but comparison with other techniques for assessing root production would still be recommended.

16.6 ROOT PHENOLOGY AND GROWTH

16.6.1 Root observation chambers

Continuous nondestructive assessments of root growth have been practiced under field conditions for decades. This involves observations of root growth against glass or rigid plastic barriers. These barriers can be as simple as a vertical glass pane buried in the soil with a hole or trench alongside so that an observer can see the roots which are growing against the glass (e.g. Ares, 1976). More permanent installations of varying size, termed root observation chambers, rhizotrons or root laboratories, have been built. These facilities and their installation and use have been described in considerable detail by Boehm (1979) which obviates the need for a lengthy description here. The panes of glass or acrylic plastic sheets are usually inclined slightly from the vertical in order to intercept downward-growing roots. Sometimes they are installed in a garden or field that is subsequently planted, or sometimes they are installed near existing established plants with a thin layer of back-filled soil between the panes and the intact soil profile. Precautions are usually taken to thermally insulate the panes and to minimize exposure of the roots to light coming through the panes.

Mapping the individual root growth, making morphological and phenological observations and scoring roots by grid line intercepts provide different measures which can be useful in addressing questions such as how long do individual root elements grow, how long do intact root hairs persist, when do new roots turn color indicating suberization, or at what depths are roots growing at different times of year (e.g. Caldwell and Fernandez, 1975). Apart from manual observations and mapping approaches, time-lapse photography has also been employed (Head, 1965).

16.6.2 Root periscopes

An analogous approach to root growth observation is to use a root periscope peering through a glass or acrylic tube (either round or square in cross-section) which has been previously inserted into the soil. These are sometimes called minirhizotrons. The periscope might be as simple as a tilted mirror lowered into the glass tube and viewed from above with a magnifying glass (Boehm, 1974). The simplicity and economy of such a design has been preserved; however, improvements have been incorporated such as fiber-optic illumination and image magnification using a small telescope in a recent design of Richards (1984) (Fig. 16.4). More elaborate and, of course, more costly approaches involve the use of cameras coupled to endoscopes (Sanders and Brown, 1978; Vos and Groenwold, 1983) or even a miniaturized color video camera (Upchurch and Ritchie, 1984). (The color video system is commercially available from Circon Corp., Santa Barbara, CA, USA.) Use of the more elaborate optical systems provides the opportunity to record images on film or video tape for future processing. Use of a simple stereoscope viewer with sequential photographs taken through an endoscope (van Noordwijk *et al.*, 1985) or the color video camera provide very detailed and clear images, even with the resolution of root hairs.

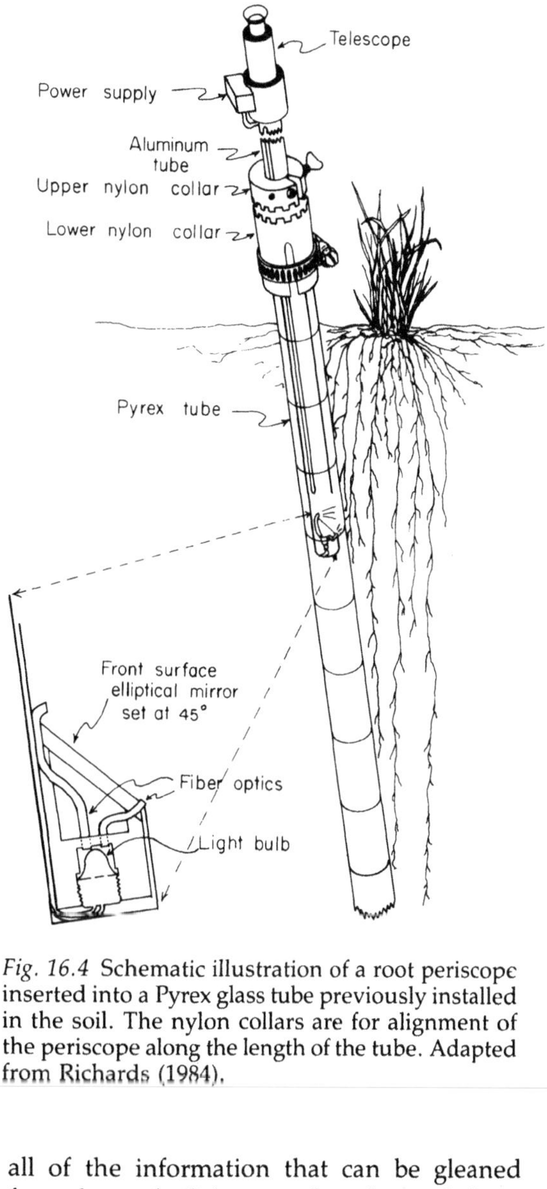

Fig. 16.4 Schematic illustration of a root periscope inserted into a Pyrex glass tube previously installed in the soil. The nylon collars are for alignment of the periscope along the length of the tube. Adapted from Richards (1984).

Since installation of tubes is a much less expensive undertaking than even the simplest rhizotrons, the periscope approach offers the possibility of much more replication and also flexibility in placement location. Providing the optical resolution is sufficient, essentially all of the information that can be gleaned from large facilities can be obtained from minirhizotrons.

Apart from root growth and phenology observations, information from minirhizotrons has also been used to estimate rooting density in the bulk soil. The relatively small obstruction to root growth caused by the presence of periscope tubes, at least those of round cross-section, afford some opportunity to extrapolate root interceptions by the tube to bulk soil rooting density. It has long been recognized that for large barriers such as rhizotron glass panes, usually the number of roots encountering a unit area of the observation pane substantially exceeds the number of roots that pass through an analogous plane in the undisturbed soil (Boehm, 1979). Comparisons of periscope observations and rooting density obtained by core excavations have been made (e.g. Sanders and Brown, 1978; Upchurch and Ritchie, 1983) and show that with sufficient sampling a good relationship can be obtained between periscope and core determinations of rooting density below the upper soil layers. (In the shallower soil depths, the periscopes provide a poor estimate.) If the minirhizotrons are installed at an angle of 45°, the estimates are much better than for vertical minirhizotrons (Bragg and Cannell, 1983). Roots encountering vertical tubes tend to grow along the tubes rather than simply to bypass the tube and continue in the same trajectory. This appears to be largely avoided by the angled minirhizotron tubes. One paper even advocates placing tubes horizontally (Meyer and Barrs, 1985).

Upchurch (1985) has developed relationships between interceptions of roots with periscope tubes and bulk rooting density. This involves the probability of intersections with a population of tubes based on root growth trajectories. The approach was shown to be successful if quite large sample sizes (numbers of tubes) are used. Another result of his approach is an assessment of root growth orientations.

Another method of improving estimates of rooting density involves reduction of the size of the periscope tube so as to minimize the size of the obstacle to root growth. Itoh (1985) used tubes of 6–15 mm in diameter with potted plants to show that decreasing tube size results in progressively less underestimation of rooting density.

16.6.3 General considerations

The choice of barrier materials for rhizotrons and periscopes is discussed in the literature to a limited extent. Yet, we have found little convincing evidence that would argue for a particular material. Glass was first used, but acrylic plastics have been used extensively. Taylor and Boehm (1976) tested acrylic plastic windows in a root observation chamber. They found that root densities within 2 mm of the plastic–soil interface were substantially greater than in the bulk soil. They concluded that the acrylic plastic windows were satisfactory for comparative and phenological measurements, but were not as satisfactory as glass windows for estimating rooting density. However, they do not report in this study a comparable set of measurements with glass windows. Apparently, they were drawing these conclusions simply on the basis of earlier experiments. They did observe that the adhesion between soil and the acrylic plastic windows they were using was poorer than that which they normally found between glass and soil. In contrast to the findings of Taylor and Boehm (1976), Voorhees (1976) reported that roots grew more slowly when against acrylic plastics. He suggested that this may involve the hydrophobic or electrical charge characteristics of acrylic plastics. Unfortunately, as in the study of Taylor and Boehm, the comparisons were made with root growth in barrier-free soil rather than with root growth along a glass–soil interface. While the issue of barrier materials remains an important consideration, more careful, controlled studies are needed to evaluate the materials effectively.

An approach to root growth observations that avoids use of barriers immediately next to growing roots involves a soil monolith perforated by a matrix of small horizontal tunnels. This has been termed a 'perforated soil system' by Bosch (1984) or a 'perforon' by Maas and Gubbels (1986). A soil monolith is either constructed by compacting soil layers or by taking an intact monolith from the soil profile. The monolith is subsequently perforated by boring holes horizontally and the entire volume of soil is then enclosed in a specialized container with access ports to each tunnel and transparent viewing windows. The roots are observed as they grow into or across the perforations by a viewing device rather than to observe root growth against the viewing windows. Such a device would avoid the problems of barrier–soil interfaces and also provides a more three-dimensional view of a root system. The difficulties in preparing these monoliths will probably limit its widespread use. Although the device avoids barriers, questions remain as to the degree to which root growth in the tunnels of the monolith represents root growth in the bulk soil.

With present technology, exposure of roots to some light in these various root observation systems is necessary in the viewing or photographing process. Light has been shown to inhibit or otherwise alter root growth (Voorhees, 1976; Boehm, 1979), but light dose–response relationships have not been elucidated that would be useful in situations as are encountered in root observation chambers or periscope tubes. The best advice that can be presently offered would be to minimize unnecessary light exposure and to realize that some influence of light is probably inevitable. A 'safe light', i.e. a source of illumination that, because of low intensity or spectral composition, can be considered to have no influence on roots is unlikely to be found. Recent demonstrations that growth of *Avena* coleoptiles in physiologically sensitive condition can be influenced by as little as 0.1 nmol photons m^{-2} (Shinkle and Briggs, 1985) encourages one to dismiss the notion of 'safe lights' for plant growth studies.

16.7 ROOT SYSTEM FUNCTION

Direct observation or recovery of roots is typically very expensive in labor and time. Extrapolation of root parameters derived from the analysis of relatively few samples to address questions about ecosystem function is very tenuous. Consequently, a number of indirect methods to assess root system function have been developed. Root activity is inferred from measurements such as the seasonal pattern and depth distribution of soil water extraction, nutrient and isotopically labeled element accumulation in shoots and recovery of soil biota typically associated with roots and rhizosphere processes.

16.7.1 Water uptake: isotope indicators of source water

Soil water can be available at considerable depth in landscape positions where water accumulates, in high-filtration soils or where ground water occurs (Chaney, 1981). In these instances, techniques that can separate the relative contribution of surface soil moisture derived from precipitation from that of deep soil reserves are necessary to understand root function more fully for plants with deep root systems. Determination of the natural abundance of ^{18}O and deuterium (D) isotopes in the transpiration stream of plants may be used to identify the sources of water available to plants.

The oxygen and hydrogen in plant xylem water is derived from soil water and no significant isotopic fractionation is associated with water uptake by roots. Isotopic fractionation processes affect the ^{18}O and D isotopic abundances of precipitation and ground waters such that these waters can be significantly different in isotopic composition. When these conditions occur, it is possible to determine the relative contribution of surface soil and deep soil water sources to the plant from an analysis of the isotopic composition of xylem water.

The natural abundances of ^{18}O and D in precipitation vary by season in temperate climates, a result of seasonal temperature variation (Dansgaard, 1964). Precipitation during cooler winter months has a lower δD than summer precipitation. White *et al.* (1985) determined the relative contribution of ground water to the water budget of tree species growing in an eastern forest (Fig. 16.5). In low-lying areas, plant xylem water had an isotopic signature near that of the available ground water (δD of −65) during periods when the surface soil was dry indicating that surface roots provided a minor source of water to the plant. Shortly after precipitation events, the δD in the xylem water shifted toward the precipitation isotope signature (δD of −20) indicating that the surface soil had become the major source of water to the plant.

In a similar manner, $\delta^{18}O$ measurements of xylem water should also distinguish between surface and deep water sources having different isotopic ratios. Allison and Hughes (1983) examined the ^{18}O and D abundances to depths of 15 m in soil profiles of *Eucalyptus* shrub and *Eucalyptus* shrub areas recently converted to wheat in Australia. Isotope ratios during the summer differed from the weighted annual precipitation average by −9 $\delta^{18}O$ and −40 δD. This was due to isotopic fractionation associated with water loss from the soil surface and water vapor movement in the soil profile. During winter, the isotope ratio of the surface soil in the cropped land increased indicating infiltration of precipitation with a higher δD and $\delta^{18}O$. However,

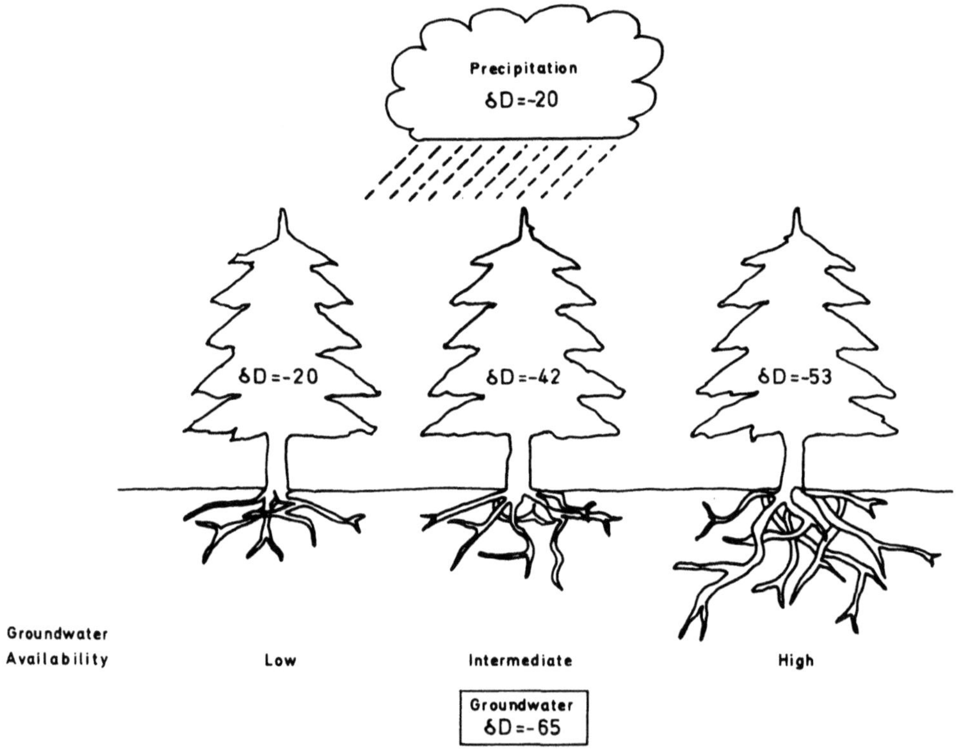

Fig. 16.5 Measurement of the natural abundance of deuterium (δD) as an indicator of water sources to plants. The δD of sap for white pine (*Pinus strobus*) trees is shown from sites having low, intermediate and high ground water availability 3 days after a rainfall which had a δD of −20. Plants without access to ground water have a δD value comparable to recent rainfall while plants with access to ground water have δD values intermediate between the two water sources. Data are from White *et al.* (1985).

there was little isotopic shift near the surface for the mallee shrubs suggesting most of the precipitation was intercepted by the plant canopies and did not infiltrate the soil. Their results show that surface and deep soil water have different isotopic abundances, which is the condition necessary to identify plant sources of water from xylem water isotope measurements.

Measurement of the D and ^{18}O abundance of soil water or xylem are made using a stable isotope ratio mass spectrometer. For D, water is reduced to H_2 by passing the sample over uranium turnings at high temperature (DeNiro, 1978). To determine $\delta^{18}O$, the water sample is equilibrated with a small volume of CO_2 followed by repeated cryogenic distillation to free the CO_2 which has come to isotopic equilibrium with the sample (Epstein and Mayeda, 1953) (see also Chapter 13).

Radioactive isotopes can also be used to assess root activity. Lewis and Burgy (1964) injected tritiated water into wells as deep as 38 m in a California oak woodland. Leaf water was extracted and analyzed for tritium activity using scintillation counting techniques. Plant uptake was detected indicating root activity to at least 38 m in this woodland.

16.7.2 Water uptake: soil water depletion

The pattern of water depletion from the soil profile is an indicator of water uptake and, therefore, root activity. Water loss from soil results from evaporation and root uptake. Evaporation is significant near the soil surface (upper 1 m) but is not a significant pathway for water loss from greater depths. Thus, water depletion from subsurface soil can be attributed to root activity. There are many approaches for measuring soil water content including direct and indirect methods (Gardner, 1965; Hanks and Ashcroft, 1980). These methods are discussed in greater detail in Chapter 3.

Changes in soil water attributable to root activity can be determined by coring and gravimetric determination of soil water content. Soil samples should be sealed immediately after collection to avoid water loss and should be dried thoroughly at 105°C. This approach is time consuming and many cores may be required in heterogeneous soils. However, close agreement between changes in gravimetric water content and changes in root density have been reported (Boehm *et al.*, 1977).

The methods most useful for assessing root activity are those where a measuring device is placed in the soil and soil water status can be measured nondestructively during the growing season. Soil psychrometers, soil moisture resistance blocks and the neutron probe meet these requirements.

Soil water potential can be determined by measuring changes in the resistance of electrodes embedded in porous blocks made of gypsum or fiberglass (Gardner, 1965). A correlation of soil water potential with soil water content is needed if plant water use is to be determined. Arkley (1981) used this method with fiberglass moisture-temperature cells to study water extraction patterns by a mixed-conifer forest in southern California. The sensors performed well for the entire 3.5-year study. Sensors constructed of gypsum are less resistant to the damaging effects of soil water and salts. Care should be taken to ensure good contact with the soil during installation.

Thermocouple psychrometers provide information on the water potential of a soil by measuring the vapor pressure (Brown, 1970; Savage and Cass, 1984; Wiebe *et al.*, 1971). Small thermocouple psychrometers which can be inserted into the soil are available commercially (Wescor Inc. and J.D. Merrill, both in Logan, UT, USA). Psychrometers with screen cages have numerous advantages over those with ceramic cups. It is very important to minimize thermal gradients within the psychrometer since psychrometers are very sensitive to small temperature differences (Wiebe *et al.*, 1977). Despite this problem psychrometers have been used successfully in desert soils to study soil water extraction (Schlesinger *et al.*, 1987). Psychrometers have also been used to indirectly assess water transfer between soil layers through roots (Richards and Caldwell, 1987).

Soil water content can be measured using a neutron moisture meter (neutron probe) by the neutron attenuation technique (Gardner, 1965; Greacen, 1981). A source of fast neutrons is lowered into an access tube in the soil. Fast neutrons, generated by an americium–beryllium source, that encounter hydrogen associated with soil water become thermalized (slow neutrons). A proportion of these are scattered back to the probe and are counted by a detector adjacent to the source of the fast neutrons. In most soils, water is the only appreciable source of H. Thus the number of slow neutrons reaching the detector is related to volumetric soil water content by a calibration curve. However, soils high in organic matter and B, Fe or Cl give erroneous results. Furthermore, readings in the upper 20 cm of soil are not reliable since back-scattering is reduced. The advantage of this approach is that repeated measurements can be made on a relatively large soil volume. The neutron access tubing of aluminum or PVC should be installed with a minimum of soil disturbance. Rambal (1984) examined seasonal water extraction to 4.5 m in a *Quercus coccifera* evergreen scrub system for seven consecutive years using the neutron probe technique. He developed a set of calibration curves assuming different volumes of rock in the soil to correct for the high variability in rock distribution. Repeated measurements of the same soil profiles provided detailed information on seasonal water extraction and hence root activity.

For installation of neutron probe access tubes to 10 m depths, a portable rotary three-way combination drill works well in the rugged terrain of southern California (Osborne and Pelishek, 1961). The soil can be brought up by wet or dry vacuum, depending on soil water content.

Other means of nondestructive soil moisture measurement are also available including tensiometers (Gardner, 1965), fiber-optic probes (Alessi and Prunty, 1986) and time-domain reflectometry (Topp and Davis, 1985). Fiber-optic probes are at present quite new and have not, to our knowledge, been used in the field. Time-domain reflectometry has been used in the field, but we feel still requires further evaluation for specific purposes. Probes may be installed vertically or horizontally (Fig. 16.6) and have the advantage of providing integrated values along the length of the probes. Air gaps next to the probes in the soil do, however, present a problem.

Perhaps the newest technology to emerge with the potential to examine the *in situ* distribution of roots in soil and to assess directly the flow of water within a root system is proton (^1H) NMR imaging (Bottomley *et al.*, 1986). Systems developed for medical research create images of static and radiofrequency magnetic fields that can be employed to obtain maps of mobile water distribution within the plant. Bottomley *et al.* (1986) examined potted *Vicia faba* plants using ^1H NMR. They evaluated several common potting media and a field soil for image distortion and signal loss. The greatest distortion was found with the finest textured material, a field soil (fine–loamy siliceous, thermic Typic Hapludult). Better results were obtained with coarse potting media such as peatlite and perlite. Uptake and movement of water within the plant were successfully determined using a $CuSO_4.5H_2O$ solution with Cu^{2+} as a tracer (Cu^{2+} is paramagnetic) and with water application alone to a plant growing in a dry pot. The advantage of the NMR technique is that it is entirely nondestructive and uses nonionizing radiation which can permeate most soil media and does not have any known effects on plant

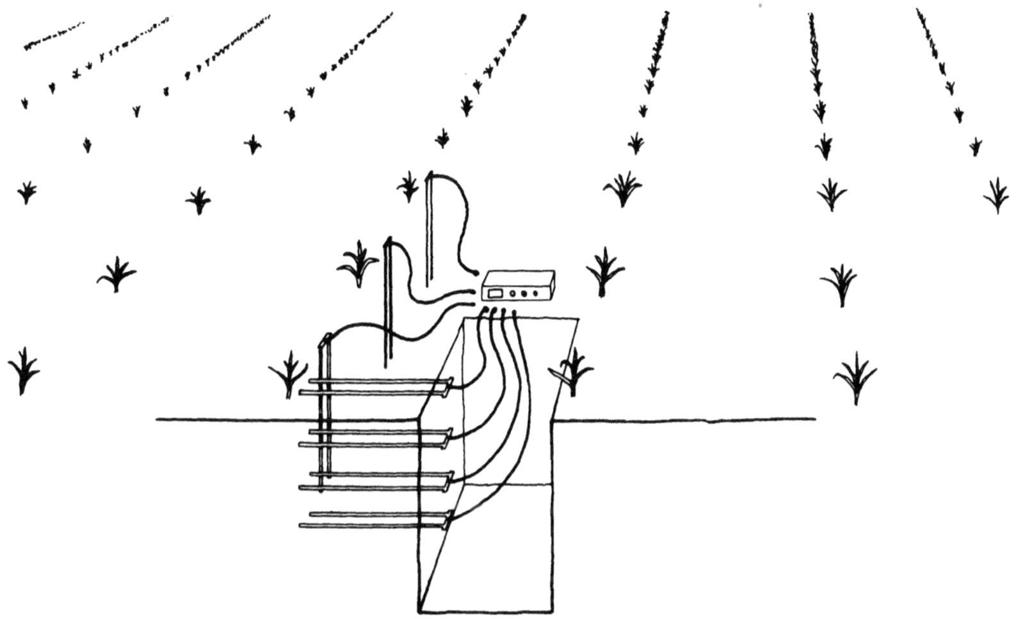

Fig. 16.6 Diagram of vertical and horizontal installation of time-domain reflectometry probes in an agricultural field. The small case represents the electronic package for determination of dielectric constants. Adapted from Topp and Davis (1985).

function. Direct sensing of the soil environment in the field will require considerable advancement in imaging technology. However, early laboratory studies indicate that this is a promising approach and may ultimately develop into a field technique for studying surface roots.

Computer-assisted tomography (CAT) to X-ray attenuation measurements is widely used in medicine. It appears to be promising in the determination of water depletion zones near individual roots in soil (Hainsworth and Aylmore, 1986).

16.7.3 Using rare chemical and radioactive tracers to assess root activity

As mentioned in the previous section (16.7.1), and will be discussed later (16.7.5), the natural abundance of stable isotopes can be used to infer sources of water or nitrogen for plant uptake. However, tracers applied to plants or soils can also be useful tools. Tracers have been frequently used to assess the extent of plant root systems by labeling a plant and sampling the soil volume for the tracer, or, conversely, by labeling the soil in different locations and determining how much of the label appears in different plants. In either case, one assumes that the activity of the roots and the distribution of the label in the root system is reasonably uniform and, furthermore, that the label is reasonably immobile in the soil, but remains available for plant uptake. Of course, the tracer should have a negligible effect on plant metabolism. Many general aspects of such labeling approaches have been discussed elsewhere (Boehm, 1979; Nye and Tinker, 1977).

The goals of such experiments are often simply to determine where active roots are located in the soil. However, questions can also be addressed concerning where different species in a mixture obtain their resources

(Fitter, 1986), the competitive ability of species for certain nutrients (see Section 16.7.6) or transfer of mineral nutrients from one plant to another (Chiariello *et al.*, 1982).

The most commonly used tracers are radioactive isotopes, especially ^{32}P (Nye and Tinker, 1977). Stable isotopes, especially ^{15}N, have also been employed (see Section 16.7.4) and elements normally not abundant in most soils such as lithium, boron, rubidium and strontium, have also been used as tracers (Boehm, 1979; Fitter, 1986). Some of these, such as lithium and boron, can have effects on plant metabolism even in small quantities.

Sometimes, only qualitative information is being sought with tracers, such as whether or not roots of an individual plant have accessed a tracer in a labeled soil location, or whether a tracer has been transferred from one plant to another. In such cases, the uniformity of tracer uptake by roots or translocation of the tracer within the plant is not of great concern. However, when more quantitative information is desired, the assumptions concerning uniformity can become a major limitation in the interpretation of results. Roots of the same plant differ in absorptive capacity and plants allocate different elements preferentially to different organs and tissues of different developmental status. Furthermore, if a particular root accesses a tracer placed in a certain soil location, this root may preferentially serve only a segment of the shoot system. Therefore to obtain quantitative information from tracer studies, it is necessary to sample all plant parts adequately and establish total tracer pool quantities in plants. The problem is analogous to estimating total labile carbon reserves in plants. It is not sufficient to only know concentrations of sugars and starches in different plant parts, but the mass of these organs must also be established in order to estimate pools. Obtaining such information for total tracer pools can be a sizable undertaking. If it cannot be adequately conducted, one must be cautious in the interpretation of results.

General aspects of tracer labeling in soils and plants are covered in Boehm (1979). Recovery and counting of radioactive isotopes is treated in Wang *et al.* (1975), and Fitter (1986) gives techniques for tracers such as strontium and lithium.

16.7.4 Nitrogen uptake

Nitrogen is the single element which most frequently limits plant productivity in natural systems (Clark and Rosswall, 1981). Numerous studies have examined N uptake by plants by applying nitrogenous compounds to soil and measuring plant N uptake and growth response relative to nonfertilized plants. The nitrogen applied to the soil can be enriched or depleted in the stable ^{15}N isotope and can be applied to the soil at various depths and distances from the plants of interest. Incorporation of the isotopic label into aboveground tissues is evidence of root activity at the particular location or depth. Small amounts of highly enriched ^{15}N can be applied to the soil to minimize the stimulating effects of N on plant growth. This approach can be used to monitor N movement and uptake in natural systems.

There are several difficulties in using applications of stable N isotopes to soil for studying root function. Most soils have only a small fraction of total soil N in solution. This creates an isotopic dilution problem when adding labeled N to the system. Unlabeled ^{14}N constantly enters the soil solution by mineralization of organic matter and by exchange with other cations on the organic and mineral soil particles. Labeled ^{15}N may leave the soil solution due to many factors besides plant uptake, including immobilization by soil microbes and cation exchange. Consequently, the isotopic composition of the plant available N pool tends to decline with time when ^{15}N-enriched compounds are added to soil. Since the spatial and temporal variability in soil N distribution and availability are high, it is not possible to label the entire rooting zone of a plant uniformly in the field. This, coupled with the high mobility of N (especially NO_3^-) in soil, complicates the interpretation of stable N isotope studies in plant–soil systems.

Techniques for N isotope analysis of plant and soil material are similar. Nitrogen in the sample must be converted to N_2 for analysis by mass spectrometry (Bremner, 1965a). This can be accomplished by standard Kjeldahl digestion (wet oxidation) followed by steam distillation or microdiffusion procedures to recover the plant or soil N as NH_4^+. Ammonium is then reduced to N_2 by reaction with sodium hypobromite in the absence of air and introduced into a stable isotope mass spectrometer. Alternatively, the Dumas method (dry oxidation) can be employed (Bremner, 1965b). For this method, the sample is heated to a high temperature with a copper oxide catalyst and the gases liberated are purified over hot Cu to reduce oxides of N to N_2 and then over copper oxide to reduce CO to CO_2. Nitrogen isotope analysis can also be accomplished by NMR and emission spectroscopy, but these approaches are not frequently used in field studies.

Nitrogen isotopes have been widely used in agricultural studies to investigate fertilizer use efficiency and symbiotic N_2 fixation (Hauck and Bystrom, 1970). Nitrogen tracers have been less frequently employed in ecological studies. The relatively high costs of ^{15}N-enriched materials and of the instrumentation to determine N isotopic composition have limited the use of N isotopes in ecological field studies. However, recent advances in mass spectrometry including the development of automated sampling systems have decreased the per sample cost of isotopic analyses. These advances and an increased interest in the use of stable isotopes in ecological research (Ehleringer *et al.*, 1986) have led to the establishment of several commercial laboratories which can determine the isotopic abundance of samples at reasonable cost (e.g. Isotope Services Inc., Los Alamos, New Mexico, USA) (McInteer and Montoya, 1980).

16.7.5 Symbiotic nitrogen fixation

Some plants can utilize atmospheric N_2 by forming a symbiotic association with soil organisms. Legumes develop root nodules infected with *Rhizobium* and *Bradyrhizobium* while nonlegumes are nodulated by *Frankia*. Methods to assess symbiotic N_2 fixation in the field have been reviewed (Burris, 1974; Bergersen, 1980a; Knowles, 1980; LaRue and Patterson, 1981; Rennie and Rennie, 1983; Silvester, 1983; Hauck and Weaver, 1986; Shearer and Kohl, 1986). Traditional methods to assess root nodule activity require quantitative nodule recovery and an assay for nitrogenase activity by the acetylene reduction method (Hardy *et al.*, 1973) or exposure of the nodules to $^{15}N_2$ and isotopic analysis of the nodules and/or plant tissues for the incorporation of ^{15}N (Bergersen, 1980b).

Nitrogenase, which reduces atmospheric N_2 to NH_3, also reduces acetylene to ethylene. Thus, the exposure of root nodules or a root system to acetylene with the subsequent analysis of the evolved ethylene by gas chromatography is a sensitive and inexpensive assay of nitrogenase activity. However, extrapolation of acetylene reduction activity (ARA) to the quantity of nitrogen fixed by a root system is difficult. Theoretically, three moles of ethylene are evolved for every mole of N_2 fixed. This ratio has been shown to vary and the conversion factor for a particular system should be determined using $^{15}N_2$ (Bergersen, 1980b).

To quantify N_2 fixation by measuring ARA, it is necessary to determine nodule mass per plant or per unit area. This is difficult even for shallow rooted herbaceous plants and is virtually impossible for deeply rooted woody plants since symbiotic activity may occur at several meters depth (Virginia *et al.*, 1986). Nodule activity is known to vary diurnally and seasonally. Furthermore, it is not feasible to recover nodules associated with deep roots or from rocky soils under most conditions.

Another approach is available when certain conditions are met. Measurements of the natural abundance of ^{15}N in soil and plant tissues can provide information on the N sources available to the plant and may be an indicator of root depth. Most soils have a higher ^{15}N abundance than atmospheric N_2 (Shearer *et al.*, 1978) ranging from −5 to +15 $\delta^{15}N$. Soil ^{15}N enrichment results from isotopic fractionation during N transformations in soil which usually favor the movement and loss of the lighter ^{14}N isotope (Delwiche and Steyn, 1970). Since soil N and the atmosphere often differ significantly in natural ^{15}N abundance, it is possible to determine from measurements of plant ^{15}N abundance the relative importance of these two sources of N to plants that are capable of symbiotic N_2 fixation.

Several studies have demonstrated that N_2-fixing plants can be distinguished from nonfixing plants growing on the same site based on natural ^{15}N abundance (Shearer and Kohl, 1978; Delwiche *et al.*, 1979; Virginia and Delwiche, 1982; Shearer *et al.*, 1983). Successful application of the method requires that soil N have a significantly different ^{15}N abundance from the atmosphere and that isotopic fractionation effects associated with the uptake and translocation of N within the plant do not alter or mask the differences between fixing and nonfixing plants. Since the ^{15}N abundance of available soil N is difficult to determine, the ^{15}N abundance of non-N_2-fixing control or reference plants is determined as a proxy for soil ^{15}N abundance. The reference plants integrate temporal and spatial variation in soil isotopic composition. Ideally the fixing and reference plants should be of similar growth form and have comparable rooting structure (Virginia and Delwiche, 1982).

The difference in the ^{15}N abundance between the N_2-fixing and nonN_2-fixing reference plants growing on a site is directly proportional to the amount of N fixed by the symbiotic plant, assuming no isotopic fractionation during N_2 fixation. However, often there is a small but measurable isotopic fraction effect associated with the fixation of N_2. The fractional contribution of fixed N to the plant (FN_{dfa}, or 'fixed N derived from the atmosphere') can be calculated from an isotope dilution expression:

$$FN_{dfa} = (\delta^{15}N_o - \delta^{15}N_t)/(\delta^{15}N_o - \delta^{15}N_a)$$

where $\delta^{15}N_o$ is the ^{15}N abundance of a suitable nonfixing control plant, $\delta^{15}N_t$ is the ^{15}N abundance of the N_2-fixing plant and $\delta^{15}N_a$ is the ^{15}N abundance of fixed N after it has been incorporated in plant tissue. The value of $\delta^{15}N_a$ can be obtained by growing the plant in an N-free medium.

Shearer *et al.* (1983) have successfully applied the natural abundance method to assess N_2 fixation by the deeply rooted desert woody legume *Prosopis glandulosa* where nodulation occurs at depths greater than 4 m (Virginia *et al.*, 1986) (Fig. 16.7). The natural abundance method is the only feasible way to study deep root function in such systems. The chief advantage of this method is that an analysis of easily collected aboveground tissues provides information on the symbiotic activity of the root system. Variation within plants in natural ^{15}N abundance has been observed (Shearer *et al.*, 1980, 1983; Virginia *et al.*, 1984). However, leaves are typically close to the mean isotopic ratio for the entire plant and are the tissue of choice for natural ^{15}N abundance comparisons. This method also provides a time-integrated measure of symbiotic activity by the plant.

Since the natural variation in ^{15}N abundance between fixing and nonfixing plants is typically small (< 10δ ^{15}N), care must be taken during sample preparation to avoid isotopic alteration due to incomplete digestion or cross contamination during distillation. Analysis requires an isotope ratio mass spectrometer with a dual collector to measure the $^{14}N^{15}N$ and $^{14}N^{14}N$ peaks simultaneously.

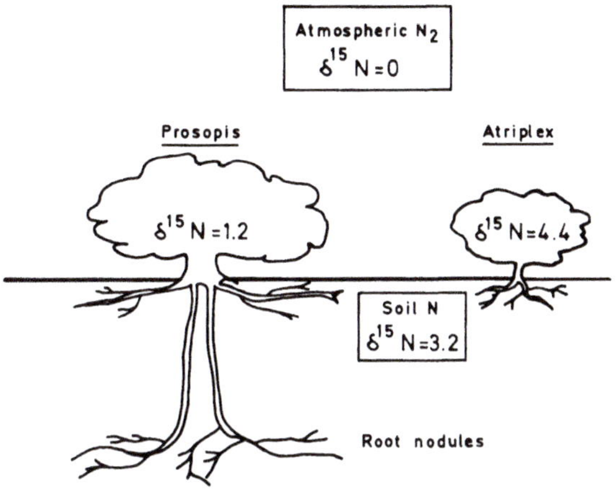

Fig. 16.7 Application of measurements of natural ^{15}N abundance to detect symbiotic N_2 fixation by plants. The $\delta^{15}N$ of the N_2-fixing woody legume mesquite (*Prosopis glandulosa*) and the nonN_2-fixing reference species *Atriplex polycarpa* are shown for Harper's Well in the California Sonoran Desert. The $\delta^{15}N$ of mesquite leaves is between the $\delta^{15}N$ of the atmosphere, soil N and that of the reference species indicating that symbiotic N_2 fixation is a significant source of N to mesquite at this site. Based on data of Shearer *et al.* (1983).

Natural ^{15}N abundance may provide information on rooting depth. Chamise (*Adenostoma fasciculatum*) is a deeply rooted shrub which is common in the chaparral of southern California. The $\delta^{15}N$ of chamise seedlings the first year following fire in Sequoia National Park, California was comparable to shallow-rooted plants growing on the site. (Chamise resprouts having an established deep-root system from before the fire had a ^{15}N comparable to other deeply rooted species as expected.) By the third year after the fire, chamise seedlings and resprouts had an ^{15}N abundance indicating deep rooting for both (Virginia *et al.*, 1989). These data suggest that chamise seedlings reach the rooting depth of mature plants within 3 years after fire. In other systems, the ^{15}N abundance of nonfixing plants has been related to plant growth form, with ^{15}N abundance decreasing with a presumed increase in rooting depth (Virginia and Delwiche, 1982). This relationship deserves further study since indirect methods to assess root distribution in rocky soils are greatly needed.

16.7.6 Use of multiple tracers

Combinations of tracers applied simultaneously or sequentially to soils or plants can be a powerful and efficient approach to questions concerning the location and timing of nutrient acquisition and competition between plants. For example, by applying three different tracers ($SrCl_2$, RbCl and LiCl) at different soil depths, Fitter (1986) was able to determine the depth of major root activity for different species in a grassland at various times of year. Similarly, Christians *et al.* (1981) were able to determine the relative proportion of phosphorus acquired by plants from the soil and from a thatch layer by applying ^{32}P and ^{33}P in these source locations. Such problems can be approached using only one tracer. An example is the work of Goodman and Collison (1982) who labeled replicate monocultures and mixtures of clover and grass at different depths with ^{32}P to determine depth profiles and competitive ability for phosphorus uptake. Use of a single tracer, however, depends in such cases on the availability of a very uniform set of replicate plots as well as an ambitious sampling program. Thus, the advantages of a multiple tracer approach are clear. On the other hand, the assumption that must be met in the multiple tracer approach is that the plant has the same affinity for absorbing and allocating the different tracers used. This assumption is much easier to meet with isotopes of the same element than it is with different elements.

Since isotopes of the same element such as ^{32}P and ^{33}P would also be translocated similarly within the plant, there is no need to determine total tracer pool sizes in order to address many questions as long as one adequately accounts for specific preferential root–shoot associations (see Section 16.7.2).

By determining ratios of ^{32}P and ^{33}P in small tissue samples, one can adequately assess the proportionate contribution from the two sources of P represented by the two isotopes. If only small tissue samples are needed, the same individual plants can be sampled through time which provides another dimension to the assessment. In the case of ^{32}P and ^{33}P, their relatively short half-lives limit this assessment to approximately 50 days following labeling.

A dual-phosphorus-isotope approach was used to determine the competitive effectiveness of different species with a common indicator plant by determining ^{32}P and ^{33}P ratios in tissues of the indicator plant (Fig. 16.8) (Caldwell *et al.*, 1985). A dual isotope approach also allows greater resolution statistically. Further aspects of dual-isotope approaches are discussed in Caldwell and Eissenstat (1987).

16.7.7 Root 'porometers' and chambers for field use

Determining metabolic activity of individual intact roots under field conditions is challenging, to say the least. Not surprisingly, little progress has been made in this area. Roots are often excised in the field and then immediately subjected to incubation and measurements of nutrient absorption capacity (Chapin, 1974), respiratory activity (Holthausen and Caldwell, 1980) or mitotic indexing (Dunsworth and Kumi, 1982). However, little has been done with roots still intact with the remainder of the root system. We have found only two examples.

Glavac and Ebben (1986) described a small root chamber that enclosed a small segment of a root system with soil (Fig. 16.9). This root system segment, which is still attached to the remainder of the root system, is subjected in

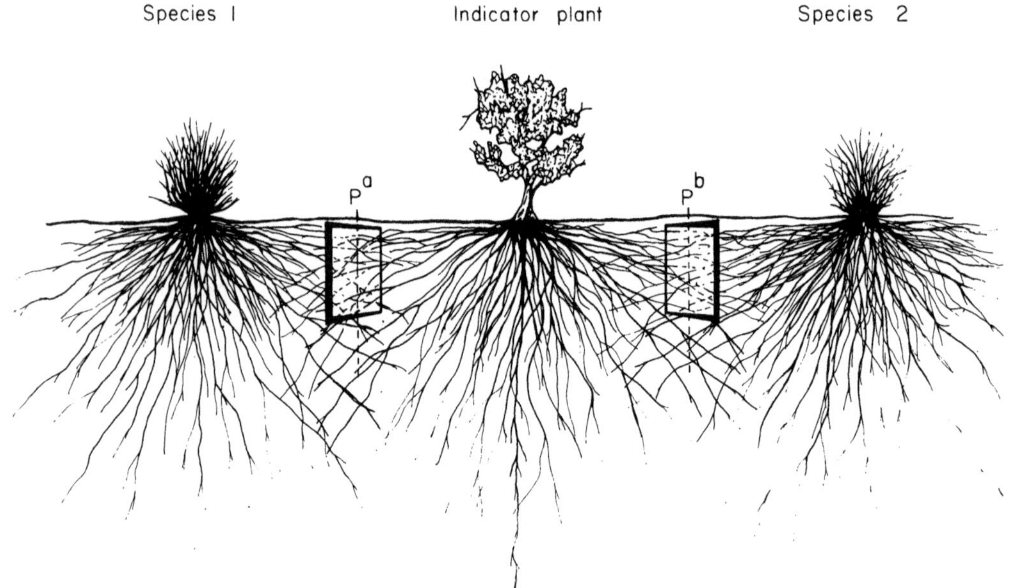

Fig. 16.8 Experimental design for assessing the ability of an indicator plant to acquire phosphorus radioisotopes from two test species. The two isotopes, ^{32}P and ^{33}P, are placed as orthophosphoric acid in a row of small vertical holes previously punched into the soil. The isotope solutions injected into the row of holes form a plane of isotope in the soil interspace between each test species and the indicator plant. Small tissue samples of the indicator plant are then assessed for ratios of the isotopes. Adapted from Caldwell and Eissenstat (1987). See also Caldwell *et al.* (1985).

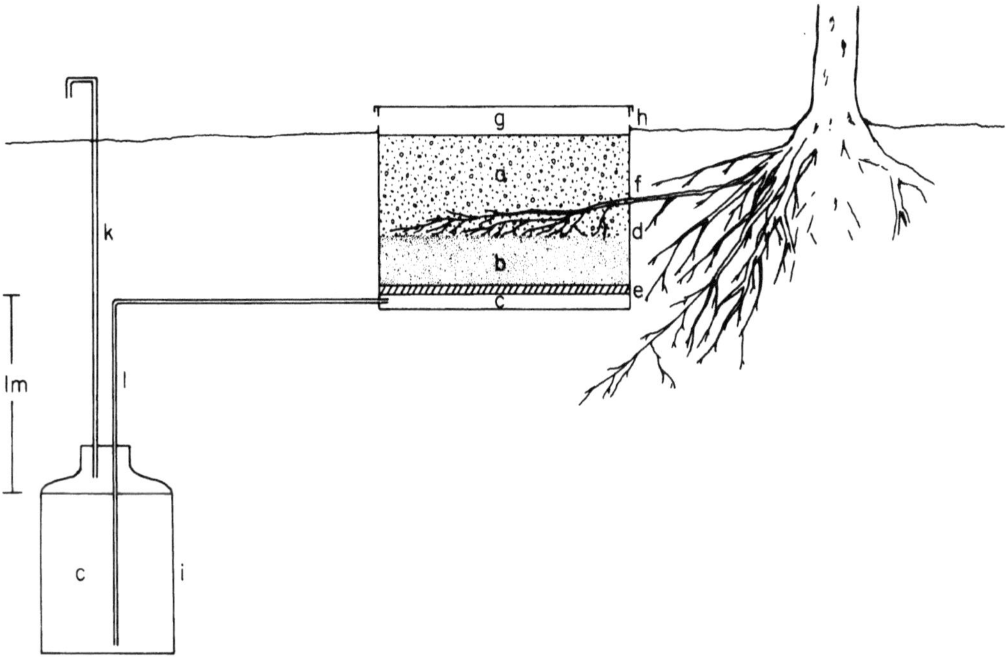

Fig. 16.9 A root chamber used to assess the reaction of roots to different growing media and soil chemicals in the field. (a) Soil with intact tree root, (b) a nonbuffered small-grain sand, (c) nutrient solution, (d) fine PVC mesh netting (1 mm^2 pore diameter), (e) ceramic plate, (f) side opening, (g) air ventilation space, (h) opening for ventilation, (i) flask as nutrient solution reservoir, (k) pressure equilibration tube and (l) interconnecting tubing. Adapted from Glavac and Ebben (1986).

the chamber to different nutrient and toxic chemical conditions by a system involving a PVC net, washed sand, a ceramic plate and a remote solution reservoir. Capillary action in the ceramic plate in equilibrium with the remainder of the chamber contents results in a constant supply to the root system of these nutrients and chemicals. This approach has been primarily used to observe morphological and tissue changes following 4-week exposures to different chemical environments. This is particularly germane to the European forest decline problem.

Root respiration has been determined under field conditions with a porometer system (Sisson, 1983). Individual roots, free of soil and still intact with the remainder of the root system, have been placed in a temperature-controlled root porometer for CO_2 exchange measurements. Although the roots must still be perturbed somewhat in order to brush away the soil and place them in the porometer, this approach appears to be the only presently feasible way to determine respiration of intact roots in the field. There are several field approaches to assessing root and soil respiration together for unperturbed systems. These include determination of CO_2 profiles within the soil, or assessing soil CO_2 efflux from the soil surface using static- or dynamic-chamber methods, or aerodynamic methods of computing CO_2 flux above the soil surface. These approaches were compared by de Jong *et al.* (1979). None of these techniques, however, allows the separation of root and soil respiration.

16.8 ROOT ASSOCIATIONS

16.8.1 Mycorrhizae

The roots of most plants form a mutualistic association with fungi known as mycorrhizae (Mosse *et al.*, 1981). Mycorrhizal fungi act to increase absorption of nutrients by increasing the effective exploratory surface area in addition to affecting plant physiological activity. Mycorrhizal plants have been shown to have greater tolerance to drought, toxic metals, saline soil, root pathogens, high soil temperatures and adverse pH than plants not infected with mycorrhizal fungi (Harley and Smith, 1983). Also, mycorrhizal associations can alter root growth patterns and the mycorrhizal fungi may consume a significant portion of the plant's photosynthate. Because of these important effects on root functioning, it is frequently necessary to assess the mycorrhizal status of plant roots in ecological studies. Members of most terrestrial plant families usually form vesicular–arbuscular mycorrhizae (VAM). Roots infected with VAM fungi show little morphological change and staining is usually required to detect infection. The primary cortex of the root is invaded by inter- and intracellular aseptate fungi of the Endogonaceae. Methods to assess VAM activity usually involve clearing and staining of roots (Kormanik and McGraw, 1982). Reproductive structures (spores) can be directly extracted from soils and counted using a variety of methods. Overall reproductive potential can be estimated by the most probable number (MPN) approach (Porter, 1979). A series of serial dilutions (to 10^{-1}, 10^{-2} or 10^{-3}) of the soil of interest are made using a sterile soil or media as the diluent. A test plant is grown in these soil dilutions and the roots are examined for the presence or absence of mycorrhizal infection. The most probable number of VAM propagules in the sample soil (number/100 g of soil) can be calculated using statistical tables.

Methods for mycorrhizal research have been compiled by Schenck (1982). Roots can be collected using any number of approaches outlined in this chapter (Sections 16.2.2 and 16.2.3). The objective of most studies is to establish if the plant is mycorrhizal and to quantify the extent of this symbiosis within the root system. Studies examining the development of mycorrhizae as related to plant growth may require enumeration of fungal structures such as vesicles, arbuscules, spores and hyphae.

The method of Phillips and Hayman (1970) for clearing and staining roots for VAM determination has become widely adopted. Roots, washed free of soil, are killed and fixed in a FAA solution (formalin–aceto–alcohol). The roots are cleared of cellular contents by treatment with KOH and the VA mycorrhizae are stained with acid fuchsin or a variety of other stains such as Trypan Blue, Sudan IV and Cotton Blue. The Phillips and Hayman (1970) procedure requires the use of highly toxic phenols or chloral hydrate. Kormanik *et al.* (1980) have modified this procedure to eliminate the use of toxic phenols, but this results in reduced clarity of internal structures.

After staining, mycorrhizal activity can be determined by several approaches (Giovannetti and Mosse, 1980). The grid-line intersect method can be used to provide information on total root length and the portion of root length colonized by VAM (Ambler and Young, 1977). Roots are spread on a Petri dish having a marked grid and are viewed with a dissection scope. The presence or absence of mycorrhizal structures is scored at the grid intersections. Root length is determined by the Newman method (Section 16.3). Alternatively, 1 cm long root segments may be mounted on slides and examined microscopically at 100–250 ×. Root sections are scored for the presence of structures of interest and results expressed as percent of VAM root segments. This method can be useful when attempting to quantify structural aspects of the mycorrhizal association since a higher magnification is generally employed. Recent modifications have employed scoring random line intersects at high magnification (e.g. Allen, 1983) on small root segments for fine root systems.

Although clearing and staining procedures are the most commonly used for studying VAM, other methods have been developed. Chitin is a component of the hyphal wall of VAM fungi but is virtually absent from plant tissues. Hepper (1977) developed a colorimetric method to assess VAM infection based on the conversion of chitin to glucosamine. Bethlenfalvay *et al.* (1981) compared histological techniques and the chitin method to measure mycorrhizal infection in soybean. This method may work in potting media without organic matter. However, in soil, amino acids and chitin breakdown residues can interfere with the detection of VAM chitin (Pacovski and Bethlenfalvay, 1982). Also, numerous other root-inhabiting organisms (eg. fungi and nematodes) are composed of chitin which interferes with estimates. In general, we do not recommend chitin assays for VAM in soil.

Ames *et al.* (1982) have reported that VAM arbuscules autofluoresce under ultraviolet light. Other VAM structures such as mycelium and spores do not. Only arbuscles could be detected by this method. This method does have the advantage that the same root examined for VAM infection can also be used for biochemical or nutrient analysis. (Clearing and staining assays are destructive.)

External and internal VAM hyphae can be estimated, but the techniques are labor intensive and require considerable experience in identifying the structures of interest. These include estimating hypha length using membrane filter procedures (e.g. Allen and MacMahon, 1985) and intensive internal sectioning and electron microscopy to estimate internal mass (Toth and Toth, 1982).

The presence and number of VAM mycorrhizal spores and other propagules (soil-borne vesicles, mycelia, infected root segments) in a soil can be of interest to indicate potential reestablishment and reproduction of VAM. Daniels and Skipper (1982) listed methods for extracting and quantifying VAM propagules from soil. Propagules can be removed from soil by a variety of methods (Smith and Skipper, 1979). Elutriators often can be used for this process, and soil biota (e.g. nematodes) can be examined in the same sample (Byrd *et al.*, 1976). However, soil physical conditions and spore types can influence the extraction efficiency of various methods and a variety of techniques should always be initially evaluated (Ianson and Allen, 1986).

If the viability of overall inoculum (consisting of spores, root fragments, etc.) is of interest, indirect MPN soil dilution tests are appropriate. Field soil is mixed with sterile soil as the diluent at 10^{-3} to 10^{-5}. Seedlings of the desired test plant are grown for 6 weeks or more in the soil. Roots are then harvested and scored for the presence of VAM fungi. The MPN of viable inoculum can be determined from statistical tables.

Ectomycorrhizal fungi form a mantle around the surface of plant roots in many tree species, e.g. most conifers, Fagaceae, Dipterocarpaceae and a number of Leguminosae and Myrtaceae, and penetrate the root cortex intercellularly forming a network-like structure, the Hartig net. Infection produces a gross change in root morphology which often can be detected visually. Infected short roots typically show thickening and branching (Harley and Smith, 1983; Schenck, 1982). Before ectomycorrhizal infection can be assessed, roots must be cleaned of adhering soil. Soil cores may require soaking in deflocculating agents to remove soil and organic materials from the roots. The objectives of the study will determine the quantification method (Grand and Harvey, 1982). Root segments can be placed in subjective infection categories based on visual examination to determine the intensity of infection. Infection is expressed as the percent of short roots infected with mycorrhizal fungi. Alternatively, mycorrhizal tips can be counted. Relationships relating tip number to tip weight, surface area or volume can be derived to evaluate mycorrhizal biomass and other ecologically meaningful parameters.

16.8.2 Nitrogen-fixing symbionts

Plants capable of symbiotic N_2 fixation support specialized root structures termed nodules. The nodules contain the N_2-fixing endosymbiont. Often it can be difficult to determine if a root swelling is actually an N_2-fixing nodule. For example, *Artemisia ludoviciana* was considered a symbiotic plant based on the morphological presence of nodule-like structures (Farnsworth and Hammond, 1968). Later more detailed examination revealed that the structures were insect galls (Wullstein and Harker, 1982). Nodules can be identified by assaying for nitrogenase activity (Section 16.7.5). The nodule can be sectioned, stained and examined microscopically for the presence of infected cells (Dart, 1975).

Nodulation may also be confirmed by isolating the endosymbiont. Procedures for isolating rhizobia from legume nodules have been reviewed by Vincent (1970) and Date (1982). Rhizobia can be isolated from surface-sterilized nodules by streaking crushed nodule contents on yeast extract mannitol agar plates. Reinoculation of the host plant with the pure rhizobial culture and subsequent plant growth in an N-free medium is required to confirm that the isolated organism was a rhizobial strain.

A high population density of rhizobia in a soil is indirect evidence of host plant root activity and nodulation at a particular soil depth. Estimation of the total number of rhizobia in a soil is difficult. The most probable number (MPN) soil dilution technique can provide an estimate of host nodulating rhizobia in a soil (Vincent, 1970). Plants are inoculated with aliquots of a serial dilution series of the soil under study. The population density of nodulating rhizobia in the soil sample can be calculated from the proportion of test plants forming nodules at each dilution.

Nonlegumes are nodulated by *Frankia* actinomycetes. The first successful isolation of *Frankia* was reported by Callaham *et al.* (1978). The procedure has been refined but is considerably more complicated than rhizobia isolation and culture. Isolation methods have been reviewed by Burggraaf *et al.* (1981). Methods usually include sucrose sedimentation of the root nodule homogenate and suspension in the top layer of a double-agar layer system. Benson (1982) described a simple method to isolate the endophyte of *Alnus incana* spp. *rugosa* based on rapid filtration and transfer of vesicles to a simple mineral medium.

16.9 CONCLUDING THOUGHTS

Study of plant root systems in the field is clearly difficult, often imprecise, fraught with variability and liable to misleading results. Yet, a variety of approaches are available and new techniques, primarily indirect (e.g. Section 16.7), continue to emerge. Bringing several approaches to bear on the same questions will allow strengths and weaknesses of different techniques and measures to compensate one another. Only 8 years ago, Boehm (1979) concluded that 'revolutionary new root-study methods are not in sight at the present time.' Much has happened since that time and an optimistic view of the possibilities for study of root systems can clearly be taken.

ACKNOWLEDGEMENTS

Information and very helpful review comments from J.H. Richards, M.F. Allen, and D.M. Eissenstat are gratefully acknowledged. Many of the ideas and some of the techniques for root studies resulted from research funded by the National Science Foundation Ecosystem Studies Program (M.M.C.: BSR-8207171 and BSR-8705492; R.A.V.: BSR-8216814 and BSR-8506807).

REFERENCES

Aber, J.D., Melillo, J.M., Nadelhoffer, K.J., McClaugherty, C.A. and Pastor, J. (1985) Fine root turnover in forest ecosystems in relation to quantity and form of nitrogen availability: a comparison of two methods. *Oecologia*, **66**, 317–21.

Alessi, R.S. and Prunty, L. (1986) Soil-water determination using fiber optics. *Soil Sci. Soc. Am. J.*, **50**, 860–3.

Allen, M.F. (1983) Formation of vesicular-arbuscular mycorrhizae in *Atriplex gardneri* (Chenopodiaceae): seasonal response in a cold desert. *Mycologia*, **75**, 773–6.

Allen, M.F. and MacMahon, J.A. (1985) Impact of disturbance on cold desert fungi: Comparative microscale dispersion patterns. *Pedobiologia*, **28**, 215–24.

Allison, G.B. and Hughes, M.W. (1983) The use of natural tracers as indicators of soil-water movement in a temperate semi-arid region. *J. Hydrol.*, **60**, 157–73.

Ambler, J.R. and Young, J.L. (1977) Technique for determining root length infected by vesicular-arbuscular mycorrhizae. *Soil Sci. Soc. Am. J.*, **41**, 551–6.

Ames, R.H., Ingham, E.F. and Reid, C.P.P. (1982) Ultraviolet-induced autofluorescence of arbuscular mycorrhizal root infections: An alternative to clearing and staining methods for assay infections. *Can. J. Microbiol.*, **28**, 351–5.

Ares, J. (1976) Dynamics of the root system of blue grama. *J. Range Manag.*, **29**, 208–13.

Arkley, R.J. (1981) Soil moisture use by mixed conifer forest in a summer-dry climate. *Soil Sci. Soc. Am. J.*, **45**, 423–7.

Belford, R.K., Rickman, R.W., Klepper, B. and Allmaras, R.R. (1986) Studies of intact shoot–root systems of field-grown winter wheat. I. Sampling techniques. *Agron. J.*, **78**, 757–60.

Benson, D.R. (1982) Isolation of *Frankia* strains from alder actinorhizal root nodules. *Appl. Environ. Microbiol.*, **44**, 461–5.

Bergersen, F.J. (ed.) (1980a) *Methods for Evaluating Biological Nitrogen Fixation*, John Wiley and Sons, New York.

Bergersen, F.J. (1980b) Measurement of nitrogen fixation by direct means. In *Methods for Evaluating Biological Nitrogen Fixation* (ed. F.J. Bergersen), John Wiley and Sons, New York, pp. 65–110.

Bethlenfalvay, G.J., Pacovsky, R.S. and Brown, M.S. (1981) Measurement of mycorrhizal infection in soybeans. *Soil Sci. Soc. Am. J.*, **45**, 871–5.

Boehm, W. (1974) Mini-rhizotrons for root observations under field conditions. *Zeitschr. Acker-Pflanzenbau*, **140**, 282–7.

Boehm, W. (1977) Development of soybean root systems as affected by plant spacing. *Zeitschr. Acker-Pflanzenbau*, **144**, 103–12.

Boehm, W. (1979) *Methods of Studying Root Systems*, Springer-Verlag, New York.

Boehm, W., Maduakor, H. and Taylor, H.M. (1977) Comparison of five methods for characterizing soybean rooting density and development. *Agron. J.*, **69**, 415–19.

Bosch, A.L. (1984) A new root observation method: the perforated soil system. *Acta Oecol., Oecol. Plant.*, **5**, 61–74.

Bottomley, P.A., Rogers, H.H. and Foster, T.H. (1986) NMR imaging shows water distribution and transport in plant root systems *in situ*. *Proc. Natl. Acad. Sci. U.S.A.*, **83**, 87–9.

Bragg, P.L. and Cannell, R.Q. (1983) A comparison of methods, including angled and vertical minirhizotrons, for studying root growth and distribution in a spring oat crop. *Plant Soil*, **73**, 435–40.

Bremner, J.M. (1965a) Isotope-ratio analysis of nitrogen in nitrogen-15 tracer investigations. In *Methods of Soil Analysis. Part 2. Chemical and Microbiological Properties. Agronomy No.* 9 (eds C.A. Black *et al.*), American Society of Agronomy, Madison, pp. 1256–86.

Bremner, J.M. (1965b) Total nitrogen. In *Methods of Soil Analysis. Part 2. Chemical and Microbiological Properties. Agronomy No.* 9 (eds C.A. Black *et al.*), American Society of Agronomy, Madison, pp. 1149–78.

Brown, R.W. (1970) Measurement of soil water potential with thermocouple psychrometers: construction and application. USDA Forest Service Research Report INT-80.

Burggraaf, A.J.P., Quispel, A., Tak, T. and Valstar, J. (1981) Methods of isolation and cultivation of *Frankia* species from actinorhizas. *Plant Soil*, **61**, 157–68.

Burris, R.H. (1974) Methodology. In *The Biology of Nitrogen Fixation* (ed. A. Quispel), Elsevier, Amsterdam, pp. 9–33.

Byrd, D.W. Jr., Barker, K.R., Ferris, H., Nusbaum, C.J., Griffin, W.E., Small, R.H. and Stone, C.A. (1976) Two semi-automatic elutriators for extracting nematodes and certain fungi from soil. *J. Nematol.*, **8**, 206–12.

Caldwell, M.M. (1987) Competition between root systems in natural communities. In *Root Development and Function* (eds P.J. Gregory, J.V. Lake and D.A. Rose), Cambridge University Press, Cambridge, pp. 167–85.

Caldwell, M.M. and Camp, L.B. (1974) Belowground productivity of two cool desert communities. *Oecologia*, **17**, 123–30.

Caldwell, M.M. and Eissenstat, D.M. (1987) Coping with variability: Examples of tracer use in root function studies. In *Plant Response to Stress – Functional Analysis in Mediterranean Ecosystems* (eds J.D. Tenhunen, F. Catarino, O.L. Lange and W.C. Oechel), Springer-Verlag, Heidelberg, pp. 95–106.

Caldwell, M.M., Eissenstat, D.M., Richards, J.H. and Allen, M.F. (1985) Competition for phosphorus: Differential uptake from dual-isotope-labeled soil interspaces between shrub and grass. *Science*, **229**, 384–6.

Caldwell, M.M. and Fernandez, O.A. (1975) Dynamics of Great Basin shrub root systems. In *Environmental Physiology of Desert Organisms* (ed. N.F. Hadley), Dowden, Hutchinson & Ross, Stroudsburg, PA, pp. 38–51.

Caldwell, M.M., Richards, J.H., Manwaring, J.H. and Eissenstat, D.M. (1987) Rapid shifts in phosphate acquisition show direct competition between neighbouring plants. *Nature, London*, **327**, 615-16.

Caldwell, M.M., White, R.S., Moore, R.T. and Camp, L.B. (1977) Carbon balance, productivity and water use of cold-winter desert shrub communities dominated by C_3 and C_4 species. *Oecologia*, **29**, 275–300.

Callaham, D., Torrey, J.G. and Del Tredici, P. (1978) Isolation and cultivation *in vitro* of the actinomycete causing root nodulation in *Comptonia*. *Science*, **199**, 899–902.

Chaney, W.R. (1981) Sources of water. In *Water Deficits and Plant Growth* (ed. T.T. Kozlowski), Academic Press, New York, Vol. VI, pp. 1–47.

Chapin, III, F.S. (1974) Morphological and physiological mechanisms of temperature compensation in phosphate absorption along a latitudinal gradient. *Ecology*, **55**, 1180–98.

Chiariello, N., Hickman, J.C. and Mooney, H.A. (1982) Endomycorrhizal role for interspecific transfer of phosphorus in a community of annual plants. *Science*, **217**, 941–3.

Christians, N.E., Karnok, K.J. and Logan, T.J. (1981) Root activity in creeping bentgrass thatch as measured by ^{32}P and ^{33}P. *Commun. in Soil Sci. Plant Anal.*, **12**, 765–74.

Clark, F.E. and Rosswall, T. (eds) (1981) *Terrestrial nitrogen cycles. Processes, ecosystem strategies and management impacts. Ecol. Bull.*, Stockholm No. **33**, 714 pp.

Costigan, P.A., Rose, J.A. and McBurney, T. (1982) A microcomputer based method for the rapid and detailed measurement of seedling root systems. *Plant Soil*, **69**, 305–9.

Daniels, B.A. and Skipper, H.D. (1982) Methods for the recovery and quantitative estimation of propagules from soil. In *Methods and Principles of Mycorrhizal Research* (ed. N.C. Schenck), American Society of Phytopathology, St. Paul, MN, pp. 29–35.

Dansgaard, W. (1964) Stable isotopes in precipitation. *Tellus*, **16**, 436–68.

Dart, P.J. (1975) Legume root nodule initiation and development. In *The Development and Function of Plant Roots* (eds J.G. Torrey and D.T. Clarkson), Academic Press, New York, pp. 467–506.

Date, R.A. (1982) Collection, isolation, characterization and conservation of *Rhizobium*. In *Nitrogen Fixation in Legumes* (ed. J.M. Vincent), Academic Press, New York, pp. 95–109.

Delwiche, C.C. and Steyn, P.L. (1970) Nitrogen isotope fractionation in soils and microbial reactions. *Environmental Science and Technology*, **4**, 927–35.

Delwiche, C.C., Zinke, P.J., Johnson, C.M. and Virginia, R.A. (1979) Nitrogen isotope distribution as a presumptive indicator of nitrogen fixation. *Bot. Gaz.*, **140**, 65–9.

DeNiro, M.J. (1978) The effects of different methods of preparing cellulose nitrate on the determination of D/H ratios of non-exchangeable hydrogen of cellulose. *Earth Planet. Sci. Lett.*, **54**, 177–85.

Drew, M.C. and Saker, L.R. (1980) Assessment of a rapid method, using soil cores, for estimating the amount and distribution of crop roots in the field. *Plant Soil*, **55**, 297–305.

Dunsworth, B.G. and Kumi, J.W. (1982) A new technique for estimating root system activity. *Can. J. For. Res.*, **12**, 1030–2.

Ehleringer, J.R., Rundel, P.W. and Nagy, K.A. (1986) Stable isotopes in physiological ecology. *TREE*, **1**, 42–5.

Epstein, S. and Mayeda, T. (1953) Variation of ^{18}O content of waters from natural sources. *Geochim. Cosmochim. Acta*, **4**, 213–24.

Fabiao, A., Persson, H.A. and Steen, E. (1985) Growth dynamics of superficial roots in Portuguese plantations of *Eucalyptus globulus* Labill, studied with a mesh bag technique. *Plant Soil*, **83**, 233–42.

Farnsworth, R.B. and Hammond, M.W. (1968) Root nodules and isolation of endophyte on *Artemisia ludoviciana*. *Proc. Utah Acad. Sci.*, **45**, 182–8.

Fitter, A.H. (1982) Morphometric analysis of root systems: application of the technique and influence of soil fertility on root system development in two herbaceous species. *Plant, Cell Environ.*, **5**, 313–22.

Fitter, A.H. (1986) Spatial and temporal patterns of root activity in a species-rich alluvial grassland. *Oecologia*, **69**, 594–9.

Fusseder, A. (1983) A method for measuring length, spatial distribution and distances of living roots *in situ*. *Plant Soil*, **73**, 441–5.

Fusseder, A. (1985) Verteilung des Wurzelsystems von Mais im Hinblick auf die Konkurrenz um Makronahrstoffe. *Zeitschr. Pflanzenernaehr. Bodenkunde*, **148**, 321–34.

Gardner, W.H. (1965) Water content. In *Methods of Soil Analysis. Part 1. Physical and Mineralogical Properties including Statistics of Measurement and Sampling. Agronomy No. 9* (eds C.A. Black *et al.*), American Society of Agronomy, Madison, pp. 82–127.

Giovannetti, M. and Mosse, B. (1980) An evaluation of techniques for measuring vesicular-arbuscular mycorrhizal infection in roots. *New Phytol.*, **84**, 489–500.

Glavac, V. and Ebben, U. (1986) Die Wurzelkammer, eine einfache Einrichtung zur experimentellen Nachprufung der Bodentoxizität an ausgewachsenen Baumen im Freiland. *Angew. Bot.*, **60**, 95–102.

Goodman, P.J. and Collison, M. (1982) Varietal differences in uptake of ^{32}P labelled phosphate in clover plus ryegrass swards and monocultures. *Ann. Appl. Biol.*, **100**, 559–65.

Grand, L.F. and Harvey, A.E. (1982) Quantitative measurement of ectomycorrhizae on plant roots. In *Methods and Principles of Mycorrhizal Research* (ed. N.C. Schenck), American Society of Phytopathology, St. Paul, MN, pp. 157–64.

Greacen, E.L. (1981) *Soil Water Assessment by the Neutron Method*, CSIRO, Australia.

Hackett, C. (1968) A study of the root system of barley. I. Effects of nutrition on two varieties. *New Phytol.*, **67**, 289–99.

Hainsworth, J.M. and Aylmore, L.A.G. (1986) Water extraction by single plant roots. *Soil Sci. Soc. Am. J.*, **50**, 841–8.

Hanks, R.J. and Ashcroft, G.L. (1980) *Applied Soil Physics. Soil Water and Temperature Applications*, Springer-Verlag, Berlin and Heidelberg, p. 159.

Hansson, A.C. and Andren, O. (1986) Belowground plant production in a perennial grass ley (*Festuca pratensis* Huds.) assessed with different methods. *J. Appl. Ecol.*, **23**, 657–66.

Hansson, A. and Steen, E. (1984) Methods of calculating root production and nitrogen uptake in an annual crop. *Swed. J. Agric. Res.*, **14**, 191–200.

Hardy, R.W.F., Burns, R.C. and Holsten, R.P. (1973) Applications of the measurement of the acetylene-ethylene assay for measurement of nitrogen fixation. *Soil Biol. Biochem.*, **5**, 47–81.

Harley, J.L. and Smith, S.E. (1983) *Mycorrhizal Symbiosis*, Academic Press, New York.

Hauck, R.D. and Bystrom, M. (1970) *^{15}N. A Selected Bibliography for Agricultural Scientists*, Iowa State University Press, Ames, Iowa.

Hauck, R.D. and Weaver, R.W. (eds) (1986) Field measurement of dinitrogen fixation and denitrification. *Soil Sci. Soc. Am. Spec. Publ.*, No. 18, Madison, Wisconsin.

Head, G.C. (1965) Studies of diurnal changes in cherry root growth and nutational movements of apple root tips by time-lapse cinematography. *Ann. Bot.*, **29**, 219–24.

Hellmers, H., Horton, J.S., Juhren, G. and O'Keefe, J. (1955) Root systems of some chaparral plants in southern California. *Ecology*, **36**, 667–78.

Hepper, C. (1977) A colorimetric method for estimating vesicular-arbuscular mycorrhizal infection in roots. *Soil Biol. Biochem.*, **9**, 15–18.

Hoffmann, A. and Kummerow, J. (1978) Root studies in the Chilean matorral. *Oecologia*, **32**, 57–69.

Holthausen, R.S. and Caldwell, M.M. (1980) Seasonal dynamics of root system respiration in *Atriplex confertifolia*. *Plant Soil*, **55**, 307–17.

Ianson, D.C. and Allen, M.F. (1986) The effects of soil texture on extraction of vesicular-arbuscular mycorrhizal fungal spores from arid sites. *Mycologia*, **78**, 164–8.

Itoh, S. (1985) *In situ* measurement of rooting density by microrhizotron. *Soil Sci. Plant Nutr.*, **31**, 653–6.

Jong, E. de, Redmann, R.L. and Ripley, E.A. (1979) A comparison of methods to measure soil respiration. *Soil Sci.*, **127**, 300–6.

Joslin, J.D. and Henderson, G.S. (1984) The determination of percentages of living tissue in woody fine root samples using triphenyltetrazolium chloride. *For. Sci.*, **30**, 965–70.

Kelley, J.M., VanDyne, G.M. and Harris, W.F. (1974) Comparison of three methods of assessing grassland productivity and biomass dynamics. *Am. Midland Nat.*, **92**, 357–69.

Kelley, O.J., Hardman, J.A. and Jennings, D.S.

(1947) A soil-sampling machine for obtaining two-, three-, and four-inch diameter cores to a depth of six feet. *Soil Sci. Soc. Am. Proc.*, **12**, 85–7.

Knowles, R. (1980) Nitrogen fixation in natural plant communities and soils. In *Methods for Evaluating Biological Nitrogen Fixation* (ed. F.J. Bergersen), John Wiley and Sons, New York, pp. 557–82.

Knievel, D.P. (1973) Procedure for estimating ratio of living to dead root dry matter in root core samples. *Crop Sci.*, **13**, 124–6.

Kormanik, P.P., Bryan, W.C. and Schultz, R.C. (1980) Procedures and equipment for staining large numbers of plant roots for endomycorrhizal assay. *Can. J. Microbiol.*, **26**, 536–8.

Kormanik, P.P. and McGraw, A.-C. (1982) Quantification of vesicular-arbuscular mycorrhizae in plant roots. In *Methods and Principles of Mycorrhizal Research* (ed. N.C. Schenck), American Society of Phytopathology, St. Paul, MN pp. 37–47.

Kummerow, J. (1981) Structure of roots and root systems. In *Mediterranean-Type Shrublands* (eds F. di Castri, D.W. Goodall and R.L. Specht), Elsevier Science Publishers, Amsterdam, pp. 269–88.

Kummerow, J., Krause, D. and Jow, W. (1977) Root systems of chaparral shrubs. *Oecologia*, **29**, 163–77.

Lamont, B. (1983) Root hair dimensions and surface/volume/weight ratios of roots with the aid of scanning electron microscopy. *Plant Soil*, **74**, 149–52.

LaRue, T.A. and Patterson, T.G. (1981) How much nitrogen do legumes fix? *Adv. Agron.*, **34**, 15–38.

Lewis, D.C. and Burgy, R.H. (1964) The relationship between oak tree roots and groundwater in fractured rock as determined by tritium tracing. *J. Geophys. Res.*, **69**, 2579–88.

Litav, M. and Harper, J.L. (1967) A method for studying spatial relationships between the root systems of two neighboring plants. *Plant Soil*, **26**, 389–92.

Maas, M.V.M. and Gubbels, M.E.M.N. (1986) Root development in *Robinia pseudoacacia*: Observation in a perforon root box system. *Abstr. 2nd Int. Legume Conf. Missouri Botanical Garden*, St. Louis, pp. 53–4.

McGowan, M., Armstrong, M.J. and Corrie, J.A. (1983) A rapid fluorescent-dye technique for measuring root length. *Exp. Agric.*, **19**, 209–16.

McInteer, B.B. and Montoya, J.G. (1980) Automation of a mass spectrometer for nitrogen stable isotope analysis, U.S. Department of Energy Report No. LA-UR-80-245, Los Alamos, New Mexico.

Marsh, B.a'B. (1971) Measurement of length in random arrangements of lines. *J. Appl. Ecol.*, **8**, 265–7.

Marshall, J.D. and Waring, R.H. (1985) Predicting fine root production and turnover by monitoring root starch and soil temperature. *Can. J. For. Res.*, **15**, 791–800.

Meyer, W.S. and Barrs, H.D. (1985) Non-destructive measurement of wheat roots in large undisturbed and repacked clay soil cores. *Plant Soil*, **85**, 237–47.

Melhuish, F.M. and Lang, A.R.G. (1969) A new technique for estimating diameter, total length and surface area of roots grown in soil. In *Root Growth* (ed. W.J. Whittington), Butterworths, London, pp. 397–8.

Milchunas, D.G., Laurenroth, W.K., Singh, J.S., Cole, C.V. and Hunt, H.W. (1985) Root turnover and production by ^{14}C dilution: implications of carbon partitioning in plants. *Plant Soil*, **88**, 353–65.

Mosse, B., Stribley, D.P. and LeTacon, F. (1981) Ecology of mycorrhizae and mycorrhizal fungi. *Adv. Microb. Ecol.*, **5**, 137–210.

Newman, E.I. (1966) A method for estimating the total length of root in a sample. *J. Appl. Ecol.*, **3**, 139–45.

Nye, P.H. and Tinker, P.B. (1977) *Solute Movement in the Soil–root System*. University of California Press, Berkeley, 342 pp.

Osborne, J.F. and Pelishek, R.E. (1961) Installing deep neutron tubes. *Agric. Eng.*, **42**, 611–12.

Ottman, M.J. and Timm, H. (1984) Measurement of viable plant roots with the image analyzing computer. *Agron. J.*, **76**, 1018–20.

Pacovsky, R.S. and Bethlenfalvay, G.J. (1982) Measurement of the extraradical mycelium of a vesicular-arbuscular mycorrhizal fungus in soil by chitin determination. *Plant Soil*, **68**, 143–7.

Pagliai, M., LaMarca, M. and Lucamante, G. (1983) Micromorphometric and micromorphological investigation of a clay loam in viticulture under zero and conventional tillage. *J. Soil Sci.*, **34**, 391–403.

Persson, H. (1979) Fine-root production, mortality and decomposition in forest ecosystems. *Vegetatio*, **41**, 101–9.

Persson, H. (1980) Spatial distribution of fine-root growth, mortality and decomposition in a young Scots pine stand in Central Sweden. *Okios*, **34**, 77–87.

Peterson, B.J., Howarth, R.W. and Garritt, R.H. (1985) Multiple stable isotopes used to trace the flow of organic matter in estuarine food webs. *Science*, **227**, 1361–3.

Phillips, J.M. and Hayman, D.S. (1970) Improved procedures for clearing and staining parasitic and vesicular-arbuscular mycorrhizal fungi for rapid assessment of infection. *Trans. Br. Mycol. Soc.*, **42**, 421–38.

Porter, W.M. (1979) The 'most probable number' methods for enumerating infective propagules of vesicular arbuscular mycorrhizal fungi in soil. *Austr. J. Soil Res.*, **17**, 515–19.

Rambal, S. (1984) Water balance and pattern of root water uptake by a *Quercus coccifera* evergreen scrub. *Oecologia*, **62**, 18–25.

Rennie, R.J. and Rennie, D.A. (1983) Techniques for quantifying N_2-fixation in association with nonlegumes under field conditions. *Can. J. Microbiol.*, **29**, 1022–35.

Richards, D.F., Gaubran, J.H., Garwoli, W.N. and Doly, M.W. (1979) A method for determining root length. *Plant Soil*, **52**, 69–76.

Richards, J.H. (1984) Root growth response to defoliation in two *Agropyron* bunchgrasses: field observations with an improved root periscope. *Oecologia*, **64**, 21–5.

Richards, J.H. and Caldwell, M.M. (1987) Hydraulic lift: substantial nocturnal water transport between soil layers by *Artemisia tridentata* roots. *Oecologia*, **73**, 486–9.

Ringrose-Voase, A.J. and Bullock, P. (1984) The automatic recognition and measurement of soil pore types by image analysis and computer programs. *J. Soil Sci.*, **35**, 673–84.

Robinson, D. and Rorison, I.H. (1983) A comparison of the responses of *Lolium perenne* L., *Holcus lanatus* L. and *Deschampsia flexuosa* (L.) Trin. to a localized supply of nitrogen. *New Phytol.*, **94**, 263–73.

Rowse, H.R. and Phillips, D.A. (1974) An instrument for estimating the total length of root in a sample. *J. Appl. Ecol.*, **11**, 309–14.

Rumbaugh, M.D., Clark, D.H. and Pendry, B.M. (1988) Determination of root mass ratios in alfalfa-grass mixtures using near-infrared reflectance spectroscopy. *J. Range Manag.*, **41**, 488–90.

Runge, M. (1983) Physiology and ecology of nitrogen nutrition. In *Encyclopedia of Plant Physiology, Vol. 12C Physiological Plant Ecology III. Responses to the chemical and biological environment.* (eds O.L. Lange, P.S. Nobel, C.B. Osmond and H. Ziegler). Springer-Verlag, Berlin, pp. 163–200.

St. John, T.V., Coleman, D.C. and Reid, C.P.P. (1983) Growth and spatial distribution of nutrient-absorbing organs: selective exploitation of soil heterogeneity. *Plant Soil*, **71**, 487–93.

Sanders, J.L. and Brown, D.A. (1978) A new fiber optic technique for measuring root growth of soybeans under field conditions. *Agron. J.*, **70**, 1073–6.

Savage, M.J. and Cass, A. (1984) Measurement of water potential using *in situ* thermocouple hygrometers. *Adv. Agron.*, **37**, 73–126.

Schenck, N.C. (ed.) (1982) *Methods and Principles of Mycorrhizal Research*, American Society of Phytopathology, St. Paul, MN.

Schlesinger, W.H., Fonteyn, P.J. and Marion, G.M. (1987) Soil moisture content and transpiration in the Chihuahuan Desert of New Mexico. *J. Arid Environ.*, **12**, 119–26.

Shearer, G. and Kohl, D.H. (1978) ^{15}N abundance in N-fixing and non-N-fixing plants. In *Mass Spectrometry in Biochemistry and Medicine* (ed. A. Frigerio), Plenum Press, New York, Vol. 1, pp. 605–22.

Shearer, G.B. and Kohl, D.H. (1986) N_2-fixation in field settings: Estimations based on natural ^{15}N abundance. *Austr. J. Plant Physiol.*, **13**, 699–756.

Shearer, G.B. Kohl, D.H. and Chien, S.H. (1978) The nitrogen-15 abundance in a wide variety of soils. *Soil Sci. Soc. Am. J.*, **42**, 899–902.

Shearer, G.B., Kohl, D.H. and Harper, J.E. (1980) Distribution of ^{15}N among plant parts of nodulating and non-nodulating isolines of soybeans. *Plant Physiol.*, **66**, 57–60.

Shearer, G., Kohl, D.H., Virginia, R.A., Bryan, B.A., Skeeters, J.L., Nilsen, E.T., Sharifi, M.R. and Rundel, P.W. (1983) Estimates of N_2-fixation from variation in the natural abundance of ^{15}N in Sonoran Desert ecosystems. *Oecologia*, **56**, 365–73.

Shierlaw, J. and Alston, A.M. (1984) Effect of soil compaction on root growth and uptake of phosphorus. *Plant Soil*, **77**, 15–28.

Shinkle, J.R. and Briggs, W.R. (1985) Physiological mechanism of the auxin-induced increase in light sensitivity of phytochrome-mediated growth responses in *Avena* coleoptile sections. *Plant Physiol.*, **79**, 349–56.

Silvester, W.B. (1983) Analysis of N_2 fixation. In *Biological Nitrogen Fixation in Forest Ecosystems. Foundations and Applications* (eds J.C. Gordon and C.T. Wheeler), Junk, Boston, pp. 172–212.

Sims, P.L. and Singh, J.S. (1978) The structure and function of ten western North American grasslands. II. Intraseasonal dynamics in primary producer compartments. *J. Ecol.*, **66**, 547–72.

Singh, J.S. and Coleman, D.C. (1973) A technique for evaluating functional root biomass in grassland ecosystems. *Can. J. Bot.*, **51**, 1867–70.

Singh, J.S., Lauenroth, W.K., Hunt, H.W. and Swift, D.M. (1984) Bias and random errors in

estimators of net root production: A simulation approach. *Ecology*, **65**, 1760–4.
Sisson, W.B. (1983) Carbon balance of *Yucca elata* Engelm. during a hot and cool period *in situ*. *Oecologia*, **57**, 352–60.
Smith, G.W. and Skipper, H.D. (1979) Comparison of methods to extract spores of vesicular-arbuscular mycorrhizal fungi. *Soil Sci. Soc. Am. J.*, **43**, 722–5.
Smucker, A.J.M., McBurney, S.L. and Srivastava, A.K. (1982) Quantitative separation of roots from compacted soil profiles by the hydro-pneumatic elutriation system. *Agron. J.*, **74**, 500–3.
Svejcar, T.J. and Boutton, T.W. (1985) The use of stable carbon isotope analysis in rooting studies. *Oecologia*, **67**, 205–8.
Svoboda, J. and Bliss, L.C. (1974) The use of autoradiography in determining active and inactive roots in plant production studies. *Arct. Alp. Res.*, **6**, 257–60.
Taylor, H.M. and Boehm, W. (1976) Use of acrylic plastic as rhizotron windows. *Agron. J.*, **68**, 693–4.
Tennant, D. (1975) A test of modified line intersect method of estimating root length. *J. Ecol.*, **63**, 995–1001.
Toth, R. and Toth, D. (1982) Quantifying vesicular-arbuscular mycorrhizae using a morphometric technique. *Mycologia*, **74**, 182–7.
Topp, G.C. and Davis, J.L. (1985) Measurement of soil water content using time-domain reflectometry (TDR): A field evaluation. *Soil Sci. Soc. Am. J.*, **49**, 19–24.
Upchurch, D.R. (1985) Relationship between observations in mini-rhizotrons and true root length density, Ph.D. dissertation, Texas Technical University.
Upchurch, D.R. and Ritchie, J.R. (1983) Root observations using a video recording system in mini-rhizotrons. *Agron. J.*, **75**, 1009–15.
Upchurch, D.R. and Ritchie, J.R. (1984) Battery-operated color video camera for root observations in mini-rhizotrons. *Agron. J.*, **76**, 1015–17.
van Noordwijk, M., de Jager, A. and Floris, J. (1985) A new dimension to observations in minirhizotrons: A stereoscopic view on root photographs. *Plant Soil*, **86**, 447–53.
Viehmeyer, F.J. (1929) An improved soil-sampling tube. *Soil Sci.*, **27**, 147–52.
Vincent, J.M. (1970) *A Manual for the Practical Study of Root-nodule Bacteria*, IBP Handbook No. 15, Blackwell Science Publication, Oxford.
Virginia, R.A., Baird, L.M., LaFavre, J.S., Jarrell, W.M., Bryan, B.A. and Shearer, G. (1984) Nitrogen fixation efficiency, natural ^{15}N abundance, and morphology of mesquite (*Prosopis glandulosa*) root nodules. *Plant Soil*, **79**, 273–84.
Virginia, R.A. and Delwiche, C.C. (1982) Natural ^{15}N abundance of presumed N_2-fixing and non-N_2-fixing plants from selected ecosystems. *Oecologia*, **54**, 317–25.
Virginia, R.A., Jarrell, W.M., Rundel, P.W., Shearer, G. and Kohl, D.H. (1989) The use of variation in the natural abundance of ^{15}N to assess symbiotic N_2-fixation by woody plants. In *Stable Isotopes in Ecological Research* (eds P.W. Rundel, J.R. Ehleringer and K.A. Nagy), Ecological Studies vol. 68, Springer-Verlag, New York, pp. 375–94.
Virginia, R.A., Jenkins, M.B. and Jarrell, W.M. (1986) Depth of root symbiont occurrence in soil. *Biol. Fertil. Soils*, **2**, 127–30.
Vogt, K.A., Grier, C.C., Gower, S.T., Sprugel, D.G. and Vogt, D.J. (1986) Overestimation of net root production: a real or imaginary problem? *Ecology*, **67**, 577–9.
Voorhees, W.B. (1976) Root elongation along a soil-plastic container interface. *Agron. J.*, **68**, 143.
Vos, J. and Groenwold, J. (1983) Estimation of root density by observation tube and endoscope. *Plant Soil*, **74**, 295–300.
Wang, C.H., Willis, D.L. and Loveland, W.D. (1975) *Radiotracer Methodology in the Biological, Environmental, and Physical Sciences*, Prentice Hall, Englewood Cliffs, New Jersey, 480 pp.
White, S.W.C., Cook, E.R., Lawrence, J.R. and Broecker, W.S. (1985) The D/H ratios of sap in trees: Implications for water source and tree ring D/H ratios. *Geochim. Cosmochim. Acta*, **49**, 237–46.
Wiebe, H.H., Brown, R.W. and Barker, J. (1977) Temperature gradient effect on *in situ* hygrometer measurement of water potential. *Agron. J.*, **69**, 933–9.
Wiebe, H.H., Campbell, G.S., Gardner, W.H., Rawlins, S.L., Cary, J.G. and Brown, R.W. (1971) Measurement of plant and soil water status. *Utah Agric. Exp. Sta. Bull.*, 484.
Wullstein, L.H. and Harker, A. (1982) Nonconfirmation of nodulation in *Artemisia ludoviciana*. *Am. J. Bot.*, **69**, 160–2.

17

Field methods used for air pollution research with plants

William E. Winner and Carol S. Greitner

17.1 INTRODUCTION

Recent awareness of the regional distribution patterns of air pollutants such as ozone, oxides of sulfur and nitrogen, and their transformation products, has led physiological ecologists to consider these chemical agents as environmental stresses. From this perspective, important ideas have emerged regarding the value of air-quality-monitoring data as well as the transfer and application of research approaches used for traditional stress physiology studies to studies of plant responses to air pollutants. Development of these ideas suggests that degradation of air quality may reduce plant productivity, fitness and capacity to resist other environmental stresses. Physiological ecologists performing field work in areas with elevated air pollutant levels need to be aware of how their research can contribute to this area of environmental science and how these agents may be affecting their study plants. This awareness should include an understanding of air pollution sources and chemistry, some responses of plants to air pollutants, and methods used to measure the effects of air pollutants on plants.

In this article we discuss the uptake of pollutants by vegetation and how this can be estimated. However, our main focus is on the techniques that physiological ecologists use to assess the responses of plants to air pollutants. The techniques discussed are principally those used in field studies and include cuvettes, systems for gaseous fumigation experiments and systems for manipulating/simulating rainfall. Emphasis on techniques for field research precludes discussion of important laboratory/greenhouse methods (Hogsett *et al.*, 1987), but is consistent with the theme of this book and the tendency for scientists to work on field-grown plants whenever possible. In addition, we refer the reader to discussion of air pollution sources, concentrations and distribution (Soderlund and Svensson, 1976; National Research Council, 1977, 1978, 1986; Zajac and Grodinska, 1982; Galloway *et al.*, 1984; Jacob *et al.*, 1985; Whitmore, 1985; US Environmental Protection Agency, 1986; Miller *et al.*, 1987).

We have further attempted to summarize air pollution exposures that are known to affect the physiology, growth and foliar appearance of tobacco, radish, *Populus* species and white pine (Tables 17.1 and 17.2). We

Table 17.1 Responses of plants to O_3 exposures

	Species	*O_3 (ppm)*	*Duration*	*Effect*	*Reference*
Foliar injury	Tobacco	0.05	4 h	5% foliar injury	Tingey *et al.* (1973)
	Radish	0.15	6 h d^{-1}, 5 d	13% foliar injury	Beckerson and Hofstra (1979)
	Populus species (poplar and aspen)	0.041	12 h d^{-1}, 23 w	1333% increase in leaf drop	Mooi (1980)
	White pine	0.10	6 h	20% of sensitive clones injured	Houston (1974)
Physiological response	Tobacco	0.30	2 h	48% reduction in chlorophyll content	USEPA (1978)
	Radish	0.17	Continuous for 5 w	Foliage produced in O_3 adapted to O_3; hypocotyl growth suppressed	Walmsley *et al.* (1980)
	Populus species (poplar and aspen)	0.085–0.125	5.5 h d^{-1}, 6–7 d w^{-1}, 6 w	Photosynthesis inhibited by up to 40%	Reich and Amundson (1985)
	White pine	0.10–0.14	7 h d^{-1}, 3 d w^{-1}, 12 w	10% reduction in photosynthesis	Reich and Amundson (1985)
Growth response	Tobacco	0.05	7 h d^{-1}, 5 d w^{-1}, 4 w	1% reduction in leaf dry weight	Tingey and Reinert (1975)
	Radish	0.05	8 h d^{-1}, 5 d w^{-1} 5 w	50% reduction in root growth	Tingey *et al.* (1971)
	Populus species (poplar and aspen)	0.15	12 h d^{-1}, 50 d	Height growth suppressed threefold; leaf area and weight and total weight suppressed	Jensen (1981)
	White pine	No data available			

chose SO_2 and O_3 and these plant species because they have been studied extensively and they provide examples of air pollution exposures known to influence vegetation. Although concentrations and durations of exposures are not entirely comparable between experiments, these tables show plants to be somewhat more sensitive to O_3 than to SO_2. No trends in species sensitivity are apparent, but these species are commonly studied because they are known to be sensitive to pollutants. Finally, although these exposures may not be definitive for field sites because some of the reports are from laboratory/greenhouse studies, they indicate that low concentrations of these gases over short time periods are of potential significance to vegetation. The effects of air pollutants on plants are reviewed in numerous volumes, including Smith (1981), Guderian (1985), McLaughlin (1985), Winner *et al.* (1985a) and Reuss and Johnson (1986). These reviewers summarize research findings on the effects of various air pollutants on plant biochemistry, physiology, growth and ecology and consider both crop and forest species.

17.2 STUDIES OF AIR POLLUTION ABSORPTION

Physiological ecologists have research approaches which are useful in analyzing the effects of air pollutants on plants. Some research has focused on determining the absorption rates of gaseous pollutants by foliage. This determination extends the same concepts and techniques used for measuring

Table 17.2 Responses of plants to SO_2 exposures

	Species	*SO_2 (ppm)*	*Duration*	*Effect*	*Reference*
Foliar injury	Tobacco	1	4 h	13% leaf area injured	Tingey *et al.* (1971)
	Radish	0.15	6 h d^{-1}, 5 d	0.9% leaf area injured	Beckerson and Hofstra (1979)
	Populus species (poplar and aspen)	0.2	24 h	1 clone: up to 33% of area of older leaves necrotic	Kimmerer and Kozlowski (1981)
	White pine	0.05–0.15	6 h	100% of sensitive clones injured	Houston (1974)
Physiological response	Tobacco	2	0.5 h	Tobacco with high levels of abscisic acid: transpiration decreased to 20–65% of initial level	Kondo *et al.* (1980)
	Radish	0.15	5 d	Stomatal opening	Beckerson and Hofstra (1979)
	Populus species (poplar and aspen)	0.20	8 h	Decreased diffusive conductance	Kimmerer and Kozlowski (1981)
	White pine	0.05	2 h	Photosynthesis reduced 10–27%	Eckert and Houston (1980)
Growth response	Tobacco	0.05	5 h d^{-1}, 5 d w^{-1}, 4 w	Shoot weight reduced 22%	Tingey and Reinert (1975)
	Radish	0.05	8 h d^{-1}, 5 d w^{-1}, 5 w	Plant fresh weight reduced 15%, root dry weight reduced 17%	Tingey *et al.* (1971)
	Populus species (poplar and aspen)	0.25	12 h d^{-1}, 50 d	Suppression of: height, leaf area and weight, total weight	Jensen (1981)
	White pine	0.025	6 h	Reduced needle elongation	Houston (1974)

CO_2 and water vapor fluxes between the leaf and air to also include gaseous pollutants such as O_3, NO_2 and SO_2. These studies generally use leaf/branch chambers for both fumigations and gas flux rate measurements. The analysis of gas exchange not only provides a continuous record of CO_2, H_2O and air pollutant fluxes but also allows for the calculation of partial pressures of these gases in the leaf mesophyll.

Air pollution absorption studies that have allowed measurement of pollutant uptake by cuvettes and leaves have been conducted with gaseous pollutants including sulfur dioxide (SO_2) (Black and Unsworth, 1979a; Winner and Mooney, 1980a,b), nitrogen oxides (NO_x) (Rogers *et al.*, 1977) and ozone (O_3) (Taylor *et al.*, 1982). To illustrate how uptake determinations can be made for cuvettes and leaves, SO_2 can be used as a model pollutant (Table 17.3). Total SO_2 uptake by a leaf in a cuvette system is determined by measuring the difference in SO_2 concentrations entering and leaving the cuvette and multiplying this concentration by the air flow rate (Taylor and Tingey, 1979; Winner and Mooney, 1980a,b). The product gives the total SO_2 uptake rate by both the leaf and cuvette interior. The uptake rate of the empty cuvette must be subtracted from this value to determine the total leaf uptake rate. Once leaf uptake rates are known, integration of these values over the exposure period gives total SO_2 uptake by the leaf during exposure.

Table 17.3 Equations used in air pollution flux calculations

1. SO_2 absorption rate by an empty cuvette:
   $J_{cuvette} = ([SO_2]_{in} - [SO_2]_{out}) \times \text{flow rate} \times \text{cuvette interior area}^{-1}$
2. SO_2 flux on a leaf area basis:
   $J_{total\ leaf} = \{([SO_2]_{in} - [SO_2]_{out}) \times \text{flow rate} \times \text{leaf area}^{-1}\} - J_{cuvette}$
3. SO_2 absorption rate through stomata:
   $J_{absorbed} = [SO_2]_{out} \times g_{SO_2}$
4. SO_2 adsorption rate to leaf surface:
   $J_{adsorbed} = J_{total\ leaf} - J_{absorbed}$
5. SO_2 conductance value determined from H_2O conductance value:
   $g_{SO_2} = g_{H_2O} \times (D_{SO_2} \text{ in air} \times D_{H_2O} \text{ in air}^{-1})$

$J_{cuvette}$ = flux rate in mol m^{-2} s^{-1}
cuvette interior area = m^2
$[SO_2]_{in}$ = SO_2 concentration entering the cuvette in mol m^{-3}
$[SO_2]_{out}$ = SO_2 concentration leaving the cuvette in mol m^{-3}
flow rate = m^3 s^{-1}
$J_{total\ leaf}$ = flux rate in mol m^{-2} s^{-1}
leaf area = m^2
$J_{absorbed}$ = flux rate in mol m^{-2} s^{-1}
g_{SO_2} = stomatal conductance for SO_2 in m s^{-1}
$J_{adsorbed}$ = flux rate in mol m^{-2} s^{-1}
g_{H_2O} = stomatal conductance for water vapor in m s^{-1}
D_{SO_2} = diffusivity for SO_2 in air
D_{H_2O} = diffusivity for H_2O in air

Division of the total SO_2 uptake rate of the leaf by the leaf area gives the total uptake rate on a leaf area basis.

Total SO_2 uptake by leaves can be partitioned between that adsorbed on to leaf surfaces, such as cuticle, and that absorbed through stomata (Black and Unsworth, 1979b; Winner and Mooney, 1980a, b; Table 17.3). Since stomata provide the principal path of entry of SO_2 and other gaseous pollutants, into the mesophyll where metabolism is subsequently altered, it is important to quantify air pollution absorption through stomata. Foliar SO_2 absorption rates are calculated by measuring stomatal conductance to water vapor, converting these values to SO_2 conductance values and multiplying SO_2 conductance by the SO_2 concentration around the leaf (the concentration leaving the cuvette). As with total SO_2 uptake, absorption rates can be integrated over the exposure period to determine total SO_2 absorbed during a period of interest.

Equations used to calculate SO_2 adsorption to the cuticle, SO_2 flux through stomata and total SO_2 uptake rates are reviewed by Winner *et al.* (1985a,b). That review also presents a discussion of methods for calculating the stomatal conductance for any gas using the stomatal conductance value for water vapor and a diffusivity value for the other gas.

Application of the equations of SO_2 flux analysis requires the assumptions that SO_2 internal partial pressure values in leaf mesophyll are near zero and that water vapor fluxes are driven by the same thermodynamic laws governing SO_2 fluxes. Although there have been no direct measures of SO_2 concentrations in leaf mesophyll tissues, Black and Unsworth (1979c) used gas-exchange studies with *Vicia faba* to show that SO_2 internal concentrations are near zero. They found that the ratio of calculated SO_2 flux/ambient SO_2 concentration was linearly related to stomatal conductance with a slope of one, suggesting SO_2 internal concentration is low. More work to verify SO_2 internal concentrations and to clarify thermodynamic processes driving SO_2 fluxes is needed, as well as studies to extend this analytical approach to other pollutants such as O_3.

Although analysis of gas fluxes between the leaf and air were made from concepts emerging from cuvette studies, cuvettes are not necessary to calculate gaseous air pollution absorption by plants. The assumptions above being true, air pollution absorption can be calculated from foliar conductance values (including stomatal and boundary layer components) and air pollution concentrations measured for leaf surfaces in any environment. Procedures for determining stomatal and boundary layer conductances are discussed in Chapters 5 and 8 respectively.

SO_2 adsorbed by the leaf surface, determined by subtracting SO_2 absorbed through stomata from total SO_2 uptake by the leaf (Winner and Mooney, 1980a,b), may or may not have biological significance. The movement of adsorbed SO_2 into mesophyll tissues depends on cuticular conductance to SO_2, cuticular integrity and period of time required for SO_2 to become oxidized to the sulfate oxidation state. Although sulfate is known to be less toxic than sulfite, little is known about the potential for sulfite and sulfate in the cuticle to affect leaf metabolism directly. These ions may also be dissolved in precipitation and enter the plant via foliar absorption or absorption by roots upon moving from foliage to soils.

The approach of using leaf conductance values and ambient air pollution concentrations to calculate air pollution absorption is referred to as the water vapor surrogate approach. Although other model approaches exist for estimating deposition velocities of gases and particles to surfaces are known (Fowler, 1985), these approaches have not been thoroughly developed in cuvette studies. Attempts to measure tissue level increases in sulfur content following fumigation are fraught with difficulty associated with (1) the necessity of thoroughly washing SO_2 from leaf surfaces in order to calculate absorbed SO_2, (2) the possible translocation of absorbed S out of fumigated leaves, (3) the possible reduction of absorbed SO_2 and subsequent emission of H_2S from leaves, and (4) technical difficulties in precise determination of pre- and post-fumigation S content of tissue. Even if all of these problems are overcome, some air pollutants such as O_3 may not change the chemical composition of foliage so as to reflect absorption. Therefore attempts to calculate air pollution flux rates on the basis of stomatal conductance and ambient air pollution concentrations have conceptual and technical advantages over other methods.

17.3 AIR POLLUTION INSTRUMENTATION

Continuous, automated instrumental methods for monitoring SO_2, O_3 and NO_x have been developed and are commercially available. These instruments can differ in their mode of operation (Table 17.4) and performance (US Environmental Protection Agency, 1983). Instrument specifications, including lag time of the reading, accuracy and precision, can differ between manufacturers and should be thoroughly investigated before selecting an instrument.

17.3.1 SO_2 monitors

The two most commonly used SO_2 analyzers operate either by flame photometry or by fluorescence spectroscopy. Flame photometric analyzers operate by burning the compound to be monitored in a flame and measuring the emitted light energy. When an S-containing gas is burned in an H_2 flame, S_2 is formed and raised to an excited state which then

Table 17.4 Examples of air pollution monitors and calibrators

Instrument	*Principle*	*Manufacturers*
SO_2 monitor	Flame photometry	CSI[a]
	Fluorescence spectroscopy	CSI, Dasibi[b], TECO[c]
SO_2 calibrator	Permeation tube	CSI, Metronics[d], Sensidyne[e], TECO
O_3 monitor	Chemiluminescence	Bendix[f], CSI
	UV photometry	Dasibi, Mast Development Corp.[g]
O_3 calibrator	UV photometry	CSI, Dasibi
NO_2 monitor	Chemiluminescence	Bendix, CSI
NO_2 calibrator	Permeation tube	Bendix, Metronics, TECO
	Gas phase titration	CSI

[a] Columbia Scientific Instruments, Austin, TX, USA.
[b] Dasibi Corp., Glendale, CA, USA.
[c] Thermo Electron Corp., Hopkinton, MA, USA.
[d] Metronics, Santa Clara, CA, USA.
[e] Sensidyne (formerly Bendix), Largo, FL, USA.
[f] Many monitors and calibrators still in use but no longer manufactured.
[g] Mast Development Corp., Davenport, IA, USA.

emits light peaking at 394 nm. The light passes through a filter which selects for the wavelength of strongest emission and is detected by a photomultiplier tube. The tube provides an electrical analog signal to the monitor's electronic components. The intensity of the light is proportional to the concentration of excited S_2, which is in turn proportional to the original amount of S-containing gas in the air sample.

Advantages of flame photometric monitors include low maintenance, high sensitivity, fast response and selectivity for S compounds. The major disadvantages are the need to filter out other S-containing gases in the sample stream to make the analysis specific for SO_2, and the hazards involved with handling the highly flammable H_2 gas.

Fluorescence analyzers utilize a radiation source to excite compounds of interest which then emit fluorescent light as they return to the ground state. Some of the light energy is dissipated through vibration and rotation, so that the emitted light is of lower energy and longer wavelength than the original light source. UV radiation at 190–230 nm from a lamp is absorbed by SO_2 molecules, and the emitted fluorescent light is detected by a photomultiplier tube. Two types of UV lamps are used by different manufacturers: a pulsed source (ThermoElectron Corp.) and a continuous source (CSI, Dasibi). Response times of fluorescent spectrometers are slower than those of flame photometers, but they are relatively insensitive to temperature and flow variations, and do not require hazardous reagents such as H_2.

17.3.2 O_3 monitors

Two methods are commonly used for measuring O_3 concentrations. In the first, the chemiluminescent reaction with ethylene (C_2H_4), O_3 in the air reacts with C_2H_4 to form activated formaldehyde which emits light with a peak at 440 nm. This light is detected by a photomultiplier tube. The 440 nm peak is specific to this reaction, and its intensity is proportional to the O_3 concentration. This method is dependable because the reaction (and therefore the equipment) is simple and there is no interference from other air pollutants. The explosive nature and expense of C_2H_4 are the major drawbacks of this procedure.

UV photometry uses an application of the Beer–Lambert law, in which the difference between the transmittance of radiation through a gas sample with and without O_3 is related to the concentration of O_3. UV photometric detectors operate by passing a monochromatic beam of light at 254 nm produced by a mercury lamp through a cell containing O_3-free air, then through a cell with O_3. The amount of light absorbed by the sample with O_3 depends upon the number of O_3 molecules in the light beam, and the difference in absorption of UV radiation is monitored by a UV detector as a measure of O_3 concentration. These monitors are fast, stable, require low maintenance and are safe to operate.

17.3.3 NO_2 monitors

The primary method of NO_2 analysis involves a chemiluminescent reaction with O_3. NO_2 is reduced to NO in a converter and reacts with O_3 generated within the monitor. Light with a peak at 1200 nm (unique to this reaction) is produced, and its intensity is proportional to the concentration of NO. Any NO in the sample passes unchanged through the converter, so this reading gives the concentration of NO plus NO_2 (total NO_x). When the converter is bypassed, NO alone is measured and by cycling between these two modes, NO_2 is measured by subtraction. The chemiluminescent monitor has a relatively fast response time, with minimal interference from other air pollutants, except where PAN (peroxyacetyl nitrates) levels are high. Proper flow of O_3 and air sample are crucial in this procedure.

17.3.4 Calibration

Air-pollution-monitoring programs require calibration efforts to verify that instruments are functioning correctly and giving accurate measurements. Methods are available from instrument suppliers and have also been documented by the US Environmental Protection Agency (1983). Common to all methods is the procedure of generating known gas concentrations that are supplied to the monitor and compared with monitor readings to document any inaccuracy and guide calibration adjustments. Common practice with air pollution monitors is to check their zero reading with pollution-free air, and to check their upscale span with a calibration gas close in concentration to levels used in experiments. Such zero/span check calibration exercises should be performed weekly to define instrument drift and to detect changes in performance. Criteria, such as a 10% drift in zero or span, should be established to define conditions for a multipoint calibration exercise resulting in adjustments of zero and/or span setting. Accurate records of zero and span readings as well as adjustments should be kept.

SO_2 analyzers are most commonly calibrated with a permeation tube. This is a tube of an inert semipermeable material such as Teflon which is filled with liquid SO_2. When the tube is maintained at constant temperature, SO_2 gas escapes at a constant rate reflecting temperature and tube length. The rate of SO_2 emission is determined by sequential weighings which measure the loss of SO_2 over time. Different concentrations of SO_2 are generated by changing carrier gas flow rates around the tube and supplying these gas mixtures to the monitor. Another calibration procedure involves titrating SO_2 from a cylinder with sodium thiosulfate in order to verify the SO_2 concentration in the cylinder. Cylinders with known concentrations of SO_2 in N_2, traceable to official standards, are used with mass flow controllers to generate a range of SO_2 concentrations for calibration.

O_3 is an unstable gas and it cannot be containerized for calibration. O_3 must be generated at the calibration site, usually by passing O_2 through UV light of 185 nm. The concentration of O_3 produced can be varied by moving an aluminum shield across the source lamp, thereby controlling the amount of radiation to which the air is exposed. A stable flow of O_2, a transformer to assure constant voltage to the lamp and a UV photometer are required for O_3 monitor calibration.

NO_2 analyzers may be calibrated with a NO_2 permeation tube using the same principles as for SO_2. In addition, the US Environmental Protection Agency (1983) has approved a gas phase titration procedure in which known quantities of O_3 are generated and added to excess NO from a pressurized cylinder. As the O_3 reacts with the NO, the decrease in NO concentration observed is equivalent to the amount of NO_2 produced and supplied to the monitor.

17.3.5 Monitoring nongaseous air pollutants

Aerosols are particles in the air measuring from 10^{-2} to about 5 μm in diameter. Particles larger than 5 μm can be considered dust. Both dust and aerosols can settle as dry deposition and can be important forms of chemical inputs to ecosystems. Measurements of aerosol and dust concentrations in air are most commonly made with high-volume air samplers, although low-volume samplers, dichotomous samplers, Nephelometers and impact collectors can be used with similar results (Giever, 1976; Kretzschmer and Pauwels, 1982). High-volume air samplers pull air through filter packs shielded with a roof to prevent collection by particle settling. Air pulled under the shield is passed through a series of increasingly fine filters that allow aerosols and dust to be separated by size class and then analyzed for chemical content. Filters can also be treated with chemicals to absorb gases. Whether particles or gases are retained on filters, the concentration in air is calculated on the basis of air flow rate through the sampling train. High-volume air samplers are easy and economical to operate, but only supply measurements of air quality averaged over the sampling period. Thus this method works best for air pollutants which are uniformly dispersed over time rather than for those that are distributed episodically. In addition, the analysis of filter packs can be tedious and involves complex chemical and mathematical techniques (West, 1976).

17.3.6 Monitoring wet deposition

The chemical analysis of wet deposition requires the collection of rain, fog and snow. Rain can be collected as a bulk sample using a funnel that drains rain into a storage vessel. A bulk sample not only includes rain, but also chemicals deposited on the funnel between rains that are then washed into the storage vessel by precipitation.

Wet deposition collectors are now designed with two storage vessels and a cover which moves between the two storage vessels (Galloway and Likens, 1976). The cover shields one vessel during fair weather, but is activated to move and shield the other vessel when moisture is present. Thus one vessel collects dry deposition and the other vessel collects precipitation. These collectors are often equipped with a tipping bucket rain gauge to provide a record of rainfall rates during sampling periods. One such wet/dry collector is made by Aerochem Metrics, Miami, FL, USA. More than 150 of these types of collectors are now used in the National Atmospheric Deposition Program/National Trends Network. This network of monitoring stations has operated with standardized methods of collecting and analyzing rain samples since 1978. The collectors are located across all 50 states of the US and parts of Canada and Mexico and provide a valuable data base resource for those needing information on rainfall rates and chemistry, including pH, conductivity and content of anions and cations (NADP/NTN Coordinator's Office,

Natural Resource Ecology Laboratory, Colorado State University, Fort Collins, CO 80523, USA).

Two major forms of fog collectors exist, passive and active. Passive collectors include screens on which the droplets are collected (used by Johnson *et al.*, 1987). Fog can actively be collected using filters which are swung or spun through fog (Dollard and Unsworth, 1983; Johnson *et al.*, 1987; Waldman *et al.*, 1987). Obtaining fog droplet size distributions, droplet concentrations or deposition rates from either active or passive collectors is difficult. Nonetheless, samples can be obtained for chemical analysis.

Rain and fog passing through forest and crop canopies can be chemically modified by contact with foliage (Lindberg *et al.*, 1986). Differences between precipitation and throughfall chemistry result when ions are either washed from or are retained by leaf surfaces. The design of throughfall collectors is often similar to bulk precipitation collectors, consisting of a funnel draining into a storage device. Ideally, these collectors should be equipped with caps during dry periods to prevent collection of dry deposition.

Samples of precipitation and throughfall which are collected must be handled with care to prevent contamination, the growth of micro-organisms and chemical changes such as degassing or changes in oxidation state of unstable ions (Galloway and Likens, 1978). Common procedures for handling samples include collecting them on a weekly basis, covering them, storing them as cold as possible and analyzing them as soon after collection as possible.

17.4 CUVETTES

17.4.1 Design and materials

Cuvettes used as fumigation chambers have a number of features in common. They are transparent to allow foliage to receive light. An air circulation scheme is usually constructed to both mix air within the cuvette and to move air through the cuvette. Mixing air within the cuvette is important to reduce the size of the boundary layer around the foliage and to be sure all gases, including air pollutants injected into the cuvette, are uniformly mixed. Air flow through the cuvette allows the gas concentrations in the cuvette to be monitored and controlled and allows the measurement of CO_2, water vapor and air pollution fluxes between the leaf and air. It is important to note that, as with CO_2 and water vapor, the air pollution concentration leaving a cuvette or chamber is assumed to be the ambient concentration around the leaf. Measuring air pollution concentrations in the air streams both entering and leaving cuvettes and fumigation chambers is useful, as discussed earlier in this chapter, because the concentration difference can be used to calculate total air pollution uptake within the cuvette. Chapter 11 discusses the general design of cuvettes for gas exchange studies and the control of cuvette environment.

Cuvettes and fumigation chambers should be built so that air pollution adsorption to tubing, chamber walls and valves is minimized. If adsorption is high, the task of controlling ambient levels in the cuvette becomes more difficult. For example, early in a fumigation, air pollution concentrations entering the cuvette must be higher than the desired ambient levels because the hardware of the cuvette is an air pollution sink. As the fumigation proceeds, the strength of the chamber sink decreases as its potential surfaces for adsorption become saturated and enter into a state of equilibrium with the air. Air pollution supply rates to the cuvette must be reduced during this period to attain constant ambient fumigation levels. During the period when the cuvette is a strong sink, its adsorption may be larger than uptake by the leaf making estimates of the leaf sink strengths more difficult. Unfortunately, rates of air pollution adsorption on to hardware tend to increase with increases in humidity or decreases in temperature. Humidity is important because it may change as air pollutants alter leaf conductance and transpiration rates. Finally, when the fumigation is terminated, air pollution adsorbed to the cuvette and associated hardware will be released into pollution-free air. This process of degassing results in imprecise end times for experiments and necessitates flushing cuvettes for long periods with clean air before the next experiment can be started. The degree to which system adsorption of air pollution makes controlled fumigations difficult can be accentuated when experiments require constantly altering gas concentrations to represent more closely the fluctuations that generally occur under field conditions.

Some procedures which can help reduce adsorption of air pollutants to chambers and cuvettes include:

1. Use only Teflon, stainless-steel or glass materials for construction. More specifically, use Teflon tubing, glass windows and valves with interior parts of stainless steel or Teflon. This last suggestion applies to metering valves, valves for stream selection and solenoid valves;
2. Avoid using materials with surfaces of Tygon, Plexiglass, brass or aluminum. Do not use Tygon tubing, Plexiglass windows or valves of brass or aluminum;
3. Use stainless-steel or Teflon compression fittings in all air pollution supply lines and sampling lines. Do not build lines by slip fitting tubing into Tygon sleeves and using hose clamps;
4. Use flow meters/rotameters with glass columns and stainless-steel metering valves;
5. Do not allow water vapor condensation to occur within the cuvette or lines which handle air pollutants. This may mean applying heat tapes to lines carrying air with high humidity.
6. Keep the surfaces of the cuvette and all supply and sampling lines clean of dust, debris, plant tissue, etc;
7. Keep air pollution dispensing and sampling lines to the shortest length possible;
8. Check the air pollution adsorption characteristics of the empty cuvette or fumigation chamber often. If adsorption to this hardware increases dramatically, sample air systematically at different points to verify the new sink and either clean or replace the part.

These suggestions are only provided as general procedures for reducing air pollution adsorption on to the cuvette and other hardware. Totally eliminating adsorption of pollutants to hardware is not possible but investment in time and materials to reduce the adsorption is well worth while.

Air filtration is another important aspect of cuvette and chamber fumigation systems. Air should be filtered as part of its conditioning prior to entering the cuvette. Dust can be filtered with Teflon membranes or fiberglass filter packs. Ambient pollutants should also be eliminated so that the only pollutants in the cuvette or chamber are those chosen and released by the scientist. Activated charcoal can be used to filter many pollutants such as O_3 (suppliers include Barnebey-Cheney Co., Columbus, OH and Cambridge Filter Corp., Syracuse, NY, USA). However, gaseous pollutants such as SO_2 and NO are not efficiently captured in charcoal but can be filtered with chemical absorbants (e.g. Purafil, Purafil Inc., Atlanta, GA, USA). Finally, air should be filtered again after use to avoid polluting a laboratory or field research site.

17.4.2 Cuvettes in air pollution studies

Cuvettes in current use are generally designed to monitor gas flux on a leaf area basis for an individual leaf, or leaf part, while it is still attached to the plant. Whole plant and multiple plant chambers are also used for gas-exchange studies. Cuvettes must have gas-handling systems and monitors for water vapor, CO_2 and possibly air pollutants. Integration of these components constitutes a gas-exchange system. See Chapter 11 for a discussion of cuvette design.

Some cuvette systems used in air pollution experiments are commonly operated independently from exposures, i.e. plant gas-exchange rates are determined before or after fumigations but fumigations do not occur within the cuvette. Cuvettes in this case can be used to document the effects of gaseous air pollution exposures (e.g. McLaughlin *et al.*, 1979; Biggs and Davis, 1980). Other workers (e.g. Kimmerer and Kozlowski, 1981; Olszyk and Tibbitts, 1981; Norby and Kozlowski, 1981, 1982) have taken porometer readings on plants during exposures to air pollutants in fumigation chambers. This approach allows the dynamic responses of leaves to air pollutants to be measured. Some systems exist in which air pollutants can be introduced directly into the cuvette. These systems constitute gas-exchange/fumigation systems and can also be used to determine the dynamics of air pollution absorption and physiological responses of leaves (Black and Unsworth, 1979a; Taylor and Tingey, 1979;

Table 17.5 Performance evaluation for cuvette/gas-exchange systems

	Cuvette type[a]			
	1	2	3	4
Evaluation of operational characteristics				
Closed air flow system	Y[b]	N	Y	Y
Flow-through air flow system	N	Y	Y	N
Temperature control	N	Y	Y	N
Humidity control	N	Y	Y	N
Clip-on cuvette	Y	N	Y	Y
Whole-leaf cuvette	Y	Y	Y	Y
Short-term measurements only	Y	N	N	Y
Long-term measurement capability	N	Y	Y	N
Diagnostic gas-exchange capability	N	Y	Y	Y
Air pollution fumigation capability	N	Y	Y	N
Classification of some models				
Diffusive resistance porometer (numerous models)	X			
Analytical Developmental Co.[c] Delta T System	X			
LI-COR steady-state porometer (1600)[d]			X	
LI-COR photosynthesis system (6000 and 6200)[d]	X			
Siemans-Allis Inc. cuvette systems[e]		X		
Data Design Group system (PACsys 9900)[f]			X	
Analytical Development Co. system (LCA-2)[c]		X		
Armstrong Enterprises system (custom design)[g]		X	X	
Walz system (custom design)[h]		X		

[a] See text.
[b] Y = yes, N = no for this type of cuvette.
[c] Hoddesdon, Herts, UK.
[d] Lincoln, NE, USA.
[e] Cherry Hill, NJ, USA; Karlsruhe, W. Germany.
[f] La Jolla, CA, USA.
[g] Palo Alto, CA, USA.
[h] Effeltrich, W. Germany.

Winner and Mooney, 1980a,b; Hallgren *et al.*, 1982; Noble and Jenson, 1983; Atkinson *et al.*, 1986).

There are four basic approaches to air pollution–gas-exchange studies in cuvettes (Table 17.5). All of these cuvette systems are commercially available; however, many scientists develop their own equipment or modify commercial units.

Type 1. *Closed cuvettes* are used for sampling rates of gas absorption or emission from leaves. These cuvettes are not suitable for use as fumigation chambers because they lack temperature control and should be on leaves for short (1–2 min) periods. These cuvettes are generally used in environments that contain gaseous pollutants. These cuvettes have been used to capture C_2H_4 (Tingey *et al.*, 1976) and H_2S (Winner *et al.*, 1982) emitted from leaves.

Type 2. *Differential flow-through cuvettes* are used for continuously monitoring gas absorption or emission from leaves. An early example of this type was developed in the late 1960s at Brookhaven National Laboratory, New York, USA, and is diagrammed in Fig. 17.1 (from Smith, 1981). Newer cuvettes are more suitable fumigation chambers because they can be built with a high degree of environmental control and can be attached to the leaves for periods of several days. Monitoring and control of all important environmental factors allow study of dynamic leaf responses to gaseous pollutants such as SO_2

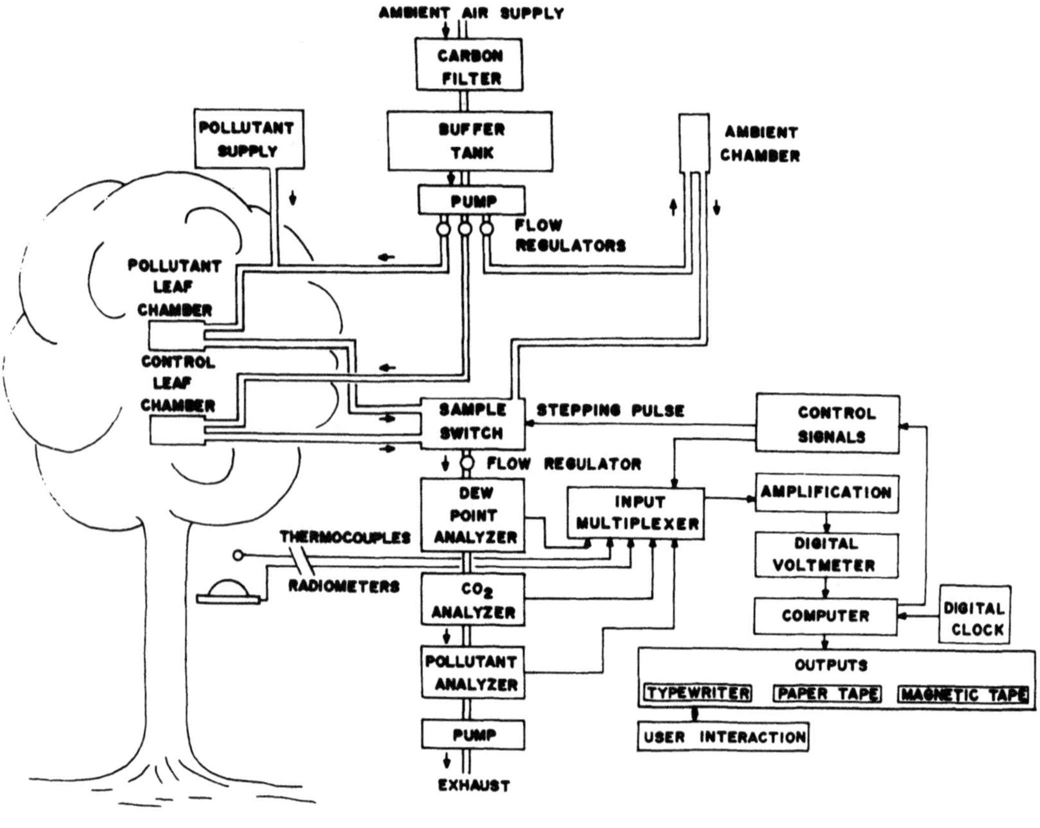

Fig. 17.1 Schematic diagram of the cuvettes, pollutant exposure equipment and data-acquisition system in the Brookhaven fumigation technique (from Smith, 1981).

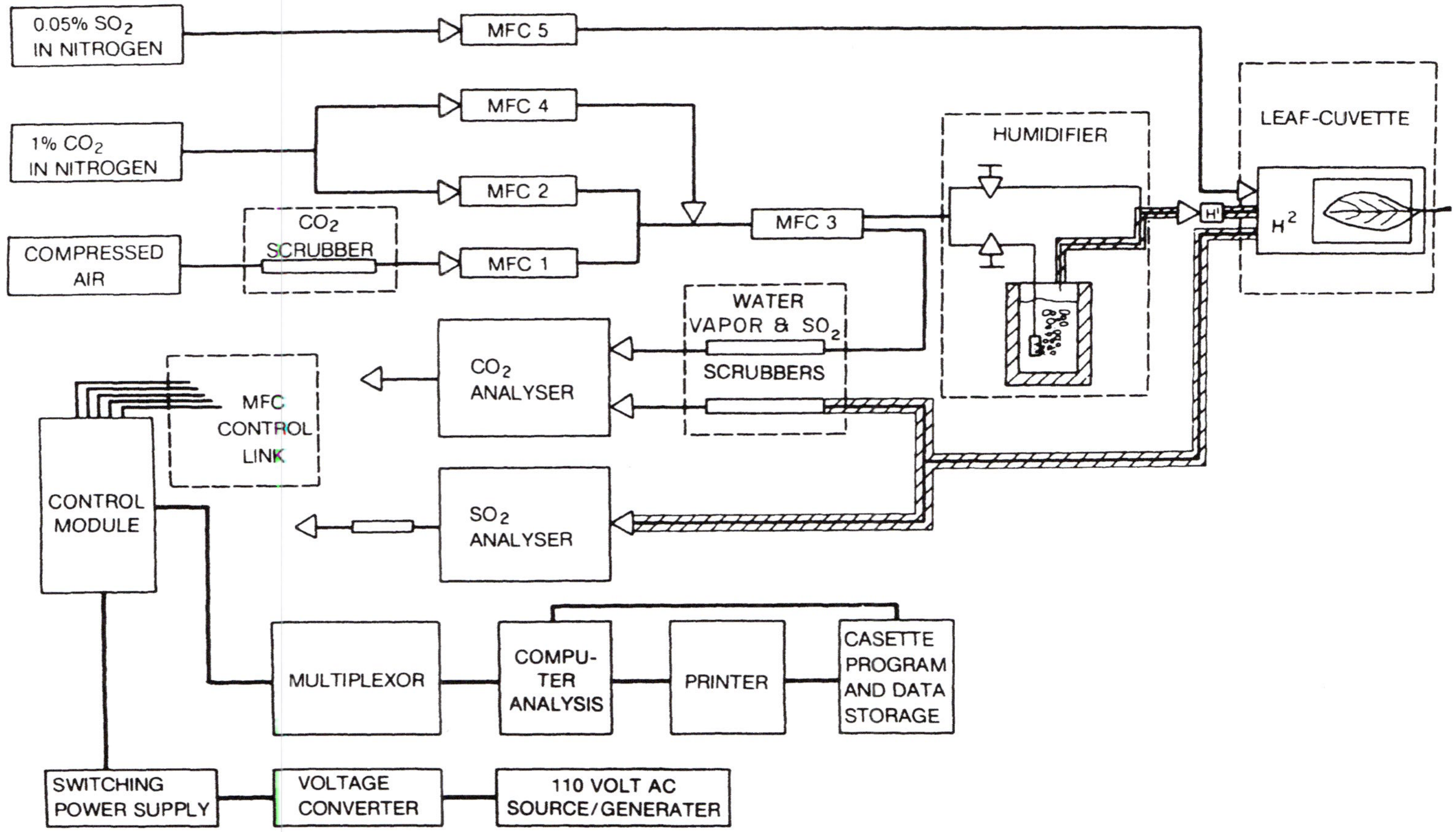

Fig. 17.2 Schematic overview of the Armstrong Enterprises (Palo Alto, CA, USA) portable gas-exchange–fumigation system showing the cuvette, humidifier, air-mixing equipment, mass flow controllers (MFC 1–5), gas analyzers, planned computer interfacing, insulated (///) and uninsulated (–) gas flow lines and wiring (MFC control link to control module) (Atkinson *et al.*, 1986).

under constant conditions so that results from many experiments can be compared (Winner *et al.*, 1982). The interaction of gaseous pollutants with environmental factors can also be studied (Norby and Kozlowski, 1981, 1982; Jones and Mansfield, 1982). These systems can be used to determine the effects of gaseous pollutants on the responses of leaves to wide ranges of light, CO_2 and vapor pressure deficit. The resultant response curves are useful for diagnosis of air pollution-caused changes in quantum yield and intrinsic photosynthetic capacity of mesophyll cells.

Type 3. *Null-balance* closed-loop gas-exchange systems equipped with flow-through cuvettes are usually designed for environmental control and can be used as fumigation chambers in long-term experiments. Units commercially available without temperature control can be used to survey gas-exchange rates for leaves of plants in air pollution exposure chambers or in other polluted environments. The Armstrong Enterprises system (Palo Alto, CA, USA; Fig. 17.2; Atkinson *et al.*, 1986) can be operated either in the differential flow-through or null-balance closed-loop mode.

Type 4. *Cuvettes for radioisotope studies* can help show the effects of air pollutants on photosynthesis, phloem and transport functions, and to help quantify air pollution absorption rates and translocation within plants. For example, $^{14}CO_2$ and $^{11}CO_2$ can be introduced into cuvettes containing leaves previously fumigated with SO_2 or O_3. Patterns of isotope distribution for pollution-treated leaves can be compared with patterns observed for control leaves to reveal the effects of treatments on photosynthesis and patterns of carbon translocation. $^{35}SO_2$ can also be introduced into cuvettes to quantify SO_2 absorption, metabolic fates of SO_2 and the translocation of SO_2 and its metabolites. Thus cuvettes are generally useful for studies with radioisotopes which can be introduced in gaseous form.

17.5 FIELD FUMIGATION SYSTEMS AND APPROACHES

17.5.1 Field fumigation chambers

(a) Open top chambers

Air pollutants such as O_3 and SO_2 are often dispersed over large regions. For example, O_3 is known to occur throughout much of eastern North America and southern California during much of the growing season (Fig. 17.3). Since there are no pollution-free 'control' areas, the regional distribution of air pollutants makes their impact on plants difficult to assess. In addition, environmental differences between sites, such as changes in soil types and contrasts in day length, temperature and water availability, make it difficult to compare air pollution impact data from two sites. Open-top fumigation chambers offer an approach for monitoring growth responses of plants to air pollutants at a single field site (Heagle *et al.*, 1979). Plants are typically raised in these chambers for a full growing season and then assessed for air pollution-caused injury.

Open-top chambers are cylinders of clear plastic which are about 3 m in diameter and 2.5 m high and were originally designed in the early 1970s (Heagle *et al.*, 1973; Mandl *et al.*, 1973). Clear plastic sides are stretched over an aluminum framework. A large fan located outside the chamber blows ambient air through a duct and into a plenum surrounding the base of the chamber. The air is vented through holes in the plenum into the cylinder, flows up through the chamber and exits through the open top. Positive pressure from the fan minimizes the amount of ambient air which might spill into the chamber through the open top. High winds can result in the incursion of ambient air, with its pollutants, into the top of the chamber resulting in the loss of treatment control. The ingress of ambient pollutants into open-top chambers can be reduced by the use of a frustrum and/or an internal collar (Buckenham

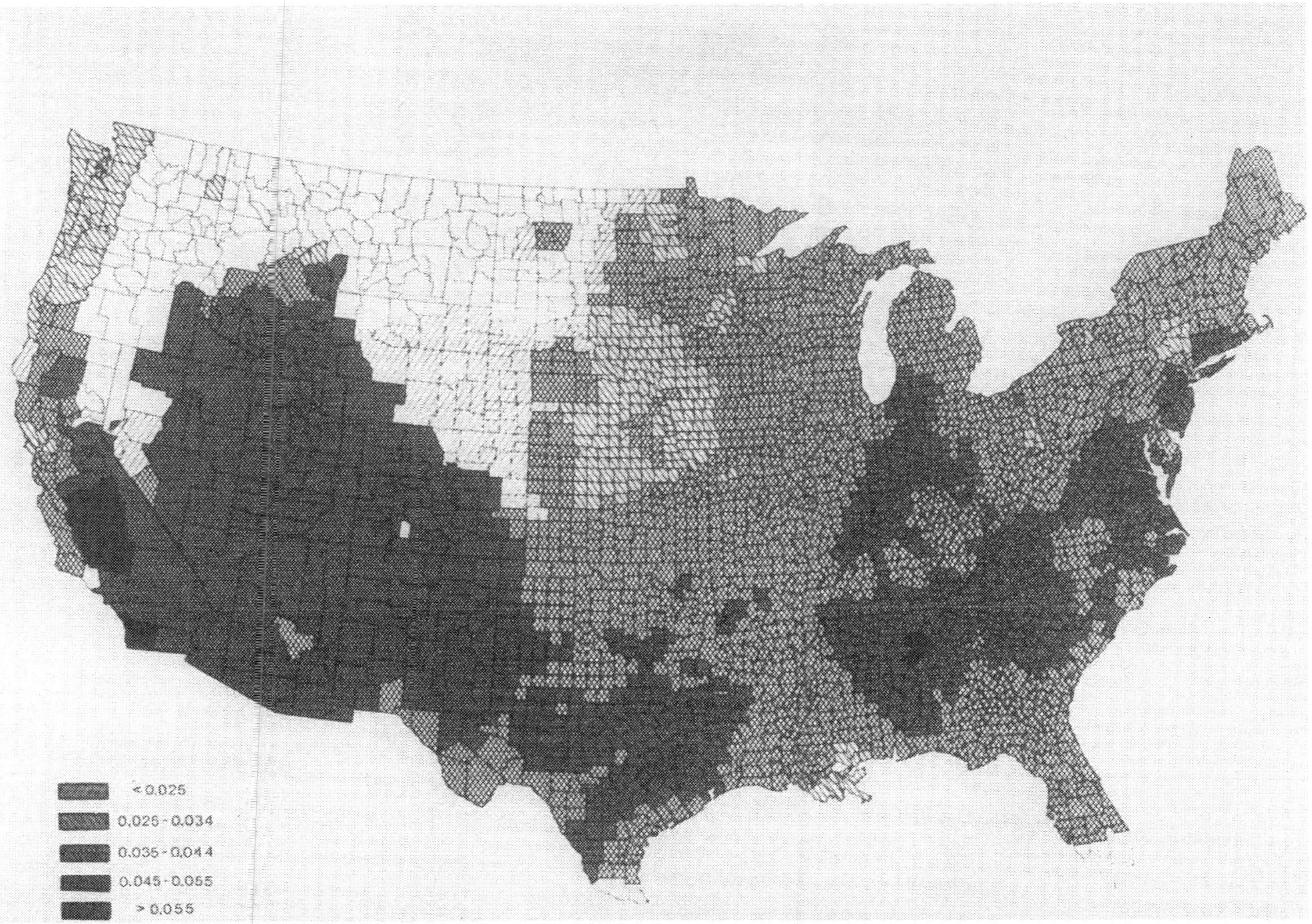

Fig. 17.3 Seasonal 7 h mean O_3 concentrations in the contiguous 48 states of the US in 1982 (courtesy of D.T. Tingey, US EPA, Corvallis, OR).

Fig. 17.4 Photograph of a modified open-top chamber. The modifications include a frustrum to reduce ingression of ambient air, a rain cap and netting over the open portion of the chamber to exclude insects. (Courtesy of W.E. Hogsett, US EPA, Corvallis, OR.)

et al., 1981; Fig. 17.4). The basic design of the open-top chamber has been modified to a rectangular form large enough to accommodate moderate-sized trees (Fig. 17.5).

Air pumped through these chambers can be treated in order to manipulate concentrations of pollutants in the chamber. Untreated unfiltered air will contain ambient pollutants. Charcoal filters installed at the blower can be used to scrub ambient pollutants and provide a control (except for precipitation) treatment. This filtered treatment is usually considered the experimental control although some gases such as NO and SO_2 are not efficiently scrubbed with charcoal. At sites where the ambient O_3 concentration is above 0.06 ppm, charcoal filters only reduce O_3 levels to about 0.025 ppm, so that even the control treatment is not pollution free. Finally, pollutants such as SO_2, O_3 and NO_x from artificial sources can be added to the air going into the chamber. The filtration and supplementation of pollutants offer a full range of pollution treatments for experiments in which plants are raised for an entire growing season. For example, O_3 exposures can be designed for constant daily concentrations or can be computer controlled to yield treatments designed to test hypotheses about plant responses to changing pollution levels.

One approach to using these chambers is to plant a field crop using appropriate agronomic processes and then to position open-top chambers over the plants. These chambers can also be located at permanent stations and

Fig. 17.5 Photograph of open-top chamber designed for fumigating larger woody plants. (Courtesy of R. Kohut, Boyce Thompson Institute.)

plants placed directly into them. Three chambers are commonly used to replicate a treatment and experiments usually have four air pollution treatments. Thus 12 chambers are commonly used to complete one experiment.

Differences in environmental factors between the chamber and the open field can be large because the environment inside the chamber has patterns of air movement which differ from those in the field. Light, air temperature, humidity and soil water availability for plants inside the chamber may also differ from the conditions for plants outside the chambers. For example, the environment in open-top chambers in Wisconsin had a 20% reduction in light, 2° C temperature increase and 10% decrease in evaporative loss compared with the neighboring non-chamber environment (Olszyk *et al.*, 1980). The environmental modifications caused by the chambers differ between sites, and within or between days at the same site. This occurs because the extent of environmental modification is related to the magnitude of solar radiation, ambient humidity, amount of vegetation in chambers, and other physical and biological factors (Unsworth, 1982). To measure the chamber effect, open-air plots are commonly planted. Three circles the diameter of the chambers are prepared and planted as for chambers but without the framework, plastic or blowers. Plant growth in these plots is compared with that in the chambers receiving ambient air.

Biotic agents, such as insects, fungi, bacteria, and viruses may enter the chambers and cause visible foliar injury or changes in plant growth which might be construed as air pollution damage. Environmental stresses, such as excessive heat and light, drought and nutrient deficiencies can also cause visible injury to leaves which could be misinterpreted as air pollution injury. In addition, environmental stresses and disease agents can influence foliar injury symptoms. Visible foliar injury is caused by air pollutants reacting with compounds found in the leaf. As environmental stresses cause the amount of water, nitrogen, phenols, etc. in leaves to change, the type of visible injury which develops from absorbed air pollution may also change.

Precipitation may also cause complications in an open-top chamber experiment. Rainfall may be unusually high or low during the season and may not fall uniformly across all chambers or within a chamber. Rainfall chemistry is also an uncontrolled variable which may affect plants.

(b) Enclosed flow-through chambers

Field fumigation chambers have been built that are enclosed and therefore have no problems with ingress of ambient air pollutants. All air is filtered and pollutants are added as needed for specific experimental designs. Differences in amounts of ambient rainfall within and between chambers are eliminated because of the enclosure and plants are uniformly irrigated with the chosen solution. These chambers commonly are 2.2–2.5 m in diameter and have transparent walls (Ashenden *et al.*, 1982; Musselman *et al.*, 1985; Fig. 17.6). Enclosed field chambers must have rapid air turnover rates to prevent overheating. Closed chambers are generally located in fixed positions and house potted plants, so they are similar to greenhouses with large air-handling systems. These chambers can be equipped with lights and other devices to minimize environmental gradients in chambers and to help reliably reconstruct environmental conditions from experiment to experiment.

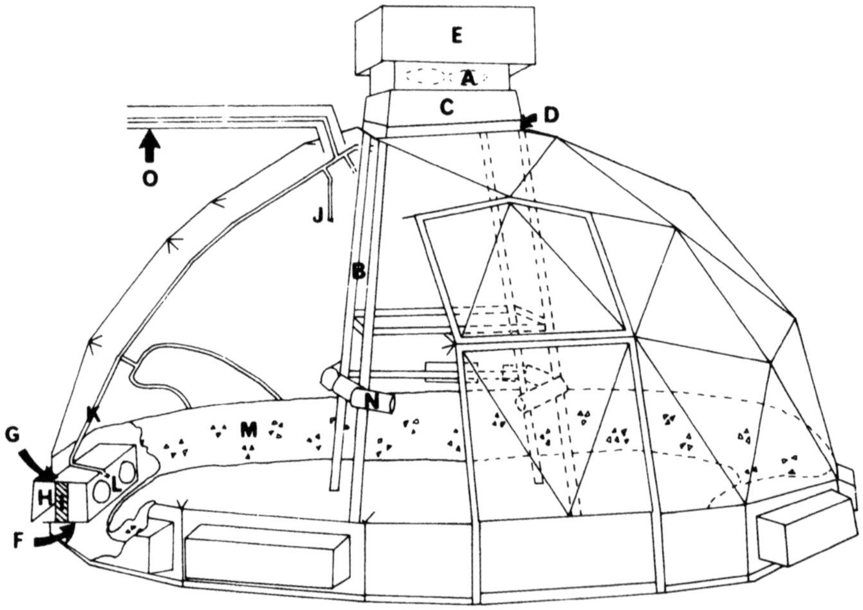

Fig. 17.6 Diagram of an enclosed chamber used by Ashenden *et al.* (1982). A, extractor fan; B, angle iron frame; C, aluminum sheet; D, rubber collar; E, cowl; F, open-ended box; G, upper side of box; H, cowls; I, charcoal filter; J, condensation trap; K, supply pipe (pollutant); L, polythene sheet baffle over filter box; M, mixing tunnel; N, tangential fan; O, lagged drainpipe.

17.5.2 Rainfall fog/simulators

Field studies aimed at assessing the effects of rainfall chemistry on crops have involved excluding ambient rain from plants and generating simulated rain. Simulated rainfall solutions are prepared by diluting premixed solutions of rainfall concentrate containing various ions with deionized water. The supply of deionized water at field sites must be large enough to provide simulated rain in the same quantity as that falling as ambient rain. Rainfall exclusion over field crops has been achieved by erecting transparent roofs over large areas of fields. In some cases the roofs are permanently located (Irving and Miller, 1981); in other cases they are located on tracks and roll forward to cover plants only during periods of rain (Evans *et al.*, 1982). Although both permanently fixed and retractable rainfall exclusion devices effectively eliminate rainfall, these two systems have important differences. In general, the retractable cover systems allow normal dry deposition processes to take place between rain storms. Permanent roofs may alter the quantity and quality of light reaching the plants. The patterns of dry deposition under permanently fixed roofs are difficult to assess, but edge effects near the roof line are apt to be large. Rainfall under exclusion devices is applied with nozzle systems and usually involves piping large quantities of simulated rain solutions to dispensing units. The piping and nozzles are usually located under, and attached to, the roof. Nozzles must be carefully located to insure uniform coverage of rain over treatment areas and to prevent the drift of small droplets intended for one treatment zone into another. Since these systems involve plants and nozzles at fixed locations, experiments must be designed to account for treatment errors.

Experiments involving only rainfall exclusion roofs, whether permanently fixed or moveable, have some limitations. Since O_3 is almost always present in ambient air at such field sites, growth responses detected in experiments might be due not to acid rain alone, but rather to combinations of acid rain and O_3. The inability to factor out the role of O_3 in these experiments also points to the problem of lack of control for other environmental variables, and the inability to achieve such control with modest add-on development.

17.5.3 Systems for controlling both wet and dry deposition

Ambient rain can pose problems in field experiments involving gaseous pollutants because rainfall quantities, rates and chemistry may not be uniform either across a field site or even within a chamber. One approach for eliminating rainfall from open-top chambers is to equip chambers with retractable covers (Fig. 17.7a). The gap between the top of the chamber and the cover allows a normal pattern of air movement through the chamber even when the covers are positioned over the chambers. Although the covers are transparent, they do reduce light levels for plants and for this reason retractable covers for open-top chambers are preferred over permanent covers. Retractable covers may operate by responding to sensors which detect rain (rain sensor, Wong Laboratories, Cincinnati, OH, USA) or by requiring site operators to swing the covers into place manually. The motorized system may be less reliable than the more labor-intensive practice of positioning covers by hand. Irrigation must be provided for plants regardless of which type of rain exclusion device is used.

Rainfall simulators have been developed for open-top chambers equipped with rainfall exclusion covers (Johnston *et al.*, 1986; Fig. 17.7b). This innovation allows the control of both gaseous pollutants and rainfall chemistry in open-top chamber experiments. The addition of rainfall as an experimental variable in open-top chamber experiments which are

a

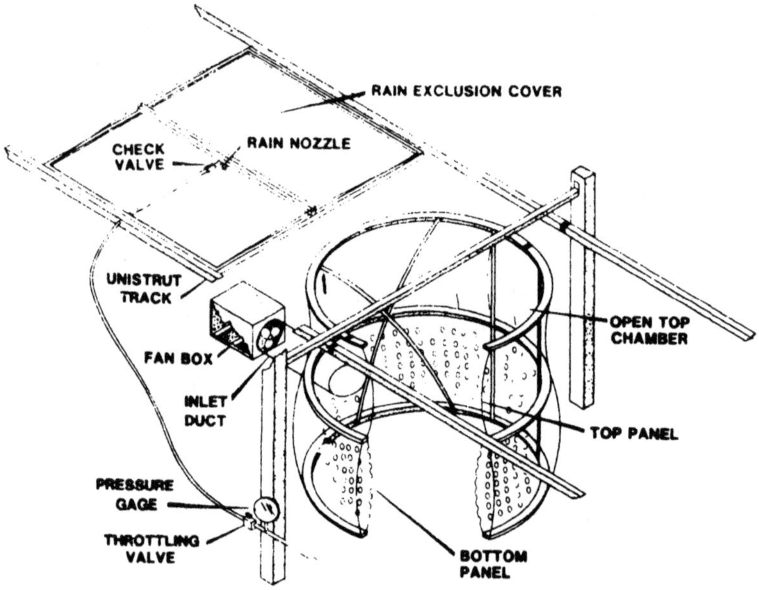

b

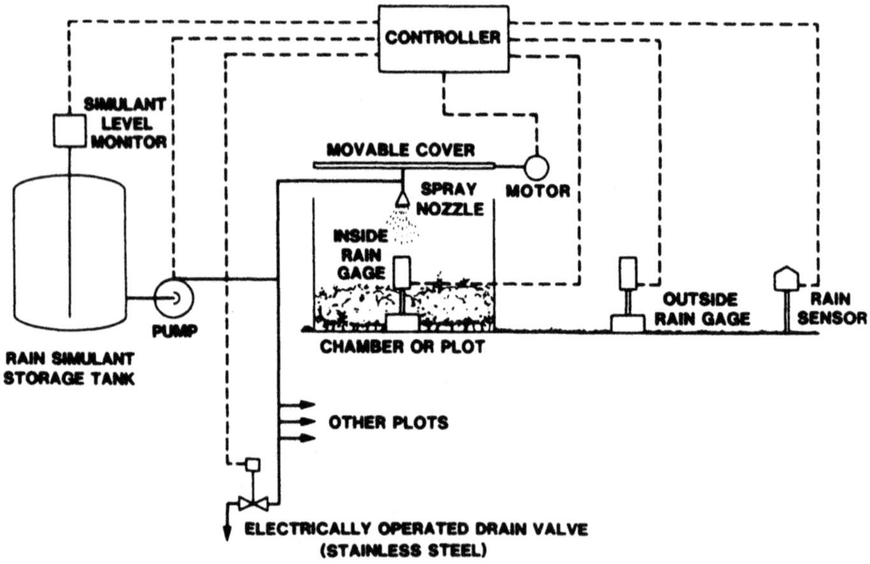

Fig. 17.7 (a) Detail of an open-top chamber with rain exclusion cover (Johnston *et al.*, 1986). (b) Illustration of the control circuitry for the automatic operation of rain exclusion and simulant distribution systems for an open-top chamber (Johnston *et al.*, 1986).

also testing the effects of a gaseous pollutant, such as O_3, increases the number of chambers needed for a study. A multifactorial experiment involving four pH levels and four levels of O_3 and replicated three times would require 48 chambers.

Equipment has also been developed to dispense fog directly in open-top chambers (Musselman *et al.*, 1985). The chemistry, droplet size and duration of fogging period can be controlled by regulating the pressure at which solutions are delivered to fog-making nozzles. Stainless-steel nozzles are available from Bete Fog Nozzle, Greenfield, MA, USA and Mee Ind., San Gabriel, CA, USA. In the system developed by Musselman and colleagues (1985), the simulated fog solution is pressurized to 100 psi resulting in fog droplets 20 μm in diameter. The deposition of fog on to leaf surfaces in fields is in part dependent upon winds which drive the droplets into the canopy. Since horizontal winds are absent from open-top chambers, this component of the deposition process is lacking, and may be reproducible only in wind tunnels. Such tunnels are not only used to test patterns of air movements around field chambers but also modified for sophisticated analysis of air pollution deposition on to plant leaf surfaces (Ligotke *et al.*, 1986).

17.5.4 Chamberless systems

Owing to inherent limitations of field chambers, efforts to develop chamberless exposure systems have been underway since the mid-1970s. The earliest and simplest experimental outdoor fumigations were releases of pure gases from single point sources for treatment of lichens (Skye, 1968). Lack of control resulted in large fluctuations in concentration within the target area.

The zonal air pollution system (ZAPS) has been the most frequently tested and utilized chamberless system. It was developed independently in the mid-1970s in France (deCormis *et al.*, 1975) and the US (Lee and Lewis, 1976, 1978). The basic components of the ZAPS are a pollutant source (usually a cylinder of SO_2), a compressor or blower, a series of aluminum pipes with holes drilled at regular intervals (which constitutes the air pollution dispensing manifold and is suspended above the target area), and pollutant sampling lines and monitor.

The pollutant is injected into the air stream provided by the compressor, travels through the pipes and is released through the small holes. Vertical pipes have been used to release gas at 0.5 and 2 m above the ground to create a relatively homogeneous distribution (deCormis *et al.*, 1975), but horizontal pipes with holes about 0.8 m above the ground have also been employed (Lee and Lewis, 1978). 'Hot spots' can occur close to the source SO_2 outlets, but within 1 m of the pipe concentrations become comparable to those of locations half-way between delivery lines (Lee and Lewis, 1978).

The frequency distributions of median SO_2 concentrations were approximately log normal in Lee and Lewis's (1978) system, which they claimed is typical around actual point sources. These researchers (Lee and Lewis, 1978) were able to maintain different SO_2 distributions on four plots. SO_2 concentrations in chamberless fumigation systems may be neither normally nor log normally distributed, but intermediate. Miller *et al.* (1980) controlled the concentrations in their plots to prevent them from exceeding certain limits, and they felt that this caused an observed departure from log normality. The continuous fluctuation of the SO_2 concentrations during the fumigations was considered advantageous (Miller *et al.*, 1980) because it made the exposures more similar to that of actual SO_2 releases from a point source.

The development of the ZAPS was advanced by introducing computer control of SO_2 release (Greenwood *et al.*, 1982). The gas was emitted from a line source placed sufficiently far upwind of the main treatment areas so that the gradient of mean concen-

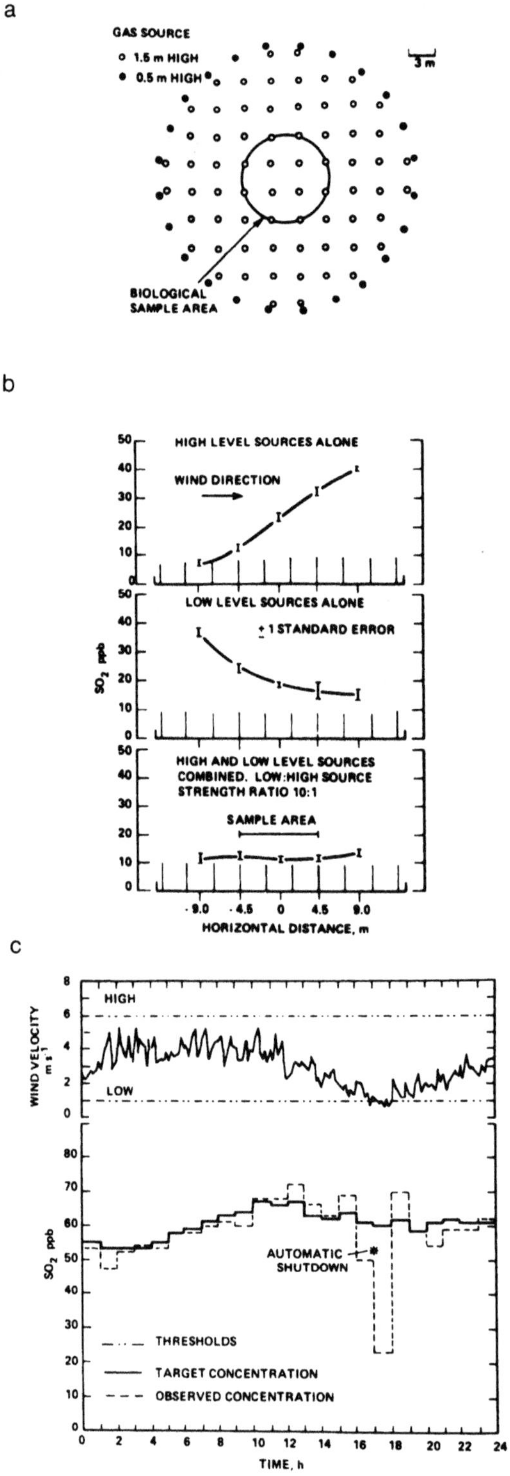

tration across the area was small compared with the range of short-term concentrations to which plants were exposed. SO_2 was monitored along the upwind, center and downwind edge of the treatment zone. The computer program was designed to maintain the mean at 0.1 ppm above background, with adjustments made if the concentration differed by more than 0.02 ppm from the projected value.

McLeod *et al.* (1985) adopted deCormis *et al.*'s (1975) concept of releasing SO_2 from two heights in the canopy and combined this with computer control of gas release. Low-level sources emitting SO_2 at 0.5 m surrounded high-level sources perforated at 1.5 m above the ground in a circular pattern (Fig. 17.8a,b). The center of each plot was the area monitored. The delivery system was designed to limit the SO_2 concentration in the release pipes to 1% or less in an effort to avoid hot spots. When the wind velocity dropped below 1 m s^{-1}, the SO_2 tank was automatically shut off and turned back on when the wind speed rose again. Observed SO_2 concentrations closely followed target concentrations, except when wind speeds were very low and the SO_2 was shut off (Fig. 17.8c).

Another type of chamberless fumigation system is the linear gradient system (Shinn *et al.*, 1977; Laurence *et al.*, 1982; Reich *et al.*, 1982). The system consists of three perforated polyethylene tubes connected to a blower and pollutant-mixing chamber for exposure of four rows of plants. The frequency and size of the exit holes generate a gradient of pollutant levels down the length of tubes. Because the tubes are laid along the ground, the gas is released near the base of the plants

Fig. 17.8 Schematic diagram of a chamberless air pollution fumigation system showing the pattern of high- and low-level sources (a), concentrations measured when high- and low-level sources are used alone and in combination (b) and target and observed SO_2 concentrations at different wind speeds (c) (McLeod *et al.*, 1985).

and is blown up through the canopy, reversing the usual vertical pollutant concentration profile seen in ambient air. An extensive air pollution sampling and monitoring system is required. Advantages of the system include: a graded series of exposures can be conducted in a small area, plants may be grown according to accepted agronomic practices, the cost is relatively low and more than one air pollutant can be applied.

Chamberless systems allow more plants to grow under natural conditions where the environmental modification is the addition of air pollutants. Crops can be planted into local soil and cultivated, fertilized, etc. according to common agricultural procedures. Native species in a relatively undisturbed ecosystem can be studied. Free movement of animals, pollinators and disease agents can take place. Light intensity, wind and temperature are unaffected by the system, and plant growth and morphology are normal. Flexibility is built into the ZAPS so that pipes can be moved, holes can be drilled at different intervals, and the pollutant concentration in the pipes altered to achieve desired concentrations and exposure periods. Both ZAPS and linear gradient setups simulate the type of exposure that would occur from a point source in ambient air.

One drawback with chamberless fumigation systems is that environmental factors other than air pollution treatments may have effects on plants which are not uniform across the study area. Such effects may be difficult to either detect or consider when attempting to measure air pollution impacts. For example, insects, diseases, hail or rainfall may affect only part of the field. Unless these variables are continuosly monitored, undetected or uncontrolled factors may be mistaken for either enhanced or reduced air pollution damage. Another drawback is that chamberless systems can only add the air pollutant treatments to the background pollution already in the air, thereby precluding studies of an air pollutant at levels lower than those of ambient conditions. Thus, these systems are best suited for sites where concentrations of ambient pollutants are known to be low. For the ZAPS especially, fumigations can only be conducted when the wind is blowing from the proper direction; otherwise control plots may be exposed. Certain wind speeds and sudden changes in wind directions may cause variations in pollutant concentration within a plot. It is more difficult to characterize doses in the chamberless systems than in chambers; the mean alone is not enough to describe the dose the plants have received (Miller *et al.*, 1980).

Undoubtedly, further refinements to ZAPS and linear gradient systems will be made, and new styles of open-air fumigation systems will be developed in the future. When parallel experiments are carried out in both chambers and chamberless systems, the combined knowledge gained will be of great value in furthering our understanding of air pollutant impacts on plants.

17.5.5 Gradient studies

One approach for studying responses of plants and plant communities to air pollutants is to develop studies along gradients of air pollution concentrations and exposure periods. The advantage of exploiting such gradients is that air pollution exposures occur naturally and do not require construction of fumigation equipment. The most successful studies along air pollution stress gradients usually involve a relatively small geographical area (less than 200 km^2), a uniform vegetation type, a uniform topography and a single air pollution source with a well-defined pattern of plume dispersal. Attempts to study plants along regional gradients of O_3 and acid deposition stress may provide interesting data concerning air pollution impacts. However, these studies must be carefully designed to understand how all pollutants, environmental variables and species distributions change along transects which may be greater than 200 km in length and span one or several states.

One example of research at a well-defined study site is located near Fox Creek, Alberta, Canada, where a natural gas refinery is a point source for SO_2 (Winner and Bewley, 1978). SO_2 produced from the refinery is dispersed by prevailing winds from the northwest creating SO_2 concentration gradients. Fifteen plots in ecologically similar stands were selected and canopy coverage for each understory species, in each stand, was measured using standard synecological techniques. The stands were located progressively farther downwind and upwind from the refinery. Comparison of the coverage of each species on a stand-by-stand basis allowed determination of species tolerant of SO_2, species sensitive to SO_2, and weedy species which invaded the disturbed community. Species differing in SO_2 tolerance were then selected for physiological studies designed to determine why such differences occurred (Winner and Bewley, 1983). Similar approaches can be used by physiological ecologists to exploit further plant species found along air pollution concentration gradients.

17.6 SUMMARY

Studies of plant responses to air pollutants now include analysis of physiological, growth, and ecological processes for pollutants deposited in air and precipitation over vast regions. These experiments require well-conceived designs that exploit a variety of research facilities. Such facilities include cuvettes that can be used to define physiological responses of plants to air pollutants, to characterize air pollution absorption rates and to help understand the influence environmental factors have on plant responses to air pollutants. Field fumigation techniques may utilize either chamber or chamberless designs. The chamber systems result in environmental modifications whereas chamberless systems may have problems with environmental heterogeneity across field sites. Most research to date has focused on approaches for fumigating plants with O_3, SO_2 and NO_2; however, fumigation systems have more recently been built that also allow control of rainfall rates and chemistry. Studies using stable isotopes, radioisotopes, chlorophyll fluorescence, plant water relations and energy budget analysis techniques will be important in the future. Air pollution studies can embrace an array of environmental and biological factors and focus on plant processes, thereby providing physiological ecologists with excellent opportunities to extend their work into this important area of plant science.

REFERENCES

Ashenden, T.W., Tabner, P.W., Williams, P., Whitmore, M.E. and Mansfield T.A. (1982) A large-scale system for fumigating plants with SO_2 and NO_2. *Environ. Pollut. (Ser. B)*, **3**, 21–6.

Atkinson, C.J., Winner, W.E. and Mooney, H.A. (1986) A field portable gas exchange system for measuring carbon dioxide and water vapour exchange rates of leaves during fumigation with SO_2. *Plant, Cell Environ.*, **9**, 711–19.

Beckerson, D.W. and Hofstra, G. (1979) Response of leaf diffusive resistance of radish, cucumber, and soybean to O_3 and SO_2 singly or in combination. *Atmos. Environ.*, **13**, 1263–8.

Biggs, A.R. and Davis, D.D. (1980) Stomatal response of three birch species exposed to varying acute doses of SO_2. *J. Am. Soc. Hort. Sci.*, **105**, 514–16.

Black, V.J. and Unsworth, M.H. (1979a) A system for measuring effects of sulphur dioxide on gas exchange of plants. *J. Exp. Bot.*, **114**, 81–8.

Black, V.J. and Unsworth, M.H. (1979b) Effects of low concentrations of sulphur dioxide on net photosynthesis and dark respiration of *Vicia faba*. *J. Exp. Bot.*, **30**, 473–83.

Black, V.J. and Unsworth, M.H. (1979c) Resistance analysis of sulphur dioxide fluxes to *Vicia faba*. *Nature, London*, **282**, 68–9.

Buckenham, A.H., Parry, M.A., Whittingham, C.P. and Young, A.T. (1981) An improved open-topped chamber for pollution studies on crop growth. *Environ. Pollut. (Ser. B)*, **2**, 475–82.

deCormis, L., Bonte, J. and Tisne, A. (1975)

Technique experimentale permettant l'etude de l'incidence sur la vegetation d'une pollution par le dioxyde de soufre appliquee en permanence et a'dose subnecrotique. *Pollut. Atmos.*, **66**, 103–7.

Dollard, G.J. and Unsworth, M.H. (1983) Field measurements of turbulent fluxes of wind-driven fog drops to a grass surface. *Atmos. Environ.*, **17**, 775–80.

Eckert, R.T. and Houston, D.B. (1980) Photosynthesis and needle elongation response of *Pinus strobus* clones to low level sulfur dioxide exposures. *Can. J. For. Res.*, **19**, 357–61.

Evans, L.S., Levin, K.F., Cunningham, E.A. and Patti, M.J. (1982) Effects of simulated acidic rainfall on yields of field-grown crops. *New Phytol.*, **91**, 429–41.

Fowler, D. (1985) Deposition of SO_2 onto plant canopies. In *Sulfur Dioxide and Vegetation: Physiology, Ecology, and Policy Issues* (eds W.E. Winner, H.A. Mooney, and R.A. Goldstein), Stanford University Press, Stanford, CA, pp. 389–402.

Galloway, J.N. and Likens, G.E. (1976) Calibration of collection procedures for the determination of precipitation chemistry. *Water, Air, Soil Pollut.*, **6**, 241–58.

Galloway, J.N. and Likens, G.E. (1978) The collection of precipitation for chemical analysis. *Tellus*, **30**, 71–82.

Galloway, J.N., Likens, G.E. and Hawley, M.E. (1984) Acid precipitation: natural versus anthropogenic components. *Science*, **226**, 829–31.

Giever, P.M. (1976) Particulate matter sampling and sizing. In *Air Pollution. Vol. III. Measuring, Monitoring, and Surveillance of Air Pollution* (ed. A.C. Stern), Academic Press, New York, pp. 3–50.

Greenwood, P., Greenhalgh, A., Baker, C. and Unsworth, M. (1982) A computer-controlled system for exposing field crops to gaseous air pollutants. *Atmos. Environ.*, **16**, 2261–6.

Guderian, R. (ed.) (1985) *Air Pollution by Photochemical Oxidants*, Springer-Verlag, New York.

Hallgren, J.E., Linder, S., Richter, A., Troeng, E. and Granat, L. (1982) Uptake of SO_2 in shoots of Scots Pine: field measurements of net flux of sulphur in relation to stomatal conductance. *Plant, Cell Environ.*, **5**, 75–83.

Heagle, A.S., Body, D.E. and Heck, W.W. (1973) An open-top field chamber to assess the impact of air pollution on plants. *J. Environ. Qual.*, **2**, 365–72.

Heagle, A.S., Philbeck, R.B., Rogers, H.H. and Letchworth, M.B. (1979) Dispensing and monitoring ozone in open-top field chambers for plant-effects studies. *Phytopathology*, **69**, 15–20.

Hogsett, W.E., Olszyk, D., Ormrod, D.P., Taylor G.E. Jr., and Tingey, D.T. (1987) *Air Pollution Exposure Systems and Experimental Protocols*, Vols. 1 and 2. US Environmental Protection Agency, No. EPA/600/3-87/037a and b.

Houston, D.B. (1974) Responses of selected *Pinus strobus* clones to fumigation with sulfur dioxide and ozone. *Can. J. For. Res.*, **4**, 65–8.

Irving, P.M. and Miller, J.E. (1981) Productivity of field-grown soybeans exposed to acid rain and sulfur dioxide alone and in combination. *J. Environ. Qual.*, **10**, 473–8.

Jacob, D.J., Waldman, J.M., Munger, J.W. and Hoffman, M.R. (1985) Chemical composition of fog water collected along the California coast. *Environ. Sci. Technol.*, **19**, 730–46.

Jensen, K.F. (1981) Growth analyses of hybrid poplar cuttings fumigated with ozone and sulphur dioxide. *Environ. Pollut. (Ser. A)*, **26**, 243–50.

Johnson, C.A., Sigg, L. and Zobrist, J. (1987) Case studies on the chemical composition of fogwater: the influence of local gaseous emissions. *Atmos. Environ.*, **21**, 2365–74.

Johnston, J.W., Jr., Shriner, D.S. and Abner, C.H. (1986) Design and performance of an exposure system for measuring the response of crops to acid rain and gaseous pollutants in the field. *J. Air Pollut. Control Assoc.*, **36**, 894–9.

Jones, T. and Mansfield, T.A. (1982) The effect of SO_2 on growth and development of seedlings of *Phleum pratense* under different light and temperature environments. *Environ. Pollut. (Ser. A)*, **27**, 57–71.

Kimmerer, T.W., and Kozlowski, T.T. (1981) Stomatal conductance and sulfur uptake of five clones of *Populus tremuloides* exposed to sulfur dioxide. *Plant Physiol.*, **67**, 990–5.

Kondo, N., Maruta, I. and Sugahara, K. (1980) Effects of sulfite and pH on abscisic acid-dependent transpiration and on stomatal opening. *Plant Cell Physiol.*, **21**, 817–28.

Kretzschmer, J.G. and Pauwels, J.B. (1982) Comparison between six different instruments to determine suspended particulate matter levels in ambient air. In *Atmospheric Pollution, 1982: Proceedings of the 15th International Colloquium, UNESCO*, Paris, France (ed. M.M. Benarie), Elsevier Scientific Publishing Co., New York, pp. 265–72.

Laurence, J.A., MacLean, D.C., Mandl, R.H., Schneider, R.E. and Hansen, K.S. (1982) Field tests of a linear gradient system for exposure of row crops to SO_2 and HF. *Water, Air, Soil Pollut.*, **17**, 399–407.

Lee, J.J. and Lewis, R.A. (1976) Field experimental component. In *The Bioenvironmental Impact of a*

Coal-fired Power Plant, 1st Interim Report (eds J.J. Lee and R.A. Lewis), US Environmental Protection Agency, Corvallis, OR, EPA-600/3-76-002, pp. 95–101.

Lee, J.J. and Lewis, R.A. (1978) Zonal air pollution system: design and performance. In *The Bioenvironmental Impact of a Coal-fired Power Plant, 3rd Interim Report* (eds J.J. Lee and R.A. Lewis), US Environmental Protection Agency, Corvallis, OR, EPA-600/3-78-021, pp. 322–44.

Ligotke, M., Cataldo, D., van Voris, P. and Novich, C. (1986) Analysts use wind tunnel to study particle behavior in the environment. Battelle Research and Development, Pacific Northwest Laboratories, Richland, WA.

Lindberg, S.E., Lovett, G.M., Richter, D.D. and Johnson, D.W. (1986) Atmospheric deposition and canopy interaction of major ions in a forest. *Science*, **231**, 141–5.

Mandl, R.H.,. Weinstein, L.H., McCune, D.L. and Keveny, M. (1973) A cylindrical open-topped chamber for exposure of plants to air pollutants in the field. *J. Environ. Qual.*, **2**, 371–6.

McLaughlin, S.B. (1985) Effects of air pollution on forests: A critical review. *J. Air Pollut. Contr. Assoc.*, **35**, 512–34.

McLaughlin, S.B., Shriner, D.S., McConathy, R.K. and Mann, L.K. (1979) The effects of SO_2 dosage kinetics and exposure frequency on photosynthesis and transpiration of kidney beans (*Phaseolus vulgaris* L.). *Environ. Exp. Bot.*, **19**, 179–91.

McLeod, A.R., Fackrell, J.E. and Alexander, K. (1985) Open-air fumigation of field crops: criteria and design for a new experimental system. *Atmos. Environ.*, **19**, 1639–49.

Miller, D.R., Byrd, J.E. and Perone, M.J. (1987) The source of Pb, Cu, and Zn in fogwater. *Water, Air, Soil Pollut.*, **32**, 329–40.

Miller, J.E., Sprugel, D.G., Muller, R.N., Smith, H.K. and Xerikos, P.B. (1980) Open-air fumigation system for investigating sulfur dioxide effects on crops. *Phytopathology*, **70**, 1124–8.

Mooi, J. (1980) Influence of ozone on the growth of two poplar cultivars. *Plant Dis.*, **64**, 772–3.

Musselman, R.C., Sterrett, J.L. and Granett, A.L. (1985) A portable fogging apparatus for field or greenhouse use. *HortSci.*, **20**, 1127–9.

National Research Council (1977) *Ozone and Other Photochemical Oxidants*, National Academy of Sciences, Washington, DC.

National Research Council (1978) *Sulfur Oxides*, National Academy of Sciences, Washington, DC.

National Research Council (1986) *Acid Deposition: Long Term Trends*, National Academy of Sciences, Washington, DC.

Noble, R.D. and Jensen, K.F. (1983) An apparatus for monitoring CO_2 exchange rates in plants during SO_2 and O_3 fumigation. *J. Exp. Bot.*, **34**, 470–5.

Norby, R.J. and Kozlowski, T.T. (1981) Relative sensitivity of three species of woody plants to SO_2 at high or low exposure temperature. *Oecologia*, **51**, 33–6.

Norby, R.J. and Kozlowski, T.T. (1982) The role of stomata in sensitivity of *Betula papyrifera* Marsh. seedlings to SO_2 at different humidities. *Oecologia*, **53**, 34–9.

Olszyk, D.M. and Tibbitts, T.W. (1981) Stomatal response and leaf injury of *Pinus strobus* L. with SO_2 and O_3 exposures. *Plant Physiol.*, **67**, 539–44.

Olszyk, D.M., Tibbitts, T.W. and Hertzberg, W.M. (1980) Environment in open-top field chambers utilized for air pollution studies. *J. Environ. Qual.*, **9**, 610–15.

Reich, P.B. and Amundson, R.G. (1985) Ambient levels of ozone reduce net photosynthesis in tree and crop species. *Science*, **230**, 566–70.

Reich, P.B., Amundson, R.G. and Lassoie, J.P. (1982) Reduction in soybean yield after exposure to ozone and sulfur dioxide using a linear gradient exposure technique. *Water, Air, Soil Pollut.*, **178**, 29–36.

Rogers, H.H., Jeffries, H.E., Stahel, E.P., Heck, W.W., Ripperton, L.A. and Witherspoon, A.M. (1977) Measuring air pollutant uptake by plants: a direct kinetic technique. *J. Air Pollut. Contr. Assoc.*, **27**, 1192–7.

Reuss, J.O. and Johnson, D.W. (1986) *Acid Deposition and the Acidification of Soils and Waters*, Springer-Verlag, New York.

Shinn, J.H., Clegg, B.R. and Stuart, M.L. (1977) A minimum field fumigation method for exposing plants to controlled gradients of air pollution levels. Preprint of paper presented at Department of Energy, Division of Biomedical and Environmental Research Workshop, Ecological and Agricultural Effects of Coal Combustion, Oak Ridge, TN.

Skye, E. (1968) Lichens and air pollution. *Acta Phytogeogr. Suec.*, **52**, 1–123.

Smith, W.H. (1981) *Air Pollution and Forests – Interactions between Air Contaminants and Forest Ecosystems*, Springer-Verlag, New York, Heidelberg and Berlin.

Soderlund, R. and Svensson, B.H. (1976) The global nitrogen cycle. In *Nitrogen, Phosphorus and Sulphur – Global Cycles* (eds B.H. Svensson and R. Soderlund), Scope Report No. 7, Ecological Bulletin No. 22, Royal Swedish Academy of Sciences, pp. 23–73.

Taylor, G.E., Jr. and Tingey, D.T. (1979) A gas-exchange system for assessing plant performance in response to environmental stress. Environmental Protection Agency, US. Report No. 600/3-79-108.

Taylor, G.E., Jr., Tingey, D.T. and Ratsch, H.C. (1982) Ozone flux in *Glycine max* (L.) Merr.: Sites of regulation and relationships to leaf injury. *Oecologia*, **53**, 179–86.

Tingey, D.T., Heck, W.W. and Reinert, R.A. (1971) Effect of low concentrations of ozone and sulfur dioxide on foliage, growth and yield of radish. *J. Am. Soc. Hort. Sci.*, **96**, 369–71.

Tingey, D.T. and Reinert, R.A. (1975) The effect of ozone and sulphur dioxide singly and in combination on plant growth. *Environ. Pollut.*, **9**, 117–25.

Tingey, D.T., Reinert, R.A., Dunning, J.A. and Heck, W.W. (1973) Foliar injury responses of eleven plant species to ozone/sulfur dioxide mixtures. *Atmos. Environ.*, **7**, 201–8.

Tingey, D.T., Standley, C. and Field, R.W. (1976) Stress ethylene evolution: a measure of ozone effects on plants. *Atmos. Environ.*, **10**, 969–74.

United States Environmental Protection Agency (1978) *Air Quality Criteria for Ozone and Other Photochemical Oxidants*, US Environmental Protection Agency, Environmental Criteria and Assessment Office, Research Triangle Park, NC, EPA report no. EPA-600/8-78-004. Available from NTIS, Springfield VA; PB80-124753.

United States Environmental Protection Agency (1983) *Air Pollution Training Institute Course* 464, *Analytical Methods for Air Quality Standards, Student Manual*, 2nd edn, Research Triangle Park, NC, EPA 450/2-81-018b.

United States Environmental Protection Agency (1986) *Air Quality Criteria for Ozone and Other Photochemical Oxidants*, US Environmental Protection Agency, Research Triangle Park, NC, EPA/600/8-84/020 cf.

Unsworth, M.H. (1982) Exposure to gaseous pollutants and uptake by plants. In *Effects of Gaseous Air Pollution in Agriculture and Horticulture* (eds M.H. Unsworth and D.P. Ormrod) Butterworth Scientific, London, pp. 43–63.

Waldman, J.M., Jacob, D.J., Munger, J.W. and Hoffmann, M.R. (1987) Pollutant deposition in radiation fog. In *The Chemistry of Acid Rain* (eds R.W. Johnson, G.E. Gordon, W. Calkins and A.Z. Elzerman), American Chemical Society, Washington, DC, pp. 250–7.

Walmsley, L., Ashmore, M.R. and Bell, J.N.B. (1980) Adaptation of radish *Raphanus sativus* L. in response to continuous exposure to ozone. *Environ. Pollut. (Ser. A)*, **23**, 165–77.

West, P.W. (1976) Analysis of inorganic particulates. In *Air Pollution. Vol. III. Measuring, Monitoring, and Surveillance of Air Pollution* (ed. A.C. Stern), Academic Press, New York, pp. 51–97.

Winner, W.E. and Bewley, J.D. (1978) Contrasts between bryophyte and vascular plant synecological responses in an SO_2-stressed white spruce association in central Alberta. *Oecologia*, **33**, 311–25.

Winner, W.E. and Bewley, J.D. (1983) Photosynthesis and respiration of feather mosses fumigated at different hydration levels with SO_2. *Can. J. Bot.*, **61**, 1456–61.

Winner, W.E., Koch, G. and Mooney, H.A. (1982) Ecology of SO_2 resistance: IV. Predicting metabolic responses of fumigated shrubs. *Oecologia*, **52**, 16–22.

Winner, W.E. and Mooney, H.A. (1980a) Ecology of SO_2 resistance: I. Effects of fumigation on gas exchange of deciduous and evergreen shrubs. *Oecologia*, **44**, 290–5.

Winner, W.E. and Mooney, H.A. (1980b) Ecology of SO_2 resistance: II. Photosynthetic changes of shrubs in relation to SO_2 absorption and stomatal behavior. *Oecologia*, **44**, 296–302.

Winner, W.E., Mooney, H.A. and Goldstein, R.A. (eds) (1985a) *Sulfur Dioxide and Vegetation: Physiology, Ecology, and Policy Issues*, Stanford University Press, Stanford, CA.

Winner, W.E., Mooney, H.A., Williams, K. and Von Craemmerer, S. (1985b) The effects of SO_2 on photosynthesis. In *Sulfur Dioxide and Vegetation: Physiology, Ecology, and Policy Issues* (eds W.E. Winner, H.A. Mooney and R.A. Goldstein), Stanford University Press, Stanford, CA, pp. 118–32.

Zajac, P.K. and Grodinska, K. (1982) Snow contamination by heavy metals and sulphur in Cracow agglomeration (southern Poland). *Water, Air, Soil Pollut.*, **17**, 269–80.

Appendix

Table A1 The Système International d'Unités (SI) system of units

	Unit	*Symbol*	*Definition*
Basic units			
length	meter	m	
mass	kilogram	kg	Defined
time	second	s	in terms of
thermodynamic temperature	kelvin	K	physical
amount of substance	mole (Avogadros number)	mol	standards*
electric current	ampere	A	
luminous intensity	candela	cd	
Derived units			
force	newton	N	1 kg m/s^2
energy	joule	J	1 N·m
pressure	pascal	Pa	1 N/m^2
power	watt	W	1 J/s
volume	cubic meter	m^3	m^3
density	kilogram/cubic meter	kg m^{-3}	kg/m^3
Celsius temperature	degree Celsius	°C	K
frequency	hertz	Hz	s^{-1}
voltage (electromotive force)	volt	V	1 W/A
resistance (electric)	ohm	Ω	1 V/A
capacitance (electric)	farad	F	1 A·S/V
concentration	mole/cubic meter	mol m^{-3}	mol/m^3
specific heat	joule/kilogram kelvin		J/kg·K
thermal conductivity	watt/meter kelvin		W/m·K

* Weast, R.C. (ed.) (1975) *Handbook of Chemistry and Physics*, 56th edn, C.R.C. Press, Cleveland.

Table A2 Prefixes to units for powers of ten

Magnitude	*Symbol*	*Prefix*	*Magnitude*	*Symbol*	*Prefix*
10^{-1}	d	deci*-	10^1	da	deka-*
10^{-2}	c	centi*-	10^2	h	hecto*-
10^{-3}	m	milli-	10^3	k	kilo-
10^{-6}	μ	micro-	10^6	M	mega-
10^{-9}	n	nano-	10^9	G	giga-
10^{-12}	p	pico-	10^{12}	T	tera-

* Not recommended. Acceptable units are multiples of 10^3 or 10^{-3}.

Table A3 Values of physical constants

Quantity	*Symbol*	*Value*	*Unit*
Standard temperature	T_o	273.15	K
Standard pressure	p_o	101.325	kPa
Gas constant for an ideal gas	R	8.31441	$J\ mol^{-1}\ k^{-1}$
Avogadros number	N_A	6.022×10^{23}	molecules mol^{-1}
Molar volume of an ideal gas at standard temperature and pressure	V_m	2.241×10^{-2}	$m^3\ mol^{-1}$
Apparent molecular weight of air		28.964×10^{-3}	$kg\ mol^{-1}$
Molecular weight of water		18.016×10^{-3}	$kg\ mol^{-1}$
Molecular weight of carbon dioxide		44.01×10^{-3}	$kg\ mol^{-1}$
Ratio of densities of dry air and water vapor		0.622	dimensionless
Stefan–Boltzmann constant	σ	5.6696×10^{-8}	$W\ m^{-2}\ K^{-4}$
Planck's constant	h	6.6262×10^{-34}	J s
Wein's constant		2897	μm K
Velocity of light	c	2.998×10^8	ms^{-1}
Gravitational constant	G	6.6720×10^{-11}	$N\ m^2\ kg^{-2}$
Specific heat of dry air at 100 kPa pressure	c_p	1.012×10^3	$J\ kg\ K^{-1}$

Table A4 Conversion factors

Pressure	
1 bar	100 kPa
1 atmosphere	101.325 kPa
1 pound per square inch (PSI)	6.89 kPa
1 mm Hg (0°C)	0.13332 kPa
1 mm H_2O (4°C)	9.8064×10^{-3} kPa
Energy	
1 calorie	4.184 J
1 watt hour	3.60×10^3 J
1 watt s	1 J
1 erg	10^{-7} J
1 dyne cm	10^{-7} J
Power	
1 calorie s^{-1}	4.184 W
1 calorie min^{-1}	6.973×10^{-2} W
1 British thermal unit s^{-1}	1.0544×10^3 W
1 erg s^{-1}	10^{-7} W
1 kg m s^{-1}	0.10197 W
Force	
1 dyne	10^5 N
kilogram force	9.80665 N
pound force	4.44822 N
Flux	
1 calorie $cm^{-2}\ min^{-1}$	697 W m^{-2}
1 langley min^{-1}	697 W m^{-2}
1 erg $cm^{-2}\ s^{-1}$	10^{-3} W m^{-2}
1 mg $CO_2\ dm^{-2}\ hour^{-1}$	0.631 μmol $CO_2\ m^{-2}\ s^{-1}$

continued

1 ng $CO_2\ cm^{-2}\ s^{-1}$	0.227 μmol $CO_2\ m^{-2}\ s^{-1}$
1 mg $CO_2\ m^{-2}\ s^{-1}$	22.7 μmol $CO_2\ m^{-2}\ s^{-1}$
1 mg $H_2O\ m^{-2}\ s^{-1}$	0.055 mmol $H_2O\ m^{-2}\ s^{-1}$
1 g $H_2O\ dm^{-2}\ hour^{-1}$	1.542 mmol $H_2O\ m^{-2}\ s^{-1}$
1 kg $H_2O\ m^{-2}\ hour^{-1}$	15.42 mmol $H_2O\ m^{-2}\ s^{-1}$
1 μeinstein $m^{-2}\ s^{-1}$	1 μmol photons $m^{-2}\ s^{-1}$
Area	
1 square foot	0.092903 m^2
1 square inch	0.6452×10^{-3} m^2
1 acre	4046.86 m^2
1 hectare	10 000 m^2
Volume	
1 cubic foot	0.0283168 m^3
1 cubic inch	0.16387×10^{-4} m^3
1 cubic centimeter	10^{-6} m^3
1 gallon (US)	0.3785×10^{-2} m^3
1 gallon (British)	0.4546×10^{-2} m^3
Length	
1 inch	2.54×10^{-2} m
1 foot	0.3048 m
1 mile (US statute)	1609.344 m
Mass	
1 pound	0.45359 kg
1 ounce	2.834×10^{-2} kg
1 ton (short, 2000 pounds)	907.18 kg
1 ton (metric)	1000 kg

Table A5 Properties of air from −5 to 45°C at 100 kPa pressure where appropriate

t (°C)	*Density* Dry ($kg\ m^{-3}$)	Saturated ($kg\ m^{-3}$)	*Thermal conductivity, dry air* ($W\ m^{-1}\ K^{-1}$)	*Dry air* Thermal diffusivity ($mm^2\ s^{-1}$)	Kinematic viscosity ($mm^2\ s^{-1}$)	*Binary diffusion coefficients in air* Water vapor ($mm^2\ s^{-1}$)	CO_2 ($mm^2\ s^{-1}$)
−5	1.316	1.314	0.0240	18.3	12.9	20.5	12.4
0	1.292	1.289	0.0243	18.9	13.3	21.2	12.9
5	1.269	1.265	0.0246	19.5	13.7	22.0	13.3
10	1.246	1.240	0.0250	20.2	14.2	22.7	13.8
15	1.225	1.217	0.0253	20.8	14.6	23.4	14.2
20	1.204	1.194	0.0257	21.5	15.1	24.2	14.7
25	1.183	1.169	0.0260	22.2	15.5	24.9	15.1
30	1.164	1.145	0.0264	22.8	16.0	25.7	15.6
35	1.146	1.121	0.0267	23.5	16.4	26.4	16.0
40	1.128	1.096	0.0270	24.2	16.9	27.2	16.5
45	1.110	1.068	0.0274	24.9	17.4	28.0	17.0

Table A6 Properties of water at temperatures from 0 to 45°C

t (°C)	*Density* ($g\ cm^{-3}$)	*Latent heat of vaporization* ($J\ kg^{-1} \times 10^{-3}$)	*Psychrometric constant* ($kPa\ K^{-1}$)	*Kinematic viscosity* ($mm^2\ s^{-1}$)	*Solubility of in water** CO_2 ($m^3\ m^{-3}$)	O_2 ($m^3\ m^{-3}$)	*Thermal conductivity* ($W\ m^{-1}\ K^{-1}$)
0	0.99987	2513	0.646	1.79	1.713	0.0489	0.552
5	0.99999	2501	0.649	1.52	1.424	0.0429	
10	0.99973	2489	0.652	1.31	1.194	0.0380	0.578
15	0.99913	2477	0.656	1.14	1.019	0.0342	
20	0.99862	2465	0.659	1.01	0.878	0.0310	0.599
25	0.99823	2454	0.661	0.89	0.759	0.0283	
30	0.99707	2442	0.665	0.80	0.665	0.0261	0.615
35	0.99567	2430	0.668	0.72	0.592	0.0244	
40	0.99224	2418	0.672	0.66	0.530	0.0231	0.632
45	0.99025	2394	0.675	0.60	0.479	0.0219	

* Volume of gas contained in volume of water when pressure of gas is 101.3 kPa. From Henry's law the amount of gas dissolved in water is proportional to its partial pressure. Thus to calculate the concentration (*C*):

$$C = p_N \,.\, \alpha / v$$

where p_N is the partial pressure of gas $_N$, α is the solubility coefficient from the table, and v is the volume. The presence of salts depresses the solubility slightly.

Table A7 Saturation vapor pressure in kPa over water for temperatures from 0 to 50°C in 0.1°C increments*

t °C	0.0	0.1	0.2	0.3	0.4	0.5	0.6	0.7	0.8	0.9
0	0.611	0.615	0.620	0.624	0.629	0.633	0.638	0.643	0.647	0.652
1	0.657	0.661	0.666	0.671	0.676	0.681	0.686	0.690	0.695	0.700
2	0.705	0.711	0.716	0.721	0.726	0.731	0.736	0.742	0.747	0.752
3	0.758	0.763	0.768	0.774	0.779	0.785	0.790	0.796	0.802	0.807
4	0.813	0.819	0.824	0.830	0.836	0.842	0.848	0.854	0.860	0.866
5	0.872	0.878	0.884	0.890	0.897	0.903	0.909	0.915	0.922	0.928
6	0.935	0.941	0.948	0.954	0.961	0.967	0.974	0.981	0.988	0.994
7	1.001	1.008	1.015	1.022	1.029	1.036	1.043	1.050	1.058	1.065
8	1.072	1.079	1.087	1.094	1.102	1.109	1.117	1.124	1.132	1.140
9	1.147	1.155	1.163	1.171	1.179	1.187	1.195	1.203	1.211	1.219
10	1.227	1.235	1.244	1.252	1.261	1.269	1.277	1.286	1.295	1.303
11	1.312	1.321	1.329	1.338	1.347	1.356	1.365	1.374	1.383	1.392
12	1.402	1.411	1.420	1.430	1.439	1.449	1.458	1.468	1.477	1.487
13	1.497	1.507	1.517	1.527	1.537	1.547	1.557	1.567	1.577	1.587
14	1.598	1.608	1.619	1.629	1.640	1.650	1.661	1.672	1.683	1.693
15	1.704	1.715	1.726	1.738	1.749	1.760	1.771	1.783	1.794	1.806
16	1.817	1.829	1.841	1.852	1.864	1.876	1.888	1.900	1.912	1.924
17	1.937	1.949	1.961	1.974	1.986	1.999	2.012	2.024	2.037	2.050
18	2.063	2.076	2.089	2.102	2.115	2.129	2.142	2.156	2.169	2.183
19	2.196	2.210	2.224	2.238	2.252	2.266	2.280	2.294	2.308	2.323
20	2.337	2.352	2.366	2.381	2.396	2.411	2.426	2.441	2.456	2.471
21	2.486	2.501	2.517	2.532	2.548	2.563	2.579	2.595	2.611	2.627
22	2.643	2.659	2.675	2.692	2.708	2.725	2.741	2.758	2.775	2.792
23	2.809	2.826	2.843	2.860	2.877	2.895	2.912	2.930	2.947	2.965
24	2.983	3.001	3.019	3.037	3.056	3.074	3.092	3.111	3.130	3.148
25	3.167	3.186	3.205	3.224	3.243	3.263	3.282	3.302	3.321	3.341
26	3.361	3.381	3.401	3.421	3.441	3.462	3.482	3.503	3.523	3.544
27	3.565	3.586	3.607	3.628	3.649	3.671	3.692	3.714	3.736	3.758
28	3.780	3.802	3.824	3.846	3.869	3.891	3.914	3.937	3.959	3.982

Table A7 continued

t°C	0.0	0.1	0.2	0.3	0.4	0.5	0.6	0.7	0.8	0.9
29	4.005	4.029	4.052	4.075	4.099	4.123	4.147	4.171	4.195	4.219
30	4.243	4.267	4.292	4.317	4.341	4.366	4.391	4.417	4.442	4.467
31	4.493	4.518	4.544	4.570	4.596	4.622	4.649	4.675	4.702	4.728
32	4.755	4.782	4.809	4.836	4.864	4.891	4.919	4.947	4.974	5.003
33	5.031	5.059	5.087	5.116	5.145	5.174	5.203	5.232	5.261	5.290
34	5.320	5.350	5.380	5.410	5.440	5.470	5.500	5.531	5.562	5.593
35	5.624	5.655	5.686	5.718	5.749	5.781	5.813	5.845	5.877	5.910
36	5.942	5.975	6.008	6.041	6.074	6.107	6.141	6.174	6.208	6.242
37	6.276	6.311	6.345	6.380	6.414	6.449	6.484	6.520	6.555	6.591
38	6.626	6.662	6.698	6.735	6.771	6.808	6.845	6.881	6.919	6.956
39	6.993	7.031	7.069	7.107	7.145	7.183	7.222	7.261	7.299	7.338
40	7.378	7.417	7.457	7.497	7.537	7.577	7.617	7.658	7.698	7.739
41	7.780	7.821	7.863	7.905	7.946	7.988	8.031	8.073	8.116	8.159
42	8.202	8.245	8.288	8.332	8.375	8.419	8.464	8.508	8.553	8.597
43	8.642	8.687	8.733	8.778	8.824	8.870	8.916	8.963	9.010	9.056
44	9.103	9.151	9.198	9.246	9.294	9.342	9.390	9.439	9.487	9.536
45	9.585	9.635	9.684	9.734	9.784	9.835	9.885	9.936	9.987	10.038
46	10.089	10.141	10.193	10.245	10.297	10.350	10.403	10.456	10.509	10.562
47	10.616	10.670	10.724	10.778	10.833	10.888	10.943	10.998	11.054	11.110
48	11.166	11.222	11.279	11.336	11.393	11.450	11.507	11.565	11.623	11.682
49	11.740	11.799	11.858	11.917	11.977	12.037	12.097	12.157	12.218	12.278
50	12.340	12.401	12.462	12.524	12.587	12.649	12.712	12.775	12.838	12.901

* Values in the table were calculated using the Goff–Gratch formulation for saturated water vapor pressure:

$$\log_{10} e_w = -7.90298(T_s/T - 1) + 5.02808 \log_{10}(T_s/T) - 1.3816 \times 10^{-7}(10^{11.344(1-T/T_s)} - 1) + 8.1328 \times 10^{-3}(10^{-3.49149(T_s/T-1)} + \log_{10} e_{ws}$$

where e_w is the saturated vapor pressure (mbar), T_s is the steam point temperature (372.16 K), T is the temperature (K) and e_{ws} is the saturation vapor pressure of pure water at the steam point temperature (1013.246 mbar). The values in the table were converted to kPa (10mbar=1kPa). For more details see List 1947, Smithsonian Meteorological Tables. (Goff, J.A. and Gratch, S. (1946) *Trans. Amer. Soc. and Vent. Eng.*, **52**, 95).

Table A8 Vapor pressure as a function of dry-bulb temperature (t_d) and the depression of the wet-bulb temperature

	Depression of wet-bulb temperature (Δt), °C									
	0	1	2	3	4	5	6	7	8	9
t_d, °C	Vapor pressure, kPa									
1	0.656									
2	0.705	0.589								
3	0.757	0.638	0.522							
4	0.812	0.690	0.571	0.455						
5	0.871	0.745	0.623	0.504	0.388					
6	0.934	0.804	0.678	0.556	0.437	0.321				
7	1.001	0.867	0.737	0.611	0.489	0.370	0.255			
8	1.071	0.933	0.799	0.670	0.544	0.422	0.303	0.188		
9	1.147	1.004	0.866	0.732	0.602	0.477	0.354	0.236	0.121	
10	1.226	1.079	0.936	0.798	0.665	0.535	0.409	0.287	0.169	0.054
11	1.311	1.159	1.012	0.869	0.731	0.597	0.468	0.342	0.220	0.102
12	1.401	1.243	1.091	0.944	0.802	0.664	0.530	0.401	0.275	0.153
13	1.496	1.333	1.176	1.024	0.876	0.734	0.596	0.463	0.333	0.208
14	1.597	1.428	1.265	1.108	0.956	0.809	0.667	0.529	0.395	0.266
15	1.703	1.529	1.360	1.197	1.040	0.888	0.741	0.599	0.461	0.328
16	1.816	1.635	1.461	1.292	1.130	0.973	0.821	0.674	0.532	0.394
17	1.935	1.748	1.567	1.393	1.224	1.062	0.905	0.753	0.606	0.464
18	2.062	1.867	1.680	1.499	1.325	1.157	0.994	0.837	0.685	0.539
19	2.195	1.993	1.799	1.612	1.431	1.257	1.089	0.926	0.769	0.618
20	2.336	2.127	1.925	1.731	1.544	1.363	1.189	1.021	0.859	0.702
21	2.485	2.267	2.058	1.857	1.663	1.476	1.295	1.121	0.953	0.791
22	2.641	2.416	2.199	1.990	1.789	1.595	1.408	1.227	1.053	0.885
23	2.807	2.573	2.348	2.131	1.922	1.720	1.527	1.340	1.159	0.985
24	2.981	2.738	2.504	2.279	2.062	1.853	1.652	1.458	1.271	1.091
25	3.165	2.913	2.670	2.436	2.211	1.994	1.785	1.584	1.390	1.203
26	3.359	3.096	2.844	2.601	2.367	2.142	1.925	1.717	1.516	1.322
27	3.563	3.290	3.028	2.775	2.532	2.299	2.074	1.857	1.648	1.448
28	3.777	3.494	3.221	2.959	2.707	2.464	2.230	2.005	1.789	1.580
29	4.003	3.708	3.425	3.152	2.890	2.638	2.395	2.162	1.937	1.720
30	4.241	3.934	3.639	3.356	3.083	2.821	2.569	2.326	2.093	1.868
31	4.490	4.171	3.865	3.570	3.287	3.015	2.753	2.500	2.258	2.024
32	4.752	4.421	4.102	3.796	3.501	3.218	2.946	2.684	2.432	2.189
33	5.028	4.683	4.352	4.033	3.727	3.432	3.149	2.877	2.615	2.363
34	5.317	4.958	4.614	4.282	3.964	3.658	3.363	3.080	2.808	2.546
35	5.621	5.248	4.889	4.544	4.213	3.895	3.589	3.294	3.011	2.739
36	5.939	5.551	5.178	4.820	4.475	4.144	3.826	3.520	3.225	2.942
37	6.273	5.869	5.481	5.109	4.750	4.406	4.075	3.756	3.450	3.156
38	6.623	6.203	5.800	5.412	5.039	4.681	4.337	4.005	3.687	3.381
39	6.990	6.553	6.133	5.730	5.342	4.970	4.611	4.267	3.936	3.618
40	7.374	6.920	6.483	6.064	5.660	5.273	4.900	4.542	4.198	3.867
41	7.776	7.304	6.850	6.413	5.994	5.591	5.203	4.831	4.473	4.129
42	8.197	7.706	7.234	6.780	6.344	5.924	5.521	5.134	4.761	4.403
43	8.638	8.127	7.636	7.164	6.710	6.274	5.855	5.452	5.064	4.692
44	9.099	8.568	8.057	7.566	7.094	6.640	6.204	5.785	5.382	4.995
45	9.581	9.028	8.497	7.987	7.496	7.024	6.570	6.134	5.715	5.312

	Depression of wet-bulb temperature (Δt), °C									
	10	*11*	*12*	*13*	*14*	*15*	*16*	*17*	*18*	*19*
t_d, °C	*Vapor pressure, kPa*									
12	0.035									
13	0.086									
14	0.141	0.019								
15	0.199	0.074								
16	0.261	0.132	0.006							
17	0.327	0.193	0.064							
18	0.397	0.259	0.126							
19	0.471	0.329	0.192	0.059						
20	0.550	0.404	0.262	0.124						
21	0.634	0.483	0.336	0.194	0.057					
22	0.723	0.566	0.415	0.268	0.127					
23	0.817	0.655	0.499	0.347	0.201	0.059				
24	0.917	0.749	0.587	0.431	0.280	0.133				
25	1.023	0.849	0.682	0.520	0.363	0.212	0.066			
26	1.135	0.955	0.781	0.614	0.452	0.295	0.144			
27	1.254	1.067	0.887	0.713	0.546	0.384	0.228	0.077		
28	1.379	1.186	0.999	0.819	0.646	0.478	0.316	0.160	0.009	
29	1.512	1.311	1.118	0.931	0.751	0.578	0.410	0.249	0.092	
30	1.652	1.444	1.243	1.049	0.863	0.683	0.510	0.342	0.181	0.025
31	1.800	1.583	1.375	1.175	0.981	0.795	0.615	0.442	0.274	0.113
32	1.956	1.731	1.515	1.307	1.106	0.913	0.727	0.547	0.374	0.207
33	2.121	1.887	1.663	1.447	1.239	1.038	0.845	0.659	0.479	0.306
34	2.294	2.052	1.819	1.594	1.378	1.170	0.970	0.777	0.591	0.411
35	2.477	2.226	1.983	1.750	1.526	1.310	1.102	0.902	0.709	0.522
36	2.670	2.409	2.157	1.915	1.682	1.457	1.242	1.034	0.833	0.640
37	2.873	2.601	2.340	2.088	1.846	1.613	1.389	1.173	0.965	0.765
38	3.087	2.804	2.533	2.271	2.019	1.777	1.545	1.320	1.105	0.897
39	3.312	3.018	2.736	2.464	2.202	1.951	1.709	1.476	1.252	1.036
40	3.549	3.243	2.949	2.667	2.395	2.133	1.882	1.640	1.408	1.184
41	3.798	3.480	3.174	2.880	2.598	2.326	2.065	1.813	1.572	1.339
42	4.059	3.729	3.411	3.105	2.811	2.529	2.257	1.996	1.745	1.503
43	4.334	3.990	3.659	3.341	3.036	2.742	2.460	2.188	1.927	1.676
44	4.622	4.265	3.921	3.590	3.272	2.967	2.673	2.391	2.119	1.858
45	4.925	4.553	4.195	3.851	3.521	3.203	2.898	2.604	2.322	2.051

$\Delta t = t_d - t_w$ where t_d and t_w are the dry and wet-bulb temperatures respectively, calculated from:

$$e = e_s - 0.000660(1+0.00115 t_w)\, p\, \Delta t, \text{ for } p = 100 \text{ kPa};$$

e is the vapor pressure, e_s is the saturated vapor pressure at t_d.

Table A9 Conversion factors to convert conductance in units of mm s^{-1} to units of mol m^{-2} s^{-1}. To use, find the value* corresponding to the leaf temperature (columns) and pressure (rows) and multiply it by the conductance in mm s^{-1}

P *kPa*	*t* °C 5	10	15	20	25	30	35	40	45
102	0.0435	0.0427	0.0420	0.0413	0.0406	0.0399	0.0393	0.0386	0.0380
101	0.0439	0.0432	0.0424	0.0417	0.0410	0.0403	0.0397	0.0390	0.0384
100	0.0444	0.0436	0.0428	0.0421	0.0414	0.0407	0.0400	0.0394	0.0388
99	0.0448	0.0440	0.0433	0.0425	0.0418	0.0411	0.0405	0.0398	0.0392
98	0.0453	0.0445	0.0437	0.0430	0.0422	0.0415	0.0409	0.0402	0.0396
97	0.0457	0.0449	0.0442	0.0434	0.0427	0.0420	0.0413	0.0406	0.0400
96	0.0462	0.0454	0.0446	0.0439	0.0431	0.0424	0.0417	0.0411	0.0404
95	0.0467	0.0459	0.0451	0.0443	0.0436	0.0429	0.0422	0.0415	0.0408
94	0.0472	0.0464	0.0456	0.0448	0.0440	0.0433	0.0426	0.0419	0.0413
93	0.0477	0.0469	0.0461	0.0453	0.0445	0.0438	0.0431	0.0424	0.0417
92	0.0482	0.0474	0.0466	0.0458	0.0450	0.0442	0.0435	0.0428	0.0422
91	0.0488	0.0479	0.0471	0.0463	0.0455	0.0447	0.0440	0.0433	0.0426
90	0.0493	0.0484	0.0476	0.0468	0.0460	0.0452	0.0445	0.0438	0.0431
89	0.0499	0.0490	0.0481	0.0473	0.0465	0.0457	0.0450	0.0443	0.0436
88	0.0504	0.0495	0.0487	0.0478	0.0470	0.0463	0.0455	0.0448	0.0441
87	0.0510	0.0501	0.0492	0.0484	0.0476	0.0468	0.0460	0.0453	0.0446
86	0.0516	0.0507	0.0498	0.0490	0.0481	0.0473	0.0466	0.0458	0.0451
85	0.0522	0.0513	0.0504	0.0495	0.0487	0.0479	0.0471	0.0464	0.0456
84	0.0528	0.0519	0.0510	0.0501	0.0493	0.0485	0.0477	0.0469	0.0462
83	0.0535	0.0525	0.0516	0.0507	0.0499	0.0490	0.0483	0.0475	0.0467
82	0.0541	0.0532	0.0522	0.0513	0.0505	0.0496	0.0488	0.0481	0.0473
81	0.0548	0.0538	0.0529	0.0520	0.0511	0.0503	0.0494	0.0487	0.0479
80	0.0555	0.0545	0.0535	0.0526	0.0517	0.0509	0.0501	0.0493	0.0485

* The conversion factors are strictly correct only under isothermal conditions. Where leaf temperature is different from air temperature use the leaf temperature to minimize the error. When leaf and air temperatures differ as they usually do, the error should under most circumstances be no larger than 1–2%.

Table A10 Atmospheric pressure and the ratio of the atmospheric pressure (p) to standard pressure (p_o) as a function of elevation

Elevation *meters*	*Pressure* *kPa*	*Ratio* p/p_o
0	101.32	1.0000
100	100.13	0.9882
200	98.95	0.9765
300	97.77	0.9649
400	96.61	0.9535
500	95.46	0.9421
600	94.32	0.9309
700	93.19	0.9197
800	92.08	0.9087
900	90.97	0.8978
1000	89.87	0.8870
1100	88.79	0.8763
1200	87.72	0.8657
1300	86.65	0.8552
1400	85.60	0.8448
1500	84.56	0.8345
1600	83.52	0.8243
1700	82.50	0.8142
1800	81.49	0.8042
1900	80.49	0.7943
2000	79.49	0.7846
2100	78.51	0.7749
2200	77.54	0.7653
2300	76.58	0.7558
2400	75.63	0.7464
2500	74.68	0.7371
2600	73.75	0.7278
2700	72.82	0.7187
2800	71.91	0.7097
2900	71.00	0.7008
3000	70.11	0.6919
3100	69.22	0.6832
3200	68.34	0.6745
3300	67.47	0.6659
3400	66.61	0.6574
3500	65.76	0.6490
3600	64.92	0.6407
3700	64.09	0.6325
3800	63.26	0.6244
3900	62.45	0.6163
4000	61.64	0.6083

Atmospheric pressures were calculated from the relationship

$$P = 101.325\,[1-(2.2569\ 10^{-5}\ Z)]^{5.2553}$$

where Z is the elevation in meters.

Table A11 Radiation (watts m^{-2}) emitted from a blackbody at temperatures from −20 to 70°C

t °C	*0*	*1*	*2*	*3*	*4*	*5*	*6*	*7*	*8*	*9*
					Watts m^{-2}					
−20	234	238	241	245	249	253	257	261	265	269
−10	273	277	282	286	290	294	299	303	308	312
0	317	322	326	331	336	341	346	351	356	361
10	366	371	377	382	387	393	398	404	409	415
20	421	426	432	438	444	450	456	462	468	475
30	481	487	494	500	507	514	520	527	534	541
40	548	555	562	569	576	584	591	598	606	613
50	621	629	637	645	652	660	669	677	685	693
60	702	710	719	727	736	745	754	763	772	781
70	790	799	808	818	827	837	846	856	866	876

Calculated from the Stefan–Boltzmann equation:

$$E = \sigma T^4$$

where E is the emitted radiation (watts m^{-2}), σ is the Stefan–Boltzmann constant (5.6696×10^{-8} W m^{-2} K^{-4}) and T is the surface temperature (K).

Table A12 Composition of the atmosphere

*Constituent**		*Molecular weight*	*Mole fraction*	*mol* m^{-3}	*kg* m^{-3}
Nitrogen	N_2	28.015	.7809	34.86	0.6977
Oxygen	O_2	32.000	.2095	9.35	0.2992
Argon	A	39.944	.0093	0.41	0.0164
Carbon dioxide	CO_2	44.101	.000340†	1.518×10^{-2}	0.6681×10^{-3}

* Minor trace gases are not listed.

† Average annual value measured at Manna Loa, Hawaii in 1983. It varies seasonally and is currently increasing at a rate of about 0.0000015 $year^{-1}$ (Gates, D.M. (1983) An overview, In *CO_2 and Plants* (ed. E.R. Lemon) Westview Press, Boulder, CO).

Table A13 Water vapor pressures and relative humidities in equilibrium with saturated aqueous salt solutions*

Temperature °C	*LiCl*	$CaCl_2$	$MgCl_2$	$Mg(NO_3)_2$	*NaCl*	*KCl*
			Water vapor pressure (kPa)			
5	0.1192		0.2936	0.5113	0.6615	0.7654
10	0.1576		0.4114	0.7027	0.9288	1.0657
15	0.2036	0.6079	0.5677	0.9523	1.2876	1.4656
20	0.2606	0.7658	0.7222	1.2732	1.7630	1.9898
25	0.3532	0.9181	1.0368	1.6757	2.3856	2.6705
30	0.4739	0.9443	1.3741	2.1782	3.1913	
35	0.6287	1.1773	1.8001	2.7962		
			Relative humidity (%)			
5	13.7		33.7	58.6	75.9	87.8
10	12.8		33.5	57.2	75.7	86.8
15	11.9	35.6	33.3	55.8	75.5	85.9
20	11.1	32.7	33.0	54.5	75.4	85.1
25	11.2	29.0	32.7	52.9	75.3	84.3
30	11.2	22.2	32.3	51.3	75.2	
35	11.2	20.9	32.0	49.7		

* Source: Achenson, D.T. (1965) Vapor pressures of saturated aqueous salt solutions, In *Humidity and moisture measurement and control in science and industry*, Vol 3 (eds A. Wexler and W.A. Wildhack), Reinhold, New York.

Table A14 Water potentials of sodium chloride solutions at temperatures from 5 to 35°C*

Temperature °C	*Molality* 0.05	0.2	0.4	0.6	0.8	1.0	1.2	1.4	1.6	1.8	2.0
						Water potential, MPa					
5	−0.218	−0.852	−1.69	−2.54	−3.40	−4.27	−5.16	−6.07	−7.00	−7.94	−8.92
10	−0.222	−0.868	−1.73	−2.59	−3.47	−4.37	−5.27	−6.21	−7.16	−8.13	−9.13
15	−0.226	−0.884	−1.76	−2.64	−3.54	−4.45	−5.39	−6.35	−7.33	−8.33	−9.36
20	−0.230	−0.900	−1.79	−2.69	−3.61	−4.55	−5.51	−6.49	−7.49	−8.52	−9.57
25	−0.234	−0.915	−1.82	−2.74	−3.68	−4.64	−5.62	−6.62	−7.65	−8.70	−9.78
30	−0.238	−0.930	−1.86	−2.79	−3.75	−4.73	−5.73	−6.75	−7.81	8.80	9.98
35	−0.242	−0.946	−1.88	−2.84	−3.81	−4.82	−5.83	−6.88	−7.95	−9.04	−10.16

* Source: Lang, A.R.G. (1967) Osmotic coefficients and water potential of sodium chloride solutions from 0 to 40°C. *Australian Journal of Chemistry*, **20**, 2017–23.

Table A15 Glucose Equivalents, construction costs and biosynthetic efficiencies of some non-nitrogenous plant constituents are given below. Glucose Equivalents (GE′) were calculated from the elemental compositions of the compounds (see Chapter 15). Pathway costs were calculated using biochemical pathways. See Williams *et al.* (1987) for details. Values for alternate pathways were averaged. Biosynthetic efficiencies were calculated as GE′/cost

Compound	*Molecular mass* ($g\ mol^{-1}$)	*GE′* ($g\ glu\ g^{-1}$)	*Cost* ($g\ glu\ g^{-1}$)	*Biosynthetic efficiency* (E_B)
Organic acids				
pyruvate	88	0.852	0.852	1.000
malate	134	0.672	0.690	0.973
citrate	192	0.703	0.716	0.982
oxaloacetate	132	0.568	0.587	0.968
fumarate	116	0.776	0.798	0.973
succinate	118	0.890	0.869	1.024
oxalate	90	0.167	0.378	0.441
Carbohydrates				
glucose	180	1.000	1.000	1.000
fructose	180	1.000	1.028	0.973
mannose	180	1.000	1.028	0.973
galactose	180	1.000	1.056	0.947
lactose	342	1.053	1.082	0.973
cellulose	(162)	1.111	1.173	0.947
hemicellulose*	–	1.132	1.296	0.874
hemicellulose†	–	1.132	1.205	0.939
sucrose	342	1.053	1.096	0.960
starch	(162)	1.111	1.173	0.947
ribose-5-P	230	0.652	0.671	0.973
erythrose-4-P	200	0.600	0.642	0.935
pinitol	194	1.160	1.302	0.890
Fatty acids and triglycerides				
caprylic acid	144	2.290	2.412	0.950
capric acid	172	2.440	2.576	0.947
lauric acid	200	2.548	2.694	0.946
myristic acid	228	2.629	2.783	0.945
palmitic acid	256	2.693	2.852	0.944
stearic acid	285	2.744	2.908	0.944
oleic acid	283	2.711	2.984	0.908
linoleic acid	280	2.677	3.060	0.875
linolenic acid	278	2.642	3.138	0.842
glyceryl tripalmitate	807	2.696	2.852	0.946
Other non-nitrogenous compounds				
rubber	(68)	3.085	3.318	0.930
diplacol (resin)	440	1.943	2.310	0.841
Lignin				
coumaryl alcohol radical	149	2.063	2.431	0.849
coniferyl alcohol radical	179	1.885	2.488	0.758‡
sinapyl alcohol radical	209	1.758	2.528	0.695‡
Monoterpenes				
camphor	152	2.663	2.868	0.928
pinene	136	3.085	3.431	0.899
Tannins (condensed)				
procyanidin subunit	288	1.471	1.591	0.924
prodelphinidin subunit	304	1.431	1.558	0.919

* Hemicellulose with residue composition reported by Bauer *et al.* (1973) *Plant Physiol.*, **51**, 174–87. Myo-inositol pathway used.

† As above but with pathways employing dehydrogenases.

‡ S-adenosyl methionine involved in biosynthetic pathway.

Table A16 Glucose Equivalents, construction costs and biosynthetic efficiencies of some nitrogen-containing plant constituents are given below. Glucose Equivalents (GE′), pathway costs and biosynthetic efficiencies were calculated as in Table A15. The nitrogen substrate forms used for the various calculations are indicated

Compound	*Molecular mass ($g\ mol^{-1}$)*	*GE′ ($g\ glu\ g^{-1}$)* NH_4^+	NO_3^-	*Cost ($g\ glu\ g^{-1}$)* NH_4^+	NO_3^-	*Biosynthetic efficiency (E_B)* NH_4^+	NO_3^-
Amino acids							
alanine	89	1.011	1.685	1.072	1.746	0.943	0.965
arginine	174	0.948	2.327	1.230	2.609	0.770	0.892
aspartate	133	0.677	1.128	0.736	1.187	0.919	0.950
asparagine	132	0.682	1.591	0.855	1.764	0.797	0.902
glutamate	147	0.918	1.326	0.972	1.380	0.945	0.961
glutamine	146	0.925	1.746	1.013	1.835	0.913	0.952
glycine	75	0.600	1.400	0.689	1.489	0.870	0.940
isoleucine	131	1.717	2.174	1.863	2.321	0.921	0.937
leucine	131	1.717	2.174	1.784	2.242	0.962	0.970
lysine	146	1.438	2.259	1.460	2.282	0.985	0.990
ornithine	132	1.250	2.158	1.392	2.300	0.898	0.938
phenylalanine	165	1.818	2.181	1.994	2.358	0.911	0.925
proline	115	1.434	1.956	1.574	2.095	0.911	0.933
serine	105	0.714	1.286	0.842	1.413	0.848	0.910
threonine	119	1.008	1.512	1.120	1.625	0.900	0.930
tryptophan	204	1.691	2.279	1.943	2.530	0.870	0.901
tyrosine	181	1.574	1.906	1.733	2.065	0.908	0.923
valine	117	1.538	2.050	1.588	2.100	0.969	0.976
Protein							
zein	–	1.453	2.030	1.741	2.318	0.835	0.876
zein†	–	1.457	2.178	1.762	2.484	0.827	0.877
hordein	–	1.384	1.967	1.675	2.259	0.827	0.871
hordein†	–	1.388	2.158	1.697	2.467	0.818	0.875
Nucleic acid							
AMP	347	0.649	1.513	0.877	1.741	0.740	0.869
GMP	363	0.579	1.405	0.817	1.644	0.708	0.855
UMP	324	0.695	1.065	0.801	1.172	0.867	0.909
CMP	323	0.697	1.254	0.819	1.376	0.851	0.911
dAMP residue	313	0.767	1.726	1.053	2.012	0.728	0.858
dGMP residue	329	0.684	1.596	0.979	1.891	0.699	0.844
dTMP residue	304	0.938	1.332	1.058	1.453	0.886	0.917
dCMP residue	289	0.831	1.454	1.003	1.626	0.828	0.894
Other nitrogen-containing compounds							
protoporphyrin	563	2.001	2.428	2.324	2.751	0.861	0.883
Cyanogenic glucosides							
dhurrin	311	1.350	1.543	1.678	1.871	0.805	0.825
prunasin	295	1.474	1.678	1.821	2.024	0.810	0.829
Alkaloids							
nicotine	162	2.221	2.961	2.686	3.427	0.827	0.864‡
caffeine	194	1.005	2.242	1.661	2.898	0.607	0.774‡
N storage and transport							
allantoin	158	0.190	1.709	0.820	2.339	0.231	0.731
N-acetyl arginine	216	1.041	2.152	1.268	2.379	0.821	0.905
N-acetyl ornithine	174	1.293	1.982	1.400	2.090	0.923	0.948
sarcosine	89	1.011	1.685	1.303	1.977	0.777	0.853‡
betaine	117	1.538	2.050	2.087	2.600	0.737	0.789‡

All glu, gln, asp and asn assumed to be glu and asp, respectively.
† All glu, gln, asp and asn assumed to gln and asn, respectively.
‡ S-adenosyl methionine involved in biosynthetic pathway.

Index

Note: Abbreviations in subentries of the index:

CAM Crassulacean acid metabolism
IRGA Infrared gas analyzer
NIR Near-infrared radiation
PAR Photosynthetically active radiation

All other abbreviations are listed within the index.

GPSR Compliance

The European Union's (EU) General Product Safety Regulation (GPSR) is a set of rules that requires consumer products to be safe and our obligations to ensure this.

If you have any concerns about our products, you can contact us on ProductSafety@springernature.com

In case Publisher is established outside the EU, the EU authorized representative is:

Springer Nature Customer Service Center GmbH
Europaplatz 3
69115 Heidelberg, Germany

Batch number: 10800581

Printed by Printforce, the Netherlands